Regulated Streams
Advances in Ecology

Regulated Streams

Advances in Ecology

Edited by
John F. Craig
Freshwater Institute
Winnipeg, Manitoba, Canada

and
J. Bryan Kemper
Alberta Environment
Edmonton, Alberta, Canada

PLENUM PRESS • NEW YORK AND LONDON

Library of Congress Cataloging in Publication Data

International Symposium on Regulated Streams (3rd: 1985: Edmonton, Alta.)
  Regulated streams: advances in ecology / edited by John F. Craig and J. Bryan Kemper.
      p. cm.
  Includes bibliographies and index.

  ISBN-13: 978-1-4684-5394-2   e-ISBN-13: 978-1-4684-5392-8
  DOI: 10.1007/978-1-4684-5392-8

  1. Stream ecology—Congresses. 2. Rivers—Regulation—Environmental aspects—Congresses.
I. Craig, John F. II. Kemper, J. B. III. Title.
QH541.5.S7I57 1985
574.5'26323—dc19   87-16596
                                                                      CIP

Proceedings of the Third International Symposium on Regulated Streams, sponsored by Alberta
Environment and Alberta Water Resources Commission, held August 4–8, 1985, in Edmonton,
Alberta, Canada

© 1987 Plenum Press, New York
A Division of Plenum Publishing Corporation
233 Spring Street, New York, N.Y. 10013

Softcover reprint of the hardcover 1st edition 1987

PREFACE

The contributions and attendance at the Third International Symposium on Regulated Streams held at Edmonton, Alberta, in August, 1985 was a strong indication of the continuing interest in regulated stream ecology. The keyword to the symposium was `advances´ both in basic science and scientific management of regulated streams. The editors hope the contributions which make up this volume fulfil this premise. The meeting was truly international with representatives of thirteen countries from North America, Europe, Asia and the Southern Hemisphere.

The papers are grouped under the headings, Fish Ecology, Invertebrate Ecology, Physical Processes, Water Quality and Special Topics, although there is much overlap between sections. As far as possible, the editorial approach has been to present first articles which give an overview of the subject for that group and many of these contributions discuss the development of models. The papers that follow outline empirical data that have been collected, often in support of these models. Finally, contributions describing case histories are given. In the summary the editors have attempted to synthesize the main contributions and make some conclusions especially pertaining to the direction of future research.

The Edmonton meeting was a great success in all aspects. It was a time for the exchange of ideas both formally and informally. The participants agreed that the event was very well organised. The chairmen also did an excellent job in keeping speakers to time and in conducting their sessions efficiently.

We are extremely grateful to the referees who worked diligently in improving the scientific quality of the manuscripts.

We have been very fortunate in having the services of Mrs. Susan Hamilton to undertake the typing of this book. We thank her for all her efforts. We also acknowledge the assistance of Betty Hope through all stages of preparation.

The Fourth International Meeting on Regulated Streams is to be held at Loughborough, England, in 1988 under the direction of Dr. Roger Petts. We wish him every success.

J.F. Craig
J.B. Kemper

CONTENTS

SECTION I: FISH ECOLOGY

SECTION II: INVERTEBRATE ECOLOGY

SECTION III: PHYSICAL PROCESSES

SECTION IV: WATER QUALITY

SECTION V: SPECIAL TOPICS

SECTION I

FISH ECOLOGY

A METHOD FOR TREATMENT OF DATA FROM THE INSTREAM FLOW INCREMENTAL

METHODOLOGY FOR INSTREAM FLOW DETERMINATION

William H. Geer

Utah Division of Wildlife Resources
Salt Lake City, UT 84116, USA

INTRODUCTION

The alteration of streamflow regimes is a major cause of degradation of western U.S. trout streams, and several instream flow needs methodologies have been developed to determine both the impacts of streamflow alteration and the streamflow regimes necessary for fishery protection. Perhaps the most widely used is the Instream Flow Incremental Methodology (IFIM) of the U.S. Fish and Wildlife Service (Bovee 1982). The field application of the IFIM is described in Bovee and Milhous (1978), and the hydraulic and habitat modeling are discussed in Milhous, Wegner, and Waddle (1981). But, while IFIM users can collect field data and simulate hydraulic and habitat parameters by standardized procedures, the final analysis of habitat simulation data is not standardized or well described. Simulation output frequently is an enormous mass of refined data subject to divergent interpretation by different fishery analysts.

This paper gives stepwise procedures of IFIM data analysis that lead different analysts to similar conclusions on both the impacts of streamflow alteration and streamflow needs. They mathematically incorporate specific aquatic habitat management goals and express the effects of selected streamflow regimes as percentages of a calculated optimum habitat condition for the subject stream. The procedures permit the comparison of altered streamflow regimes to historical conditions and to stream habitat potential. Analytical procedures are demonstrated on actual IFIM data used to assess the fishery impacts of a proposed hydroelectric power project in northern Utah.

THE LOGAN CITY POWER PROJECT AND THE IMPACT AREA

In 1981, the City of Logan, Utah, applied for an appropriation of 28.3 m^3 s^{-1} from the Logan River for hydroelectric power generation. With respect to fishery resources, the pertinent aspects of the proposed project design and operation are: (1) power generation would be stable rather than peaking, (2) power water would be diverted at Second Dam on the Logan River, (3) power water would be returned to the stream at the inlet to First Reservoir, and (4) The Logan City Power Project (LCPP) would not alter the consumptive diversion regime of the Logan-Smithfield-Hyde Park (L-S-HP) Canal located between the proposed power diversion and return flow points 'Figure 1).

1

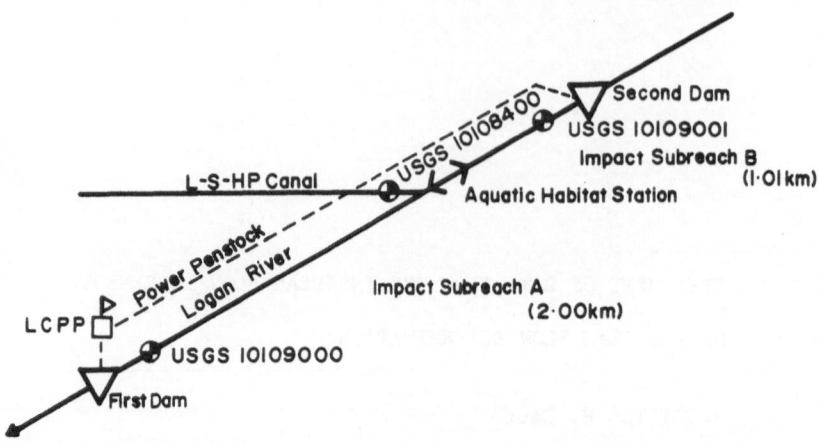

Figure 1. Diagrammatic map of the Logan City Power Project (LCPP) on the
Logan River in northern Utah; Impact Reach = Impact Subreaches
A+B.

The Logan River is tributary to the Bear River in the Great Salt Lake
drainage. The portion of the Logan River to be directly impacted by the
LCPP is the 3.01 km reach (the Impact Reach) from the inlet of First
Reservoir to Second Dam, and is comprised of two subreaches. Impact
Subreach A extends from the inlet to First Reservoir upstream 2.00 km to
the L-S-HP Canal; Impact Subreach B extends from the L-S-HP Canal 1.01 km
to Second Dam. Power water diversions and instream flow releases would be
made at Second Dam.

The subreaches differ markedly in the summer streamflow due to
consumptive diversions into the L-S-HP Canal. The U.S. Geological Survey
(USGS) has continuously recorded streamflow since 1913 in Impact Subreach A
at gage No. 10109000. Diversions into the L-S-HP Canal have been recorded
at gage No. 10108400 since 1963. The flow in Impact Subreach B (simulated
gage No. 10109001) is the sum of flows at gages No. 10109000 and
10108400. There are no perennial surface tributaries to the Logan River in
the Impact Reach. For Water Years 1970-1980, the annual mean flows in
Impact subreaches A and B were 6.97 and 7.79 $m^3 s^{-1}$,
respectively. Additional hydrological summaries are given in the
"Procedures" section.

The Impact Reach is in a U-shaped canyon with little channel
meandering. Subreach A has a mean channel gradient of 2.53%; Subreach B has
2.37%. Subreach A has been partially modified by channelization due to
highway encroachment, whereas Subreach B approximates a natural stream
channel.

PROCEDURES

This report describes only the analysis of data from the habitat
simulation model of the IFIM methodology (Milhous et al. 1981; Part VI). It
is assumed that the hydraulic simulation model accurately predicts physical
habitat components in a specified range of streamflows. In the LCPP
example, aquatic habitat data was collected in 1981 at a broad range of

flows at a single study site of nine cross-sections in Impact Subreach B (Figure 1). Habitat area values from Impact Subreach B were applied to Subreach A using an arbitrary adjustment described later.

Step 1

Summarize the historical streamflow data for the prevailing hydrological conditions in the subject stream reach. Because the development of a stream biotic community is defined largely by the prevailing hydrological conditions, it follows that an intelligent derivation of both the impacts of streamflow alteration and the recommended streamflow regime for fishery protection should be constrained by hydrologic conditions. Stream ecological conditions generally evolve under limits imposed by base and flood flows.

Base and flood flow values should be used in stream habitat studies to define the limits of extrapolation of habitat simulation predictions. The flood flow estimate sets the upper limit of habitat simulation predictions, whereas the base flow value sets the lower limit of flows recommended for fishery protection.

Annual or seasonal mean or median flows are commonly used to further characterize a stream for water development and set the upper extremes of recommended fishery flows. In these procedures, base, median, and flood flows are quantified from streamflow duration curves for a specified period of record (POR) (Searcy 1969).

LCPP Example. Monthly streamflow duration curves were prepared for two POR in Impact Subreach B. The first POR (Table 1) is Water Years 1970-1980, the period since the closure of the original UP&L power project and in which time the present fishery of wild brown trout and mountain whitefish developed. The streamflow duration statistics from that POR were used to define the upper limits of LCPP aquatic impacts and instream flow needs. Table 1 also lists the monthly mean diversions into the L-S-HP Canal. The streamflow in Impact Subreach A is the flow in Impact Subreach B minus the diversion into the L-S-HP Canal.

The second POR is Water Years 1931-1980 for Impact Subreach B, and was used to determine both the long-term base flow in Impact Subreach B and the frequency of occurrence of the recommended instream flows in a 50-year period of the historical streamflow regime. However, to shorten this report, the long-term base flow value from the second POR is reported in the data from the first POR in Table 1; the balance of the second POR is not reported here since it is not absolutely necessary to demonstrate the procedures. The base flow estimate used in the LCPP was the daily mean flow equaled or exceeded in 100% of the POR ($1.42 \text{ m}^3 \text{ s}^{-1}$ in Table 1).

The monthly flood flow values for the LCPP were taken as the daily mean flow equaled or exceeded 10% of the POR for the month in question since that value encompassed the most frequent seasonal floods. The 10-percentile exceedance flow is a bankfull value and occurred frequently enough to strongly influence the stream morphology and environment. The monthly 50-percentile exceedance (median) flows are alternatives to the monthly mean flows. A monthly flow greater than the 50-percentile exceedance value would not be recommended because it would have less than a 50% chance of occurrence under prevailing hydrological conditions.

3

Step 2

Summarize pertinent biological data and define the character and attributes of the existing community of aquatic biota for the subject stream reach.

LCPP Example. The Logan River Impact Reach is a state Class 2 sport fishery managed for wild brown trout and mountain whitefish. Fish population data are summarized in Table 2 (UDWR, unpublished data). Brown trout (*Salmo trutta*) comprised 78% of the gamefish standing crop, whereas mountain whitefish (*Prosopium williamsoni*) were 17%. Mottled sculpin (*Cottus bairdi*) are abundant and are forage for the brown trout. Benthic invertebrates are abundant and composed of Trichoptera (70%), Plecoptera (20%), and Ephemeroptera (10%).

Step 3

For the subject stream reach, set a fishery management goal that can be expressed as habitat quantity and quality. The IFIM habitat model gives predicted habitat areas as weighted usable area, or WUA (Milhous et al. 1981). A general goal should first be established for all aquatic species of concern, followed by a specific working goal for selected target species that are representative of the aquatic community and have habitat

Table 1. Streamflows ($m^3 s^{-1}$) in the LCPP impact reach in water years 1970-1980.

	Daily Mean flow Equaled or Exceeded The Following Percent of the POR			Monthly Mean Flow In L-S-HP Canal
	100	50	10	
January	1.42*	3.26	4.25	0.11
February	1.95	3.03	3.96	0.11
March	2.15	3.26	5.95	0.11
April	2.35	5.64	13.03	0.11
May	3.68	18.86	28.32	0.99
June	3.11	22.43	33.98	1.70
July	2.61	10.76	19.54	2.10
August	2.35	6.77	9.85	1.70
September	2.15	5.44	7.08	1.10
October	2.15	4.39	5.95	0.40
November	1.95	3.91	5.38	0.20
December	1.95	3.34	4.53	0.11

*Designated long-term base flow for Water Years 1931-1980.

Table 2. Population data on March 24, 1981 for fish at least 152 mm total
length in the LCPP impact reach.

	Standing Crop fish km^{-1}	kg ha^{-1}
Brown trout	1,111	132
Mountain whitefish	68	29
Mottled sculpin	392	--

Trout:Whitefish standing crop ratio = 132:29 = 4.6:1   5:1

suitability criteria compatible with IFIM analysis. An example of the
habitat suitability criteria are the probability-of-use criteria for
salmonids cited in Bovee (1978).

LCPP Example. The general management goal is to perpetuate the present
Class 2 fishery values in the Impact Reach. The specific working goal is to
retain the existing habitat quantity and quality for all life stages of
brown trout, mountain whitefish, and seven benthos species of Trichoptera,
Ephemeroptera, and Plecoptera. Brown trout and mountain whitefish are to be
retained as wild populations, with brown trout dominant over
whitefish. This dominance is to be reflected as a 5:1 weighting of brown
trout WUA over mountain whitefish WUA based on the 5:1 standing crop ratio
shown in Table 3. Likewise, benthic invertebrates are given the same 5:1
weighting over whitefish in order to maintain adequate food resources for
trout. These weighting factors are mathematically demonstrated in Step
8. Mottled sculpin are not in the working goal because habitat suitability
criteria are not yet available for them, but the provision of adequate
benthos habitat should also satisfy habitat needs of sculpin.

In addition to the 5:1 preference, a 3:1 preference in overall aquatic
habitat value is arbitrarily given to Impact Subreach B over A. This
weighting factor (demonstrated in Step 11) favors the better habitat
conditions in Subreach B over A, while allowing the use of the single study
site in Subreach B to describe the stream environment in both
subreaches. The 5:1 and 3:1 preference factors directly influence the
final quantification of LCPP impacts and the recommended fishery flow
releases.

Step 4

Identify the life stage periodicities of the target organisms selected
for the specific working goal (Step 3) and the biologically significant
periods (BSP) of the year. A BSP is a continuous period of time (e.g.,
October-November) in which a specific combination of taxa and life stages
occurs.

LCPP Example. For the LCPP target organisms, the life stage
periodicities and BSPs are listed in Table 3.

Determine the ranges of flows for which WUA values are to be predicted. These ranges depend on the time of year, the target taxa and life stages present in that period, and the range of flows occurring in that period over the POR. It usually is best to extrapolate WUA as close to $0 \text{ m}^3 \text{ s}^{-1}$ as possible considering the accuracy of the hydraulic simulation model. The monthly estimates of flood flows in Step 1 set the upper limits of WUA prediction. A monthly streamflow range delineated in this manner will permit modeling of habitat area over a sufficiently broad range so

Table 3. Monthly presence (X) of target taxa and life stages*, grouped in biologically significant periods (BSP), in the LCPP impact reach.

| | Brown Trout | | | | | Mountain Whitefish | | | | Benthic Invertebrates |
	S	I	F	J	A	S	F	J	A	L/N
BSP-1										
January		X	X	X	X		X	X	X	X
February		X	X	X	X		X	X	X	X
March		X	X	X	X		X	X	X	X
BSP-2										
April		X	X	X			X	X	X	X
May		X	X	X			X	X	X	X
June		X	X	X			X	X	X	X
BSP-3										
July		X	X				X	X		X
August		X	X				X	X		X
September		X	X				X	X		X
BSP-4										
October	X		X	X		X	X	X		X
November	X		X	X		X	X	X		X
BSP-5										
December		X	X	X			X	X	X	X

*S = spawning; I = egg incubation; F = fry; J = juvenile; A = adult; L/N = larval/nymphal.

that assessments of the effects of both low and high flows on aquatic habitat can be made. The delineated streamflow ranges should be related directly to specific hydrological conditions expected to occur in the subject stream reach.

LCPP Example. The monthly 10-percentile exceedance flows for Water Years 1970-1980 (Table 1) were arbitrarily selected as the flood flow estimates. Therefore, the WUA values for each target taxon and life stage were predicted from 0 m^3 s^{-1} to the maximum monthly 10-percentile exceedance flow occurring in the specific period of the year that the taxon and life.stage in question are present. The selected 10-percentile flow values and the expected month of occurrence are shown in Table 4.

To clarify this step, see Table 4 for brown trout spawning, which occurs in October-November (Table 3). The maximum 10-percentile exceedance flow in this period is 5.95 m^3 s^{-1}, which historically occurred in October (Table 1). Therefore, WUA predictions for brown trout spawning are made only for a flow range of 0-5.95 m^3 s^{-1}. Other taxa and life stages are treated in the same manner.

Table 4. Target taxa and life stages and the maximum monthly 10-percentile exceedance flow (m^3 s^{-1}; month of occurrence in parentheses) during periods of presence in the LCPP impact reach.

	Period of Presence	Maximum 10-Percentile Exceedance Flow
Brown Trout		
Spawning	Oct-Nov	5.95 (Oct)
Egg incubation	Dec-Mar	5.95 (Mar)
Fry	Jan-Jun	33.98 (Jun)
Juvenile	Jan-Dec	33.98 (Jun)
Adult	Jan-Dec	33.98 (Jun)
Mountain Whitefish		
Spawning	Oct-Nov	5.95 (Oct)
Fry	Jan-Jun	33.98 (Jun)
Juvenile	Jan-Dec	33.98 (Jun)
Adult	Jan-Dec	33.98 (Jun)
Benthic Invertebrates		
Larval/Nymphal	Jan-Dec	33.98 (Jun)

Step 6

For the target taxa and life stages selected in Step 3, compute WUA values in selected increments of flow within the lower and upper limits of streamflow set in Step 5.

LCPP Example. The computed WUA values for the selected taxa and life stages up to the flows specified in Table 4 are in Figures 2 and 3.

Step 7

Convert the WUA values determined in Step 6 to individual percent deviations (IPD) from optimal WUA values for each taxon and life stage. The optimum WUA for a particular taxon and life stage is the maximum areal value within the lower and upper limits of streamflow set in Step 5 (shown in the WUA graphs in Step 6). All WUA values within the designated streamflow range are converted to percent deviations (negative values) from the maximum WUA value for the subject taxon and life stage. The maximum WUA value would be 0% deviation from the optimal value.

LCPP Example. The IPD values for the selected taxa and life stages are shown in Figures 4 and 5.

Step 8

From the specific working goal (Step 3) and the taxa and life stage periodicities (Step 4), develop a series of relative weighting equations for the BSPs of the year. The relative weighting equations are used to calculate mean percent deviations (MPD) from optimal WUA from the individual percent deviations (IPD) from optimal WUA (Step 7). The MPD values are arithmetic means reflecting a balanced consideration for taxa and life stages that occur simultaneously. The MPD values simplify the mass of WUA data by converting them to single habitat index values incorporating basic management goals. The relative weighting equations should include as factors the preferences given to particular taxa or life stages as shown in Step 3.

LCPP Example. From the five BSPs in Table 3, the following five relative weighting equations were constructed using the 5:1 preference factors for brown trout and benthos from Step 3:

January, February, March:

$$MPD = [5(\text{Brown Trout I+F+J+A})+(\text{Mountain Whitefish F+J+A})+ \quad (1)$$
$$5(\text{Benthos L/N})]/28;$$

April, May, June:

$$MPD = [5(\text{Brown Trout F+J+A})+(\text{Mountain Whitefish F+J+A})+ \quad (2)$$
$$5(\text{Benthos L/N})]/23;$$

July, August, September:

$$MPD = [5(\text{Brown Trout J+A})+(\text{Mountain Whitefish J+A})+ \quad (3)$$
$$5(\text{Benthos L/N})]/17;$$

October, November:

$$MPD = [5(\text{Brown Trout S+J+A})+(\text{Mountain Whitefish S+J+A})+ \quad (4)$$
$$5(\text{Benthos L/N})]/23;$$

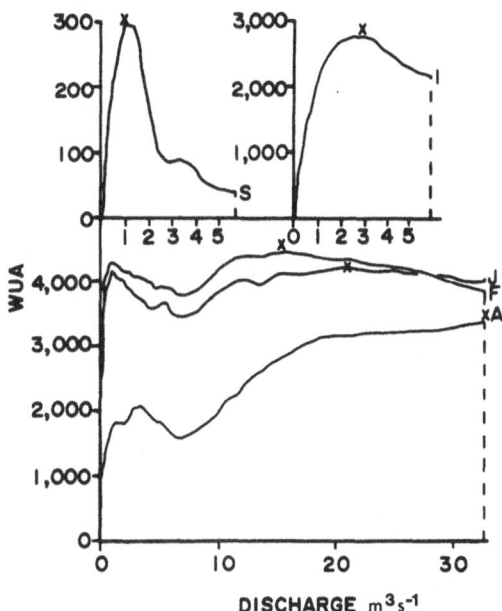

Figure 2.   WUA (m^2 $(1,000m)^{-1}$) versus streamflow (m^3 s^{-1}) for brown trout
spawning (S), egg incubation (I), fry (F), juvenile (J), and
adult (A) life stages in the LCPP Impact Reach; X marks peak
WUA.

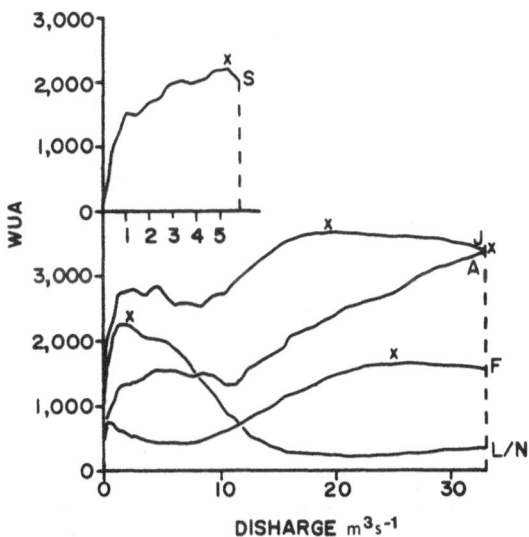

Figure 3.   WUA (m^2 $(1,000m)^{-1}$) versus streamflow (m^3 s^{-1}) for mountain
whitefish spawning (S), fry (F), juvenile (J), and adult (A)
life stages and benthic invertebrate (Ephemeroptera, Plecoptera,
and Trichoptera combined) larval and nymphal (L/N) life stages
in the LCPP Impact Reach; X marks peak WUA.

9

December:

$$MPD = [5(\text{Brown Trout I+J+A})+(\text{Mountain Whitefish J+A})+ \qquad (5)$$
$$5(\text{Benthos L/N})]/22;$$

where: MPD = mean percent deviation from optimal WUA
              at flow Q;
        S = spawning IPD at flow Q;
        I = egg incubation IPD at flow Q;
        F = fry IPD at flow Q;
        J = juvenile IPD at flow Q;
        A = adult IPD at flow Q;
      L/N = larval/nymphal IPD at flow Q; and
Denominators = the total number of life stages (expanded by
              the 5:1 preference factors) in the numerator
              of each equation.

Step 9

For each BSPs from Step 4, and for which relative weighting equations were developed in Step 8, list the maximum monthly 10-percentile exceedance flow. These values will set the upper limits of MPD determination in Step 10 (just as Step 5 values limited WUA extrapolations in Step 6). The lower flow limit should be the same as set in Step 5, which was 0 m^3 s^{-1}.

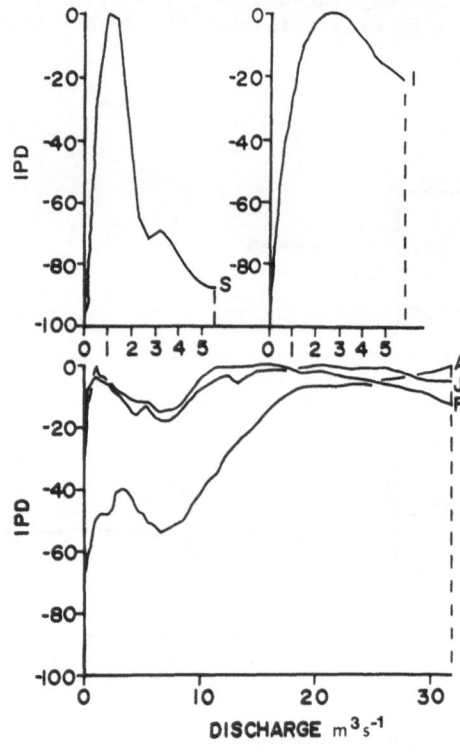

Figure 4. Individual percent deviation (IPD; %) from optimum WUA versus streamflow (m^3 s^{-1}) for brown trout spawning (S), egg incubation (I), fry (F), juvenile (J), and adult (A) life stages in the LCPP impact reach.

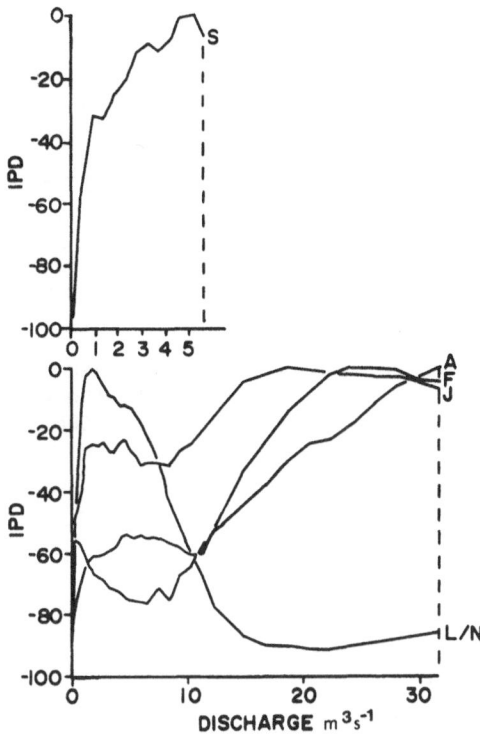

Figure 5. Individual percent deviation (IPD; %) from optimum WUA versus
streamflow ($m^3 s^{-1}$) for mountain whitefish spawning (S), fry
(F), juvenile (J), and adult (A) life stages and benthic
invertebrate larval and nymphal (L/N) life stages in the
LCPP impact reach.

Table 5. Biologically significant periods and the maximum monthly
10-percentile exceedance flows ($m^3 s^{-1}$; month of occurrence in
parentheses) in the LCPP impact reach.

Biologically Significant Period	Maximum 10-Percentile Exceedance Flow
Jan-Mar	5.95 (Mar)
Apr-Jun	33.98 (Jun)
Jul-Sep	19.54 (Jul)
Oct-Nov	5.95 (Oct)
Dec	4.53 (Dec)

11

LCPP Example. For the BSPs shown in Table 4 and Equations 1-5, the maximum monthly 10-percentile exceedance flows for Water Years 1970-1980 (Table 1) are listed in Table 5.

Step 10

For selected increments within the streamflow ranges delineated in Step 9, relative weighting equations (Step 8) are used to convert the IPD values (Step 7) to MPD values for each BSP (Step 4). If an instream flow decision is to be made from the results of a single stream study reach, and consequently from a single annual set of MPD values, go to Step 12. If a decision is to be made from multiple study reachs, go instead to Step 11, for further summarization of multiple annual sets (one per study reach) of MPD values to a final single annual set of mean MPD values for all reaches combined.

LCPP Example. Using the relative weighting equations (Step 8) and the lower and upper streamflow limits (Step 9), the IPD values in Figures 4-5 were converted to MPD values (Figure 6) for the BSPs. The Figure 6 values apply to both subreaches.

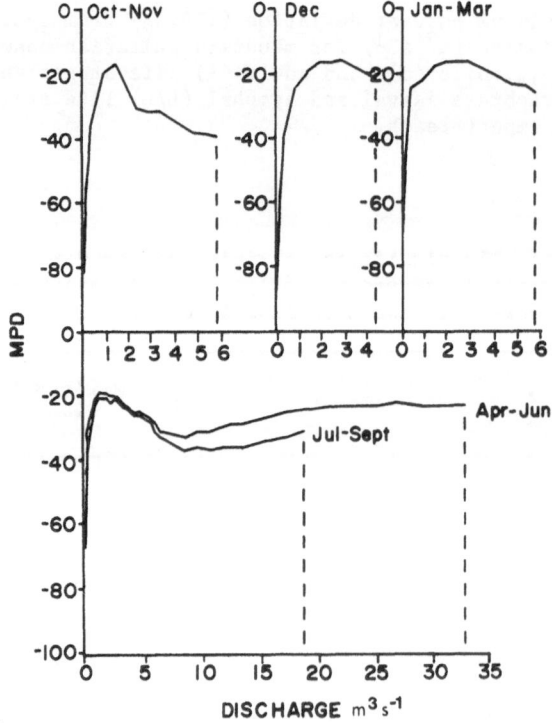

Figure 6. MPD (%) from optimal WUA versus streamflow ($m^3 s^{-1}$) for the BSPs in the LCPP impact reach.

To illustrate this, the MPD values for flows of 0.74 and 2.83 m^3 s^{-1} in July are computed. Using Equation 3 (Step 8) and the life stage IPD value for 0.74 m^3 s^{-1}, the MPD value is:

MPD = [5(Brown Trout J+A)+(Mountain Whitefish J+A)+
5(Benthic Invertebrates L/N)]/17;

MPD = [5(-6-55)+(-43-72)+5(-21)]/17 = -31%.

At 2.83 m^3 s^{-1} , the MPD value is:

MPD = [5(-9-43)+(-24-60)+5(-4)]/17 = -21%.

Step 11

This step is optional, and is used to reduce the MPD values for several individual stream reaches to a single mean MPD value for all reaches combined. A reach relative weighting equation--similar to the species relative weighting equations in Step 8--that incorporates preference factors for specific reaches is constructed. The mean MPD values are computed for the biological periods identified in Step 4 and the increments and ranges of flows set in Step 9.

LCPP Example. Using the 3:1 preference factor that favors Impact Subreach B over A (Step 3), the subreach relative weighting equation for computing a single 3:1 MPD value from individual MPD values for Subreaches A (MPD_A) and B (MPD_B) is:

$$3:1 \text{ MPD} = [(MPD_A)+3(MPD_B)]/4. \qquad (6)$$

Use of the equation is shown for July, which is in the BSP that includes August and September. Equation 3 (Step 8) is used to compute the individual MPD values for each of subreaches A and B. The same flow-versus-MPD graphs are used for both subreaches for any specific flow; that is, both reaches have an MPD value of -21% at 2.83 m^3 s^{-1} (Figure 6). But, when Subreach B has a flow of 2.83 m^3 s^{-1} in July, Subreach A would have only 0.74 m^3 s^{-1}, since the July diversion into the L-S-HP Canal averages 2.10 m^3 s^{-1} (Table 1). In Figure 6, Subreach B has an MPD value of -21%, whereas Subreach A has -31%. From equation 6, the 3:1 MPD is:

$$3:1 \text{ MPD} = [(MPD_A)+3(MPD_B)]/4 = (-31\%)+3(-21\%)]/4 = -24\%.$$

Several July values are in Table 6, whereas graphs of 3:1 MPD values for all months are in Figures 7-9.

Step 12

In this step, a theoretical best streamflow regime for aquatic habitat is prescribed from either the MPD values for single stream reaches (Step 10) or the mean MPD values for multiple stream reaches combined (Step 11). Streamflow values are selected from within ranges specified in Step 9. Within the designated flow range for the annual period of interest, the flows having the lowest MPD or mean MPD values are identified. The final flow selected as the best for the species, life stages, and management goal, and within the hydrological limitations, would be the lowest flow of those having the lowest MPD value. This flow would be the best if the stream could be managed solely for aquatic habitat optimization, given the historical hydrological conditions. The best flows will often have less than a 50% probability of occurrence under the historical streamflow

Table 6. July MPD and 3:1 MPD values (%) versus streamflow ($m^3 s^{-1}$) for the LCPP impact reach.

| Impact Subreach B | | Impact Subreach A | | Impact Reach |
Release Discharge	MPD	Predicted Flow*	MPD	3:1 MPD
0	-100	0	-100	-100
25	-38	0	-100	-54
50	-22	0	-100	-42
75	-21	1	-100	-41
100	-21	26	-35	-25
150	-25	76	-21	-24
200	-27	126	-23	-26
400	-37	326	-37	-37
700	-31	626	-33	-32

*Release discharge minus diversion into L-S-HP Canal (Table 1).

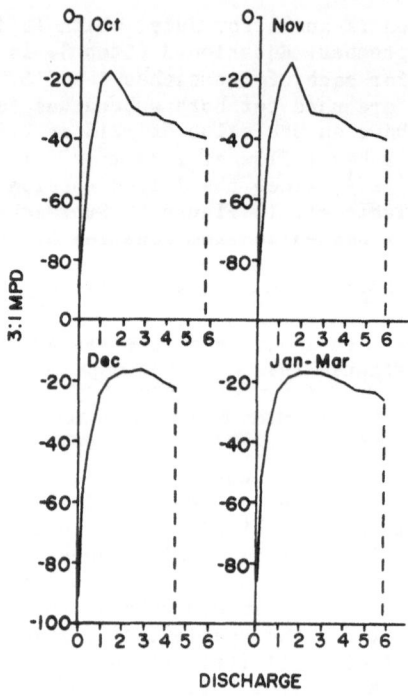

DISCHARGE

Figure 7. 3:1 MPD (%) from optimal WUA versus streamflow ($m^3 s^{-1}$) for the BSPs in the LCPP impact reach.

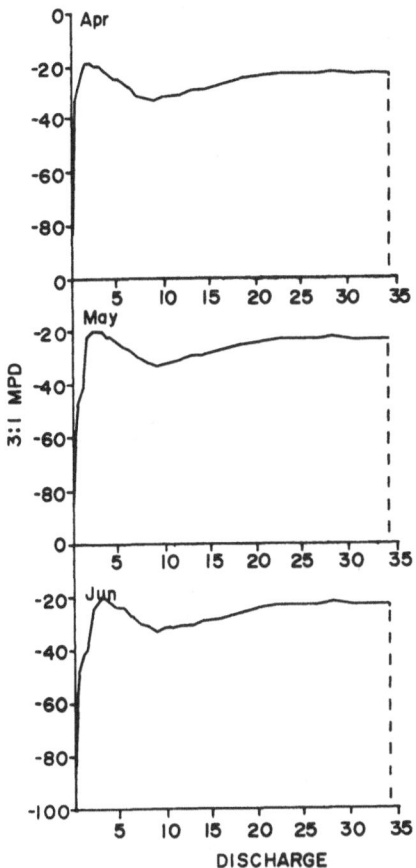

Figure 8. 3:1 MPD (%) from optimal WUA versus streamflow (m^3 s^{-1}) for the BSPs in the LCPP impact reach.

regime, and would require impoundment and regulation of downstream releases.

It must be realized that "theoretical best" refers only to values of MPD computed from WUA values, which are considerations of only stream depth, current velocity, and substrate. Many other habitat parameters not compatible with hydraulic modeling, such as fish cover and water quality, must also be considered in the determination of both streamflow alteration impacts and recommended flows.

LCPP Example. Table 7 shows the theoretical best streamflow regime for aquatic habitat. This regime does not include constraints on low or high flow recommendations.

Step 13

At this point, pre-determined constraints are applied to the results of Step 12 to determine if the theoretical best streamflow regime is a feasible recommendation. The first two constraints were determined in Step 1, in which the base flow limitation was set to protect against unreasonably low streamflow recommendations, and the mean or median flows were defined as the upper limits of recommendations. Additional constraints

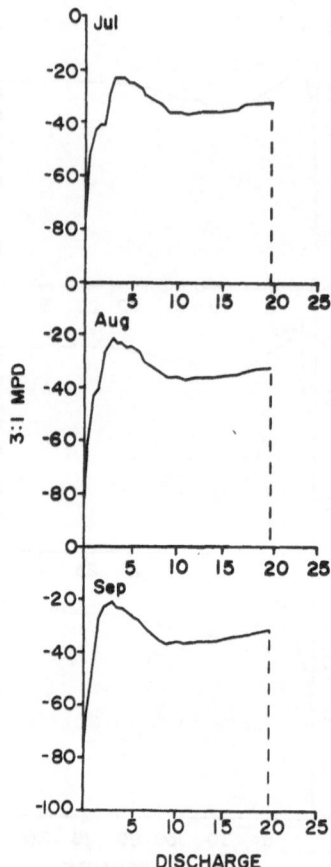

Figure 9. 3:1 MPD (%) from optimal WUA versus streamflow ($m^3 s^{-1}$) for the BSPs in the LCPP impact reach.

that should be imposed include fish cover and behavior, water quality, and other important aspects of habitat or biology.

LCPP Example. The long-term base flow value is cited in Table 1 (1.42 $m^3 s^{-1}$. From Table 8, the best flows for all months are greater than the long-term base flow.

The flows shown in Table 7 are less than the monthly median flows set in Step 1 as the upper limits of streamflow recommendation, so do not need reduction.

Cover data show that large losses of critical bank and instream fish cover in Subreach A occur at flows below 1.10 $m^3 s^{-1}$. Therefore, selection of the best 3:1 MPD value should be limited to flows at Second Dam (head of Subreach B) that provide at least 1.10 $m^3 s^{-1}$ in Subreach A. A release discharge of 3.00 $m^3 s^{-1}$ in July at Second Dam provides only 0.91 $m^3 s^{-1}$ in Subreach A (because the other 2.10 $m^3 s^{-1}$ is diverted into the L-S-HP Canal), although it is the lowest flow in the range that provides the lowest 3:1 MPD value (-23%). To maintain at least 1.10 $m^3 s^{-1}$, a release discharge of at least 3.20 $m^3 s^{-1}$ must be made, which would also provide a 3:1 MPD value of -23%. But 3.20 $m^3 s^{-1}$ would still not be the best flow, since higher releases would not only give the same 3:1 MPD value, but would

16

Table 7. Theoretical best streamflow (m^3 s^{-1}) based only on MPD values (%) in the LCPP impact reach.

	Impact Subreach B			Impact Subreach A		Impact Reach
	Release Discharge	Duration Percentage*	MPD	Predicted Flow	MPD	3:1 MPD
Jan	2.01	97	-17	1.90	-17	-17
Feb	2.01	98	-17	1.90	-17	-17
Mar	2.01	100	-17	1.90	-17	-17
Apr	1.50	100	-19	1.39	-20	-19
May	2.10	100	-19	1.10	-21	-20
Jun	3.00	100	-20	1.30	-20	-20
Jul	3.00	98	-21	0.91	-29	-23
Aug	3.00	93	-21	1.30	-22	-21
Sep	3.00	89	-21	1.90	-21	-21
Oct	1.50	100	-17	1.10	-19	-18
Nov	1.50	100	-17	1.30	-18	-17
Dec	3.00	63	-16	2.89	-16	-16

*The percent of the total days in the given month in the POR (Water Years 1931-1980) that the release discharge would have been equaled or exceeded by the historical daily mean flow.

also give better combinations of individual MPD values for each impact subreach. The best release in July would be 3.51 m^3 s^{-1} since it would: (1) give the lowest 3:1 MPD value (-23%), (2) give the best combination of individual MPD values (-23% and -22%), (3) give more than 1.10 m^3 s^{-1} in Subreach A, and (4) fall within the low and high flow limitations set in Step 1. Table 8 shows the actual best streamflow regime for the optimization of aquatic habitat in the Impact Reach.

Step 14

Compute the historical habitat conditions in terms of MPD or mean MPD values. Arithmetic means for each month are good summary statistics, but others (e.g., medians or ranges) might be better. But, both historical and development streamflow regimes should be shown in comparable terms.

LCPP Example. Historical habitat conditions are shown as mean 3:1 MPD values computed by integration and averaging of flow-related 3:1 MPD values over the hydrological period of actual fishery development (Table 1). In Step 9 (Table 5), the maximum monthly 10-percentile exceedance flow in each BSP was set as the upper limit for extrapolation of 3:1 MPD values in that BSP. Within each period, then, integration of 3:1 MPD values was performed

Table 8. Actual best streamflows (m^3 s^{-1}) based on MPD values (%) and other essential habitat parameters in the LCPP impact reach.

| | Impact Subreach B | | | Impact Subreach A | | Impact Reach |
	Release Discharge	Duration Percentage*	MPD	Predicted Flow	MPD	3:1 MPD
Jan	2.01	97	-17	1.90	-17	-17
Feb	2.01	98	-17	1.90	-17	-17
Mar	2.01	100	-17	1.90	-17	-17
Apr	1.50	100	-19	1.39	-20	-19
May	2.10	100	-19	1.10	-21	-20
Jun	3.00	100	-20	1.30	-20	-20
Jul	3.51	94	-23	1.42	-22	-23
Aug	3.00	93	-21	1.30	-22	-21
Sep	3.00	89	-21	1.90	-21	-21
Oct	1.50	100	-17	1.10	-19	-18
Nov	1.50	100	-17	1.30	-18	-17
Dec	3.00	63	-16	2.89	-16	-16

*The percent of the total days in the given month in the POR (Water Years 1931-1980) that the release discharge would have been equaled or exceeded by the historical daily mean flow.

only over the total number of days that had a daily mean flow of not more than the 10-percentile limitation. Days that had greater flows were ignored. The procedure is demonstrated for July (Table 9).

For BSP-3 (Table 3), the maximum monthly 10-percentile exceedance flow was 11.54 m^3 s^{-1} in July, which became period's upper limit for 3:1 MPD value extrapolation (Table 5). In the POR, all months of July combined had a total of 341 days, of which only 90%, or 307 days, were included in the integration/averaging computations; the remaining 34 days had daily mean flows exceeding 19.54 m^3 s^{-1}. For August and September, also in BSP-3, integration of flow-related 3:1 MPD values was also performed over the total number of days that had daily mean flows not greater than 19.54 m^3 s^{-1}.

Table 10 lists the historical monthly mean flows and mean 3:1 MPD values for the Impact Reach.

Table 9. The distribution of daily mean streamflows ($m^3 s^{-1}$) and 3:1 MPD values (%) in the LCPP impact reach in July (water years 1970-1980).

Flow Class	3:1 MPD	Days in Class
2.61- 2.80	-27	2
2.83- 3.09	-25	21
3.11- 3.37	-23	4
3.40- 3.65	-23	4
5.66- 6.20	-26	5
6.23- 6.77	-28	15
6.80- 7.33	-31	10
7.36- 8.18	-32	33
8.21- 9.03	-34	30
9.06- 9.88	-36	27
9.91-10.73	-36	20
10.76-11.86	-37	28
11.89-13.00	-36	26
13.03-14.41	-36	27
14.44-15.83	-36	21
15.86-17.25	-35	15
17.27-19.23	-34	18
19.26-20.93	-32	7(1+6*)
20.95-23.19	--	5*
23.22-25.46	--	6*
25.49-28.01	--	10*
28.03-31.12	--	7*

Total days in POR   = 341
Total days ignored in integration/averaging   = 34
Total days used in integration/averaging   = 307
Mean 3:1 MPD   = -33

*Excluded from integration because flow exceeds maximum 10-percentile exceedance value (19.54 $m^3 s^{-1}$); only 1 out of 7 days in flow class 19.26-20.93 did not exceed 19.54 $m^3 s^{-1}$, and is included in integration.

Table 10. Historical monthly mean streamflows ($m^3 s^{-1}$) and mean 3:1 MPD (%) in the LCPP impact reach (water years 1970-1980).

	Mean Flow	Mean 3:1 MPD
Jan	3.31	-18
Feb	3.14	-18
Mar	3.79	-18
Apr	7.02	-26
May	17.81	-26
Jun	22.26	-25
Jul	11.84	-33
Aug	6.85	-30
Sep	5.38	-27
Oct	4.45	-34
Nov	3.94	-34
Dec	3.45	-17

Step 15

From the target taxa and life stages (Step 3) and the WUA data (Step 6), the summary WUA statistics for both the actual best and the historical streamflow regimes are computed. Monthly mean flows might be adequate descriptors, but other statistics (base, median, or ranges of flows) may be preferable. The percent changes from historical to best flows are computed. The monthly mean 3:1 MPD values of each regime are compared.

LCPP Example. The monthly 3:1 MPD values of the historical and best streamflow regimes are shown in Table 11 for comparison.

For the historical streamflow regime monthly mean flows in both impact subreaches were determined and converted, using Figures 2 and 3, to monthly mean WUA values in each subreach for all target taxa and life stages according to their months of presence (Table 3). Then, the monthly mean WUA values were converted to annual means (Table 12).

For the actual best streamflow regime, the Table 9 monthly flow values for both impact subreaches were also converted to WUA values. The monthly mean WUA values were then converted to annual means and listed with the percent changes from the historical values in Table 12.

Table 12 shows large increases in habitat for brown trout spawning and benthos, small increases for brown trout fry and juveniles, and small to moderate losses for brown trout adults and mountain whitefish. Since

20

Table 11. A comparison of streamflow ($m^3 s^{-1}$) and 3:1 MPD values (%) of the historical and actual best streamflow regimes in the LCPP impact reach.

	Historical		Best	
	Flow*	3:1 MPD	Flow	3:1 MPD
Jan	3.31	-18	2.01	-17
Feb	3.14	-18	2.01	-17
Mar	3.79	-18	2.01	-17
Apr	7.02	-26	1.50	-19
May	17.81	-26	2.10	-20
Jun	22.26	-25	3.00	-20
Jul	11.84	-33	3.51	-23
Aug	6.85	-30	3.00	-21
Sep	5.38	-27	3.00	-21
Oct	4.45	-34	1.50	-18
Nov	3.94	-34	1.50	-17
Dec	3.45	-17	3.00	-16

*Monthly mean values.

Figures 2 and 3 show that maximum WUA occurs at different flows for different life stages, the best flows would not provide increases in all life stages. Another management goal could alter the best streamflow regime and lead to entirely different WUA means.

The designation of best streamflows is not simply a matter of selection of maximum WUA values for a desired life stage, but must include considerations for desired management results and balance with other stages and species. Some species might actually be unwanted, but inclusion of their habitat needs in the analysis keeps track of their habitat trends while maintaining emphasis on desired taxa. The losses in mountain whitefish WUA values shown in Table 12 are consistent with the management goal that brown trout would receive five times more emphasis than whitefish in the final streamflow regime. The coincident loss in brown trout adult habitat is an unavoidable consequence of the same predetermined constraints.

If only the actual best streamflow regime for aquatic habitat management is needed, the analysis would end here. If additional alternative regimes are necessary, go to Step 16.

Table 12. A comparison of annual mean WUA values ($m^2(1,000 \text{ m})^{-1}$) of the historical and actual best streamflow regimes in the LCPP impact reach.

| | Streamflow Regime | | Percent Change |
	Historical	Best	From Historical
	Brown Trout		
Spawning	74	290	+292
Egg incubation	2,706	2,682	-1
Fry	4,120	4,198	+2
Juvenile	3,748	4,010	+7
Adult	2,127	1,822	-14
	Mountain Whitefish		
Spawning	1,998	1,501	-25
Fry	758	599	-21
Juvenile	2,813	2,657	-6
Adult	1,574	1,250	-21
	Benthic Invertebrates		
Larval/Nymphal	1,637	2,197	+34

Step 16

With the comparisons of the historical and best possible streamflow regimes in Step 15, it is a simple matter to use the same data, assumptions, and constraints to develop and compare nearly any alternative streamflow regime. The MPD or mean MPD values from Steps 10 and 11 furnish the refined data necessary for flow selection, governed by the constraints dictated in Steps 1 and 13. The WUA values from Step 6 facilitate conversions of selected flows to habitat areas for further comparison.

In many streamflow alteration projects, it is necessary to prescribe alternative streamflow regimes that differ markedly from historical conditions. In many cases, habitat losses are unavoidable, but could be minimized. In others, however, habitat enhancement is possible, especially where degradation results from high runoff flows. Historical streamflows rarely provide optimal aquatic habitat, and reductions in high summer flows can lead to large habitat improvement. In fact, streamflow recommendations based on retention of only a portion of historical habitat, under the assumption that historical conditions are the best attainable, may result in unnecessary sacrifices, since a carefully selected regime of flows lower than historical may retain all of or more than historical habitat quantity and quality.

Table 13. Alternative streamflows ($m^3 s^{-1}$), based on MPD values (%) and other essential habitat parameters, in the LCPP impact reach.

| | Impact Subreach B | | | Impact Subreach A | | Impact Reach |
	Release Discharge	Duration Percentage*	MPD	Predicted Flow	MPD	3:1 MPD
Jan	1.50	100	-18	1.39	-19	-18
Feb	1.50	100	-18	1.39	-19	-18
Mar	1.50	100	-18	1.39	-19	-18
Apr	1.50	100	-19	1.39	-20	-19
May	2.10	100	-19	1.10	-21	-20
Jun	3.00	100	-20	1.30	-20	-20
Jul	3.40	95	-23	1.30	-23	-23
Aug	2.89	94	-21	1.19	-22	-22
Sep	2.29	100	-22	1.19	-23	-22
Oct	1.50	100	-17	1.10	-19	-18
Nov	1.30	100	-17	1.10	-19	-18
Dec	1.50	100	-19	1.39	-19	-19

*The percent of the total days in the given month in the POR (Water Years 1931-1980) that the release discharge would have been equaled or exceeded by the historical daily mean flow.

LCPP Example. The comparisons of historical and best flows and 3:1 MPD values (Table 11) and the graphs of flow-versus-3:1 MPD values (Figures 7-9) show that considerable flexibility exists in the selection of flows that maintain at least historical levels of aquatic habitat. In fact, large reductions in summer flows lead to apparent large increases over historical habitat because historical conditions have been limited by high current velocities due to high and prolonged runoff flows.

Table 13 shows an alternative streamflow regime that retains most of the habitat values of the actual best regime, but with lower monthly release discharges. Gains in MPD values would occur in 11 of 12 months, but the greatest gains would occur in summer. The only loss (of only 2% in the 3:1 MPD value) would be in December, but the recommended release discharge is balanced with surrounding months.

Tables 14 and 15 compare the flows, 3:1 MPD values, and annual mean WUA values of the historical and alternative streamflow regimes. Table 13 shows that the alternative regime would retain nearly all of the habitat areas of the actual best regime (Table 11). The alternative regime was

Table 14. A comparison of streamflow (m^3 s^{-1}) and 3:1 MPD values (%) of the historical and alternative streamflow regimes in the LCPP impact reach.

| | Historical | | Alternative | |
	Flow*	3:1 MPD	Flow	3:1 MPD
Jan	3.31	-18	1.50	-18
Feb	3.14	-18	1.50	-18
Mar	3.79	-18	1.50	-18
Apr	7.02	-26	1.50	-19
May	17.81	-26	2.10	-20
Jun	22.26	-25	3.00	-20
Jul	11.84	-33	3.40	-23
Aug	6.85	-30	2.89	-22
Sep	5.38	-27	2.29	-22
Oct	4.45	-34	1.50	-18
Nov	3.94	-34	1.30	-18
Dec	3.45	-17	1.50	-19

*Monthly mean values.

Table 15. A comparison of annual mean WUA values ($m^2(1,000)^{-1}$) of the historical and alternative streamflow regimes in the LCPP impact reach.

| | Streamflow Regime | | Percent Change From Historical |
	Historical	Alternative	
	Brown Trout		
Spawning	74	294	+297
Egg incubation	2,706	2,450	-9
Fry	4,120	4,237	+3
Juvenile	3,748	4,055	+8
Adult	2,127	1,800	-15
	Mountain Whitefish		
Spawning	1,988	1,500	-25
Fry	758	624	-18
Juvenile	2,813	2,579	-8
Adult	1,574	1,195	-24
	Benthic Invertebrates		
Larval/Nymphal	1,637	2,171	+33

presented as the final recommended streamflow regime for implementation with the LCPP.

ACKNOWLEDGEMENTS

Thanks are extended to Maureen M. Wilson for her assistance in the preparation and review of this manuscript.

REFERENCES

Bovee, K. D. 1978. Probability-of-use criteria for the family Salmonidae. Instream Flow Information Paper No. 4. Cooperative Instream Flow Service Group, U.S. Fish and Wildlife Service, Fort Collins, CO.
Bovee, K. D. 1982. A guide to stream habitat analysis using the instream flow incremental methodology. Instream Flow Information Paper No. 12. Cooperative Instream Flow Service Group, U.S. Fish and Wildlife Service, Fort Collins, CO.
Bovee, K. D. & Milhous, R. 1978. Hydraulic simulation in instream flow studies: theory and techniques. Instream Flow Information Paper No. 5. Cooperative Instream Flow Service Group, U.S. Fish and Wildlife Service, Fort Collins, CO.
Milhous, R., Wegner, D. L. & Waddle, T. 1981. User's guide to the physical habitat simulation system. Instream Flow Information Paper No. 11. Cooperative Instream Flow Service Group, U.S. Fish and Wildlife Service, Fort Collins, CO.
Searcy, J. K. 1969. Manual of hydrology. Part 2: Low flow techniques--flow duration curves. Water Supply Paper No. 1542-A. U.S. Geological

A CRITIQUE OF THE INSTREAM FLOW INCREMENTAL METHODOLOGY

AND OBSERVATIONS ON FLOW DETERMINATION IN NEW ZEALAND

D. Scott and C.S. Shirvell [1]

Department of Zoology
University of Otago
Dunedin, New Zealand

INTRODUCTION

One of the most popular models for simulating the effect of regulated
streamflow on fish habitat is the Instream Flow Incremental Methodology
(IFIM). This model was developed by the Instream Flow Group (IFG) and is
composed of components which simulate water temperature, water quality,
and physical habitat. The physical habitat component (PHABSIM), however,
is so frequently the only part of the method used that PHABSIM and IFIM
are often confused.

PHABSIM (and therefore IFIM) is based on several assumptions which
are not always met (Orth & Maughan 1982; Cada et al. 1983; Annear &
Conder 1984; Mathur et al. 1985). Tests relating fish biomass to amounts
of habitat based on these assumptions have given inconsistent results
(White et al. 1981; Orth & Maughan 1982; Nelson et al. 1984; Loar
1985). Because IFIM has become so widely used for water management
decisions (it is now a legal requirement in Idaho, California, Colorado,
Washington, and Oregon and is receiving increasing use in Canada and New
Zealand) there is a danger of extensive damage to natural resources if its
predictions are inaccurate. The purpose of our paper is to examine the
validity of PHABSIM's assumptions, to review the accuracy of PHABSIM's
predictions, and to discuss other methods that have been used for
determining instream flow needs in New Zealand.

Assumptions in PHABSIM

1. PHABSIM assumes that water depth, water velocity, and substrate
    size are the only physical habitat variables determining position
    choice by fish.

PHABSIM models an index of fish habitat called weighted usable area
(WUA) under different streamflows. Because PHABSIM uses only water depth,
water velocity and substrate size to calculate WUA it is implicit that the
model assumes these are the only variables determining the suitability of
physical habitat. However, other factors affect position choice by fish

---

[1]Present address: Department of Fisheries and Oceans, Rm. 207
417-2nd Ave. West, Prince Rupert, B.C., Canada V8J 1G8

and thereby determine a location's desirability. Cover is a good example of an important habitat variable imperfectly modelled by PHABSIM. While cover is well accepted as an important habitat element there is little agreement on what constitutes cover and poor understanding on how it changes with changes in streamflow. Unlike other habitat variables it is discontinuous and therefore does not fit the mathematical weighting procedure PHABSIM uses for other variables. Also no preference curves for cover have ever been developed for any species. Because cover is difficult to model, PHABSIM treats cover as though it does not change with changes in streamflow. This is false, however, when surface turbulence or water depth act as cover elements. Furthermore changes in streamflow affect access to conventional cover like logs or undercut banks. A test comparing PHABSIM's predictions of cover with locations where cover was actually present was correct for only 52% of all observations (Gowan 1984). The result of imperfect modelling of cover or inadequate inclusion of other variables important in determining habitat suitability is that PHABSIM could calculate conditions only partly relevant to fish production.

2. PHABSIM assumes that Manning's n remains constant with changes in streamflow.

PHABSIM uses either of two programs to simulate hydraulic conditions at different streamflows: IFG2 and IFG4. In IFG2, Manning's n, a coefficient of the bed roughness used in calculating the water depth, is assumed to remain constant with changes in streamflow. Because Manning's n is measured only once for the calibration streamflow, the true value of n is not known with certainty for any other streamflows simulated by the model. However, Manning's n varies from place to place and with streamflow (see Bovee & Milhous 1978; and Horton & Cochnauer 1980 for a discussion). Assuming a constant Manning's n leads to mean errors in the velocity simulations of 39% and maximum errors of 133% - 200% (Elser 1976; Bovee & Milhous 1978).

3. PHABSIM assumes that mean water velocities in individual cells change in the same way as the mean velocities for a cross-section with changes in streamflow.

The relationship

$$v = aQ^b \tag{1}$$

where v = velocity (m s^{-1})
         Q = streamflow (m^3 s^{-1})
    a and b = intercept and slope respectively
               of the log-log regression equation

is well accepted for the mean water velocity in a channel (Leopold & Maddock 1953). However, the hydraulic simulation program IFG4 uses a similar equation

$$v_i = a_i Q^{b_i} \tag{2}$$

where all the symbols have the same meaning except now for the ith cell to calculate water velocities in individual cells. This relationship is not accepted by hydraulic engineers (Bovee & Milhous 1978) and because the boundaries of individual cells are fluid rather than stationary like the stream banks, the use of this equation causes a mean error in the velocity simulations of 14% and a maximum error of 55% (Bovee, Gore & Silverman 1977; Gowan 1984).

28

4. PHABSIM -IFG2 assumes that water velocities at 6/10 of the depth affect fish preference.

Hydraulic simulations from IFG2 model water velocities at 6/10 of the depth from the surface. Because many riverine fishes reside on or near the bottom, this results in PHABSIM simulating water velocities above the depths these fish actually use. In large rivers where mean depths are greater than 1 metre, the distance between these two locations can exceed 50 cm, and the water velocity difference between these two positions can exceed 44% (Shirvell unpublished data). Because water velocity is so important in the calculation of WUA, this velocity difference can result in differences in the WUA calculation as large as 100% (Nelson 1980; Shirvell & Morantz 1983). Unless fish perceive and respond to water velocities at 6/10 of the depth, modelling habitat conditions at this position could be irrelevant to a population's biomass. As Bachman (1984) notes the position of a feeding trout is related to the vertical distribution of velocity so as to minimize energy cost.

5. PHABSIM assumes habitat preference curves can be treated as probability functions.

IFIM interprets habitat preference curves mathematically as probabilities of occurrence (Bovee & Cochnauer 1977; Bovee 1978a; Bovee 1982) but calls them suitability index curves. However IFIM interprets the modal postion of a frequency distribution curve as having a probability of one, and that points other than the mode as having probabilities of less than one. This leads to the false implication that fish will be found with certainty at locations having the modal level of occurrence of a habitat variable when in fact fish will be found with certainty only at locations within the range of their use of that habitat variable. The use of ratings as probabilities can lead to low correlation between WUA and biomass (Mathur et al. 1985). Although the IFG now discourages the name "probability-of-use" for these curves, some users still refer to them as probabilities (e.g. Glova 1982; Anderson 1984; Sheppard & Johnson 1985) and the IFG still interprets them as probabilities of occurrence (Bovee 1982, p. 176). And because PHABISM multiplies the weighting factor for two or more habitat variables together to calculate WUA, the model still manipulates these curves as if they were probability functions.

6. PHABSIM assumes habitat variables are independent in their influence on position choice.

A second consequence of multiplying the frequency of occurrence level for habitat variables together is that it implicitly assumes fish perceive the suitability of habitat variables independently. This assumption has been found to be false by Orth and Maughan (1982) and Gowan (1984). Interaction in the selection of multiple habitat variables by fish leads to practical skews in their distribution within the stream (Shirvell & Dungey 1983) which PHABSIM can not predict because it treats them as if they were independent. Interaction between habitat variables has been calculated to explain 30% of the known variation in abundance of fish when no other variables, including the habitat index WUA, were significant (Mathur et al. 1985).

7. PHABSIM assumes large areas of less than optimum habitat have the same productive capacity as small areas of optimum habitat.

PHABSIM, because it multiplies a composite weighting factor for several habitat variables times the area of stream having those conditions, implicitly assumes that (for example) 2 m^2 of half usable habitat will produce the same biomass of fish as 1 m^2 of fully usable habitat. This assumption will fail if the relationship between a fish's perception of habitat desirability and the habitat's productive capacity is nonlinear. For example, Baldes and Vincent (1969) found that brown trout (*Salmo trutta*) in an experimental flume would not use preferred habitat conditions once the area with those conditions became smaller than 0.14 m^2. In a more recent comparison of locations actually occupied by brown trout in a natural stream, Gowan (1984) found that 29% (4 cells out of 14) predicted as usable by PHABSIM were not occupied. Mathur et al. (1985) make the same point.

8.   PHABSIM assumes that areas of stream not occupied by fish are useless.

As a consequence of PHABSIM multiplying weighting factors for habitat variables together, combined with the possibility that the weighting factor for one habitat variable could be zero, it is possible for areas of stream with suitable levels for some, but not all, habitat variables to be judged as having no usable habitat. This implies that areas of the stream not actually occupied by fish can be eliminated with no effect on the fish population. When a decision is made to reduce streamflow in rivers because PHABSIM has predicted greater amounts of WUA at streamflows lower than those which occur naturally, managers are choosing to discard some productive capacity of the ecosystem for increased amounts of physical microhabitat based on the rationale that "non-occupied areas are useless". This is dangerous because it could eliminate habitat indirectly important to the fish population. Gowan (1984) for example found that 29% of a brown trout population (21 fish out of 72) occupied areas of the stream predicted as unusable by PHABSIM.

TESTS OF VALIDITY

Violation of these eight foregoing assumptions either singly or cumulatively can result in errors in the calculation of WUA. These errors combined with flaws in interpreting how habitat availability functions to determine population size can lead to incorrect conclusions about the effect of a streamflow regulation on a fish population. IFIM assumes there is a direct linear relationship between WUA and fish standing crop (Bovee 1978a, p. 345). Accepted tests of PHABSIM's validity therefore are based on the strength of correlation or regression between standing crop and the model's predictions of WUA. So far eleven studies have published 444 analyses of this relationship (Table 1). Some studies (e.g. Stalnaker 1979; Nelson et al. 1984) have found very strong relationships betwwen WUA and fish biomass, while others have found weak (e.g. Orth & Maughan 1982) or even negative (e.g. Shirvell & Morantz 1983; Loar 1985) relationships. The majority (74%) found no relationship at all.

The best relationships between weighted usable area and fish populations have been found for the biomass of populations at carrying capacity in several different streams (i.e. one data pair per stream) before a streamflow perturbation (e.g. Stalnaker 1979; Anderson 1984). Slightly less strong are relationships between a single population in one stream at several different times (i.e. multiple data pairs per single stream) (e.g. Orth & Maughan 1982; Gowan 1984). The relationship between fish numbers and weighted usable area is strongest for older age classes where usable space appears to be a factor limiting their abundance; fish numbers of young age classes whose biomass may be more related to energy availability have less strong relationships with WUA. Regardless of whether biomass or fish numbers are used, however, no relationship has been demonstrated after a streamflow perturbation. And no relationship between fish production and weighted usable area has been demonstrated either before or after a streamflow perturbation.

30

Table 1. Relationship between weighted usable area calculated by the Instream Flow Incremental Methodology and fish standing crops.

Reference	Region	Species	No. of Analyses	Strength of Relationship
Stalnaker 1979	Wyoming	brown trout (*Salmo trutta*)	1	very good, r^2=0.81
Anderson 1984	Colorado	brown trout	2	very good, r^2=0.89
Nelson et al. 1984	Pennsylvania	rock bass adults (*Ambloplites rupestris*)	2	positive, significant correlation max r^2=0.88, very good
Gowan 1984	Michigan	brown trout, 1984	1	very good, positive, significant correlation r^2=0.84
CIFSG 1979	Wyoming	brook trout (*Salvelinus fontinalis*)	1	good, r^2=0.59
Orth & Maughan 1982	Oklahoma	stoneroller (*Campostoma anomalum*) freckled madtom (*Noturus nocturnus*) orangebelly darter (*Etheostoma radiosum*)	4 4 4	good, correlation but only in summer r^2=0.47 to 0.85
Loar (Ed.) 1985	Tennessee North Carolina	brown trout	68	good; 68 out of 160 (43%) significant, positive correlation, r^2=0.41
White et al. 1981	Oregon (artificial stream)	rainbow trout (*Salmo gairdneri*) summer, autumn low streamflow	1	average, positive relationship but nonlinear
Shirvell & Morantz 1983	Nova Scotia	Atlantic salmon, juvenile (*Salmo salar*)	1	poor, negative correlation, r^2=0.1 to 0.5
Orth & Maughan 1982	Oklahoma	smallmouth bass (*Micropterus dolomieui*) juvenile adult	4 4	poor, no significant correlation r^2=0.0 to 0.5
Nelson 1980	Montana	brown trout	1	poor, Incremental Method predicted 50% optimum streamflow
Irvine et al. (in press)	New Zealand (artificial streams)	rainbow trout	8	poor, no relationship
Loar (Ed.) 1985	Tennessee	brown trout	82	poor; 10 out of 160 (6%), significant, negative correlation. r^2=0.40
Loar (Ed.) 1985	Tennessee North Carolina	rainbow trout	226 14	poor; 3 out of 240 (1%) significant, negative correlation. r^2=0. 14 out of 240 (6%), significant, positive correlation. r^2=0.45
Nelson et al. 1984	Pennsylvania	rock bass juveniles (*Ambloplites rupestris*)	2	poor, no significant correlation
White et al. 1981	Oregon (artificial stream)	rainbow trout, autumn, high streamflow	1	poor, negative correlation
Gowan 1984	Michigan	brown trout, 1983	1	poor, no relationship
Anderson 1984	Colorado	rainbow trout	2	poor, no relationship
			444	

The existence or absence of a good relationship (and therefore the success or failure of IFIM) could be due to the quality of a study's execution. This does not appear to be the explanation, however, as both strong and non-existent relationships have been found in the same study. For example Nelson et al. (1984) comment that they took extreme care to ensure their study was valid, yet they found a strong relationship for adult fish but none for juveniles. Likewise Orth and Maughan (1982) discovered the strength of the relationship varied between seasons, Gowan (1984) noted it varied between years, and Anderson (1984) found it varied between species. Studies in Table 1 indicate no consistent pattern which would allow one to predit when WUA would be a reliable determinant of fish biomass. Because a strong positive linear relationship occurs so infrequently, and appears to be so susceptible to seasonal and annual variation, it may be that the existence of a positive, significant relationship in those studies where it was found was due to chance. The commonly accepted interpretation in statistics where the existence of a relationship is rejected when it could occur by chance in 1 out of 20 occurrences suggests that too much importance may have been placed on IFIM predictions. If this interpretation were applied to the studies in Table 1, one would conclude that WUA is an unreliable predictor of fish biomass.

THE NEW ZEALAND EXPERIENCE

New Zealand is well supplied with water, over 70% of the country receiving more than 1100 mm of precipitation per year. The total annual runoff to the oceans is estimated at 400 km^3, and total consumptive use is estimated at 1.847 km^3. Table 2 shows how this consumption is distributed in relation to five other countries. However, irrigation demand is significant, particularly in the rain-shadow provinces of Otago and Canterbury where 84% of the total irrigated area is concentrated. Hydro-electricity may be a consumptive use in that it can divert water from a river for use elsewhere, and of the 62 schemes operating at least 21 involve diversion of water (Scott 1978). Against this background the response of fisheries administrations has varied from acquiescence in total diversion, as in the Southland Waiau, to expending more than 20 person years as in the Rakaia, investigating the possible consequences of diversion. Where there has been an attempt to negotiate discharges for fisheries, the responses can be classified in terms of increased complexity of evidence presented.

Table 2. Water use in New Zealand (Howie 1979).

	France	USA	Usage % UK	USSR	Japan	NZ
Agriculture	38	40	3	52	70	62
Industry	50	50	68	40	20	19
Domestic	12	10	29	8	10	19

1. Informal Approaches

The passing of the Water and Soil Act in 1967 made it easier for fisheries administrators to press for recognition of fisheries requirements in rivers subject to abstraction. By the early 1970s negotiations aimed at maintenance of flows involved experienced biologists or administrators using available flow data to arrive at flow minima and regimes that would maintain the fish population and assist angling.

The Tongariro River is a very valuable river fishery for upstream migrant rainbow trout, as well as providing recruitment for Lake Taupo. A substantial hydro-electricity development scheme involved abstraction beginning in 1973, and negotiations resulted in guaranteed flows (Richmond 1979). The mean annual discharge at the lower end of the river (Turangi) was about 53 m^3 s^{-1} and a minimum of 27.2 m^3 s^{-1} was agreed upon. The basis for this minimum was that it approximated a fairly uniform summer baseflow, thus safeguarding juvenile rearing and adult migrants (Richmond loc. cit.). It was considered that winter flows would be adequate, and Figure 1 shows the discharge before and after development. Analysis of impacts is continuing, (Richmond, pers. comm.) but it appears that there was an initial decline in the catch rate from 1973 onward, followed by a recovery (Figure 2). In addition the reduced winter flows and altered flood hydrographs may have decreased the rate at which fine sediment is moved through the system. However increased stability of discharge has improved

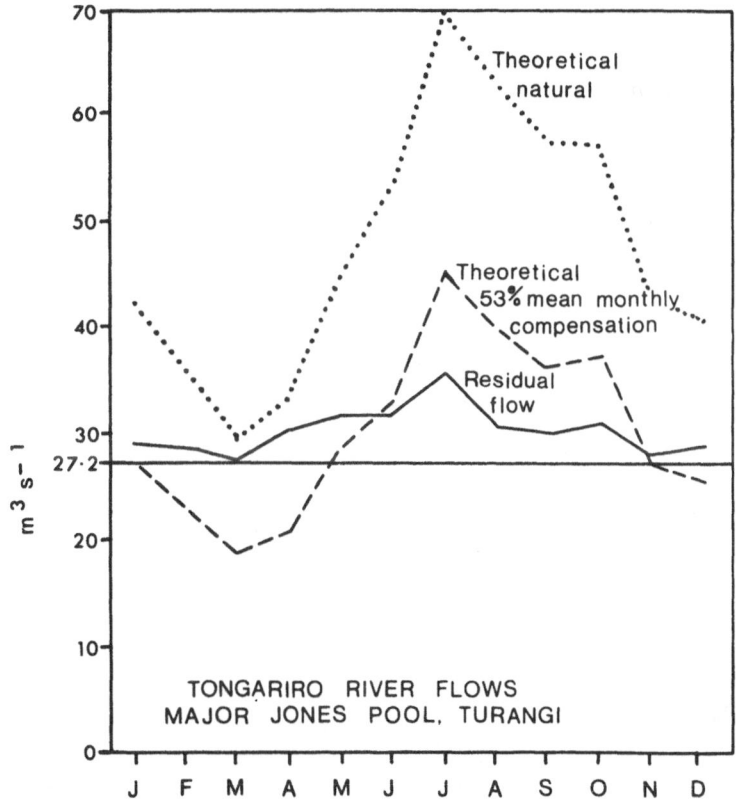

Figure 1. Seasonal flows in the Tongariro River showing flows before and after abstraction.

the habitat for juveniles, and there is no indication of a major change in the numbers of spawning migrants.

The rain fed rivers of the Canterbury Plains were prone to very low flows prior to significant abstraction, and the development of irrigation in this region has been of particular concern to fisheries. The Opihi River is typical of many smaller Canterbury rivers and supported a satisfactory population of brown trout and quinnat salmon *(Oncorhynchus tshawytscha)* (Bull 1979) prior to the development of an irrigation scheme in 1936. This scheme allowed diversion of 3.05 m^3 s^{-1}, although the actual draw-off has probably not exceeded 2.26 m^3 s^{-1} (Hardy 1972). The 50 percentile and 95 percentile (low) on the flow duration curve corresponded to 11.0 and 3.2 m^3 s^{-1} respectively, and the mean annual discharge is 30 m^3 s^{-1}, so that in an average year the scheme takes 27% of the discharge. In years of low flow the river below the irrigation intake could infiltrate into the gravel bed, so that no surface water was apparent (Hardy loc. cit). Against this background fisheries interests argued that maintenance of a continuous flow below the irrigation intake was essential and that abstraction for irrigation should be decreased progressively to achieve this. Where this action was still insufficient to maintain a continuous flow it was proposed that relief flows be achieved by a partial shutdown of abstraction for 24 hours each week. These arguments were accepted and the concept of sharing a water shortage was introduced as follows in 1973:

Opihi flow at intake	Irrigation share
5.7 - 4.0 m^3 s^{-1}	2.0 m^3 s^{-1} or less
<4.0 m^3 s^{-1}	1.14 m^3 s^{-1}

In addition when no continuous flow in the river below the intake was likely, the irrigation share was to be nil for 24 hours each week, and at times of salmon migration the period of 24 hours could be extended to 7 days (Fancourt 1974). After 5 years of operation an informed observer suggested that while the sharing concept was welcome, the level had been

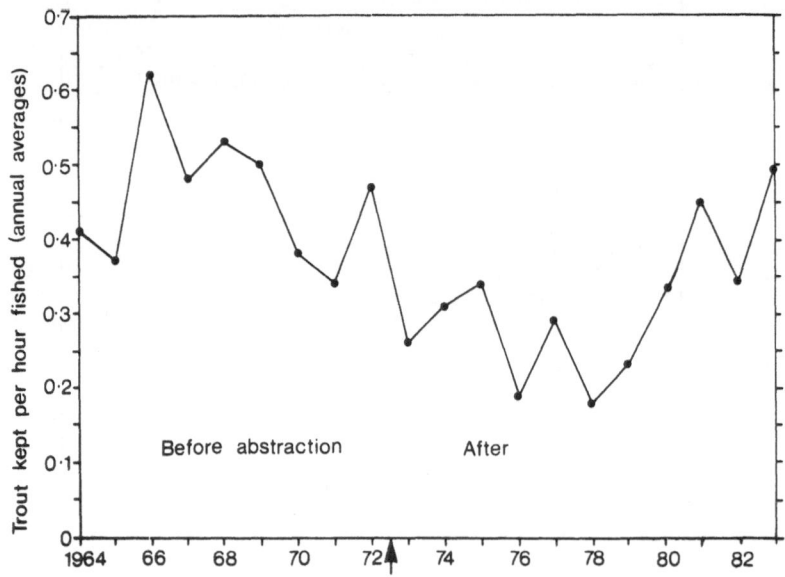

Figure 2.   Catch rate for trout in the Tongariro River.   The arrow indicates the start of the reduced flow regime.

set too low and the fishery had declined seriously (Bull 1979). A further agreement was reached aimed at reducing the frequency of low flows and came into operation in 1984 (Sagar & Jellyman 1985). This has already resulted in a more gradual flow recession.

2. Discharge Percentage Methods

The methods used have been either the Montana method (Tennant 1976) or Fraser's modification of this (Fraser 1978). Typically these methods have been applied in a context where fisheries interests have been unable either to carry out detailed studies, or to delay a judicial decision.

Deep Stream is a small (mean annual discharge 3.3 m^3s^{-1}) brown trout fishery in Otago already subject to abstraction for water supply. A further proposal to divert high flows for electricity generation was made in 1979, but it became apparent that this would involve substantial reductions in summer flows which were already low. The fisheries interests (Otago Acclimatisation Society) made a counter proposal based on 30% of the mean annual discharge. This would have resulted in a reduction of minimum summer flows (January - March) by 37% as opposed to 47% in the original proposal. The differences are shown as flow duration curves in Figure 3, and it is clear that there is a benefit at low flows from the counter proposal. The case was argued in 1981 and the decision incorporated the proposal by the Otago Society. Although a full evaluation of the fishery has not yet been made, local opinion indicates that angling is still satisfactory.

The Taieri River, in its upper reaches, is a slow meandering willow lined river with good sized brown trout (mean length of fish in angler's catch = 43-50 cm). An irrigation scheme was proposed involving withdrawal of water from the river, and a minimum flow of 0.85 $m^3 s^{-1}$ (10% of the M.A.D.) imposed without the agreement of the fisheries administration. The mean summer flows as $m^3 s^{-1}$ (1967-78) were:

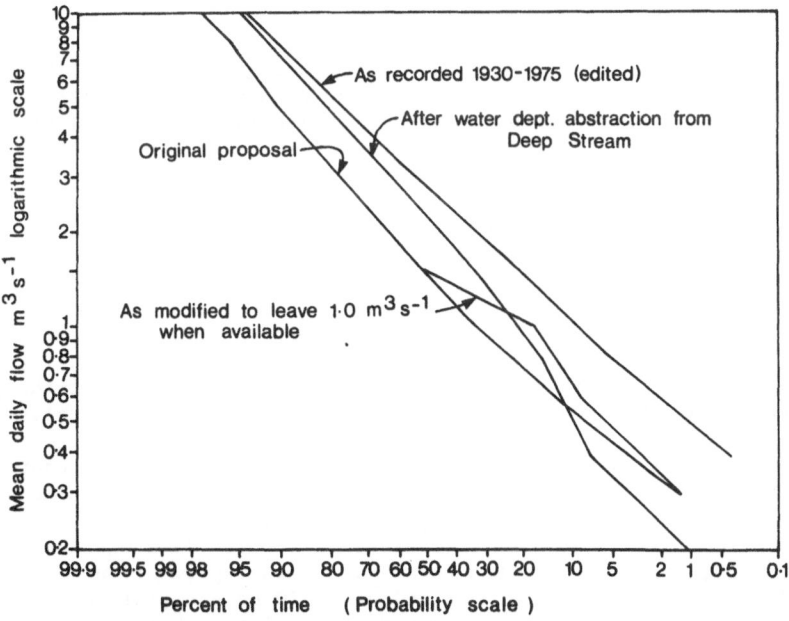

Figure 3. Flow duration curves for Deep Stream under four flow regimes.

35

Nov	Dec	Jan	Feb	March
8.4	4.0	3.8	2.1	2.6

The proponents of the scheme claimed that while the flow regime would involve the river dropping to the minimum for most of the irrigation season (September to April) extreme low summer flows would be avoided. In fact for the worst period, January–March, the proportion of daily flows less than $0.85 \text{ m}^3 \text{ s}^{-1}$ would have been 7.5%. Although the scheme was approved in 1975, it is now being reviewed.

3. Incremental Methodology

This method has been employed in at least 11 locations so far, and some of the relevant details are given in Table 3. Apart from the specific comments made above on WUA as a measure of potential physical microhabitat, it should be noted that, unlike the methods employed earlier in New Zealand, calculation of WUA does not lead directly to recommendations for a flow regime for fisheries protection (Jowett 1982; Glova 1982; Mosley 1983b). Although estimates of WUA have been used in negotiating flows, it cannot be said that they have dominated such negotiations. In all of the studies noted in Table 3, depth, velocity and substrate were measured. Object cover was not measured, although Mosley (1983a) measured water surface turbulence, and it can be argued (Bovee 1982, and this paper) that this is a significant omission. However Loar (1985) in a substantial study on streams in the Appalachians estimated instream and riparian cover. The cover estimates were not incorporated into concurrent estimates of WUA, but some negative correlations were found between trout biomass and estimates of cover, and there was a preponderance of nonsignificant correlations. In view of conflicting reports on the value of cover (Hunt 1976; Murphy et al. 1981; Hawkins et al. 1983) it would be desirable to include cover in incremental studies, particularly in relation to the size of the stream. The preference curves for fish cited in the New Zealand studies are taken from both North American (Waters 1976; Bovee 1978b) and New Zealand (Glova 1982; Shirvell & Dungey 1983) studies. In the case of invertebrates they are entirely from North American data, and this may or may not be appropriate.

The comments of the users of IFIM are of interest. Mosley (1983a) discussing braided rivers considered (p. 50) that WUA showed little consistent relationship with discharge if predictions between rivers were of interest; in extrapolation for a single river he considered that the scatter in the data made confident prediction difficult. Glova and Duncan (1985) estimated the change of WUA for several species over the range 69–$146 \text{ m}^3 \text{ s}^{-1}$ in a large braided river, the Rakaia (mean annual discharge $203 \text{ m}^3 \text{ s}^{-1}$), and concluded that for this range no change in WUA could be detected. Mosley (1983b), in a more general review, emphasized the importance of the distinction between potential usage and actual usage of habitat by fish, and noted that the relationships between many limiting factors and discharge have not yet been established. Irvine et al. (in press) examined the relation between WUA and biomass of young of the year rainbow trout in small controlled streams and could find no correlation.

The most detailed example of the use of IFIM in a judicial context is provided by the hearing on the Rakaia River. In 1983 an application was made for a National Water Conservation Order on the Rakaia by the New Zealand Acclimatisation Societies (Davis 1983). This river has the largest salmon run (20,000–30,000) in the country, and the application was supported by fisheries research equivalent to 20 person years of which about 3 person years was directed towards the relation between WUA and

Table 3. Application of IFIM in New Zealand.

Location	Derivation of WUA			Preference Curves		Application	Reference
	Predicted from 1 discharge	Interpolated or extrapolated from several discharges	Measured at 1 discharge	Fish	Food Supply		
Waitaki River			+	Bovee 1978b Waters 1976 Shirvell & Dungey 1983	Waters 1976	Not used in current flow application	Riddell & Jowett, 1983
Clutha River	+			Bovee 1978b	Waters 1976	Potential use in flow application	Jowett 1982
Tekapo River	+			Waters 1976	Waters 1976	Used in flow application	Jowett 1982
Ohau River		+		Bovee 1978b	Waters 1976	No application possible on site	Mosley 1982
Ashley River		+		Bovee 1978b Glova 1982		Little relation between WUA and discharge	Mosley 1983a
Hurunui River		+		Glova 1982		Little relation between WUA and discharge	Mosley 1983a
Rakaia River		+		Glova 1982		Little relation between WUA and discharge	Mosley 1983a
Aburiri		+		Glova 1982		Little relation between WUA and discharge	Mosley 1983a
Rakaia River		+		Glova 1982		WUA not decisive in flow negotiations	Glova & Duncan 1985
Fraser River	+			Bovee 1978b Waters 1976 Shirvell & Dungey 1983	Waters 1976	WUA not decisive in flow negotiations	Jowett 1983
Mataura River	+			Shirvell & Dungey 1983		Assessment of minimum flow	Riddell 1984
Replicate channels, Waitaki River		+		Bovee 1978b		No relation between WUA and fish biomass	Irvine, Jowett & Scott (in press)

discharge. A draft order was recommended in 1984 (Davis 1984) which reduced the minimum summer discharges substantially (Table 4 and Figure 4).

An appeal was made by the applicants against this draft (Davis 1985) on the grounds that the minima recommended were too low, and that the sharing ratio (Fig. 4) allowed too little variability in summer discharge. The evidence presented at the appeal did not rely on the relationship between WUA and discharge because it had been demonstrated (Glova & Duncan 1985) that there was no significant relation for various species of fish over the range 69-146 $m^3 s^{-1}$. Instead the evidence indicated that factors other than WUA were important in braided river channels in determining juvenile fish production and survival and the relation of fishability for anglers to discharge also indicated higher levels than those proposed. The appeal decision granted the original request of the applicants (Table 4).

4. Evaluation

The use of WUA in New Zealand as a means of predicting change in physical habitat as a consequence of flow is very recent, and it is therefore difficult to make a full evaluation. Nevertheless several points require comment in view of the increasing demand for diversion of rivers to out of stream uses. There has been insufficient attention directed to the relation between WUA and fish numbers or biomass, although Mosley (1983b) has stressed the point. The predictive value of WUA in negotiation depends on a positive relation, yet apart from the limited study of Irvine et al. (in press) which indicated no relationship, information is not available. This relationship needs to be well validated before confident predictions can be made about the usefulness of WUA. The same process of validation also provides evidence on whether or not physical habitat is limiting (Orth & Maughan 1982; Mathur et al. 1985). There are few situations in New Zealand where before and after studies on the relation between physical habitat and fish populations are feasible although

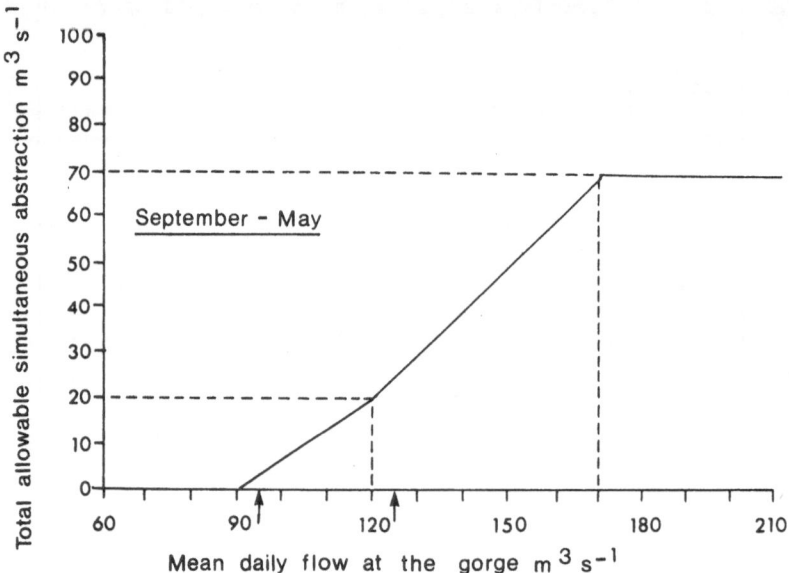

Figure 4.   Flow regime for the Rakaia proposed for the draft National Conservation Order. The arrows indicate the preexisting minima for Feb-April, the main fishing period.

Table 4. Mean and minimum monthly discharge of the Rakaia River at Rakaia Gorge 1959–82, m^3 s^{-1} (Ministry of Works and Development).

Month	Mean	Minimum	Requested by applicant
Sept	182	82	90.3
Oct	220	109	106.4
Nov	272	131	128.9
Dec	266	160	138.7
Jan	244	138	123.9
Feb	194	122	107.9
Mar	206	102	105.3
April	195	93	96.7
May	193	104	94.6

the Tongariro River and Deep Stream are possibilities. A study of this type in the Whakapapa River (Richardson & Teirney 1982) was limited by the amount of information available on fish populations before abstraction. Preference curves for indigenous fish are only in the process of being developed while preference curves for salmonids are taken largely from North America. The need to use preference curves developed in New Zealand is emphasised by the contrast between the curve for adult brown trout given by Bovee (1978b) from North American data, and by Shirvell and Dungey (1983) and Glova and Duncan (1985) for New Zealand data in respect of velocity. Preference curves for New Zealand stream invertebrates have not been developed. Winterbourn et al. (1981) have emphasised the opportunistic nature of the New Zealand lotic invertebrate fauna, and related this in part to the variability of New Zealand river ecosystems. The correlation between these invertebrates and their physical micro-habitat has not yet been determined, but an examination of Allen's findings (1951, p. 43) in respect to two important fish food items, *Deleatidium* and chironomid larvae, suggests that development of curves from local data would be preferable.

A further desirable approach is the comparison of IFIM with other methods in terms of cost benefit ratios as suggested by Mosley (1985). Campbell and Scott (1984) pointed to a phase change in behaviour of juvenile brown trout, related to a critical velocity, and considered that this phase change was well established for several salmonid species.

CONCLUSIONS

Although the use of IFIM in water management decisions is increasing, the uncritical application of the method is likely to cause difficulties. The assumptions in PHABSIM need to be carefully examined to see how they affect the usefulness of the model: variability in fish behaviour whether learned or genetic suggests that preference curves should be constructed for each river, or at least for river classes. The estimation of WUA is not an end in itself, and unless the relation of this index to fish biomass is well validated, it is difficult to see how it can be used decisively at the judicial level.

Experience in New Zealand so far does not indicate that IFIM is invariably more efficient than simpler methods, and while Mosley (1983b) has suggested that the incremental approach is the best available framework for assessing instream flow needs, it might be more helpful to regard IFIM as part of a framework that is still being developed.

REFERENCES

Allen, K. R. 1951. The Horokiwi Stream. N.Z. Marine Dept., Fish. Bull. 10:1-231.
Anderson, R. M. 1984. Fish flow investigations. In: Stream fisheries investigations. (Ed. by J.R. Ruch). pp. 1-25. Colorado Division of Wildlife, Fort Collins, Colorado.
Annear, T. C., and Conder, A. L. 1984. Relative bias of several fisheries instream flow methods. N. Amer. J. Fish. Mgt. 4: 531-539.
Bachman, R. A. 1984. Foraging behaviour of free-ranging wild and hatchery brown trout in a stream. Trans. Amer. Fish. Soc. 113: 1-32.
Baldes, R. J. and Vincent, R. V. 1969. Physical parameters of microhabitats occupied by brown trout in an experimental flume. Trans. Amer. Fish. Soc. 98: 230-238.
Bovee, K. D. 1978a. The incremental method of assessing habitat potential for coolwater species, with management implications. Amer. Fish. Soc. Spec. Pub. 11: 340-346.
Bovee, K. D. 1978b. Probability of use criteria for the family salmonidae. Instream Flow Information Paper, 4. FWS/OBS - 78/07. 80 p.
Bovee, K. D. 1982. A guide to stream habitat analysis using the instream flow incremental methodology. Instream Flow Information Paper, 12. FWS/OBS - 82/26. 248 p.
Bovee, K. D., and Cochnauer, T. 1977. Development and evaluation of weighted criteria, probability-of-use curves for instream flow assessments: fisheries. Instream Flow Information Paper 3. FWS/OBS - 77/63. 39 p.
Bovee, K. D., Gore, J., and Silverman, A. J. 1977. Field testing and adaptation of a methodology to measure instream values in the Tongue river, Northern Great Plains Region. U.S. Environmental Protection Agency, Office of Energy Activities Contract 68-01-2653.
Bovee, K. D., & Milhous, R. 1978. Hydraulic simulation in instream flow studies: theory and techniques. Instream Flow Information Paper 5. FWS/OBS-78/33. 130 p.
Bull, J. 1979. Opihi River abstraction and effects. In: The effects of water abstraction on fisheries. (Ed. by D. Scott). pp. 79-82. National Water Protection Committee of Acclimatisation Societies. Dunedin.
Cada, G. F., Sale, M. J., Cushman, R. M., and Loar, J. M. 1983. A test of a biological assumption underlying hydraulic-rating methods for assessing instream flow needs. Proceedings of Waterpower 1983, An International Conference on Hydropower. Knoxville, Tennessee. Oak Ridge National Laboratory publication. 9 p.
Campbell, R. N. B. and Scott, D. 1984. The determination of minimum discharge for 0+ brown trout (*Salmo trutta* L.) using a velocity response. N.Z. J. Mar. & Freshw. Res. 18:1-11.
CIFSG (Cooperative Instream Flow Service Group). 1979. Draft 3 year completion report. USFWS, Fort Collins, Colorado. 168 p.
Davis, S. F. 1983. Rakaia River hearing. Freshwater Catch 21: 2.

Davis, S. F. 1984. Conservation order for the Rakaia. Freshwater Catch **23**: 3-4.

Davis, S. F. 1985. Rakaia Conservation Order Report. Freshwater Catch **26**: 3.

Elser, A. A. 1976. Use and reliability of water surface profile program data on a Montana prairie stream. In: Instream flow needs. (Ed. by J. F. Orsborn and C. H. Allman) pp. 496-504. American Fisheries Society, Bethesda.

Fancourt, T. L. 1974. Co-operation leads to successful resource planning. Soil and Water **20**: 25-27.

Fraser, J. C. 1978. Suggestions for developing flow recommendations for instream uses of New Zealand streams. Water and Soil Misc. Pub. **6**: 1-13.

Glova, G. J. 1982. Fishery impact evaluation - application of the incremental method. In: River flows: conflicts of water use. (Ed. by R. H. S. McColl). pp. 16-27. Water and Soil Misc. Pub. 47.

Glova, G. J. and Duncan, M. J. 1985. Potential effects of reduced flows on fish habitats in a large braided river, New Zealand. Trans. Amer. Fish. Soc. **114**: 165-181.

Gowan, C. 1984. The impacts of irrigation water withdrawals on brown trout (*Salmo trutta*) and two species of benthic macroinvertebrates in a typical southern Michigan stream. M.Sc. thesis. Michigan State University, East Lansing, MI, USA.

Hardy, C. J. 1972. A submission and report on fisheries use and requirements in the Opihi river in relation to water abstraction. Report of the South Canterbury Acclimiatisation Society. 67 p.

Hawkins, C. P., Murphy, M. L., Anderson, N. H., and Wilzbach, M. A. 1983. Density of fish and salamanders in relation to riparian canopy and physical habitat in streams in northwestern United States. Can. J. Fish. Aquat. Sci. **40**: 1173-1185.

Horton, W.D., and Cochnauer, T. 1980. Instream flow methodology evaluation, biological criteria determination, and water quantity needs for selected Idaho streams. Idaho Department of Fish and Game, Stream Evaluation Project Phase II. 101 p.

Howie, R. 1979. Distribution of abstraction. In: The effects of water abstraction on fisheries. (Ed. by D. Scott). pp. 5-11. National Water Protection Committee of Acclimatisation Societies. Dunedin.

Hunt, R. L. 1976. A long term evaluation of trout habitat and its relation to improving management-related research. Trans. Amer. Fish. Soc. **105**: 361-364.

Irvine, J. R., Jowett, I. G., and Scott, D. (in press). A test of the instream flow incremental methodology for underyearling rainbow trout (*Salmo gairdneri* Richardson) in man-made New Zealand streams. N. Z. J. Mar. Freshw. Res.

Jowett, I. G. 1982. The incremental approach to instream flow needs, New Zealand case studies. In: River low flows: conflicts of water use. (Ed. by H. S. McColl). pp. 9-15. Water and Soil Misc. Pub. 47.

Jowett, I. G. 1983. Fraser River Hydrology and Fish Habitat. Ministry of Works and Development, Wellington. 18 p.

Leopold, L. B. and Maddock, T. Jr. 1953. The hydraulic geometry of stream channels and some physiographic implications. USGS Professional Paper 252. 57 p.

Loar, J. M. (ed). 1985. Application of habitat evaluation models in southern Appalachian trout streams. Oak Ridge National Laboratory, Environmental Sciences Division, Pub. No. 2383. 310 p.

Mathur, D., Bason, W. H., Purdy, Jr., E. J., and Silver, C. A. 1985. A critique of the instream flow incremental methodology. Can. J. Fish. Aquat. Sci. **42**: 825-831.

Mosley, M. P. 1982. Analysis of the effect of changing discharge on channel morphology and instream uses in a braided river, Ohau River, New

Zealand. Water Resources Research **18**: 800-812.

Mosley, M. P. 1983a. Response of braided rivers to changing discharge. J. Hydrol. (N.Z.) **22**: 18-67.

Mosley, M. P. 1983b. Flow requirements for recreation and wildlife in New Zealand rivers - a review. J. Hydrol. (N.Z.). **22**: 152-174.

Mosley, M. P. 1985. River channel inventory, habitat, and instream flow assessment. Progress in Physical Geography **9**: 494-523.

Murphy, M. L., Hawkins, C. P., and Anderson, W. H. 1981. Effects of canopy modification and accumulated sediment on stream communities. Trans. Amer. Fish. Soc. **110**: 469-478.

Nelson, F. A. 1980. Supplement to evaluation of four instream flow methods: applied to four trout rivers in southwest Montana. Montana Department of Fish, Wildlife and parks report to USFWS. Contract No. 14-16-0006-78-046. Bozeman, Montana. 55 p.

Nelson, P. C., Bovee, K. D., and Mayhew, D. A. 1984. Component testing of the Instream Flow Incremental Methodology: A case study of experimental design in natural stream systems. Presented at the 35th Annual AIBS meeting, International Society for Ecological Modelling Session on Populations and Environment. Colorado State University, Fort Collins, Colorado.

Orth, D. J. and Maughan, O. E. 1982. Evaluation of the incremental methodology for recommending instream flows for fishes. Trans. Amer. Fish. Soc. **111**: 413-445.

Richardson, J., and Teirney, L. D. 1982. The Whakapapa River: a study of a trout fishery under a modified flow regime. Fisheries Environmental Report **22**, 70 p. N. Z. Ministry of Agriculture and Fisheries.

Richmond, C. J. 1979. Some effects of water abstraction on the fisheries of the Central North Island. In: The effects of water abstraction on fisheries. (Ed. by D. Scott). pp. 73-78. National Water Protection Committee of Acclimatisation Societies. Dunedin.

Riddell, D. C. 1984. Surface water hydrology of the Mataura River catchment. Southland Catchment Board, Invercargill. 124 p.

Riddell, D. C. and Jowett, I. G. 1983. Lower Waitaki habitat survey. Ministry of Works and Development, Wellington. 12 p.

Sagar, P. and Jellyman, D. 1985. Regional modifications to waterways Part VI - South Canterbury. Freshwater Catch **26**: 19-24.

Scott, D. 1978. Small hydro schemes and fisheries. N. Z. Engineering. **33**: 178-179.

Sheppard, J. D., and Johnson, J. H. 1985. Probability-of-use for depth, velocity, and substrate by subyearling coho salmon and steelhead in Lake Ontario tributary streams. N. Amer. J. Fish. Mgt. **5**: 277-282.

Shirvell, C. S. and Dungey, R. G. 1983. Microhabitats chosen by brown trout for feeding and spawning in rivers. Trans. Amer. Fish. Soc. **112**: 355-367.

Shirvell, C. S., and Morantz, D. L. 1983. Assessment of the instream flow incremental methodology for Atlantic salmon in Nova Scotia. Transactions of the Canadian Electrical Association, Engineering and Operating Division. **22**, 83-HJ-108. 22 p.

Stalnaker, C. B. 1979. The use of habitat structure preferenda for establishing flow regimes necessary for maintenance of fish habitat. In: The Ecology of Regulated Streams. (Ed. by J. V. Ward and J. A. Stanford). pp. 321-337. Plenum Press, New York.

Tennant, D. L. 1976. Instream flow regimens for fish, wildlife, recreation and related environmental resources. In: Instream flow needs. (Ed. by J. F. Orsborn & C. H. Allman) pp. 359-373. American Fisheries Society, Bethesda.

Waters, B. F. 1976. A methodology for evaluating the effects of different streamflows on salmonid habitat. In: Instream Flow Needs. (Ed. by J. F. Orsborn & C. H. Allman). pp. 254-266. American Fisheries Society, Bethesda.

White, R. G., Milligan, J. H., and Bingham, A. E. 1981. Effects of reduced stream discharge on fish and aquatic macroinvertebrate populations. Idaho Water and Energy Resources Research Institute, Research Technical Completion Report Project B-045-IDA. University of Idaho, Moscow, Idaho. 283 p.

Winterbourn, M. J., Rounick, J. S., & Cowie, B. 1981. Are New Zealand stream ecosystems really different? N. Z. J. Mar. & Freshw. Res. **15**: 321-328.

CONSIDERATIONS IN ASSESSING FLUSHING FLOW

NEEDS IN REGULATED STREAM SYSTEMS

Dudley W. Reiser, Michael P. Ramey,
and Thomas R. Lambert[1]

Bechtel Inc., San Francisco, California
[1]Pacific Gas and Electric Company, San Ramon, California

INTRODUCTION

It has long been recognized that the regulation of stream flows can both positively and negatively affect existing fishery habitat and fish populations. This became most apparent in the western states where natural precipitation and runoff patterns had already produced well-defined periods of low streamflow.

As a result, fisheries biologists began investigating the relationships between fishery habitat and streamflow with the ultimate goal of being able to prescribe flows necessary for the maintenance and/or enhancement of fish populations. To this end, a wide variety of methodologies for assessing the "instream flow" needs of aquatic life have been developed and used.

However, many of the problems and questions associated with water developments still remain. Such is the case with flushing flow needs. Flushing flows, so named for their effect of removing ("flushing") fine sediments from gravels, have been the focus of relatively few fisheries studies in the past.

This study was undertaken to review and summarize existing information of flushing flows and to provide a better understanding of the physical and hydraulic parameters responsive to flushing flow conditions, and how they influence the biotic community. In so doing, special emphasis was placed on defining those factors which should be considered in making flushing flow recommendations. The intent of the study was not to develop a new methodology, but rather provide a set of guidelines useful for assessing flushing flow needs.

PROBLEM DEFINITION

In regulated streams, flushing flows may serve a variety of uses including channel maintenance, riparian habitat maintenance, prevention of vegetation encroachment, and the maintenance or enhancement of fishery habitat. Discussions in this paper are limited to the latter, and are focused on important fish spawning and rearing areas.

In general, the major reason for recommending flushing flows is for removing accumulated sediments from important fishery habitats. The need for these flows results when there has been a change in the natural hydrograph of a system due to the implementation of a water development project, for example a storage reservoir or hydroelectric development. Such projects tend to eliminate peak flows of the stream thereby reducing its competency to transport sediment at those times. The net effect is that sediment in the system tends to accumulate rather than being periodically removed, as during spring runoff. With time, continued sediment deposition and aggradation can adversely affect both spawning and rearing habitat.

In regulated streams, the solution to the problem is to periodically release high flows (flushing flows) to remove and transport sediments downstream out of the habitat. Unfortunately, few methods for determining needed flows have been developed, and those that have, have not been adequately tested. As such, many recommendations today are being made subjectively without a rational basis. Furthermore, follow-up verification studies are generally not undertaken. This can result in either a waste of water (if flows recommended were in excess of transport needs), or in the continued degradation of habitat (if flows were insufficient). Both have economic ramifications with respect to the water and fishery resources. The importance of prescribing reliable and accurate recommendations is therefore obvious.

As noted by Wesche and Rechard (1980), if flushing flows are needed to remove fines and maintain channel integrity, reliable methods should first be developed for determining the magnitude, duration, and timing of such flows. Even more fundamental is the determination of the need itself, for under some conditions, flushing flows may be more detrimental than beneficial to the resource. In short, a decision must be made by the appropriate management agency regarding the best prescribed condition for the stream in question.

FLUSHING FLOW METHODS

A variety of techniques have been used for assessing sediment transport, although remarkably few have been developed specifically for prescribing flushing flow needs in streams. Those that have are inadequately tested with respect to their reliability and accuracy, and only partially respond to the overall needs of a flushing flow (i.e., magnitude, timing, duration and effectiveness).

Table 1 presents thirteen different methods which directly or indirectly have been used for assessing flushing flow needs. The majority of the methods are office techniques requiring extensive flow records. Of these, the Tennant Methodology, based on 200% of the average annual discharge, is the most widely recognized and used technique in the western states. Most other office methods are founded on the principle that a bankfull flow or dominant discharge is the channel forming flow, and therefore should be used for effectively transporting fine sediments. Of the thirteen methods, only five address the question of timing of flows, of which three include considerations for evaluating its effectiveness. The duration of the prescribed flows as addressed by nine of the methods, ranges from a period of hours to 7 days.

Many of the proposed methods are predicated on what is called regime methods. These methods assume that some flow rates such as the bankfull flow, are the dominant channel forming flows. However, a river "in regime" in general, scours in some places and deposits in others. Thus, flushing

Table 1. Summary of methodologies for assessing flushing flow needs.

Methodology Author	Type	Basis	Method Consider Flow				Comments
			Magnitude	Timing	Duration	Effectiveness	
Tennant (Montana) Method Tennant (1975; 1976)	Office (field studies recommended but not detailed)	200% Average Annual Flow	X				Requires extensive flow records; site photographs recommended.
Northern Great Plains Resource Program Method NGPRP (1974)	Office	Average Annual Flow	X				Requires extensive flow records; method not developed primarily for flushing flows.
Dominant Discharge/ Channel Morphology Method Montana Dept. Fish Wildl. Parks (1981)	Office	Dominant Discharge (1.5 year frequency peak flow QF1.5P)	X	X	X 24 h		Requires extensive flow records (9 yr); suggests a gradual rising and receeding of the flushing flow.
Estes and Orsborn Method Estes (1984) Osborn (1982)	Office	Two year average annual peak flood event - QF2P; 3 day average around QF2P 7 day average around QF2P	X		X instan. 3 day 7 day		Requires extensive flow records; flow synthesis techniques are discussed; suggests field studies for flow verification.
Hoppe Method Hoppe (1975) Hoppe and Finnel (1970)	Office	17th percentile on flow duration curve (Q17)	X		X 48 h		Requires extensive flow records; empirically developed for the Fryingpan River, Colorado - Q17 may be specific to that system.
Bed Material Transport Method Hey (1981)	Field	Threshold discharge for transport; determined using bedload tracers	X		X		Restricted to clear water systems with good visibility; several test flows required; office techniques not described.
Instream Flow Incremental Methodology (IFIM) Bovee and Milhous (1978) Bovee (1982)	Office/Field	Indirect approach: point at which WUA (on spawning curve) begins to decrease	X				Several assumptions must be made using this approach Presently, the IFIM does not directly address flushing flows; the CIFASG is reviewing approaches for integrating this into the IFIM.
Wesche Method (Wesche et al. 1977)	Field	Bankfull discharge (empirically determined) uses drainage basin similarities for estimating unmeasured systems	X	X	X 3 day	X	Approach developed on high mountain streams in Wyoming; applicability to other systems uncertain; requires flow measurements during high flow events.

Table 1 (concluded). Summary of methodologies for assessing flushing flow needs.

Methodology Author	Type	Basis	Method Consider Flow				Comments
			Magnitude	Timing	Duration	Effectiveness	
Beschta and Jackson Method (Beschta & Jackson 1979)	Office	Flow/drainage area ratio (estimated at 0.16 cms/km²); 5th percentile on flow duration curve 05)	x				Developed in small coastal streams of Oregon; approach may not be applicable on other systems; flow records required.
Effective Discharge (O'Brien 1984)	Field/Office	Effective discharge/ Bankfull discharge	x	x	x 48 h	x	Developed on Yampa River in Colorado/Utah; extensive field measurements required; Sediment discharge relationships based on field and laboratory studies; approach included a physical model of the system; requires extensive flow records.
U.S. Forest Service Channel Maintenance Flow Method (Rosgen 1982)	Office	Bankfull discharge/Dominant discharge (1.5 year recurrence interval)	x	x	x 3 day		Developed on streams in northern Wyoming; extensive flow records required; method considers a wide range of flows not just peak flows.
Incipient Motion Methodology; Meyer-Peter Muller Based Water and Environment Consultants, Inc. (1980)	Field/Office	Predicting discharge which causes incipient motion of particle; employs Meyer-Peter Muller transport formula	x	x	x 3 day		Used on streams on south-eastern Wyoming; Meyer-Peter Muller formula can provide widely varying results; assumptions used in this technique should be evaluated on a site specific basis; technique probably suitable for implementation type studies
Incipient Motion Methodology; Shields Entrainment Function	Office	Predicting discharge for incipient motion of particle; based on a Shields entrainment function	x		x		Method based on Shields parameter of 0.03; other values can also be used which would change relationships developed; technique provides an estimate of needed flow as a function of grain size, stream width, and channel slope.

flow magnitudes based on these methods will be of uncertain accuracy at best.

Perhaps the most reliable method of those presented is to observe various test flow releases (Hey 1981). Field observations such as the sampling and tagging of bed material, would be made before and after each release flow at selected locations of the stream to determine actual effectiveness. Where test flow releases cannot be made, the use of methods based on sediment transport mechanics (e.g., Meyer-Peter & Muller 1948; Einstein 1950; Parker et al. 1982) provides the most reliable approach for the determination of required flushing flow rates. Proper application of these methods requires the collection of field data such as sediment gradation, channel geometry, and channel slope. Descriptions of the application and use of sediment transport models are provided in Reiser et al. (1985), Richardson et al. (1975) and Bjornn et al. (1977).

Another approach which looks promising for assessing flushing flow needs was derived during this study. Defined as an incipient motion method, it was derived from a Shields' entrainment function (Shields' parameter estimated at 0.03), and provides a means for estimating needed flows for bed mobilization given different grain sizes and channel slopes. The required discharges are expressed as a discharge per unit stream width (Figure 1). The duration of the estimated flows can be approximated using the travel time - median bed grain size relationships shown in Figure 2. At first glance this latter figure appears in error since it implies that a stream with a steeper slope will require a longer "flush" time. However, this is correct since the magnitude of the flow required for flushing is far less for the steeper gradient stream (Figure 2), therefore requiring a longer flow duration. Using this figure, the required duration of a flushing flow for a stream with a median grain size of 50 mm and a slope of 0.005 would be 0.28 h km^{-1} of stream; a stream 32 km long would require a flow duration of 9 h. It should be noted that this approach is an oversimplification of the process of flushing fine material from gravels and that some comparison with field data would be necessary before any confidence could be placed in the method.

Overall, it can be concluded that there is no present state-of-the-art methodology or approach for prescribing flushing flow needs. Moreover, the methods which are in use today are largely untested, and may be providing unrealistic recommendations.

EVALUATION OF FLUSHING FLOW REQUIREMENTS

From the above review and discussion it is apparent that the evaluation of flushing flows should be made with proper consideration for various physical, hydraulic and biological parameters. It is the interaction of these parameters which, in part, determines the need for, and timing and magnitude of flushing flows.

Determining the Need for Flushing Flows

Fundamental in the evaluation process is an initial determination of the need for a flushing flow. An unsubstantiated "blind" recommendation and implementation of a flushing flow may actually be detrimental to the aquatic resource. In regulated stream systems, the evaluation process should commence even before a real problem is recognized. The assessment should focus on the geomorphic and hydrologic characteristics of the drainage and how they influence the biotic environment. Through this evaluation, it should become evident whether sedimentation problems are likely to occur in the drainage below the water development project.

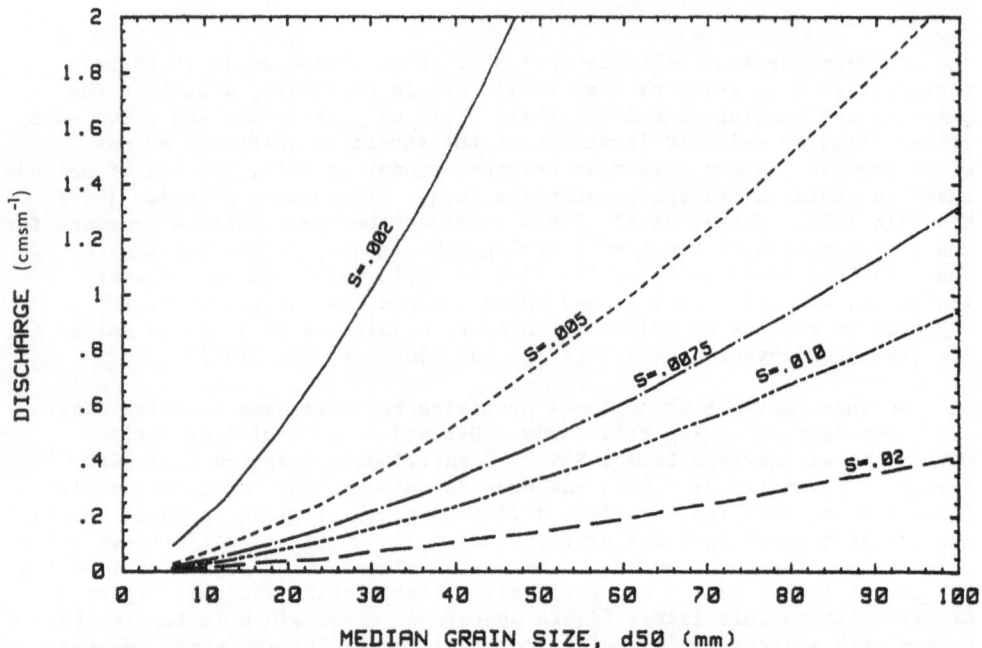

Figure 1. Critical unit discharge for bed mobilization as a function of
grain size and channel slope. Relationships derived from a
Shields' entrainment function.

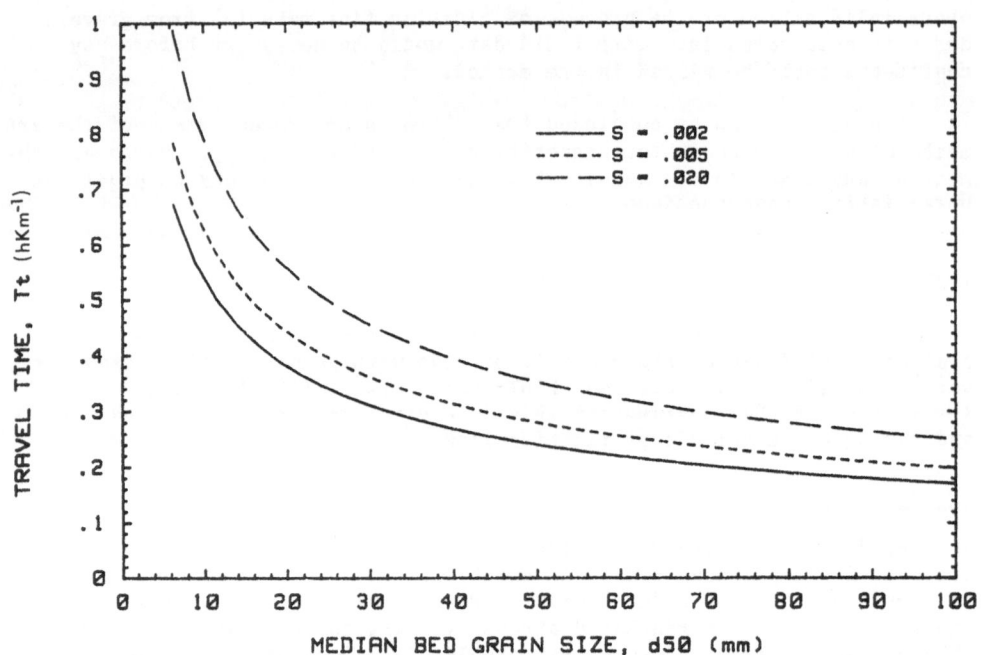

Figure 2. Time required to flush fine sediments as a function of median bed
grain size and channel slope.

Specific points for consideration include:

1. Physical location of the water development project. (Is the
    project above or below the major sediment sources in the
    drainage?).

2. Topography and geology of the project area.

3. Susceptibility of the drainage to catastrophic events (e.g.,
    landslides, storms, etc.).

4. Sensitivity of important fish species and their life history
    stages to sediment depositional effects (salmonids
    vs. centrarchids vs. catostomids, etc.).

5. Extent of man-induced activities within the drainage which may
    increase sediment recruitment (e.g., road construction, mining,
    etc.).

6. Operational characteristics of the project (important in
    determining whether the systems will be open or closed to
    upstream sediment recruitment).

As far as possible the determination of need should be based on an
objective, rather than subjective evaluation. The best approach for this
would be to establish test sections within a river reach which could be
used to monitor sediment levels. The sections should include habitats
(e.g., spawning areas, riffles, and pools) known to be used by target fish
species, and which are representative of other sections. This procedure
will define baseline sediment conditions within important habitats which
reflect unperturbed conditions. Continued monitoring of the same habitats
will permit temporal and spatial comparisons, and should delineate major
changes in sediment concentration. A variety of field techniques can be
used for monitoring sediment accumulation including:

1. Substrate core sampling and analysis (McNeil & Ahnell 1964;
    Everest et al. 1980).

2. Intergravel sediment sampling (Reiser 1983; Mahoney & Erman
    1984).

3. Visual substrate characterization and ratings (Platts et al.
    1983; O'Brien 1984).

4. Cross-sectional profiling of bed elevations (Corley & Newberry
    1982; Wesche et al. 1985; Platts et al. 1983).

5. Photographic documentation (Corley & Burmeister 1979).

6. Scour and deposition indicators (Foley 1976; Wesche et
    al. 1985).

7. Groundwater standpipes (Terhune 1958; Reiser & Wesche 1977).

8. Bedload samplers (Helley & Smith 1971; Neilson 1974).

Details on these and other methods are described in Reiser et al. (1985).
Any changes in sediment levels observed during the monitoring would need to
be evaluated with respect to potential impacts on the aquatic biota.
Designated standards or limits of sediment deposition should be established
above which a flushing flow would be required. These standards could be

based on values derived from the literature, but ideally would be developed on an individual stream or drainage basis.

Determining the Timing of Flushing Flows

When the need of a flushing flow has been established, it is equally important to determine the best time for its implementation. Important considerations in this regard include:

1. Species of fish present in the system.

2. Life history functions of important species.

3. Historical runoff period and flow availability.

In this paper, the primary concern is maintenance of aquatic biota. Hence, flow timing should be based on the life history requirements of important fishes in a system. Depending on the magnitude and duration, flushing flows may simulate a short term peaking regime with a rapid increase and decrease in discharge. Peaking flows can have deleterious effects on the aquatic resource including the dislodgement and transport of eggs (Wade & White 1978), dewatering of redds constructed during high flow periods (Reiser & White 1983; Becker et al. 1982), stranding of fish which have entered side pools that become unbridged as flows recede (Witty & Thompson 1974), and large increases in invertebrate drift (Wade & White 1978).

Ideally then, the most effective time would be that which would provide the greatest benefits, or impart the least harm to the biotic communities. This would certainly not be the case if flows were released during or after salmonid spawning. Released then, such flows could dislodge eggs and alevins and dramatically reduce recruitment potential. In contrast, flushing flows released prior to spawning should effectively remove and clean fine sediments from the substrates, and serve to enhance egg and alevin survival. Scheduled correctly, it may be possible for flushing flows to serve a dual purpose i.e., flushing fine sediments from spawning gravels, and transporting smolts downstream. Indeed, major flow releases at several Columbia River dams are specifically targeted at smolt transport. Maximization of benefits for the given water released should be a guiding principal when assessing the timing of flows.

Determining the Magnitude of Flows

The determination of the magnitude of flows is the most important, yet most difficult and least understood aspect of formulating a flushing flow recommendation. No standard method or approach has been developed for this purpose (see FLUSHING FLOW METHODS).

Certainly, the methods presented in this paper should provide some guidance in formulating recommendations. A careful review of the techniques may result in the development or adaptation of an approach which lends itself to a given problem. However, the selection of one approach over another does not guarantee any better resolution in the final recommendation.

It is of interest to note the disparity in flow recommendations which can result using two different methods. Wesche et al. (1983) noted an average difference of 60% in the flushing flows recommended on two independent studies for the same stream systems in Wyoming. The approach of Estes (1984) and Orsborn (1982) can result in as much as 600-900 percent difference in flows, when compared with recommendations derived using the

Tennant methodology (Tennant 1975). The methods which employ the
derivation of bankfull discharge and dominant discharge via office versus
field techniques would also likely vary.

In general, for studies in the planning stage, the safest approach may
be to use the technique providing the highest flow estimate. This should
be easy to determine since most of the office techniques have the same
general data requirements (see Table 1). Using this conservative approach,
water budgets and operating rules for proposed hydroelectric or water
development projects can be formulated around these needs. If refinements
are later warranted, they would likely result in a reduction rather than
increase in flows recommended.

For implementation studies which would include the development of
final recommendations for new or existing facilities, both office and field
techniques should be used. Office methods can provide an initial estimate
of needed flows which can then be refined through field evaluations.
Depending upon the project and its physical setting, field techniques can
range from the use of sediment transport mechanics to empirical assessments
of bed transport under different flow releases. The purpose of the field
component would be to verify or refine the initial recommendation as
dictated by the specific characteristics of each drainage system. Flushing
flow recommendations should be developed on a site-specific basis when
feasible.

The variability of results generated from the different methods
amplifies the importance of follow-up evaluation studies. Indeed, today
such studies remain as the only way to verify the sufficiency of a
recommendation, and furthermore provide a means to evaluate the
effectiveness of the methods themselves.

Assessing the Effectiveness of Flushing Flows

An evaluation of the effectiveness of a given discharge should be a
logical part of every flushing flow study. Only through this process can
the actual versus desired results be compared and necessary refinements
made.

In general, the same methods used for determining the need for
flushing flows would be used to assess their effectiveness. The utility of
the techniques are contingent upon their application both before, and after
a given flushing flow. In most instances however, the pre-flow assessment
should already be part of the process for determining flow need.

The selection and implementation of any assessment method should be
preceded by a review of its data collection and analysis techniques, and
its applicability to a given stream system. Where practical, special
emphasis should be made to design sampling programs for the collection of
meaningful, statistically valid data which can be factored into the
evaluation.

GUIDELINES FOR ASSESSING FLOWS

Until standard methods are developed for assessing flushing flows,
evaluations need to use an approach tailored to the specific needs and
characteristics of each stream and project. This may dictate the use of
several different office techniques to derive an initial flow estimate,
followed by detailed field studies to refine the recommendations. For
projects in the planning stage, an office approach may be all that is

Table 2. Guidelines for assessing the need for, and timing, magnitude, and effectiveness of flushing flows.

	NEED FOR	TIMING OF	MAGNITUDE OF	EFFECTIVENESS OF

CONSIDERATIONS WHEN ASSESSING:	o Physical location of project – above or below major sediment sources
	o Topography of project area- susceptibility to erosion
	o Extent of man- induced perturbations in the drainage
	o Susceptibility of drainage to catastrophic events
	o Operational characteristics of the project
	o Sensitivity of target fish species to effects of sediment deposition
		o Species of fish present in the systems (native, introduced)
		o Timing of the history functions of important species
		o Historical runoff period
		o Availability of project flows
			o Level of investigation required
			  – planning level studies
			  – implementation level studies
			o Availability of flow records
			o Availability of test flows
				o Availability and reliability of background data for defining pre-flushing flow
				o Time interval between end of flushing flow and field assessment
				o Potential influence of extraneous activities on the effectiveness of a flushing flow (e.g., sediment input from tributaries, road construction)

TECHNIQUES FOR ASSESSING:	o Establish and monitor test reaches by:
	  – substrate analysis
	  – cross sectional profiling
	  – photographic documentation
	  – scour and deposition indicators
	  – groundwater standpipes
	  – bedload samplers
	o Comparison of data with standards
	  – literature based
	  – site specific based (preferred)
	o Spot assessments made (as needed)
		o Prepare and review species life history – periodicity charts and note preferred timing release periods
		o Review historical flow records and note timing of peak flows
		o Review water budgets of project and note availability of flows
		o Adjust timing recommendations accordingly. (Timing of flows should be based on maximizing benefits for the given water released)
			o For planning level studies use appropriate office techniques for initial estimate
			o Implementation studies – refine estimates through field/laboratory investigation
			  – sediment transport models
			  – empirical assessments of bed transport
			  – physical modeling of stream reach
			o No standard approach or method presently available
			o Recommendations should be based on site specific in- formation and include estimates of flow duration
				o Pre- and post flow comparisons of substrate-sediment deposition and composition by:
				  – substrate analysis
				  – cross sectional profiling
				  – photographic documentation
				  – etc.
				o Post-flow analysis of data should be factored into necessary adjustments in recommendations

needed; implementation studies should include detailed field investigations.

Recommended guidelines for conducting flushing flow studies include:

1. The utilization of an interdisciplinary team approach. Study team members should include at a minimum a hydrologic engineer, and a fisheries biologist.

2. An initial determination of the actual need for the flushing flow should precede detailed assessments.

3. The assessment approach used should be tailored to the specific needs and characteristics of each stream and project; office and field techniques may both be required.

4. For comparative purposes, more than one method should be used for deriving flow recommendations.

5. A determination of the timing and required duration of the flow should be included as part of the assessment process.

6. Flushing flow recommendations should be stated in terms of magnitude, timing and duration.

7. Follow-up studies should be conducted to evaluate the effectiveness of the flows and allow for necessary adjustments.

A summary of guidelines including considerations and techniques for assessing the need, timing, magnitude and duration of flushing flows is presented in Table 2.

In summary, this paper has reviewed the biological and engineering considerations required for assessing flushing flows in regulated streams, and has presented brief descriptions of thirteen different techniques which have been used for making flushing flow recommendations. It is intended that this review serve as a "state-of-the-knowledge" on flushing flows and as such, be used for promoting an understanding of the complexity of the issue, and stimulating and fostering research.

REFERENCES

Becker, C. D., Neitzel, D. A. and Fickeisen, D. H. 1982. Effects of dewatering chinook salmon redds: tolerance of four developmental phases to daily dewaterings. Trans. Amer. Fish. Soc. 111: 624-637.
Beschta, R. L. and Jackson, W. L. 1979. The instrusion of fine sediments into a stable gravel bed. J. Fish. Res. Board Can. 36: 204-210.
Bjornn, T. C., Brusven, M. A., Molnau, M. P., Milligan, J., Klamt, R., Chacho, E. and Schaye, C. 1977. Transport of granitic sediment in streams and its effects on insects and fish. Univ. Idaho, For. Wild. Range Exp. Station, Bulletin No. 12, 43 pp.
Bovee, K. D. and Milhous, R. T. 1978. Hydraulic simulation in instream flow studies: theory and techniques. Instream Flow Information Paper No. 5. USDI Fish Wildl., Serv. FWS/OBS 78/33. 130 pp.
Bovee, K. D. 1982. A guide to stream habitat analysis using the instream flow incremental metholodogy. Information Paper No. 12, FWS/OBS 82/26, Fort Collins, Colorado. 248 pp.
Corley, D. R. and Burmeister, L. A. 1979. Fishery habitat survey of the South Fork Salmon River - 1979, Boise and Payette National Forests. Completion Report, U.S. Forest Service, Boise, Idaho.

Corley, D. R. and Newberry, D. D. 1982. Fishery habitat survey of the South Fork Salmon River - 1981. Boise and Payette National Forests. Completion Report, U.S. Forest Service, Boise, Idaho.

Einstein, H. A. 1950. The bedload function for sediment transportation in open channel flows. Soil Conser. Serv., Tech. Bulletin No. 1026.

Estes, C. C. 1984. Evaluation of methods for recommending instream flows to support spawning by salmon. M.S. thesis, Washington State University, Pullman, Washington, 156 pp.

Everest, F. H., McLemore, C. E. and Ward, J.F. 1980. An improved tri-tube cryogenic gravel sampler. Pacific Northwest Forest and Range Experiment Station, Portland, Oregon, PNW-350.

Foley, M. G. 1976. Scour and fill in an ephemeral stream. Proc. Third Fed. Inter. - Agency Sedimentation Conf., March 22-25, 1976. pp. 5-12.

Helley, E. J. and Smith, W. 1971. Development and calibration of a pressure-difference bedload sampler. Wash. D.C.: U.S. Dept. Interior, Geol. Survey, Water Resource Division, 18 pp.

Hey, R. D. 1981. Channel adjustment to river regulation schemes. In: Proceedings of workshop on downstream river channel changes resulting from diversions or reservoir construction. U.S. FWS., FWS/OBS -81/48. pp. 114-127.

Hoppe, R. A. 1975. Minimum streamflows for fish. Paper presented at Soils-Hydrology Workshop, USFS, Montana State Univ., Jan. 26-30. 13 pp.

Hoppe, R. A. and Finnel, L. M. 1970. Aquatic studies on the Fryingpan River, Colorado - 1969-1970. U.S. Bureau of Sport Fish and Wildlife, Division of River Basin Studies. Mimeo Rept. 12 pp.

Mahoney, D. and Erman, D. C. 1984. An index of stored fine sediment in gravel bedded streams. Water Resources Bulletin, **20** (3): 342-348.

McNeil, W. J. and Ahnell, W. H. 1964. Success of pink salmon spawning relative to size of spawning bed materials. U.S. Fish. Wildl. Service Spec. Sci. Rept. - Fish No. 469. 15 pp.

Meyer-Peter, E. and Muller, R. 1948. Formulas for bedload transport, Proceedings, Third Conference, International Association of Hydraulic Research, Stockholm, Sweden, pp. 39-64.

Montana Department of Fish, Wildlife and Parks (MDFWP). 1981. Instream flow evaluation for selected waterways in western Montana. Completion Rept. Contract No. 53-0343-0-305 submitted to U.S. Forest Service.

Neilson, D. R. 1974. Sediment transport through high mountain streams of the Idaho Batholith. M.S. Thesis, Univ. Idaho, Moscow, Idaho, 83 pp.

Northern Great Plains Resource Program. 1974. Instream needs sub-group report. Work Group C: Water. 35 pp.

O'Brien, J. S. 1984. Hydraulic and sediment transport investigation, Yampa River, Dinosaur National Monument. WRFSL Rept. 83-8.

Orsborn, J. F. 1982. Estimating spawning habitat using watershed and channel characteristics. In: Acquisition and Utilization Aquatic Habitat Inventory Information. (Ed by N. Armantrout). pp. 154-161. West. Div. Amer. Fish Soc.

Parker, G., Klingeman, P. and McLean, D. 1982. Bedload and size distribution in paved gravel bed streams. Hydraul. Div., ASCE, Vol. 108, HY4, 198 pp.

Platts, W. S., Megahan, W. F. and Minshall, G. W. 1983. Methods for evaluating stream, riparian, and biotic conditions, USDA. Gen. Techn. Rept. INT-138., Ogden, Utah, 70 pp.

Reiser, D. W. and Wesche, T. A. 1977. Determination of physical and hydraulic preferences of brown and brook trout in the selection of spawning locations. Water Resour. Ser. No. 64. Water Resour. Res. Inst. Univ. of Wyoming, 100 pp.

Reiser, D. W. and White, R. G. 1983. Effects of complete redd dewatering on salmonid egg hatching success and development of juveniles. Trans. Amer. Fish Soc. **112**: 532-540.

Reiser, D. W. 1983. Development and testing of an intergravel sampling device for quantifying fine sediments in streams. Unpublished Manuscript, Bechtel Group, Inc., San Francisco, Calif. 16 pp.

Reiser, D. W., Ramey, M. P. and Lambert, T. R. 1985. Review of flushing flows in regulated streams. Completion Rept. Pacific Gas and Electric Company, Research and Development Program, San Ramon, Calif., 97 pp + Appendices.

Richardson, E. V., Simons, D. B., Karaki, S., Mahmood, K. and Stevens, M. A. 1975. Highways in the river environment, hydraulic and environmental design considerations: training and design manual, U.S. Dept. Transportation Federal Highway Administration. U.S. Printing Office, Wash. D.C.

Rosgen, D. 1982. A procedure and rationale for securing favorable conditions of water flows on national forest system lands in northern Wyoming. Draft manuscript. Region 2 Forest Service, Fort Collins, Colorado.

Tennant, D. L. 1975. Instream flow regimens for fish, wildlife, recreations and related environmental resources. U.S. Fish and Wildlife Service Rept. Billings, Montana. 18 pp.

Tennant, D. L. 1976. Instream flow regimens for fish, wildlife, recreation and related environmental resources. In: Pro. Sym, Spec. conf. Instream Flow Needs. (Ed by T. Orsborn and C. Allman). pp. 359-373. Vol. II, Amer. Fish Soc.

Terhune, L. D. B. 1958. The Mark VI groundwater standpipe for measuring seepage through salmon spawning gravel. J. Fish. Res. Board Canada 15 (5): 1027-1063.

Wade, D. and White, R. G. 1978. Fisheries and invertebrate studies on the South Fork of the Boise River below the Anderson Ranch Dam. Annual Rept. Univ. Idaho.

Water and Environment Consultants (WEC) Inc., 1980. Flushing flow discharge evaluation for 18 streams in the Medicine Bow National Forest. Completion Rept. for Environmental Research and Technology, Fort Collins, Colorado.

Wesche, T. A., Reiser, D. W., Wichers, W. F. and Wichers, D. J. 1977. Fishery resources and instream flow recommendations for streams to be impacted by Cheyenne's proposed Phase II developments. Completion Rept. submitted to Wyoming Game and Fish Department. Wyoming Water Resources Research Institute.

Wesche, T. A. and Rechard, P. A. 1980. A summary of instream flow methods for fisheries and related research needs. Eisenhower Consortium. Bulletin No. 9. 122 pp.

Wesche, T. A., Hasfurther, V., Skinner, Q. and Hubert, W. 1983. Development of methodology to determine flushing flow requirements for channel maintenance purposes. Unpublished Manuscript, Wyoming Water Research Center, University of Wyoming.

Wesche, T. A., Hasfurther, V., Skinner, Q. and Hubert, W. 1985. Assessment of flushing flow recommendations in a steep, rough regulated tributary. This volume.

Witty, K. and Thompson, K. 1974. Fish stranding. In Anatomy of a River, Rept. of Hells Canyon Controlled Flow Task Force, Pacific Northwest River Basin Commission.

ASSESSMENT OF FLUSHING FLOW RECOMMENDATIONS IN A STEEP,

ROUGH, REGULATED TRIBUTARY

T. A. Wesche,[1] V. R. Hasfurther,[1]
W. A. Hubert,[2] and Q. D. Skinner[3]

[1]Wyoming Water Research Center
[2]Department of Zoology and Physiology
[3]Range Management Division
University of Wyoming, Laramie, Wyoming U.S.A. 82071

INTRODUCTION

Alteration of stream flow regime and sediment loading from water development activities can result in both short- and long-term changes in channel morphology and conveyance capacity. Subsequently, the condition of the aquatic habitat can be affected. In recent years, much research and development effort has been directed toward the determination of suitable instream flows to maintain fisheries habitat in regulated streams (Stalnaker & Arnette 1976; Wesche & Rechard 1980). However, there are several facets of the instream flow problem which have not been adequately investigated, one involving the recommendation of flushing flows to simulate the peak runoff hydrograph characteristics of most unregulated streams (Reiser et al. 1985).

Limited research has been conducted to develop methodology for determining the magnitude, timing, and duration of flushing flows needed to maintain channel integrity and associated habitat characteristics through the movement of sediment deposits. Of the 15 methodologies identified by Reiser et al. (1985), a majority were not designed specifically to assess flushing flows, but rather were approaches for studying sediment transport problems. The several formal methodologies currently available (Wesche et al. 1977; Environmental Research & Technology, Inc. 1980; Rosgen 1982) were developed in response to immediate management needs and are relatively untested in terms of accuracy and reliability.

During 1984, the Wyoming Water Research Center initiated a research project entitled, "Development of methodology to determine flushing flow requirements for channel maintenance purposes." Objectives of this project are to 1) document the rate of change of various channel characteristics resulting from aggradation/degradation processes under altered flow regimes; 2) quantify the physical and hydraulic properties needed to transport deposited sediment through natural channels; 3) test the predictive capabilities of existing sediment transport models against field data; and 4) develop methodology to predict conditions of flow needed to flush sediments to maintain given streams in prescribed hydraulic, physical and biologic conditions.

One stream selected for study in response to these objectives was the North Fork of the Little Snake River (North Fork), a steep, rough, regulated, headwater stream. Wesche et al. (1977) recommended both maintenance and flushing flow regimes for the North Fork in light of the proposed expansion of water diversion facilities in the drainage by the City of Cheyenne, Wyoming, as part of their Stage II water development program. Construction of Stage II began in 1983. During the late summer of 1984, intense rainfall in the construction area resulted in the deposition of a broad size range of sediments in that section of the North Fork where flushing recommendations had been made. At the request of the Wyoming Game and Fish Department and in cooperation with the United States Department of Agriculture Forest Service, the authors initiated a study of the North Fork. The objectives of this paper are to 1) describe the methods used to assess the extent of the 1984 sediment deposits; 2) present preliminary results summarizing the response of the deposited sediment to the 1985 spring runoff flow regime; and 3) evaluate the effectiveness of the 1977 flushing flow recommendations in relation to the 1984 sediment deposits.

DESCRIPTION OF STUDY AREA

The North Fork of the Little Snake River is a steep, rough, regulated tributary of the Little Snake River located in the Green River sub-basin of the Colorado River basin in southwest and southcentral Wyoming (Figure 1). The headwaters of the North Fork rise on the west slope of the Continental Divide at an elevation of 3050 meters (m) above sea level (msl) and flow southwesterly 20 kilometers (km) to the confluence with the Little Snake River at an elevation of 2130 m. Average gradient is 4.6 percent. A United States Geological Survey (U.S.G.S.) streamflow gaging station (#09251800) located 2.4 km below the study area was in operation from 1957 to 1965 and recorded a maximum discharge of 14.6 cubic meters per second (m^3 s^{-1}) on June 7, 1957. Average discharge over the period of record was 0.73 m^3 s^{-1}. Prior to initial water diversion in the mid-1960's, the North Fork hydrograph was typical of unregulated mountain streams in the central Rocky Mountain Region, with the majority of runoff occurring in the May to late-June period, as a result of the melting snowpack.

The North Fork and its tributaries support the largest known, essentially-pure, naturally-reproducing endemic population of Colorado River cutthroat trout (*Salmo clarki pleuriticus* Cope) (Binns 1977). For this reason, management of the population is a high priority for the Wyoming Game and Fish Department. Wesche et al. (1977) also report the collection of mottled sculpin (*Cottus bairdi* Girard) in the North Fork.

Transbasin diversion of water from the North Fork drainage has occurred since 1964 when the City of Cheyenne, Wyoming completed Stage I of its water development program. Approximately 9.86 x 10^6 m (8,000 acre-feet) per year have been diverted (Banner Associates, Inc. 1976). During 1983, construction began on Stage II collection facilities. When completed in 1986, a total of 28.4 x 10^6 m^3 (23,000 acre-feet) per year will be conveyed from the upper Little Snake drainage to the east slope of the Continental Divide (U.S.D.A., Forest Service, 1981).

The study area on the North Fork is located in Section 27, Township 13 North, Range 85 West at an elevation of 2615 m, within the boundaries of Medicine Bow National Forest, 2.4 km below the Stage I diversion structure. Under Stage II, this structure is being modified to increase the amount of water diverted from the North Fork proper. Within the study area boundary, a stream section 0.5 km in length, construction of a bridge and pipeline crossing was underway in the late summer of 1984 when heavy rains precipitated the sediment spill that led to the initiation of this study. Gradient through this area is 4.4 percent while the predominant natural substrate is boulders and cobbles. Wesche et al. (1977) reported a mid-July 1976 water temperature range of 12.7 to 17 °C, a total alkalinity range of 25 to 32 mg 1^{-1}, a pH of 7.1, and clear water conditions for this section of the North Fork. Standing crop estimates for Colorado River cutthroat trout ranged up to 15.6 kg ha^{-1}. Instream flow recommendations developed by Wesche et al. (1977) called for a minimum flow of 0.085 $m^3 s^{-1}$ or the natural flow, whichever is less, and a three-day annual release of 1.70 $m^3 s^{-1}$ for flushing purposes during the spring runoff period.

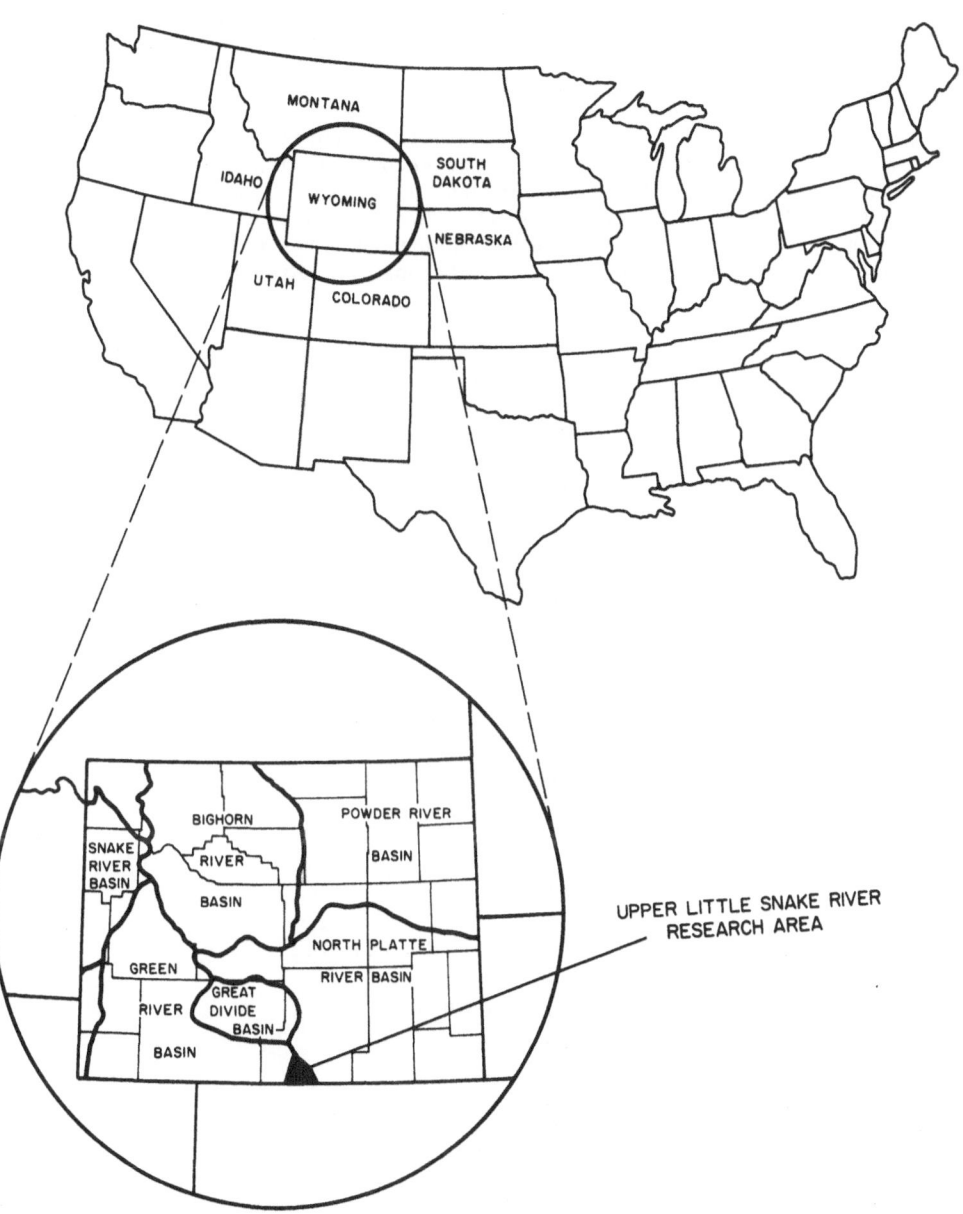

Figure 1.   Map showing location of the Upper Little Snake River Research Area.

METHODS

During the Fall of 1984, four reaches were selected for study in cooperation with personnel from the Wyoming Game and Fish Department and the U.S. Forest Service (Figure 2). Reach 1, the uppermost site, was located just above the confluence of Second Creek, approximately 400 m upstream from the North Fork bridge and pipeline crossing. Reach 1 served as the control above the construction area from which the sediment spill originated. Reaches 2, 3 and 4 were located in descending order below the North Fork crossing area and were within the zone of immediate deposition from the spill. Given the intensive nature of the sampling to be conducted, study reaches were kept short in length, with Reach 2 being the longest, 15 m. Study reaches were located close to one another to avoid compounding the access problems involved with early spring sampling in a remote, high elevation area.

Two recording streamflow gage stations were installed within the study area in early May, 1985, to monitor the spring runoff hydrograph. One station was located at Reach 1 while the second was installed at Reach 3. As no tributaries entered between Reaches 2, 3, and 4, this lower station served to define the hydrograph for the three downstream reaches. Each station consisted of a stilling well constructed from a 1.2 m length of 30 cm diameter perforated plastic pipe, a Leopold and Stevens Type F water

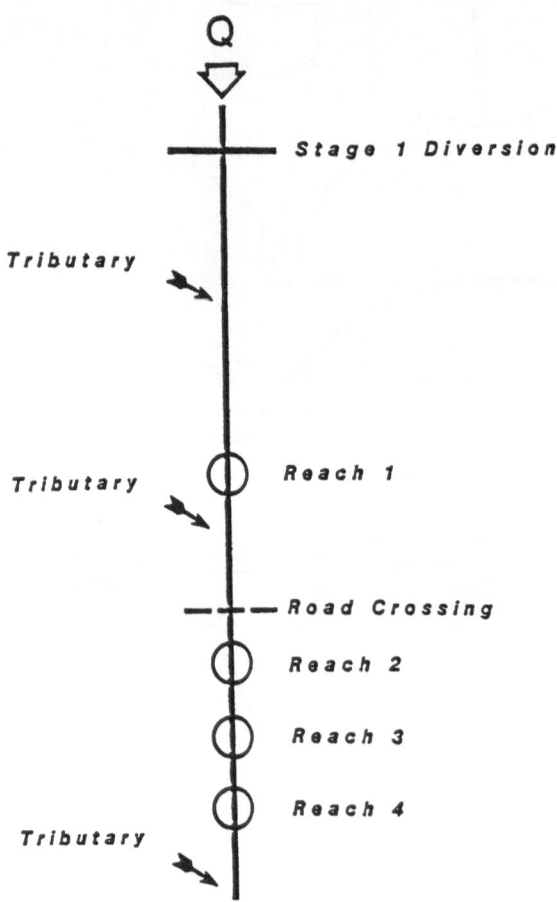

Figure 2. Study reach layout on the North Fork of the Little Snake River.

stage recorder, a steel platform on which the recorder was seated, and an outside staff gage for measuring stream stage. A rating curve for each gage station was developed following standard U.S.G.S. procedures (Buchanan & Somers 1969). Eight stage-discharge measurements were made at each station to determine the rating curves. The correlation coefficient (r) for each curve was 0.99. Recording thermographs to measure water temperature were installed in conjunction with each stream gage station. Four equally spaced cross-channel transects were established during October, 1984, within each study reach. Field data collected along these transects were used to quantify changes in response to the runoff hydrograph of 1) hydraulic characteristics, including discharge, channel width, top width, water depth, cross-sectional area, wetted perimeter, hydraulic radius, mean water velocity, bottom water velocity, and intergravel permeability; 2) bedload transport; 3) suspended sediment transport; 4) quantity and distribution of deposited sediments; and 5) quality of the deposited sediments. Given the scope of this paper, analysis will focus only on data types 4 and 5 listed above. The hydraulic and sediment transport data collected is presently undergoing analysis and will be presented in future project papers and reports.

Field sampling began in late October, 1984, was then discontinued over the winter months, and was reinitiated in early May, 1985, as spring runoff began. Sampling continued on approximately a weekly basis through early July, 1985.

The quantity of deposited sediment $[kg \ (m^2)^{-1}]$ within each study reach at each sampling time was determined by multiplying the volume of material (m^3) by its mean density $[kg \ (m^3)^{-1}]$ and dividing this product by the surface area (m^2). Volume was calculated as the product of mean depth of deposition (measured at approximately 80 locations within the reach using a round steel depth rod), reach length, and reach width. Mean density determinations were based upon the analysis of 12 core samples collected from each reach in October, 1984, May, 1985, and July, 1985, using a McNeil- Ahnell sampler (McNeil & Ahnell 1964). All cores were oven dried for at least 24 hours at 60 C to standardize weight measurements while the volume of each core was determined by water displacement.

The composition and quality of the deposited material within each reach over time was assessed by the following procedure:

1. As described above, 12 core samples were taken at each study reach at each of three sampling times. A total of 144 cores were collected (12 cores per reach times 4 reaches times 3 sampling times).

2. Particle size distribution by weight within each core sample was determined by dry-sieve analysis at the University of Wyoming's Division of Range Management Watershed Laboratory. A series of 10 sieves ranging in mesh size from 75 to 0.2 mm were used (Reiser & Wesche 1977).

3. The mean particle size distribution for each reach at each sampling time was determined by averaging the results from the 12 individual core samples. Distribution plots of particle size (mm) versus percentage (by weight) finer than the given sieve sizes were then developed.

4. Quality of the deposited material by reach over time was assessed by:

a) the median particle size read from the distribution plots described above;

b) the geometric mean particle size (d_g) calculated by the equation,

$$d_g = (d_1{}^{w_1} \times d_2{}^{w_2} \times \ldots d_n{}^{w_n}, \tag{1}$$

where d_n is the midpoint diameter of particles retained by the nth sieve and w_n is the decimal fraction by weight of particles retained on the nth sieve (Platts et al. 1983);

c) The Fredle Index (f) calculated by the equation,

$$f = \frac{d_g}{S_o} \tag{2}$$

where S_o is the sorting coefficient defined as the ratio of d75 to d25 where the particle size diameters are 75 and 25 percent finer on a weight basis of the sample (Lotspeich & Everest 1981).

RESULTS

A summary of hydraulic characteristics for each study reach is presented in Table 1. As indicated by these data, Reach 2 had the steepest gradient and subsequently the highest water velocities and shallowest water depths. Reach 4, the lowermost site, consisted primarily of pool habitat having the lowest gradient, deepest water and slowest velocities. Reaches 1 and 3 were similar in hydraulic characteristics and represented more moderate conditions.

Table 1. Mean hydraulic characteristics of the four North Fork study reaches at a low and a high discharge.

	Hydraulic Characteristics					
Reach	Discharge (m^3 s^{-1})	Top Width (m)	Cross Sectional Area (m^2)	Mean Depth (m)	Mean Velocity (m^3 s^{-1})	Water Surface Slope (percent)
#1	0.10	5.79	0.66	0.11	0.17	2.6
	1.12	6.59	1.96	0.30	0.58	---
#2	0.12	6.37	0.40	0.06	0.33	4.5
	1.83	7.19	1.90	0.27	0.97	---
#3	0.10	6.04	0.60	0.10	0.16	3.0
	2.11	7.50	2.42	0.33	0.88	---
#4	0.09	4.88	0.63	0.13	0.15	0.4
	2.86	8.55	4.45	0.53	0.68	---

Spring 1985 runoff hydrographs for the two streamflow gaging stations are presented in Figure 3. While the magnitude of the runoff was greater at the lower station due to the tributary which entered the North Fork immediately below Reach 1, timing and duration were similar. Also shown on Figure 3 is the magnitude of the flushing flow recommended by Wesche et al. (1977) for the North Fork in the vicinity of the three lower study reaches. This recommendation, 1.7 m^3 s^{-1} for a duration of 3 days, was based upon field measurement of bankfull discharge and the findings of Eustis and Hillen (1954).

Three major runoff peaks occurred during 1985 which equalled or exceeded the magnitude and duration of the recommended flushing flow (Figure 3). Each peak had a maximum instantaneous discharge of 2.97 m^3 s^{-1} while the maximum mean daily peaks ranged from 2.06 to 2.26 m^3 s^{-1}. Based upon maximum instantaneous discharge, the earliest peak lasted 3 days (May 10 to 12), the second peak extended over 8 days (May 23 to 30), and the third peak exceeded the recommended discharge on five consecutive days (June 6 to 10). A fourth peak occurred in late June during which the maximum flow approached the 1.7 m^3 s^{-1} level, but only for a portion of one day.

The quantity of deposited material within each study reach at each sampling time is presented in Figure 4. Deposition was consistently lowest in Reach 1, the upstream control, and Reach 2, the uppermost study section below the construction area. Quantities in these two reaches varied from 79 to 153 kg $(m^2)^{-1}$. The high gradient through Reach 2 probably explains the relative lack of deposition in this area. Based upon the early May and July 1985 data, Reach 2 experienced a net export of 7 kg $(m^2)^{-1}$ through the spring runoff period, while Reach 1, a moderate gradient section, realized a loss of 22 kg $(m^2)^{-1}$.

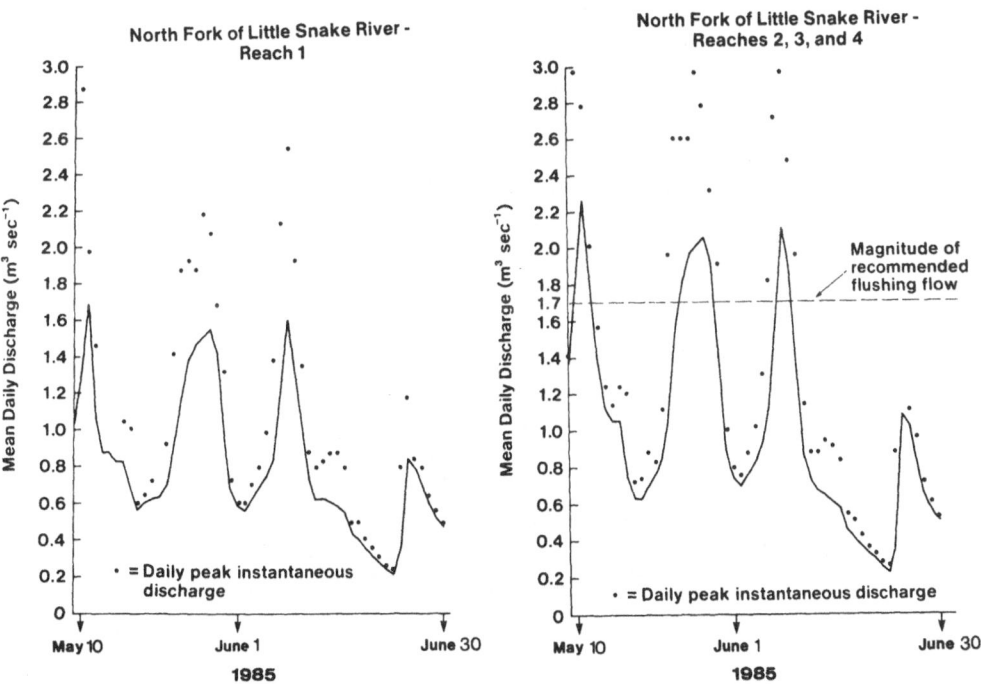

Figure 3. Spring runoff hydrographs for the two North Fork Stream gage stations.

The quantity of deposited material sampled in Reach 3 ranged from 145 to 230 kg $(m^2)^{-1}$. From early May to July 1985, a net export of 21 kg $(m^2)^{-1}$ was observed in this moderate gradient reach. The trend of the data, while greater in magnitude, paralleled that found for Reach 1, a section having similar hydraulic characteristics.

Reach 4, the lower gradient pool section, was found to have the greatest magnitude and variation of deposited material. In early May, prior to the first peak in the hydrograph, the amount of deposition present was 403 kg $(m^2)^{-1}$. The effects of the three major peak runoff events and one minor event observed during the spring of 1985 on Reach 4 are evident from Figure 4. In total, these four flushes reduced the amount of deposition to 248 kg $(m^2)^{-1}$, a net export of 155 kg $(m^2)^{-1}$.

Comparison of October 1984 and early May 1985 deposition quantities indicates considerable aggradation during the winter (Figure 4). Deposits in the moderate gradient reaches, 1 and 3, increased 52 and 20 kg $(m^2)^{-1}$, respectively, while the low gradient pool section, Reach 4, realized a net gain of 249 kg $(m^2)^{-1}$. The high gradient Reach 1 was the only study section to export material over the winter, showing a net loss of 30 kg $(m^2)^{-1}$. As the streamgage stations were not in operation during the winter and bedload transport samples could not be taken due to access limitations and heavy snow cover over the stream, the source and timing of these deposits cannot be positively identified; however, early winter thaws on the disturbed soils throughout the study area are probably responsible.

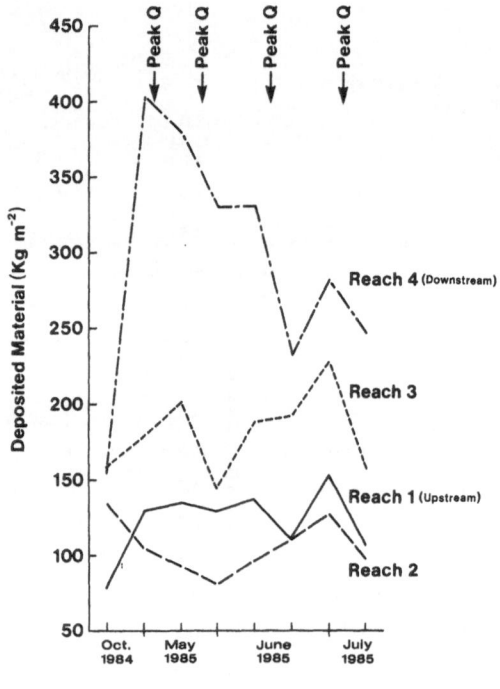

Figure 4. Comparison of deposited material in the four North Fork study reaches in relation to the spring runoff peak discharges.

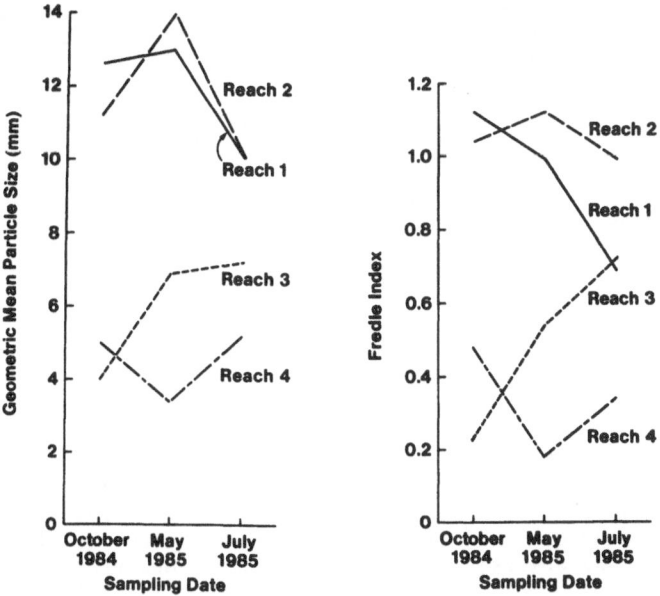

Figure 5. Quality of deposited material in the four North Fork study reaches over time.

The relative quality of deposited material in each of the study reaches over time is provided in Figure 5. As median particle size data were similar in both magnitude and variation to the geometric means, they are not presented.

The geometric mean particle size was consistently larger in Reaches 1 (range 10 to 13 mm) and 2 (10 to 14 mm) than in the lower two sections. Data for Reach 3 varied from 4.0 to 7.2 mm while the range for Reach 4 was 3.4 to 5.2 mm. Geometric means for both reaches 3 and 4 increased in response to the runoff peaks.

Fredle numbers, a spawning substrate quality index developed by Lotspeich and Everest (1981), appear to be quite low in all study reaches when compared to the preliminary relationships presented by Platts et al. (1983) between index values and percent survival-to-emergence of eggs from several salmonid species. However, the trend of our data is similar to that for geometric mean particle size and indicates improvement of deposition quality in Reaches 3 and 4 in response to the spring runoff hydrograph.

CONCLUSIONS

Based upon our analysis of the data presented, the following conclusions can be drawn:

1. Three spring runoff flushes meeting or exceeding the magnitude and duration of the recommended flushing flow for this section of the North Fork of the Little Snake River were somewhat successful in reducing the quantity of deposited material.

2. Flushing was more effective in steeper gradient reaches, while results regarding duration of the individual flushes are at present inconclusive.

3. As indicated by the Fredle Index, quality of the deposited material was very low throughout the study area.

4. Quality of deposited material showed an improving trend in response to the runoff hydrograph within the study reaches having the largest quantities of deposition.

REFERENCES

Banner Associates, Inc. 1976. Report on proposed expansion of Cheyenne's Little Snake Diversion Facilities. First section of the general report and reconnaissance report on Hog Park Reservoir enlargement. Prepared for the City of Cheyenne, Wyoming, Board of Public Utilities.

Binns, N. A. 1977. Present status of indigenous populations of cutthroat trout (*Salmo clarki*), in southwest Wyoming. Fisheries Technical Bull. No. 2, Wyoming Game and Fish Department, Cheyenne, Wyoming.

Buchanan, T. J. and Somers, W. P. 1969. Discharge measurements at gaging stations. Book 3, Chapter A8 of Techniques of Water Resources Investigations of the U.S. Geological Survey. U.S. Government Printing Office, Washington, D.C.

Environmental Research & Technology, Inc. 1980. Flushing flow discharge evaluation for 18 streams in the Medicine Bow National Forest. Report prepared for the City of Cheyenne, Wyoming, Board of Public Utilities.

Eustis, A. B. and Hillen, R. H. 1954. Stream sediment removal by controlled reservoir releases. Prog.-Fish Cult., 16(1): 30-35.

Lotspeich, F. B. and Everest, F. H. 1981. A new method for reporting and interpreting textural composition of spawning gravel. Research Note PNW-369; Pacific Northwest Forest and Range Experiment Station, Corvallis, Oregon.

McNeil, W. J. and Ahnell, W. H. 1964. Success of pink salmon spawning relative to size of spawning bed materials. U.S. Fish and Wildlife Service Special Scientific Report - Fish No. 469. 15 p.

Platts, W. S., Megahan, W. F. and Minshall, G. W. 1983. Methods for evaluating stream, riparian, and biotic conditions. U.S. Department of Agriculture Forest Service, General Technical Report INT-138, Intermountain Forest and Range Experimentation Station, Ogden, Utah.

Reiser, D. W., Ramey, M. P. and Lambert, T. R. 1985. Review of flushing flow requirements in regulated streams. Pacific Gas and Electric Co., San Ramon, California. 97 pp. + App.

Reiser, D. W. and Wesche, T. A. 1977. Determination of physical and hydraulic preferences of brown and brook trout in the selection of spawning locations. Water Resources Series No. 64. Water Resources Research Institute, University of Wyoming, Laramie, Wyoming.

Rosgen, D. 1982. A procedure and rationale for securing favorable conditions of water flows on national forest system lands in northern Wyoming. Draft manuscript. Region 2, U.S. Forest Service, Fort Collins, Colorado.

Stalnaker, C. B. and Arnette, J. L. (Eds.) 1976. Methodologies for the determination of stream resource flow requirements: An assessment. U.S. Fish and Wildlife Service, Office of Biological Services, Western Water Allocation, Washington, D. C.

United States Department of Agriculture, Forest Service. 1981. Final environmental impact statement on Cheyenne Stage II water diversion proposal. Rocky Mt. Region, Lakewood, Colorado. 235 pp. + App.

Wesche, T. A. and Rechard, P. A. 1980. A summary of instream flow methods for fisheries and related research needs. Eisenhower consortium bulletin No. 9. U.S. Government Printing Office: 1980-0-679-417/509, Washington, D.C.

Wesche, T. A., Reiser, D. W., Wichers, W. and Wichers, D. 1977. Fishery resources and instream flow recommendations for streams to be impacted by Cheyenne's proposed Phase II development. Report to Wyoming Game and Fish Department, Cheyenne, Wyoming.

SIMULATING TROUT FEEDING STATIONS IN INSTREAM FLOW MODELS

Hal A. Beecher

Washington Department of Game
600 N. Capitol Way
Olympia, Washington 98504   USA

INTRODUCTION

In recent years the Physical Habitat Simulation (PHABSIM), a portion of the Instream Flow Incremental Methodology (IFIM), has been widely used by fisheries biologists to model the relationship between stream flow (discharge) and fish habitat. PHABSIM, which is considered state-of-the-art (Loar et al. 1985), has two primary components: an hydraulic simulation and an habitat function (Stalnaker 1980; Bovee & Milhous 1978; Bovee 1982; Milhous et al. 1984).

The hydraulic simulation has several variations, but each predicts depth and velocity distributions over an array of discrete, contiguous units of streambed area ("cells") at a series of specified stream discharges. Substrate or cover values are assigned to each cell.

The habitat function, HABTAT, uses substrate or cover together with depth and velocity at a specified discharge to calculate a relative habitat value for each cell. Relative habitat value multiplied by surface area of the cell is summed for all cells to yield a weighted usable area (WUA) for each specified discharge. HABTAT uses functions called habitat preference curves or suitability of use criteria to calculate relative habitat value given depths, velocity, and substrate or cover. Habitat preference curves assign weighting or a preference factor ranging from 0 to 1 to each habitat value. For example, if the habitat preference curve is a velocity preference curve for adult rainbow trout (*Salmo gairdneri*), each water velocity is assigned a value from 0 to 1. A preference factor of 0 for velocity V_z implies that adult rainbow trout do not inhabit water with velocity V_z. A preference factor of 1 for velocity V_y implies that the adult rainbow trout inhabit water with velocity V_y as much as or more than any other velocity, all else being equal. WUA can be very sensitive to habitat preference curves.

One of the assumptions in PHABSIM as currently used is that depth, velocity, substrate and cover are adequate descriptors of physical microhabitat, provided that water quality is suitable. However, stream-dwelling trout often occupy feeding stations in which water velocity adjacent to the feeding station (Va) differs from water velocity at the feeding station (Vc), but both velocities are important (Wickham 1967; Chapman & Bjornn 1969; Griffith 1972; Everest & Chapman 1972; Campbell &

Neuner 1985). In this paper I propose a modification in the way velocity is used in HABTAT to simulate feeding stations or shear zones, and I compare a simplified version of this procedure with traditional WUA results at several sites in Washington State.

METHODS

The Washington Department of Game is responsible for recommending minimum instream flows to protect fish and wildlife habitat below hydroelectric projects or other diversion projects. Flow recommendations are based upon results of instream flow studies using PHABSIM. Instream flow studies have been conducted by project developers following guidelines developed by the Department of Game.

Velocity Preference Curve Development

Habitat preference curves are specific for individual streams as well as for each life-stage of each species (e.g., Moyle & Baltz 1985; Sheppard & Johnson 1985). For that reason the Department of Game recommends that habitat preference curve development be part of any instream flow study. Habitat preference curve development has three steps: (1) habitat mapping to determine distribution and availability of different velocities, depths, substrates, and cover types; (2) measurement of velocities, depths, substrates, and cover types used by all fish of the life-stage and species of interest; and (3) calculation of relative preference for each range of habitat values (sensu Baldridge & Amos 1981). Relative preference is calculated based on density of the target fish in each range of habitat values (see Orth et al. 1981).

In some cases conditions such as high turbidity preclude successful development of habitat preference curves. The Department of Game provides a set of habitat preference curves (called WDG curves in this paper) for use in those cases.

Feeding Station

I have used the binary criteria illustrated in Table 1 to evaluate instream flow requirements for rearing trout in a number of instream flow studies using PHABSIM in Washington during 1984 and 1985. I based the criteria in Table 1 upon incidental personal observations. Others have independently concluded that similar depth and velocity values influence rainbow trout distribution in Washington stream (Puget Sound Power & Light Company 1985; Campbell & Neuner 1985). For the primary cell (C) to be tallied as a feeding station, only one of the two adjacent cells (C-1 or C+1) needs to meet the adjacent cell criteria. I tabulated a number of feeding stations, but I did not weight feeding station according to cell area, unlike HABTAT.

Table 1. Velocity and depth criteria for a feeding station (cell C) and adjacent cells (cells C-1 and C+1) as used in feeding station analysis.

Cell	C	C-1 and/or C+1
Depth (m)	0.15	>0
Velocity (m s^{-1})	<0.3	>0.3 and >(0.15 + V_c)

72

Feeding station or focal point could be assessed by PHABSIM if the proposed procedure in Figure 1 (elaborated in Figure 2) were incorporated into HABTAT as an optional subroutine. A major advantage of using HABTAT for feeding station analysis is its ability to weight cells according to cell area. The user must specify the threshold velocity difference (ΔV min) between adjacent cells. The subroutine compares the velocity (V_c) in cell C with velocity in adjacent cells (V_{c-1} or V_{c+1}). If the difference between adjacent cells equals or exceeds ΔV min, then cell C (numbered consecutively across transect) is a potential feeding station. Evaluation of cell C by the subroutine continues as outlined in Figure 2. If the difference between adjacent cells is less than ΔV min, then cell C is not considered a feeding station.

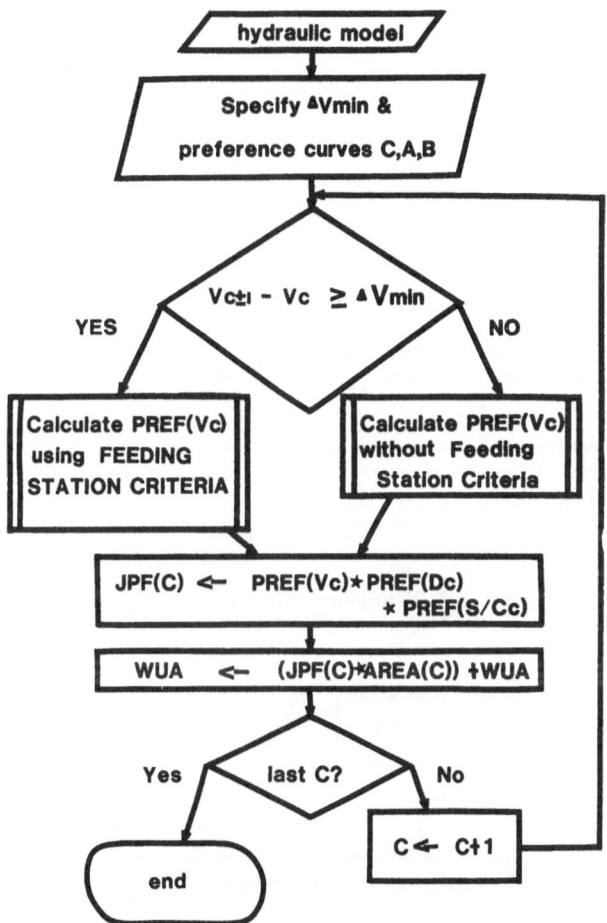

Figure 1. Proposed modification of HABTAT to consider velocity in adjacent cells. See Figure 2 for mechanics of calculation velocity preference for cell C, PREF (V_c), using feeding station criteria. User must specify threshold velocity difference (ΔV min) and velocity preference curves C (for cell being evaluated), A (for adjacent cells), and B (to be used only if non-feeding station criteria are also being used).

If the feeding station analysis is to be used, the user must provide at least two velocity preference curves to FISHFIL. Curve C is used for all C. A graphical representation of the binary criteria under C in Table 1 is an example of a simple curve. Curve A is used for adjacent cells C-1 and C+1. A graphical representation of the binary criteria under C+1 in Table 1 is an example of a simple curve A. Some cells that are not feeding stations, as characterized here, can have value as habitats for rearing fish. Consequently, a third velocity preference curve, B, is developed for areas that are not feeding stations. Assuming that non-feeding-station cells have a lesser value to rearing fish, the maximum weighting factor in velocity preference curve B is less than one.

Evaluation of feeding stations meeting the ΔV min criteria proceeds so that a joint preference factor for cell C, JPF (C), is a product of the preference factors for cell C for velocity (PREF(Vc)), depth (PREF(Dc)), and substrate or cover (PREF(S/Cc)). PREF(Vc) is the ordinate that corresponds to Vc in the preference curve C. Similarly, PREF (Va) is the

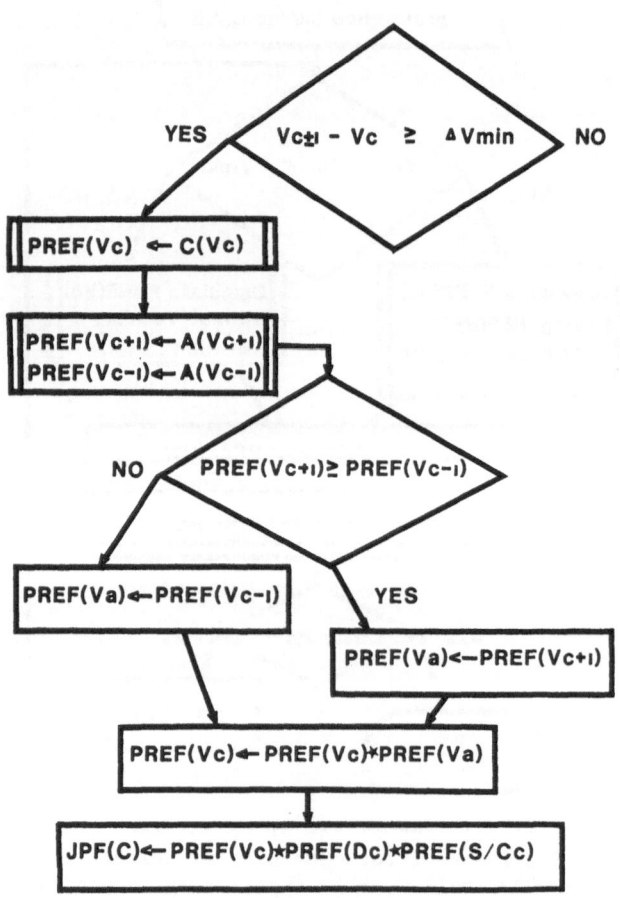

Figure 2. Mechanics for calculating velocity preference for cell C, PREF (Vc), using feeding station criteria.

higher of two adjacent cell velocity factors, PREF (Vc-1) and PREF (Vc+1), which are calculated using the preference curve A. The final PREF (Vc), which is used in calculating JPF(C), is a product of PREF(Va) and the initial PREF(Vc). Thus, velocities in both the feeding station cell C and in one adjacent cell are considered in calculation of JPF(C).

In deeper water (>75 cm) the mean of two velocity measurements, one at 0.8 depth and the other at 0.2 depth, is used as mean velocity for the water column in hydraulic simulations. Wickham (1967) observed brook trout (*Salvelinus fontinalis*) using slow water beneath faster water in a vertical feeding station. Criteria in Table 2 can be used to evaluate vertical feeding stations when velocity measurements have been made near the surface (0.2 depth) and near the bottom (0.8 depth). Vertical feeding stations can only be evaluated at measured flows because hydraulic simulations in PHABSIM generate only one velocity, a mean water column velocity, for each cell. In some cases it may be most appropriate to assume high velocities near the surface and use the criteria for 0.8 depth from Table 2 with velocity simulations for a depth near the bottom (see Milhous et al. 1984. Appendix K).

Table 2. Velocity preference criteria for vertically stratified velocities.

	Velocity (m s^{-1})
at 0.2 Depth	>0.3 and >($V_{0.8}$ + 0.15)
at 0.8 Depth	<0.3

RESULTS

Velocity Preference Curve Development

Velocity preference curves developed in Washington and Oregon streams for juvenile steelhead and for juvenile and adult rainbow trout indicate a preference for either very low velocities (near 0 m s^{-1}) or for a slightly higher velocity (near 0.3 m s^{-1}). Although each velocity preference curve is different, they fall into two distinct categories. In those preference curves showing maximum preference for velocities near 0.3 m s^{-1}, the preference factor for a velocity of 0 m s^{-1} is near 0. Examples are shown in Figure 3. Existence of two different types of curve implies two different types of optimum habitat for the same fish species. For a fish with velocity preference near 0 m s^{-1}, ideal habitat should be obtained by minimizing velocity without limiting depth or area, e.g., minimizing flow in a well-defined pool. For a fish with a preferred velocity of 0.3 m s^{-1}, a stream with uniform velocity of 0.3 m s^{-1} ought to be ideal, if the velocity preference curves are accurate and adequate to describe that part of microhabitat.

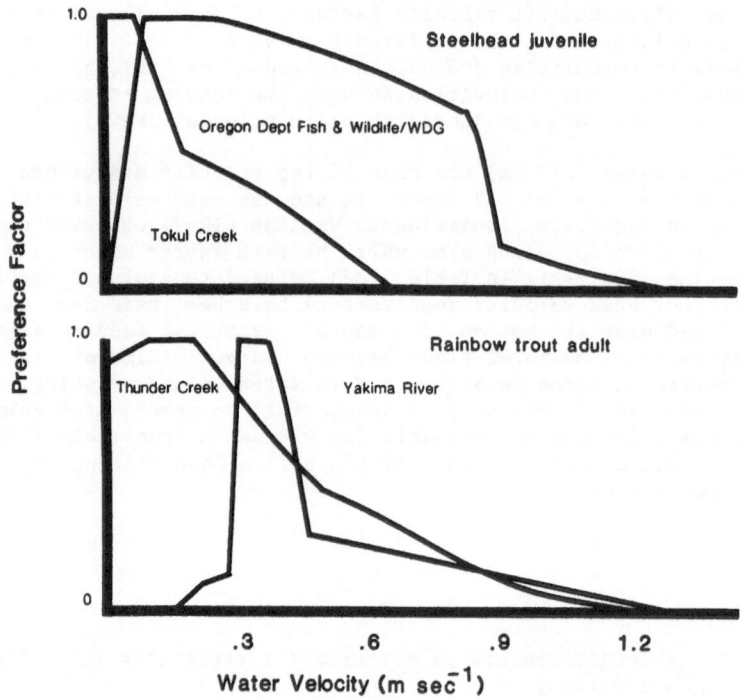

Figure 3. Examples of velocity preference curves for *S. gairdneri* used in Washington IFIM studies. Juveniles of the anadromous form (steelhead, top) and adult resident rainbow trout (bottom) show a dichotomy of velocity preference curves.

Preference for low velocity were generally found in high gradient streams in which low velocities were rare, and preferences for high velocity were generally found in lower gradient streams with extensive areas of low velocity but less area of high water velocity. Thus trout appeared to select the less available velocity. One explanation for this dichotomous preference for the less common velocity is that trout seek areas where low velocities are adjacent to higher velocities.

I reviewed preference curves and supporting data. Some of the examples in Figure 3 are smoothed versions of those presented by the developers who conducted the studies. I smoothed the right limbs of velocity preference curves where deviations from the trend over a part of the velocity range appeared to be artifacts of few observations.

Thunder Creek

At Thunder Creek, a steep (4-6%) boulder and cobble bed stream in the northern Cascade Mountains of western Washington, Puget Sound Power & Light Company (1985) conducted an instream flow study to determine operating limitations for a proposed hydroelectric project. The company developed habitat preference curves on-site (Campbell & Neuner 1985). Juvenile-adult rainbow trout habitat increased as stream flow increased within the range of flows simulated (Figure 4). In an appendix to the report, the authors questioned the validity of the study results because of a number of assumptions in PHABSIM, including habitat preference criteria (Puget Sound Power & Light Company 1985, Appendix 1). I used a feeding station analysis to respond to some of the criticisms. The feeding station analysis incorporates criteria which correspond to observations made by the authors in the appendix to their report:

a) "adult and juvenile trout selected offshore territories in protected water but immediately adjacent to high velocity water;"

b) "most adult trout were located along velocity shear lines near the head of a pool;"

c) "dominant factor determining habitat location in summer for all size classes appeared to be velocity, both faster water to provide a concentrated food source and slow water to provide cover;" and

d) "adults and juveniles did not use water less than 0.5 foot in depth."

The observations by Puget Sound Power & Light Company (1985) biologists coincide with the criteria I had developed for feeding station analysis. Despite different velocity preference curves, the two methods indicated similar relationships between flow and habitat in this study (Figure 4)

Lyre River

The Lyre River is a short, steep (4%) stream draining Lake Crescent on the northern Olympic Peninsula. Steelhead trout are the dominant fish in the study reach, but several species of salmon (*Oncorhynchus* spp.) may also use the study reach (CH2M Hill 1984).

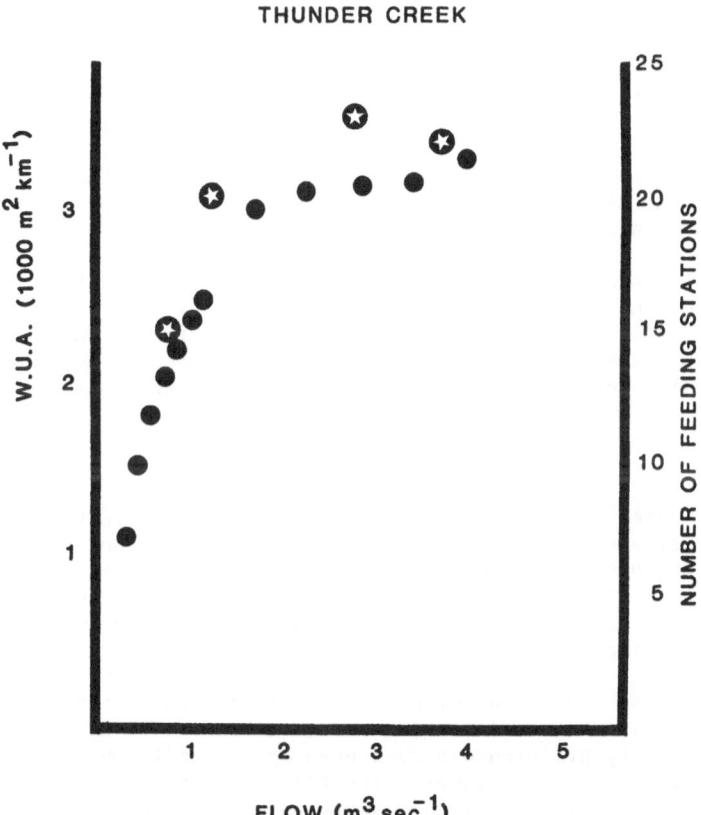

Figure 4. Trout habitat changes with flow in Thunder Creek. Trout habitat is estimated as WUA (circles) or as feeding stations (circled stars) based on measured calibration flows.

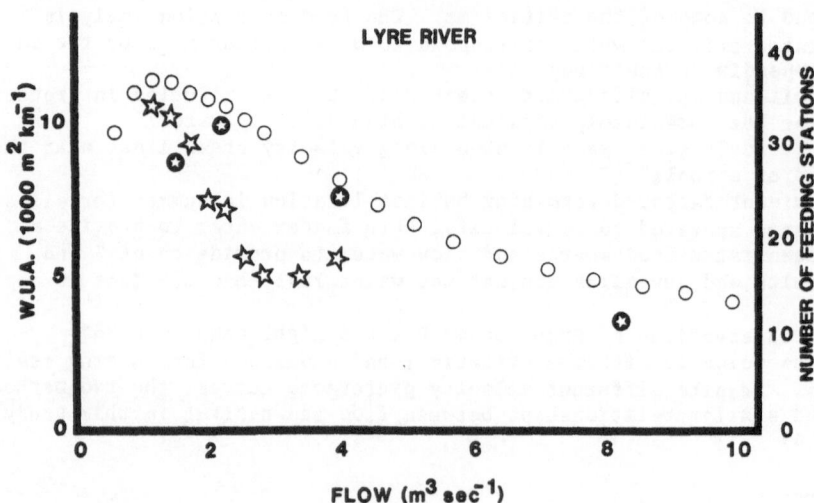

Figure 5. Steelhead habitat changes with flow in Lyre River. Steelhead habitat is estimated as WUA (circles) or as feeding stations based on measured calibration flows (circled stars) or on simulated flows (stars).

No habitat preference curves were developed on-site. Unlike the Thunder Creek instream flow study, where on-site habitat preference curve development led to the use of low velocity preference curves, the Lyre River instream flow study (CH2M Hill 1984) used the higher velocity preference curve shown on the upper right of Figure 3. This curve was provided by the Department of Game and was based on unpublished work by the Oregon Department of Fish and Wildlife.

The relationship between WUA and flow is similar to that between number of feeding stations and calibration flows (Figure 5). However, number of feeding stations at calibration flows differ from number of feeding stations at simulated flows, indicating that the hydraulic simulation does not match measured depths and velocities at calibration flows.

Tokul Creek

Weyerhaeuser Company (Barry Firth, personal communication) conducted an instream flow study at two sites on Tokul Creek, a low elevation stream used by anadromous salmonids, including steelhead and cutthroat trout (*Salmo clarki*). Both sites were bedrock channels with many boulders and some patches of gravel. The upper site had a gradient of 5.4% and the lower site had a gradient of 3.9%.

Weyerhaeuser Company developed velocity preference curves for juvenile steelhead as well as juvenile cutthroat trout and trout fry. The juvenile steelhead velocity preference curves developed on-site by Weyerhaeuser indicated a preference for lower velocities than those indicated by the WDG curve. Barry Firth (personal communication) used both the WDG curve and the on-site velocity preference curve in HABTAT runs for Tokul Creek.

At the upper site all results indicated the same trend: rearing habitat for steelhead increases as flow increases.

Results for the lower site are more interesting. Use of on-site preference curves indicates that habitat increases as flow decreases but use of WDG curves indicates that steelhead rearing habitat increases as flow increases (Figure 6). Feeding station analysis of the measured calibration flows indicates no change in habitat with change in flow, but feeding station analysis of simulated flows indicates an increase in habitat with increasing flow, similar to results with the WDG curves, illustrating again the sensitivity of feeding station analysis to hydraulic modeling.

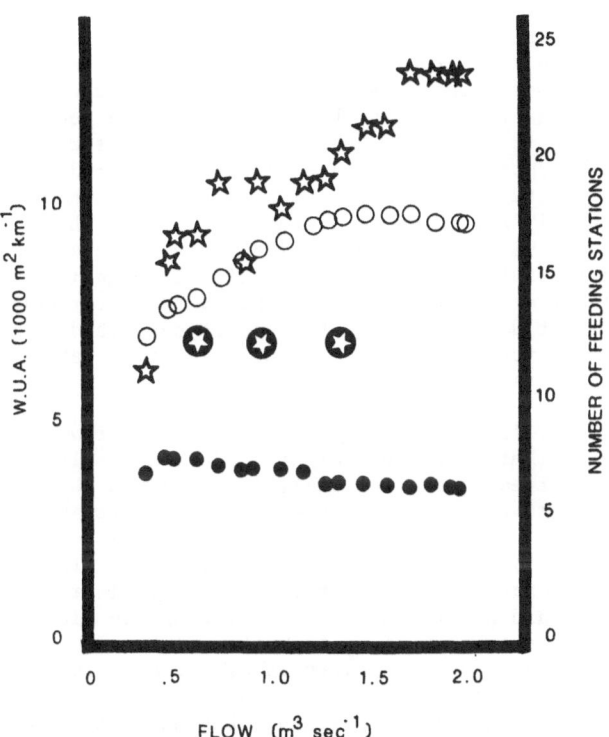

Figure 6. Use of different velocity preference criteria for juvenile steelhead at Tokul Creek indicate different relationships between habitat and flow. Solid circles are WUA based on velocity preference curve developed during IFIM study (see Figure 3 for smoothed Tokul Creek velocity preference curve). Clear circles are WUA based on velocity preference curve developed by Oregon Department of Fish and Wildlife and modified by the author (see Figure 3). Three circled stars are feeding stations determined from measured calibration data. Stars are feeding stations determined from simulated flows.

DISCUSSION

Stream-dwelling trout, including juvenile anadromous trout, frequently maintain a position in water having a relatively slow velocity adjacent to faster food-bearing water (Wickham 1967; Chapman & Bjornn 1969; Griffith 1972; Everest & Chapman 1972; Campbell & Neuner 1985). In pool-riffle or pool-cascade habitats, such conditions are often found at the upstream end of a pool, but they can occur anywhere in a stream. By occupying such a feeding station or focal point, a trout can minimize its energy expenditure while maximizing food availability. Vortices created at the shear zone between fast and slow water can capture and concentrate drifting food organisms. Neither the faster velocity of the food-bearing water nor the slower velocity of the feeding station would seem, by itself, adequate to characterize feeding station microhabitat. A characteristic of the feeding station is the combination of different adjacent velocities.

In the majority of cases, including some not presented here, feeding station analysis yields results that are consistent with results of standard HABTAT results. Both types of analysis suggest that rainbow trout and steelhead rearing habitat is limited by low flows in most Washington streams. This is also consistent with an analysis of the relationship between low flows and steelhead production (Beecher 1981).

Logical extension of the present use of HABTAT implies that ideal physical habitat would be provided by expanses of uniform velocity and uniform depth. For example, a velocity preference curve with a peak at zero velocity implies that optimum habitat would be a pool with no flow entering or leaving it. However, a combination of different water velocities in close proximity has been widely recognized as an important aspect of trout habitat in streams. Wickham (1967), Chapman & Bjornn (1969), Griffith (1972), Everest & Chapman (1972), White (1973), Campbell & Neuner (1985), and Johnson (1985) have discussed feeding stations or focal points used by trout, indicating the importance of slow water with at least a threshold depth adjacent to faster water. I have often observed trout using feeding stations as described in this report, but I have also observed trout feeding and holding in habitat that did not appear to meet the present feeding station criteria.

Use of higher velocity preference curves together with adequate cover coding and cover weighting factors might achieve a result comparable to that obtained with feeding station analysis, at least when cover such as boulders serves to reduce velocity immediately downstream.

Feeding station analysis identifies shear zones. The studies discussed in this report were conducted in high gradient trout streams to determine how salmonid habitat changes with change in flow. Shear zones are also important in other regions for other fishes. While electrofishing in large, low gradient rivers of northwest Florida, I found large concentrations of fish of many species along shear zones. Thus, the approach outlined here would be applicable beyond salmonid streams.

A drawback to the feeding station analysis is its requirement for more data in the form of additional velocity preference curves. The U.S. Fish and Wildlife Service Instream Flow and Aquatic Systems Group has recently incorporated the feeding station concept as an option in a new program, HABTAV (Dr. Robert Milhous, personal communication). HABTAV uses a single velocity preference curve along with several other single criteria, thus avoiding some of the multiplication of preference curves proposed here. Comparison of number of feeding stations at measured flows and at simulations of those same flows emphasizes the importance of hydraulic simulation. Small deviations of simulated velocities from measured velocities can change results significantly.

Stober (1984) questioned the validity of PHABSIM on large rivers (e.g., Skagit River - mean annual flow = 425± m^3 s^{-1}). His results, as well as other PHABSIM studies on rivers larger than about 100 m^3 s^{-1} mean annual flow, indicate that habitat is maximized by reducing flow below lowest flow. Use of vertical feeding station analysis (Table 2) might give a more realistic indication of habitat than using PHABSIM with mean velocity for the water column.

ACKNOWLEDGEMENTS

The three instream flow study reports discussed are public information in the files of the Washington Department of Game. This paper has benefitted from discussions with Thom Johnson (Washington Department of Game), Dr. Clair Stalnaker and Dr. Robert Milhous (Instream Flow and Aquatic Systems Group, U.S. Fish and Wildlife Service), Ron Campbell (Beak Consultants Incorporated), Mike Stempel (U.S. Fish and Wildlife Service), Jean Caldwell and John Easterbrooks (Washington Department of Fisheries), Forrest Olson (CH2M Hill), Pete Rittmueller and Michael Barclay (Cascades Environmental Services), Barry Firth and Jim Fisher (Weyerhaeuser Company), Will Sandoval (Muckleshoot Indian Tribe), Brad Caldwell (Washington Department of Ecology), Phil Hilgert (Hosey and Associates Engineering Company), Mike McDowell (Dames & Moore), and Cary Feldmann (Puget Sound Power & Light Company).

REFERENCES

Baldridge, J. E. & Amos, D. 1981. A technique for determining fish habitat suitability criteria: a comparison between habitat utilization and availability. In: Acquisition and utilization of a aquatic habitat inventory information (Ed. by N. B. Armantrout) pp. 251-258. Western Division American Fisheries Society.

Beecher, H. A. 1981. Instream flows and steelhead production in western Washington. Proc. 60th Ann. Conf. West. Assoc. Fish Wildlife Agencies, Kalispell, Montana, July 13-17, 1980, 396-410.

Bovee, K. D. 1982. A guide to stream habitat analysis using the instream flow incremental methodology. Instream Flow Information Paper No. 12, U.S. Fish and Wildlife Service FWS/OBS-82/26.

Bovee, K. D. & Milhous, R. 1978. Hydraulic simulation in instream flow studies: theory and techniques. Instream Flow Information Paper No. 5, U.S. Fish and Wildlife Service FWS/OBS-78/33.

Campbell, R. F. & Neuner, J. H. 1985. Seasonal and diurnal shifts in habitat utilized by resident rainbow trout in western Washington Cascade Mountain streams. In: Proceedings of the Symposium on Small Hydropower and Fisheries. (Ed by F. W. Olson, R. G. White, & R. H. Hamre) pp. 39-48. American Fisheries Society.

Chapman, D. W. & Bjornn, T. C. 1969. Distribution of salmonids in streams, with special reference to food and feeding. In: Symposium on salmon and trout in streams. (Ed by T. G. Northcote) pp. 153-176. H.R. Mac Millan Lectures in Fisheries, University of British Columbia, Vancouver.

CH2M Hill 1984. Lyre River instream flow assessment-prepared for Public Utility District No. 1 of Clallam County, Port Angeles, Washington, CH2M Hill, Bellevue, Washington.

Everest, F. H. & Chapman, D. W. 1972. Habitat selection and spatial interaction by juvenile chinook salmon and steelhead trout in two Idaho streams. J. Fish. Res. Board Can. 29: 91-100.

Griffith, J. S., Jr. 1972. Comparative behavior and habitat utilization of brook trout (*Salvelinus fontinalis*) and cutthroat trout (*Salmo clarki*) in small streams in northern Idaho. J. Fish Res. Board Can. **29**: 265-273.

Johnson, T. H. 1985. Density of steelhead parr for mainstem rivers in western Washington during the low flow period, 1984. Washington State Game Department, Fisheries Management Division, 85-6, Olympia.

Loar, J. M., Sale, M. J., Cada, G. F., Cox, D. K., Cushman, R. M., Eddlemon, G. K., Elmore, J. L., Gatz, A. J., Kanciruk, P., Salmon, J. A. & Vaughan, D. S. 1985. Application of habitat evaluation models in southern Appalachian trout streams. Oak Ridge National Laboratory, Environmental Sciences Divisions, Pub. No. 2383, ORNL/TM=9323.

Milhous, R. T., Wegner, D. L. & Waddle, T. 1984. User's guide to the Physical Habitat Simulation system (PHABSIM). Instream Flow Information Paper No. 11, U.S. Fish and Wildlife Service, FWS/OBS-81/43 Revised.

Moyle, P. B. & Baltz, D. M. 1985. Microhabitat use by an assemblage of California stream fishes: developing criteria for instream flow determinations. Trans. Amer. Fish. Soc. **114**: 695-704.

Orth, D. J., Jones, R. N. & Maughan, O. E. 1981. Considerations in the development of curves for habitat suitability criteria. In: Acquisition and utilization of aquatic habitat inventory information. (Ed. by N. B. Armantrout) pp. 124-133. Western Division American Fisheries Society.

Puget Sound Power & Light Company 1985. Thunder Creek instream flow study. Puget Sound Power & Light Company, Bellevue, Washington.

Sheppard, J. D. & Johnson, J. H. 1985. Probability-of-use for depth, velocity, and substrate by subyearling coho salmon and steelhead in Lake Ontario tributary streams. N. Amer. J. Fish. Manage. **5**: 277-282.

Stalnaker, C. B. 1980. The use of habitat structure preferenda for establishing flow regimes necessary for maintenance of fish habitat. In: The Ecology of Regulated Streams (Ed by J. V. Ward & J. A. Stanford) pp. 321-387. Plenum Publishing Corporation, New York.

Stober, Q. 1984. Interpretation of IFIM results. Unpublished paper presented at Pacific Fishery Biologists 46th Ann. Conf., Ocean Shores, Washington, March 19-21.

White, R. J. 1973. Stream channel suitability for coldwater fish. Proceedings of the 28th Annual Meeting of the Soil Conservation Society of America (Plants, Animals and Man), September 30 - October 3, Hot Springs, Arkansas, 61-79.

Wickham, M. G. 1967. Physical microhabitat of trout, M.S. thesis, Colorado State University, Fort Collins, Colorado. 42.

EFFECTS OF VARYING FLOWS IN MAN-MADE STREAMS

ON RAINBOW TROUT (*SALMO GAIRDNERI* RICHARDSON) FRY

J. R. Irvine
Department of Zoology
University of Otago
P.O. Box 56, Dunedin, New Zealand[1]

INTRODUCTION

The published proceedings of the two previous international symposia on regulated streams (Ward & Stanford 1979; Lillehammer & Saltveit 1984), as well as the papers presented at this third symposium, indicate the concern for understanding impacts of flow regulation on fish and other aquatic organisms. A major difficulty in the study of regulated streams is conducting studies with proper scientific controls. A survey of the papers published in the two previous symposia revealed only two studies which had the controls in space and time recommended by Green (1979). Although for many regulated rivers it is not possible to conduct impact studies with good controls, the limitations, and consequently the validity of predictions from these studies, must be realized. Impact assessments lacking control areas may record unrelated changes and attribute these to the impact. Assessments lacking a temporal control may not detect that differences existed between the control area and the impacted area before the impact occurred.

The construction in 1980 of the lower Waitaki River replicate stream channels in New Zealand allowed the possibility of conducting experiments with controls in space and time. The purpose of this paper is to present results from two experiments in which the effects of flow changes, simulating conditions below a hydroelectric peaking plant, on rainbow trout fry emigration, growth and condition, production and habitat preferences were examined.

GENERAL METHODOLOGY

Replicate Streams

A brief description of the lower Waitaki River replicate stream channels is given by Campbell et al. (1984) and Rimmer (1985) and they are described in detail by Irvine (1984). The replicate streams are approximately 32 km upstream from the mouth of the Waitaki River, South

---

[1]Present address: Department of Fisheries and Oceans, Fisheries Research Branch, West Vancouver Laboratory, 4160 Marine Drive, West Vancouver, British Columbia, Canada V7V 1N6.

Island, New Zealand (Figure 1). Stream channels were excavated in 1980 and lined to a depth of 50 cm with Waitaki River rocks which consisted of 10% gravel and 90% cobbles. The streams were approximately 100 m long, 3 m wide at their base with 2:1 side slopes and 1:900 slope. Each had four riffles about 19 m long alternating with three pools about 8 m long. Water from the Waitaki River was discharged through a 30 ha irrigation headpond into a smaller, constant level headpond, and then into the replicate streams. Stream discharge was set according to a previously determined stage-discharge relationship using slide gates at the upstream end of each stream.

Water temperature was measured during each experiment by a maximum-minimum thermometer installed at the upper end of Stream 3. Temperatures were recorded not less than once a week. Permanent staff gauges were installed at the downstream end of each stream. The length of each riffle and pool was measured and transects were established at the midpoint of each of these habitats. Stream widths were recorded at each transect. Water depths and velocities (at 0.6 times the depth) were measured at 50 cm intervals along each transect using an Ott current meter.

Fish

Collection. Fry were caught by electrofishing the streams at approximately four-weekly intervals. Seine (4 meshes cm^{-1} for youngest fry, 1.1 meshes cm^{-1} for older fry), 5 m x 1 m were first positioned

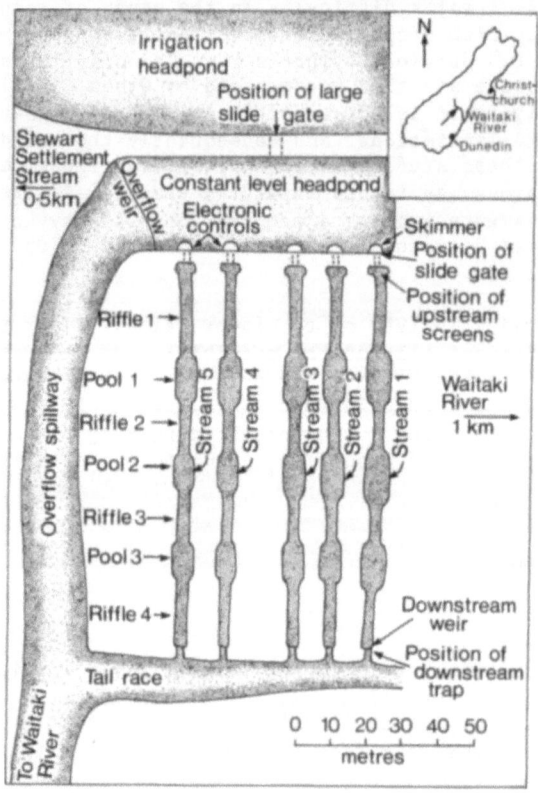

Figure 1. Diagram of the replicate stream facility (semi-diagrammatic) and the location of the Waitaki River in South Island, New Zealand.

between each riffle and pool. Depending on the water conductivity, 410-610 volts DC were used.

Electrofishing took place with the operators walking downstream and at least three successive runs were made of each stream. In Experiment 2, fry captured in riffles were kept separate from fry captured in pools and the lesser of 25 fry or the entire catch from each category were anaesthetized with ethyl-p-aminobenzoate (benzocaine) and fork lengths and wet weights were obtained. In Experiment 1, fry caught in riffles and pools were not separated and a maximum of 50 fry was weighed and measured per stream. Fry were held for at least 2 h before being released into their appropriate habitat and stream.

Downstream fry movement was monitored using inclined plane traps modified from the design of Wolf (1951). Stainless steel mesh (3.7 meshes cm^{-1}) was used for screens in the inclined planes and live boxes.

Population estimates, growth and production. Electrofishing results were analyzed using the sequential removal method of Zippen (1956) to obtain population estimates with associated 95% confidence limits. Population estimates were converted to numbers of fry per 100 m^2 to standardize for different areas in the various streams.

Growth was assumed to be isometric and condition factors were calculated for individual fry as 100 times the fry weight (g) divided by the fry length (cm) cubed.

The mathematical model of Ricker (1946) was used to estimate fry production occurring during intervals between population estimates. Instantaneous growth rates and a linear change in biomass were assumed in calculating production.

Habitat preferences. For each census, the total number of fry caught in each habitat was divided by the area of that habitat to obtain the number of fry per m^2. To facilitate comparisons between streams and dates, these densities were expressed as percentages by area. To obtain these percentages, each density estimate for each habitat was divided by the density estimate for that stream and multiplied by 100.

Pools developed at the upper end of the streams. These inlet pools were smaller and the water in them was more turbulent than downstream. Fry densities were estimated in inlet pools during Experiment 1, but subsequently upstream screens were installed below them, eliminating fry access to them, and these pools were not sampled for fry.

Statistical analyses of results. Two way model I (fixed-effects) streams-by-times analyses of variance (ANOVA) were used to analyze fry results during control periods. It was generally assumed that no interaction occurred between streams and times during control periods and so the streams and times mean squares were tested over the error (interaction) mean squares. The two null hypotheses tested were that the effects due to different streams, and the effects due to different times, were both nil.

When the discharge in some of the streams was varying, it obviously could not be assumed that interaction effects were insignificant and the computer program GLIM (General Linear Interactive Modelling) (Manly & Crosbie 1977; Baker & Nelder 1978) was used to test whether flow changes had significant effects. With GLIM it was possible to introduce an intervention factor ("fluctuation") to account for the presence or absence of fluctuating flows. "Fluctuation" was given the value of zero during the control period for streams in which flow varied later, and zero also for control streams in which flow never varied. During periods of fluctuating flows, "fluctuation" was weighted according to how long flows had been varying. GLIM performed an ANOVA and the importance of flow fluctuations was determined by observing changes to the residual sums of squares when "fluctuation" was included in the model. The use of GLIM to analyze the effects of varying flows in the replicate streams is described in more detail by Irvine (1984; 1986).

When significant values of F were obtained, Duncan's multiple range tests (DMRT) were used to compare treatment means with each other.

EXPERIMENT 1 - EFFECTS OF REGULAR SMALL CHANGES IN DISCHARGE

Experimental Design

The first purpose of this experiment was to compare the five streams during equal constant flow to provide a control in space for subsequent experiments. The second purpose was to test the hypothesis that regularly occurring changes in discharge would have no effect on rainbow trout fry production or the distribution of these fish between riffles and pools. Only Streams 2, 4 and 5 were available to test this hypothesis. Stream 2 was used as a control in time and did not vary in discharge. After the equal flow control period, fluctuations in discharge were begun in Streams 4 and 5 and fry were monitored for almost five months.

Discharge in each stream was set at 125 l s^{-1} in early October 1980. On 20 October approximately 2000 alevins per stream were injected into artificial redds constructed in Riffle 1 and on 6 November an additional 4000 late-eyed-eggs were added per stream. Although emerged fry were first observed in the streams on 25 November, it was not possible to begin monitoring downstream fry movement until 3 December.

On 17 February, manipulations of the discharge in Streams 4 and 5 were begun (Figure 2). Streams 1 and 3 were then used in a separate experiment (Campbell et al. 1984; Rimmer 1985). The discharge in Streams 4 and 5 was increased and decreased twice daily, five days per week, the time of flow changes similar to what might be found in a river regulated for hydroelectric power generation. Every morning (Monday to Friday), flows in Streams 4 and 5 were increased in four equal steps between 0630-0730 h, remained at the peak flow for 1 h, and then decreased in four steps to reach the base flow at 0930 h. The same pattern was followed in the afternoons, beginning at 1600 h and finishing at 1900 h. Stream 4 varied in discharge between 110 and 200 l s^{-1} and Stream 5 varied between 136 and 194 l s^{-1}.

Nine population estimates were made, the first three during the control period. Because of differences in population sizes between streams, on 15 January, Stream 3 was electrofished and 49 fry were transferred to Stream 2, 36 fry to Stream 4 and 41 fry to Stream 5. Beginning with the third fry census, inlet pools at the upper end of each stream were included in each census. These habitats were small at the beginning of the experiment but increased in size due to erosion. The experiment was terminated after the final population estimate on 7 July 1981.

Results

At the control flow, average water velocities ranged from 13.8 cm s^{-1} in Stream 4 to 20.1 cm s^{-1} in Stream 3 (Figure 2). Corresponding velocities in riffles were 3-5 cm s^{-1} higher, and velocities in pools were 8-13 cm s^{-1}

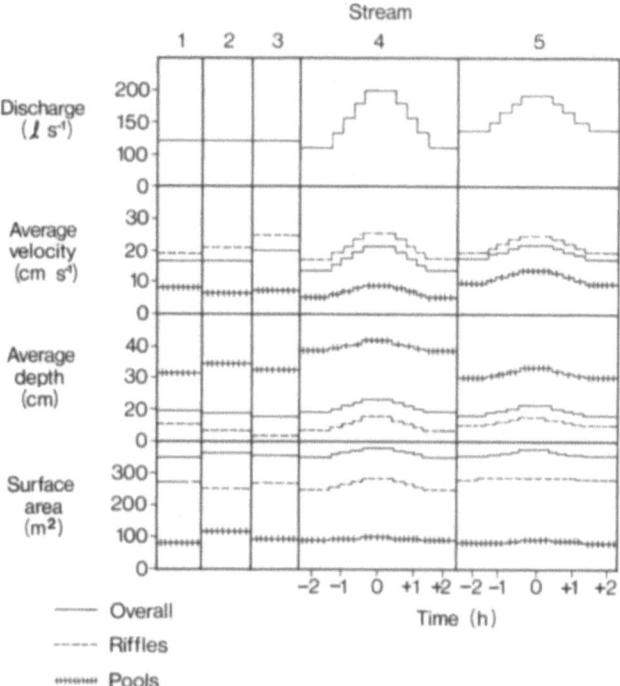

Figure 2. Physical parameters in Streams 1-5 during control flow and, for Streams 4 and 5, during fluctuating flow periods in Experiment 1. Time scale for Streams 4 and 5 in hours before (-) and after (+) the peak of each fluctuation.

lower. Average depths were similar between streams, ranging between 17.3 and 19.3 cm. Pools were 12-20 cm deeper, and riffles were 4-6 cm shallower, than stream averages. The surface area of the streams varied only between 347 and 361 m^2. Riffles comprised 68-77% of the area with pools making up the remainder.

As discharge in Streams 4 and 5 increased and decreased, so did other physical parameters (Figure 2). Average velocities in Streams 4 and 5 changed proportionately more than average depths, but peak velocities were only slightly higher than in Stream 3 at its control flow. The surface area of each stream, itself a function of average depth, varied only slightly with changes in discharge.

Peak water temperatures were 22°C in late January and early February and the minimum temperature was 4.5°C in early July. There was a gradual decrease in water temperature between mid-February and the end of the experiment.

From 3 December until flow changes were begun in Streams 4 and 5 on 17 February, between 604 (Stream 5) and 1960 (Stream 3) fry per stream were captured in the downstream traps. During this control period, fry leaving Stream 3 were significantly shorter, lighter and in poorer condition than fry leaving the other streams (DMRT; p<0.05). Emigration had virtually ceased by February and fluctuating flows had no effect.

Population densities declined rapidly in all streams from December to February (Figure 3). Effects on population densities due to different streams were significant (ANOVA; $F = 4.07$; $p<0.05$). Fry densities in Stream 3 were significantly higher than in all other streams, except Stream 4 (DMRT; $p<0.05$). There was no difference between the fry densities in Streams 1, 2, 4 and 5.

Flows began varying on 17 February in Streams 4 and 5 and in the subsequent population estimate (13 March), fry densities in Stream 4 in which discharge varied the most, were significantly greater than in the stream with the smallest range of discharge variations (5), and also greater but not significantly, than in the control stream (2) (Figure 3). Stream 4 retained the highest fry density for the duration of the experiment. When GLIM was used to determine whether these conditions were the consequence of fluctuating flows, effects due to fluctuating flows were not significant ($F = 0.15$; $p>0.50$).

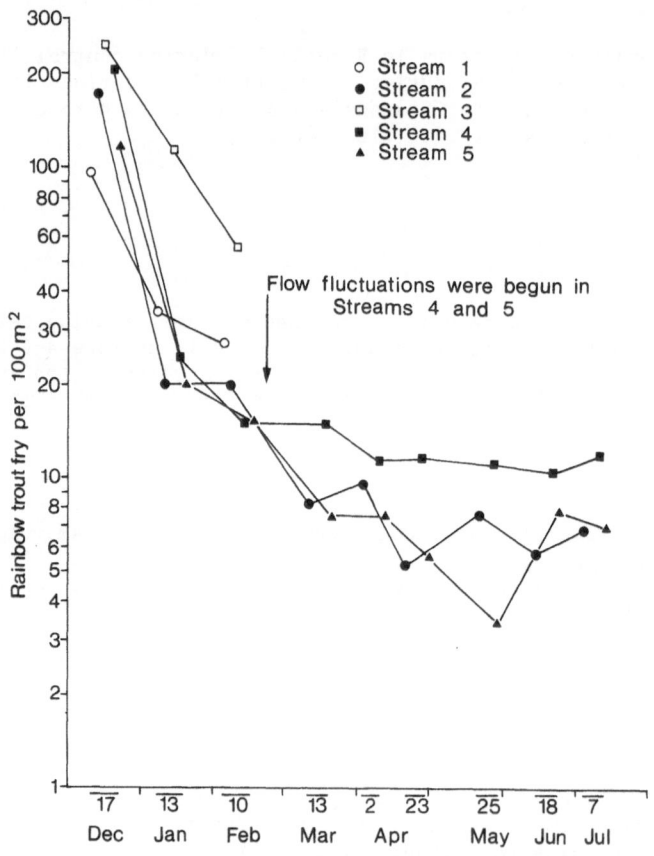

Figure 3.   Rainbow trout densities in Streams 1-5 during Experiment 1. Flow fluctuations were begun in Streams 4 and 5 on 17 February 1981.

Fry in all streams increased in length and weight during the control period (Figure 4). Effects on length due to streams were significant (ANOVA; F = 6.80; p<0.05). Stream 3 fry were significantly shorter than fry from all other streams except Stream 5 (DMRT; p<0.05). Fry weights and conditions were not significantly different in different streams.

Following the onset of flow fluctuations, fry in the control stream (2) were generally heavier and longer than those in either Streams 4 or 5 in which flows were varied, but this condition also prevailed during the control period (Figure 4). When GLIM was used to analyze the significance of fluctuating flows on fry weights and lengths, effects were not significant. Effects of fluctuating flows on condition factors were statistically significant (GLIM; F = 5.96; p<0.05), but this was probably not a real effect. Condition factors in the control stream (2) were initially lower than in the other streams. Their increase at about the time flow variations were begun in Streams 4 and 5, and no similar increase in the other streams, resulted in the test being significant.

Production during the control period ranged from 0.76-2.57 g (100 m²)⁻¹ day⁻¹ with Stream 3 having the highest production and Stream 5 the lowest production (Figure 5). A two way streams-by-dates ANOVA failed to show a significant effect due to streams. However, a DMRT indicated that production was significantly higher in Stream 3 than in Stream 5 during the control period (p<0.05). The graph of production versus time for the experimental streams (Figure 5) did not suggest that fluctuating flows had any effect on fry production, and this conclusion was confirmed by GLIM.

In December and January, riffles were the preferred habitat in all streams. By March, the importance of riffles appeared to have declined. The proportions of fry found in the inlet pools increased and from 23 April on, more than 50% of the fry caught in each stream were in these upstream

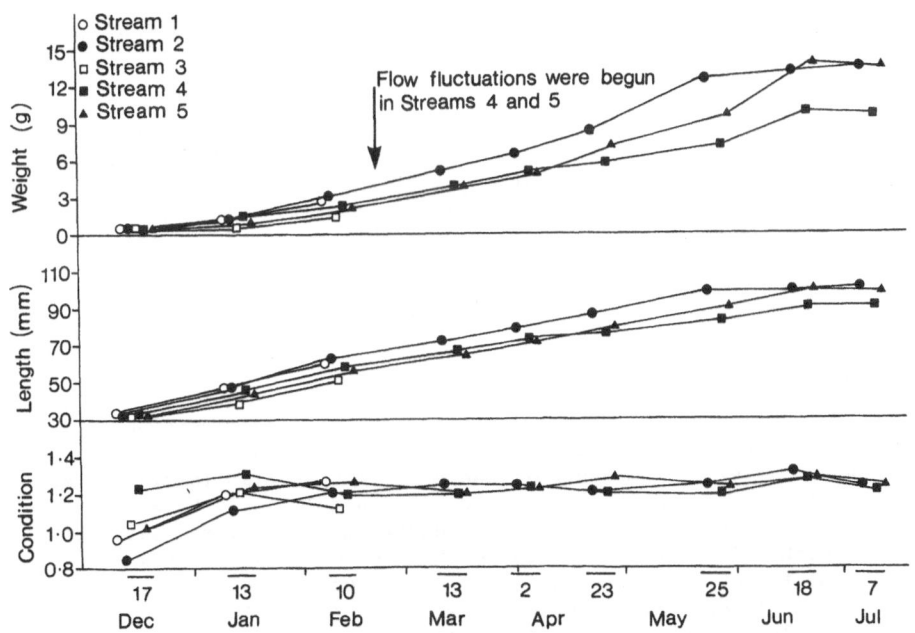

Figure 4. Mean weights, lengths and condition factors for rainbow trout fry during Experiment 1. Flow fluctuations were begun in Streams 4 and 5 on 17 February 1981.

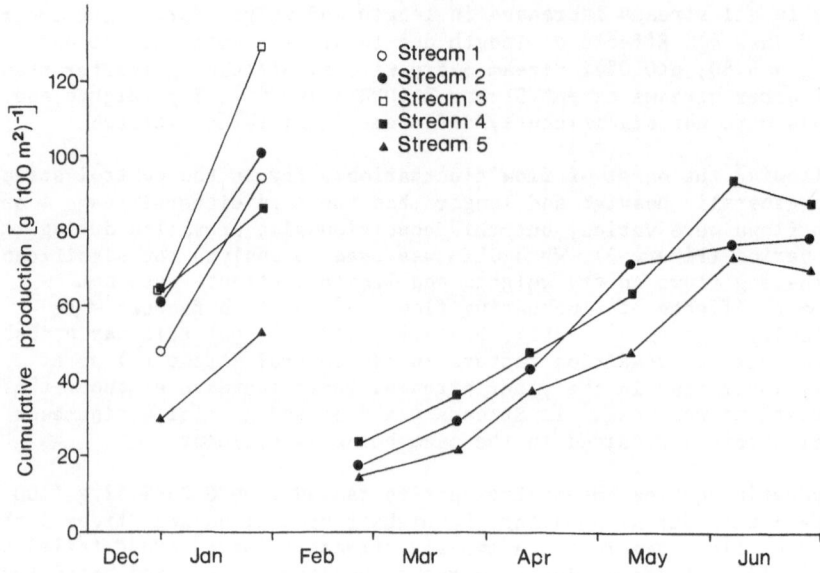

Figure 5. Cumulative rainbow trout production in Experiment 1 during the control (December–January) and fluctuating flow (February–June) periods. Flow fluctuations were begun on 17 February 1981.

pools. The same pattern occurred in the control stream (2) as in the experimental streams, suggesting that fluctuating flows had no significant effect on fry distribution.

EXPERIMENT 2 – EFFECTS OF REGULAR FIVE-FOLD CHANGES IN DISCHARGE

Experimental Design

Fluctuating flows had little effect on rainbow trout fry in Experiment 1, perhaps because the range of discharges during fluctuations was insufficient to affect the fry. In Experiment 2, the same discharge pattern was followed as in Experiment 1, except that the base flow was lowered and the peak flow was increased. As in Experiment 1, the main hypothesis tested was that regular changes in discharge would have no effect on rainbow trout fry production or the distribution of these fish between riffles and pools. In addition, an attempt was made to investigate the role of density dependent factors by stocking fry at varying densities. All five streams were used in the experiment.

Screens were installed near the upper end of the streams before Experiment 2. The screens reduced the likelihood of non-experimental fish entering the streams from upstream and experimental fish leaving in an upstream direction, and they trapped debris which might otherwise have interfered with the operation of the downstream traps. Three pairs of screens, each pair with a different mesh size, were constructed for each stream. The finest mesh screens, with 5 mm holes, slipped into a frame assembly and filtered all water entering the streams. The medium (1.2 cm) and coarse (7 cm) mesh screens leaned against the finer screens.

Two hundred and twenty-three rainbow trout fry were caught by electrofishing in nearby tributaries of the Waitaki River. The fish were

Table 1. Density of Rainbow Trout fry introduced into replicate streams on 18 February 1982, Experiment 2.

Stream	Density	Fry $(100 \ m^2)^{-1}$
1	low	9.8
2	intermediate	16.4
3	higher	36.5
4	higher	32.0
5	low	9.3

weighed and measured, and on 18 February 1982 they were released into Pool 2 of the streams to yield the densities shown in Table 1.

For the next three days, downstream-emigrating fry were returned to the streams, and from then on, downstream emigrants were recorded and permitted to escape from the system.

All streams remained at a constant discharge of 75 1 s^{-1} for a control period which lasted until 16 March. From that date the discharge in Streams 4 and 5 was increased and decreased 5-fold, twice daily, five days per week in the same pattern as in Experiment 1 (Figure 6).

Population estimates were made near the end of the control period on 3 March, and again on 14 April, 11 May and at the end of the experiment on 9 June.

Results

During control flows, average water velocities ranged from 12.6 cm s^{-1} in Stream 4 to 18.7 cm s^{-1} in Stream 2 (Figure 6). Average velocities in riffles were 2-5 cm s^{-1} faster and average velocities in pools were 10-14 cm s^{-1} slower than average stream velocities. Stream 5 had the lowest average depth (14.3 cm) and Stream 4 had the greatest (17.5 cm). Riffles were 3-7 cm shallower and pools were 14-21 cm deeper than stream averages. Surface area ranged from 307 m^2 in Stream 3 to 369 m^2 in Stream 4 with riffles constituting an average of 73% of the total area.

There were significant changes in all physical parameters in Streams 4 and 5 with changes in discharge (Figure 6). Five-fold changes in discharge resulted in a 3.1-fold change in average water velocity and a 1.6-fold change in average depth for Stream 4 and a 2.5-fold change in velocity and a 2.0-fold change in depth for Stream 5. Velocities and depths in Streams 4 and 5 at peak flows were higher than in the other streams. Changes in surface area accompanying changes in discharge were proportionately less than changes in other physical parameters.

Peak water temperatures were 20.5°C on 19 February and 12 March and the minimum temperature was 6°C at the end of the experiment. Temperatures generally declined between March and June.

Twenty fry migrated downstream during the control period (21 February - 16 March) from Stream 3, one of the higher stocking density streams.

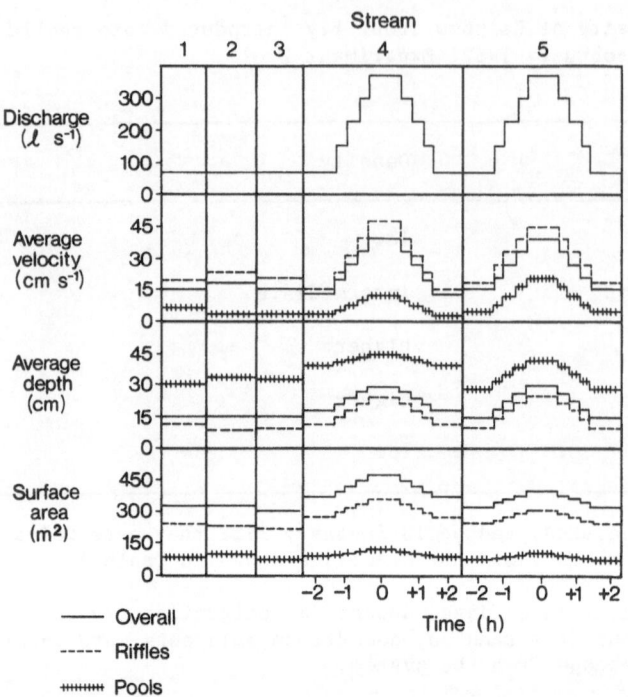

Figure 6. Physical parameters in Streams 1–5 during control flow and, for Streams 4 and 5, during fluctuating flow periods in Experiment 2. Time scale for Streams 4 and 5 during flow fluctuations in hours before (–) and after (+) the peak of each fluctuation.

Five fry left Stream 4, the other higher stocking density stream, and 3–8 fry left each of the other streams during this period. No fry left any of the streams during the period the discharge in Streams 4 and 5 was varying (17 March – 9 June) so it was concluded that flow fluctuations had no effect on downstream emigration.

Instream mortality was insignificant in both low density streams but was significant in the higher density variable flow stream. Densities in the two higher density streams were similar to the intermediate density stream by the second census (Figure 7) so results from these three streams are discussed together.

During the control period, there were no significant differences between low density streams for fry length, weight or condition (Figure 8). Fluctuating flows increased fry weight significantly (GLIM; F = 61.61; p<0.005) in the low density stream (5) but not length or condition. In the higher density streams, at the end of the experiment, fry in the stream with the discharge fluctuations (4) were longer and heavier than fry in the other streams (Figure 9), but these differences were not statistically significant. Condition factors of fry from the three higher density streams were not significantly different from each other.

Production was similar in the two low density streams during the control period, but following the onset of flow fluctuations, production appeared to occur at a faster rate in the control stream than in Stream 5

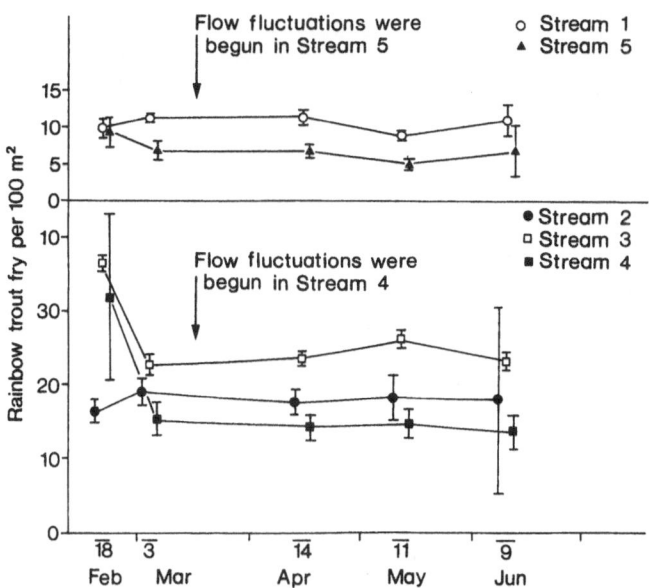

Figure 7. Density (with 95% confidence limits) of rainbow trout in low density (upper graph) and higher density (lower graph) streams during Experiment 2. Flow fluctuations were begun in Streams 4 and 5 on 16 March 1982.

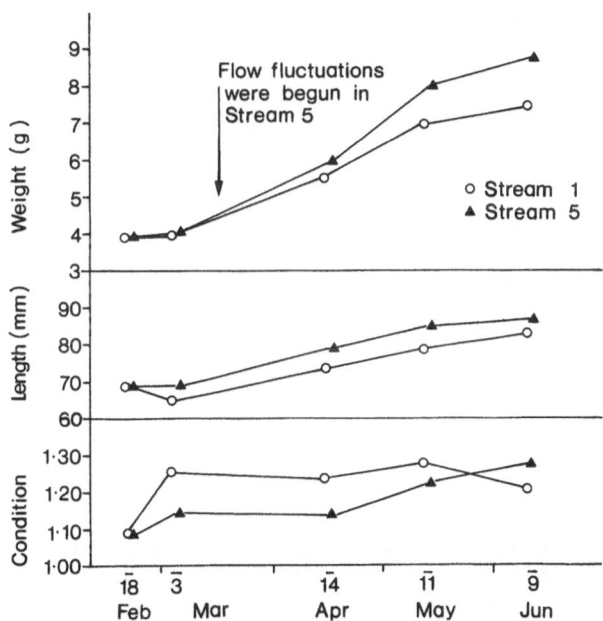

Figure 8. Mean weights, lengths and condition factors for rainbow trout fry in the two low density streams during Experiment 2. Flow fluctuations were begun in Stream 5 on 16 March 1982.

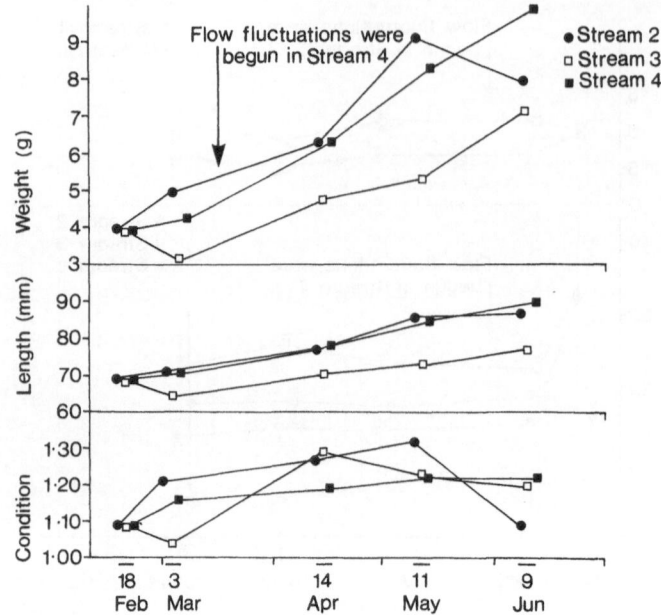

Figure 9. Mean weights, lengths and condition factors for rainbow trout
fry in the three higher density streams during Experiment 2.
Flow fluctuations were begun in Stream 4 on 16 March 1982.

(Figure 10). When GLIM was used to determine whether fluctuating flows were
responsible for this apparent difference in production [g(100 m^2)$^{-1}$ day^{-1}],
effects were not significant. Fluctuating flows did not appear to affect
fry production in the higher density streams (Figure 10), and this was
confirmed when GLIM was used to analyze production results.

Neither the presence or absence of fluctuating flows nor different fry
densities appeared to have any significant effect on habitat preferences.
The survey during the control period (3 March) indicated no consistent
pattern of habitat preferences. More fry had moved upstream of Pool 2 (into
which all fry had initially been introduced) than downstream. In all
streams there was a tendency for increasing proportions of the fry to be
found in pools as the experiment progressed. Throughout the experiment, in
every stream, the upstream riffle (Riffle 1) usually had a higher fry
density than the other riffles.

DISCUSSION

Up to five-fold flow changes occurring twice daily, five days per
week, had remarkably little effect on rainbow trout fry, illustrating how
well-adapted fry are to varying discharge. Downstream emigration was not
affected. However, in separate experiments at the replicate streams,
downstream emigration of chinook or quinnat salmon (*Oncorhynchus
tshawytscha*) fry was increased by fluctuating discharge (Irvine 1986).
Ottaway & Clarke (1981) found that more brown trout (*Salmo trutta*) fry
emigrated from stream channels after flows were increased than before.
Species differences probably exist but it is probable that fish age or size
is also important in determining whether flow changes will affect fry
emigration. The rainbow trout used in Experiments 1 and 2 were larger than

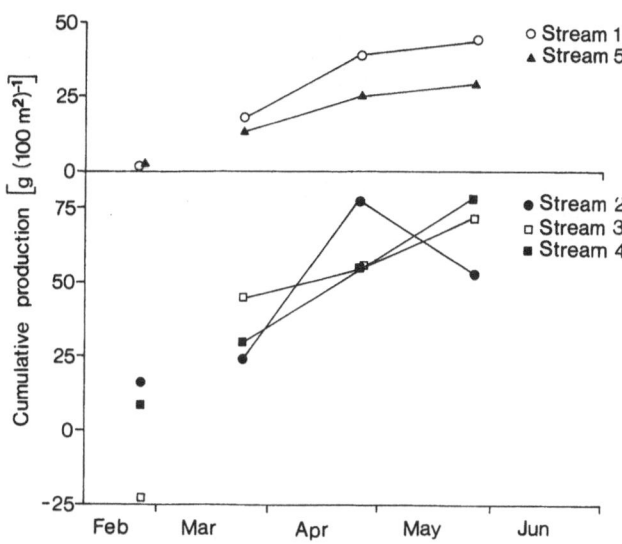

Figure 10. Cumulative rainbow trout production during Experiment 2 in low density (upper graph) and higher density (lower graph) streams during the control (February) and fluctuating flow (March-June) periods. Flow fluctuations were begun in Streams 4 and 5 on 16 March 1982.

the salmon in Irvine's (1986) study or the brown trout in Ottaway and Clarke's (1981) experiment.

Varying flows resulted in significant weight gain for rainbow trout in the low density stream in Experiment 2. Invertebrate drift densities sometimes increased during flow changes, especially after a period of stable flow (Irvine 1985), so perhaps an increased availability to trout of invertebrate prey was responsible for the trouts' weight gain. Significant effects of fluctuating flows on fish production were not found. However, as production estimates usually have large errors (Chapman 1978), it is difficult to demonstrate differences statistically.

Trout habitat preferences were similar in constant flowing and fluctuating streams. More rainbow trout were captured in pools as Experiments 1 and 2 progressed, regardless of the discharge regime. This probably represented a natural, seasonal shift to areas of slower water velocity as was found by Bustard and Narver (1975).

This study illustrates the need for good controls in flow assessment studies. Designed and constructed to be the same as the other streams, Stream 3 resembled its counterparts more closely than is possible for most comparisons of natural streams. Yet in both experiments, significant differences were found between the fry of Stream 3 and the other streams during the control period. In Experiment 1, more fry emigrated from Stream 3 and these fry were smaller and in poorer condition than fry leaving the other streams. Fry captured in Stream 3 were shorter than fry captured in the other streams, fry densities were higher and a greater fry production occurred. In Experiment 2, more fry emigrated from Stream 3 than the other streams, and yet Stream 3 supported a higher density of fry than the other streams. Fry growth rates in Stream 3 were slower, and instream mortality rates were higher, than in the two other higher density streams. Stream 3 was probably anomalous because its upstream riffle was narrow, steep and consequently fast flowing. The fast water may have been responsible for a high density of benthic invertebrates found in Stream 3 (Irvine 1984). Other workers (Mason & Chapman 1965; Slaney & Northcote 1974; Dill et al. 1981) reported a direct relationship in streams between salmonid rearing capacities and food abundance. This suggests that high invertebrate densities permitted more fry to live in Stream 3 than in the other streams. Density dependent factors then presumably reduced fry growth. In conclusion, extreme caution is required when comparing environmental results from different rivers. Seemingly minor physical differences can result in major biological changes which may render comparisons between rivers meaningless.

When one considers the natural variability in flows in streams where salmonids live, perhaps it is not surprising that trout fared as well as they did in the replicate streams with varying flows. Caution must be used when extrapolating from these experiments to effects of flow changes in larger streams. These experiments were relatively short term and effects may have shown up later. Also, fry stranding in the replicate streams was uncommon but can have a major population regulatory effect in regulated rivers. In conclusion, it is the responsibility of those investigating biological effects of varying flows to ensure that their studies are conducted with sufficient scientific rigour to demonstrate these effects conclusively.

REFERENCES

Baker, R. J. and Nelder, J. A. 1978. The GLIM system, release 3. Distributed by the Numerical Algorithms Group, 7 Banbury Road, Oxford, England.

Bustard, D. R. and Narver, D. W. 1975. Aspects of the winter ecology of juvenile coho salmon (*Oncorhynchus kisutch*) and steelhead trout (*Salmo gairdneri*). J. Fish. Res. Board Can. **32**: 667-680.

Campbell, R. N. B., Rimmer, D. M. and Scott, D. 1984. The effect of reduced discharge on the distribution of trout. In: Regulated Rivers. (Ed. by A. Lillehammer and S. J. Saltveit). pp. 407-416. Universitetsforlaget AS, Oslo, Norway.

Chapman, D. W. 1978. Production in fish populations. In: Ecology of Freshwater Fish Production. (Ed. by S. D. Gerking). pp. 5-25. Blackwell Scientific Publications, Oxford, England.

Dill, L. M., Ydenberg, R. C. and Fraser, A. H. G. 1981. Food abundance and territory size in juvenile coho salmon (*Oncorhynchus kisutch*). Can. J. Zool. **59**: 1801-1809.

Green, R. H. 1979. Sampling Design and Statistical Methods for Environmental Biologists. John Wiley & Sons, New York. 269 p.

Irvine, J. R. 1984. Effects of varying discharge on stream invertebrates and underyearling salmon and trout. Ph.D. Thesis, Zoology Department, University of Otago, Dunedin, New Zealand. 254 pp.

Irvine, J. R. 1985. Effects of successive flow perturbations on stream invertebrates. Can. J. Fish. Aquat. Sci. **42**: 1922-1927.

Irvine, J. R. 1986. Effects of varying discharge on the downstream movement of salmon fry, *Oncorhynchus tshawytscha* Walbaum. J. Fish Biol. **28**: 17-28.

Lillehammer, A. and Saltveit, S. J. (Eds.). 1984. Regulated Rivers. Universitetsforlaget AS, Oslo, Norway. 540 p.

Manly, B. F. J. and Crosbie, E. D. 1977. Examples of the use of GLIM. N.Z. Statistician **12**(1): 26-42.

Mason, J. C. and Chapman, D. W. 1965. Significance of early emergence, environmental rearing capacity, and behavioural ecology of juvenile coho salmon in stream channels. J. Fish. Res. Board Can. **22**: 173-190.

Ottaway, E. M. and Clarke, A. 1981. A preliminary investigation into the vulnerability of young trout (*Salmo trutta* L.) and Atlantic salmon (*S. salar* L.) to downstream displacement by high water velocities. J. Fish Biol. **19**: 135-145.

Ricker, W. E. 1946. Production and utilization of fish populations. Ecol. Mono. **16**: 374-391.

Rimmer, D. M. 1985. Effects of reduced discharge on production and distribution of age-0 rainbow trout in seminatural channels. Trans. Am. Fish. Soc. **114**: 388-396.

Slaney, P. A. and Northcote, T. G. 1974. Effects of prey abundance on density and territorial behaviour of young rainbow trout (*Salmo gairdneri*) in laboratory stream channels. J. Fish. Res. Board Can. **31**: 1201-1209.

Ward, J. V. and Stanford, J. A. (Eds.) 1979. The Ecology of Regulated Streams. Plenum Press, New York. 298 p.

Wolf, P. 1951. A trap for the capture of fish and other organisms moving downstream. Trans. Am. Fish. Soc. **80**: 41-45.

Zippen, C. 1956. An evaluation of the removal method of estimating animal populations. Biometrics **12**: 163-189.

SECTION II

INVERTEBRATE ECOLOGY

DEVELOPMENT AND APPLICATIONS OF MACROINVERTEBRATE

INSTREAM FLOW MODELS FOR REGULATED FLOW MANAGEMENT

James A. Gore

Faculty of Biological Sciences
University of Tulsa
Tulsa, OK 74104 USA

INTRODUCTION

Since the classic papers of Dodds and Hisaw (1924; 1925) and Nielsen (1950) and standard references on the ecology of running water ecosystems (Hynes 1970), aquatic biologists and entomologists have agreed that benthic macroinvertebrates display a wide variety of morphological and behavioral adaptations to a flowing environment. In general, benthic invertebrates have operated to take advantage of flow for feeding (Wallace & Merritt 1980) and have evolved some variation of dorso-ventral flattening to either reduce shear stress, as proposed by Steinmann (1907), Thienemann (1926), and, most recently, Statzner and Holm (1982) or to occupy a less rigorous "boundary layer" (Ambühl 1959). Since adaptations to hydraulic forces seem to be common among a wide variety of benthic invertebrate groups, it should be expected that distributions and densities of these organisms could be predicted by assessing the heterogeneity of flow characteristics within a stream reach. It is interesting to note that many aquatic biologists proceed from this basic assumption to demonstrate that distribution, density, and diversity are primarily determined by biotic interactions rather than responses to the physical environment. Recently, with increasing demands for water for energy development in the arid and semi-arid environments of the western United States, aquatic biologists and stream managers have begun to examine the impacts of altered flow regimes upon the biota of rivers and streams.

Although there remain some free-flowing rivers in North America, the majority of major river systems have been impounded at least once over the course of their flow (Stanford & Ward 1979). The most striking example of impacts to biota have been seen on the Columbia River where less than 30 percent of available habitat for salmonid spawning has remained after construction of hydroelectric facilities (Robinson 1978). With the realization of significant habitat losses to sport fisheries, the United States Fish and Wildlife Service has developed a series of protocols to predict changes in fish density and habitat available over a range of regulated discharges (Bovee & Cochnauer 1977; Bovee 1978; 1982; Bovee & Milhous 1978; Stalnaker 1979). These methods (most commonly referred to as IFG-IV or IFIM) have been adopted by most American regulatory agencies and appear to be quite effective in predicting changes to fisheries as a result of regulated flows (Stalnaker 1982). Grenney and Kraszewski (1981) have developed water quality models to complement the fish habitat and production models.

With the development of predictive models for fish, little attention has been paid to possible changes in the benthic macroinvertebrate fauna, which represent a significant portion of the diets of many game and forage fish as well as serving a crucial role in the processing of energy in lotic ecosystems. It has been assumed that macroinvertebrate responses would closely match those of the fish. However, macroinvertebrates, being considerably less mobile than fish, do not have the rapid re-invasion capabilities of fish when living in an environment of unpredictable and fluctuating discharges. Many macroinvertebrates have been shown to have narrow ranges of tolerances to changes in flow (Gore 1977). If those macroinvertebrates are a significant portion of a benthic community, a small percentage loss of fish habitat will not result in a comparable loss of macroinvertebrate populations (Figure 1). Whether or not the managed fish populations are food limited or can withstand a fairly large loss of macrocinvertebrate diet items, the imbalances to the macroinvertebrate community structure may lead to further degradation of population numbers and less efficient energy processing throughout the lotic ecosystem (Brusven et al. 1974).

In this paper I summarize the most recently proposed models to predict densities and changes in densities as a result of regulated flow. Some of these models have been developed for use with the IFG-IV programs, while others were developed as the most accurate representations of hydraulic

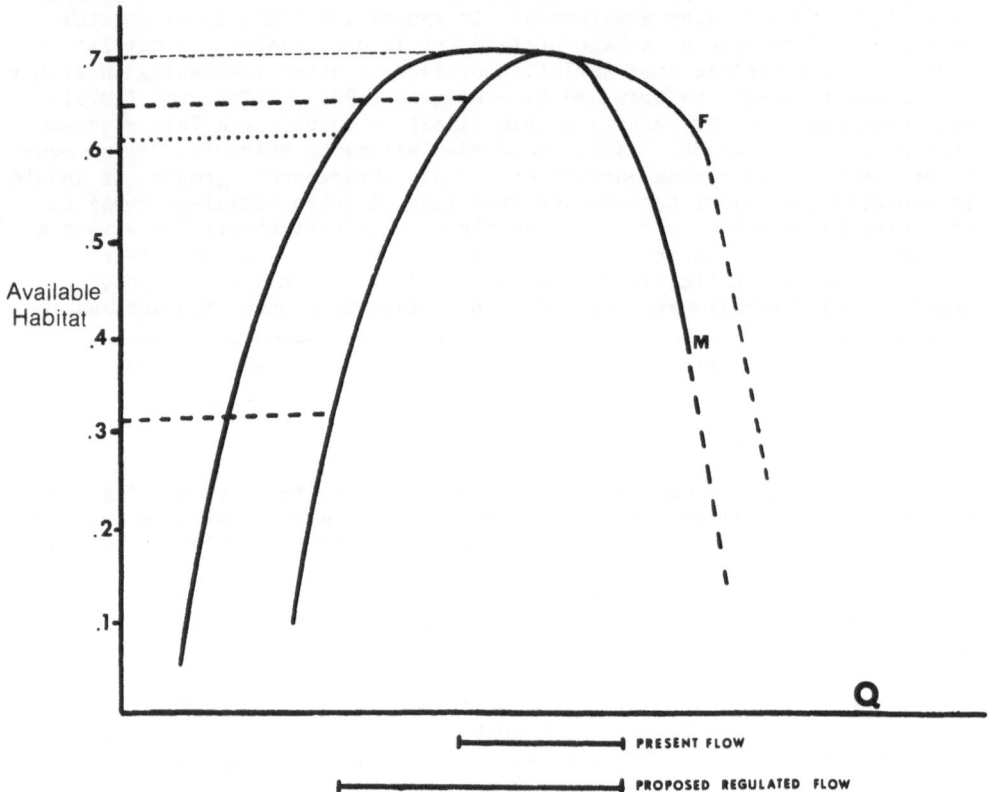

Figure 1. Alterations in available habitat for fish (F) and macroinvertebrates (M) as a function of change in flow regime (Q).

situations encountered in the stream environment. All of these models have been based upon accurate measurement of the basic stream parameters of current velocity, depth, and substrate character. These characteristics have been found to be the most useful describers of the hydrologic determinants of aquatic insect (and other macroinvertebrate) habitats (Newbury 1984) and can be used as basic values in a variety of hydraulic models (Chow 1964) as well as in the models described below.

HABITAT PREFERENCE CURVES

Gore (1978) attempted to describe the joint preferences for velocity and depth by macroinvertebrates of the Tongue River, Montana. These preferences were presented as response surfaces with the calculated centroid of velocity, depth, and highest density or diversity indicated as the optimum hydraulic environment (Figure 2). A single species, *Rhithrogena hageni* Eaton, was chosen as an indicator of maintenance of hydraulic conditions. Calculated discharge optima for the indicator were verified during drift studies when increased drift was observed at the time channel discharge continually exceeded the minimum calculated discharge level for benthic habitat (Gore 1977). Optimum hydrologic conditions could also be predicted from response surfaces plotted as density versus substrate profile and turbulence or Froude number, F, where

$$F = \frac{v2}{gd} \qquad (1)$$

and   v = mean stream velocity (cm s^{-1})
      d = stream depth (cm)
      g = acceleration due to gravity (980 cm s^{-2}).

The Froude-substrate centroids have the advantage of providing information about the distribution of macroinvertebrate densities according to substrate character, a critical physical habitat component (Minshall 1984), while retaining a hydraulic parameter which includes a factor of velocity and depth interactions.

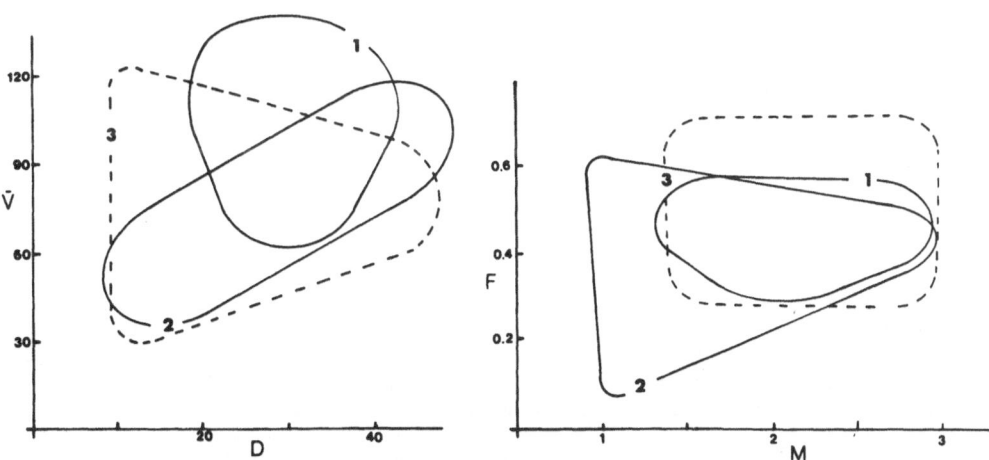

Figure 2.   Response surfaces (95% confidence ranges) for velocity (v) (cm s^{-2}) depth (d) (cm), turbulence (f), and substrate profile (M) for *Rhithrogena hageni* Eaton (1), *Symphitopsyche bifida* (Banks) (2), and *Acroneuria abnormis* (Newman). Methods are as described by Gore (1978).

Both types of response surfaces are not easily applied to the incremental instream flow methodologies since only the most preferred habitat values are determined by centroid-calculations.

In response to criticisms of the term "probability-of-use" curves and the lack of an interactive velocity-depth term in the incremental models (Patten 1979), Gore and Judy (1981) presented a series of instream flow models based on preference curves (there is, indeed, no apparent way to demonstrate a true "probability" curve) which could be used with the IFG-IV models to predict changes in macroinvertebrate habitat and density with changes in discharge.

Preference curves for each species are derived from a fit of a polynomial to a plot of cumulative mean number of individuals of a species per sample as a function of arbitrarily chosen increments of velocity, depth, and substrate character (Figure 3a). The region of greatest incremental increase in densities should represent the most preferred segment of that habitat factor. In most cases a third- or fourth-order polynomial of the forms:

$$y = C_1 + C_2x + C_3x^2 + C_4x^3 \tag{2}$$

or

$$y = C_1 + C_2x + C_3x^2 + C_4x^3 + C_5x^4 \tag{3}$$

when C_n is a constant
and x = the physical habitat parameter,

are the most appropriate fits. If the best cumulative fit is linear rather than sigmoidal, it must be concluded that the species displays equal preferences over the range of habitats sampled (in this case, the "preference curve" would be represented graphically by a horizontal line at the preference value of 1, over the range of habitat values). It should be noted that the polynomials should be fitted over the range of sample habitat parameters measured rather than the total range of habitat values found in the stream reach.

A first order derivative of the form:

$$\frac{dy}{dx} = C_2 + 2C_3x + 3C_4x^2 \tag{2$'$}$$

or

$$\frac{dy}{dx} = C_2 + 2C_3x + 3C_4x^2 + 4C_5x^3 \tag{3$'$}$$

will provide a preference curve for the indicated habitat parameter (Figure 3b) where the peak value represents the point of inflection on the fitted polynomial and is the most preferred habitat condition. When these derived curves are normalized to 1, the resulting curve will indicate a relative preference for that given habitat condition.

Once preference curves have been produced, they can be used in two fashions: 1) as indicators of changes in available habitat and 2) as predictors of density and changes in density.

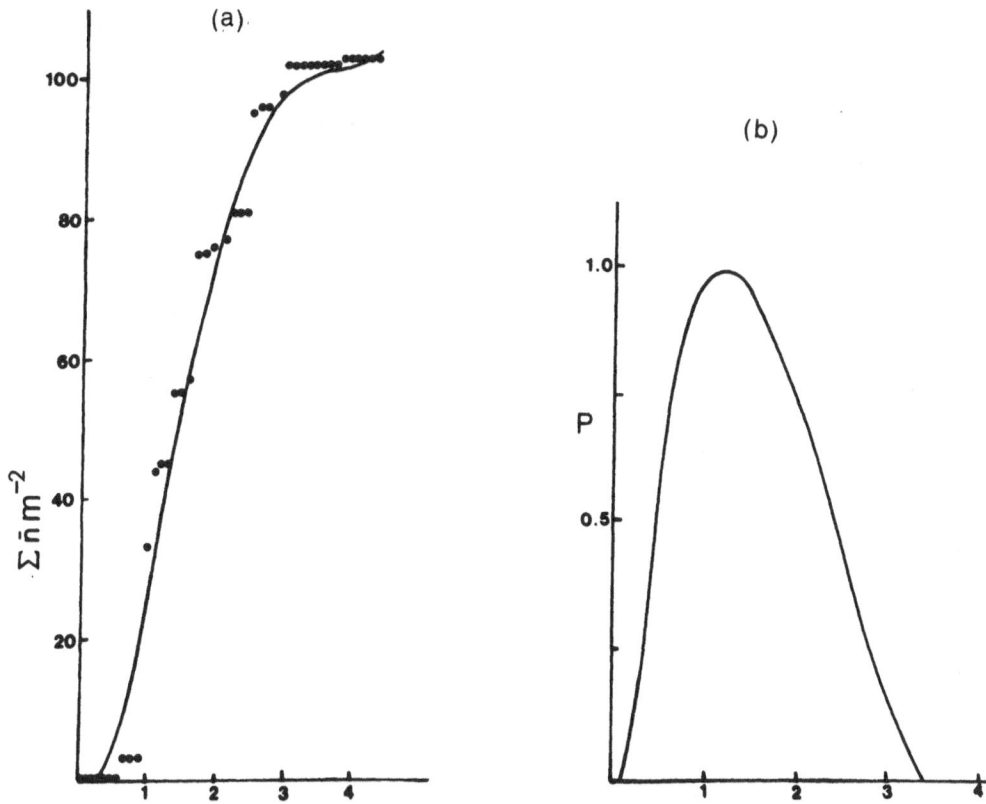

Figure 3. Fourth order polynomial fit to a plot of cumulative mean number
of individuals per m^2 per habitat increment (left, 3a). First
order derivative of the polynomial normalized to 1 produces the
preference curve (P) (left, 3b).

Weighted Usable Area

Weighted usable area (WUA) is a portion of the Physical Habitat
Simulation (PHABSIM) model of the incremental model of the U.S. Fish and
Wildlife Service (Bovee 1982). WUA provides an indication of flow and/or
flow related characteristics (relative to the preferences of a given
species) at a given discharge (Q). When WUA is plotted over the range of
discharges in a stream system, a management model (Figure 4) is produced
which will indicate habitat losses and gains under various regulated flow
regimes.

WUA is calculated as the sum of joint habitat preferences where:

$$WUA = \sum_{i=1}^{n} A_i C_i \qquad (4)$$

where
  A_i = surface area of stream element
  C_i = composite preference (or suitability)
      of the species in that stream element

with n total stream elements being examined. Composite preference is
calculated as the product of the preference values for each habitat

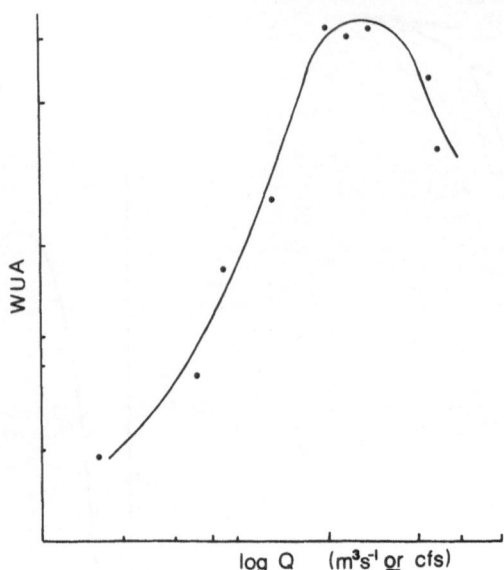

Figure 4. Weighted usable area (WUA) as a function of discharge (Q). Each
point represents a WUA value for a stream reach measured at a
single discharge value.

characteristic. For example, using velocity, substrate and depth

$$C_i = \text{velocity preference x depth preference x} \\ \text{substrate preference} \qquad (5)$$

(see Figure 5 for a numerical example). Each "stream element" is a single
grid section from a transect map of the flow characters of the stream reach
being examined (Figure 5). If the stream reach has a fairly predictable
hydrograph, PHABSIM can make accurate predictions of habitat changes based
upon transect maps measured at one low discharge period and one high
discharge period (Bovee 1982 and personal communication).

Stream managers can use WUA-discharge predictions to determine
minimum allowable discharges to maintain the indicator species or the
entire community as well as assessing the relative losses or gains in
usable habitat in comparison with those of the fish component under the
same discharge regime. It will be the decision of the stream managers
and/or regulatory agencies to set a recommended flow sequence which will
insure biotic stability.

Density Predictions

Maximum densities for a given species (that is, with a composite
preference of 1.0) can be predicted from plots of WUA and density values of
the raw data (Figure 6). A linear or curvilinear function (Gore & Judy
1981) can then be used to predict densities under variable flow
regimes. This density prediction technique has been criticized by Mathur et
al. (1985) for non-independence. However, raw data are not compared with
the derived equation to obtain statistical reliability. Plots of data drawn
from the same river system but not used in the WUA-density predictor have
been shown to be reliable (Figure 7).

A species biomass per stream element can be obtained as the product of
its density times an estimate of average biomass per individual. Stream

managers can then combine density or biomass data for each stream element
with WUA-discharge equations and obtain predictions of density or biomass
losses of critical diet items under new or proposed flow alterations.

Some Important Considerations

Although the production of preference curves and WUA predictions
appear to be relatively flexible, there are some points to be considered
which allow for the most effective use of the models.

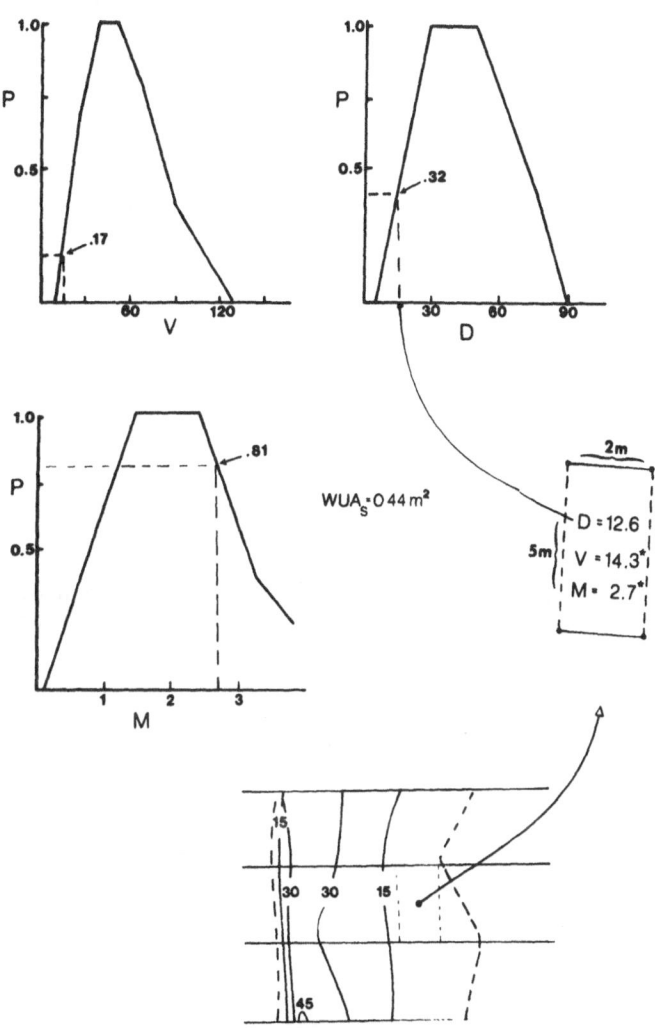

Figure 5. Weighted usable area (WUA$_s$) calculation for a single stream
element of a stream reach. WUA values are plotted from computer
generated preference curves (P = preference, range: 0-1) for the
mayfly, *Baetis hageni* Eaton. The cell is drawn from an isodepth
map of a stream reach. Mean values of depth (d) (cm), velocity
(V) (cm s^{-2}) and substrate profile (M) are derived from the four
corner points on the cell which are measurement points along the
transects used in map construction. Starred values derived from
other isoparameter projections.

Throughout this discussion, mean velocity measurements are assumed to be those velocities measured according to standard hydrologic procedures at a depth of 0.4 x depth from the substrate (Chow 1964). Although this is not a velocity encountered by benthic macroinvertebrates, it is strongly correlated to boundary layer and shear conditions (Statzner 1981a). In addition, a value of mean velocity within the water column can be used in other hydraulic applications. These models will be described later in this paper.

Substrate composition and/or roughness (profile) still need to be translated into a single effective index. Gore (1978) suggested a substrate profile measuring device which proved to be accurate when predicting flow needs (as turbulence) of invertebrate species. This method seems quite

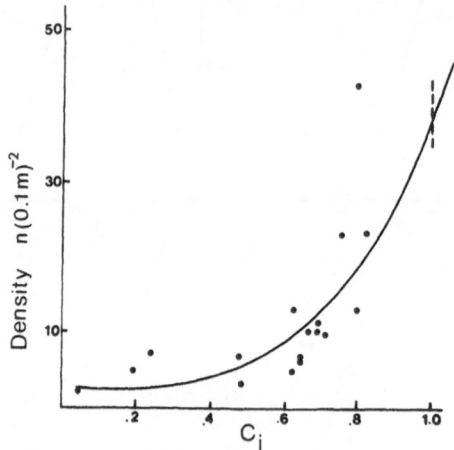

Figure 6.   Plot of density of *Baetis hageni* as a function of composite suitability of randomly chosen samples used in preference curve predictions ($r = 0.83$, $p > 0.95$).

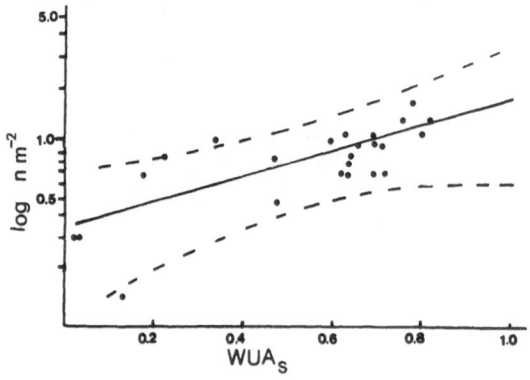

Figure 7.   Density of *Baetis hageni* as a function of WUA per element (WUA_s). Dotted lines are 95% confidence intervals of the regression line. Dots represent additional density and WUA values of samples calculated from preference curves of the same species from the same river (Tongue River, Montana), but not used to generate the regression line shown.

tedious with a large sample size. However, Statzner (personal communication) has designed modifications which decrease sampling time and appear to be cost effective. Bovee (1982) and Winget (1985) have proposed alternate numerical indices of substrate quality based upon visual observation of dominant particle sizes and amount of fine particulates in the sample area. Both indices have been shown to be effective in their respective applications but are limited to stream reaches where the substrate can be visually examined without disturbance.

It must be noted that a habitat preference curve derived for a particular species can not be universally applied to all communities in which that benthic organism exists. Since community structure is determined in at least some quasi-random mechanism (see the arguments of Gilpin & Diamond 1984; Conner & Simberloff 1984), it is not unreasonable to assume that energy dynamics, inter-, and intra-specific interactions, which affect population numbers, will vary density predictions for benthic species in any system. Thus, preference curves must be determined on a site-by-site consideration.

Finally, a great number of mathematical transformations occur in the process of raw data conversion to WUA estimates. Since the initial polynomials are fit to plots of cumulative mean number of individuals per habitat increment, a variance around each mean is implicit within the model. Differentiation and normalization of the resulting curve compounds these variances. Thus, with each conversion, variances can be expected to increase. This is especially true in the formation of the joint preference value, C_i, in equation 5, when values from several preference curves are multiplied together. Any number of habitat preference curves can be used to produce C_i values, but the fewest number of curves which provide accurate estimates will also exhibit the lowest variances. Orth and Maughan (1983) attempted to minimize these variances by log transformation of the raw data prior to construction of preference curves. These transformations were successful for predictions of the habitat preference of the water penny, *Psephenus*. Gore and Judy (1981) also presented new models which included joint velocity and depth terms in construction of preference curves. Simultaneous solution of partial differentials for v and d:

$$\frac{\partial f'}{\partial v} = a_1 + 2(a_3 v) + a_5 d = 0$$

$$\frac{\partial f'}{\partial d} = a_2 + 2(a_4 d) + a_5 V = 0$$

yielded a normalization value, N, for a new preference curve:

$$f'' = \frac{1}{N} \exp \left[-(a_1 v + a_2 d + a_3 v^2 + a_4 d^2 + a_5 vd) \right].$$

By holding depth or velocity constant, and varying the other parameter, more precise preference curves were produced. Morin et al. (1986) found that the exponential polynomial had the effect of minimizing variances and were preferred to the polynomial curves of Gore and Judy (1981) and the log transformed curves of Orth and Maughan (1983). However, it still appears that reduction in the number of preference curves is needed to increase the accuracy of WUA predictions. I offer alternate habitat predictions in subsequent sections.

Further Modifications of Preference Curves

During initial investigations of instream flow techniques, Bovee et al. (1978) suggested that a comprehensive instream flow recommendation should include a monthly assessment of flow requirements for hydrologic features such as sediment transport and ice break-up and biological features such as spawning requirements and food production. The largest of the complimentary flow requirements for each month would be included in the total annual volume commitment for use by stream managers. This procedure recognizes the variability of flow requirements during the life cycles of fish species of management concern. Although there was a suggestion of potential changes in macroinvertebrate habitat (as the food component function), these changes were not verified.

Kovalak (1978) reported size differences among six species of mayflies. Those populations of the same species living in slower currents were larger than those existing in faster waters. Differences were attributed to differential growth rates and habitat selection controlled by oxygen requirements. Body size and morphology apparently determine micro-habitat selection among some groups of macroinvertebrates. Preference for higher velocities and smaller substrate particle sizes among juveniles have been reported respectively by Rabeni (1985) for crayfish and by Gore (1983) for a pleurocerid snail. Statzner (1979) found the most distinct changes in habitat preference among instars of the hemipteran, *Aphelocheirus aestivalis*, where early instars were found in low velocity, fine substrates with late instars and adults found in high velocity, cobbled substrates. Gore (1983) produced a single complex polynomial to describe the velocity preferences of *Elimia potosiensis* (Gould) (Gastropoda: Pleuroceridae). However, it is clear the best form of instream flow analysis should be based on individual preference curves derived from instar or cohort analysis (when appropriate). Although no test has been made, to date, this form of life cycle habitat analysis could refine monthly flow allotments to better reflect macroinvertebrate habitat needs.

In some regions, an additional factor of ice formation and habitat exclusion must be considered for estimates of flows necessary to maintain invertebrate habitat. Although macroinvertebrates appear to be well adapted to environments of solid ice and frozen substrate (Olsson 1981, 1983), surface habitat exclusion and lateral movement of invertebrates to avoid adverse conditions do result in increased mortality (Olsson 1981; Logan 1963). In addition, where surface ice extends to the substrate, especially under abnormally low flow conditions, scour from break-up could result in increased mortality or reductions in local populations (Fremling 1973; Bovee et al. 1978). Johnson (1980) and Johnson et al. (1982) have produced a series of equations to estimate salmonid and macroinvertebrate habitat loss from ice cover. The amount of habitat excluded depended upon whether the habitat area was located in a pool or riffle but followed the general formula:

$$\% \text{ Excluded} = a + b (N) - c (S) - d(d_e) - e (v) \qquad (6)$$

where

a, b, c, d, and e are constants derived from field observations and

N = day number since first surface ice formation (designated as t_0)

$$S = \text{degree days of frost when } S = \int_{t_0}^{t} (-T_a)dt$$

if T_a = mean daily temperature (°C)

d_e = mean effective water depth (cm)

v = mean water velocity (cm s^{-1}) (Michel 1971).

Generally a~100, b~0.45, c~0.025, d~125, and e~35 for riffle areas where the greatest habitat loss occurred. Only a small portion of habitat was lost from pool areas (Figure 8). I suggest that these equations combined

with estimates of pool-riffle frequency will provide for better flow
management when winter ice conditions warrant these considerations.

BOUNDARY LAYER MODELS

As mentioned earlier, a number of statistical problems (increasing
variances) may be encountered within calculations of WUA from the
incremental methodology. Most important are the lack of hydraulic
simulations which describe the naturally occurring interactions of velocity
and depth. Benthic biologists have offered alternatives which use
depth-velocity-substrate interactions to describe distribution of benthic
macroinvertebrates. These models are described below.

Reynolds Number

Railsback et al. (1981) have suggested that a direct measurement of
shear stress across the substrate will be adequate to describe aquatic
insect distributions. As a ratio of turbulent shear stress to viscous shear
stress, the Reynolds number can be applied as:

$$R* = \frac{D_s^2 \ g \ R \ S}{\nu^2} \tag{7}$$

where
$\quad$ $R*$ = boundary layer value
$\quad$ D_s = particle diameter (applied as average particle
$\quad\quad\quad$ diameter in the area sampled) (cm)
$\quad$ g = gravitational acceleration
$\quad$ R = hydraulic radius (or mean depth)
$\quad$ S = bed slope
$\quad$ ν = kinetic viscosity

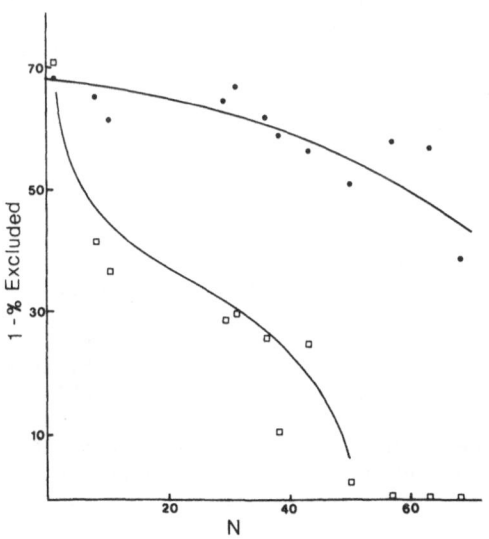

Figure 8. Habitat available as a function of days since surface ice formed
(N) for pool areas (dots) and riffle areas (open squares). Data
derived from Johnson (1980) on Wagonhound Creek, Wyoming.

Railsback et al. (1981) demonstrated a highly significant correlation between R* and measured velocities over uniform substrates in experimental channels. They note that the application of these equations is limited to areas where R* exceeds a value of 100 (most riffle habitats). Although no in situ experiments were reported, Railsback et al. (1981) suggest that the variables needed to determine R* can be determined by models contained within PHABSIM and other incremental models and could be applied to regulated flow management.

"Hydraulic Stress"

Statzner (1981a) suggested two models which describe bed roughness and associated characteristics to describe the microdistribution of benthic macroinvertebrates. Statzner transformed the Manning equation (Chow 1964):

$$V = K_s^{-1} R^{2/3} I^{1/2} \qquad\qquad (8)$$

where
$\quad$ V $\;$ = current velocity
$\quad$ K_s = channel roughness
$\quad$ R $\;$ = hydraulic radius
$\quad$ I $\;$ = slope.

The transformation yielded an equation indicating hydraulic stress (I´) applied to objects (organisms) upon the substrate where:

$$I´ = \left\{ \frac{V}{K_s \left(\dfrac{D}{2+D}\right)^{2/3}} \right\}^2 \quad X \; 1000 \qquad\qquad (9)$$

where
$\quad$ D $\;$ = water depth (cm).

This expression can be used as an indication of stress over a small substrate area (a single 0.1 m^2 macroinvertebrate sample) rather than the entire bed as expressed in the original Manning equation. Statzner (1981a) indicated that a correlated equation for laminar sub-layer thickness ($\smallint$) could be used as a descriptor of invertebrate macrohabitat:

$$\smallint' = \frac{11.5 \; \nu \; 5.75 \; \log \; \dfrac{12D}{r_p}}{U} \qquad\qquad (10)$$

where
$\quad$ $\smallint'$ = thickness in cm
$\quad$ U $\;$ = mean current velocity
$\quad$ r_p = height of bed roughness projections.

Statzner (1981b) used laminar sublayer thickness to produce reliable estimates of population size and density for the blackfly (Simuliidae), *Odagmia ornata*. Statzner suggested that laminar sublayer thickness could be used for similar predictions for other macroinvertebrates. Although he did not attempt to fit measurements of $\smallint'$ to incremental methods, Statzner did suggest the utility of sublayer thickness measurements in predicting abundances of macroinvertebrates at high and low discharges.

Statzner (1981b) used a modified substrate profile measuring device (Gore 1978) to obtain measurements of r_p. However, Statzner (personal communication) has indicated that most repeatable and consistent substrate indices (Bovee 1982; Winget 1985) should give values of $\int'$ which can be used in predictive models.

It appears that models which demonstrate the effects of turbulent flow upon shear forces are comparable to the incremental methods in their abilities to predict macroinvertebrate distributions. Indeed, Statzner (1981b) found significant relationships between Froude number calculations and substrate profile (as proposed by Gore 1978) and sublayer thickness. It may not be readily apparent as to the utility of Reynolds numbers or sublayer thickness values to the widely used IFG-IV methods. However, with the exception of temperature values used to calculate kinematic viscosity (Décamps et al. 1975), all values required for alternate "turbulence" models are also required for IFG-IV. Thus, using the preference curve methods of Gore and Judy (1981), preference curves for Froude number, Reynolds number, or sublayer thickness could be calculated. Only slight modifications of existing programs need to be made to calculate weighted usable area as a function of these more complex hydraulic characteristics. This implies that single preference curves and subsequent WUA values could replace joint preference values (C_i) and attendant variance problems as predictors of macroinvertebrate density or biomass for regulated flow recommendations.

FUTURE NEEDS

One of the primary objections to instream flow assessments for macroinvertebrates has been the apparent need for large numbers of samples for construction of adequate preference curves. In the past, I have recommended a minimum of 50 samples. This was an arbitrarily chosen number based upon experience with random sampling along transects across "typical" stream reaches. In general, that number of samples would ensure a reasonably high diversity of hydraulic values. However, stratified random samples combined with some form of sequential analysis (Resh & Price 1984; Sheldon 1985) should allow for a reduction in numbers of samples. Indeed, preliminary results seem to indicate that the minimum number of effective samples, if taken randomly from identifiable strata, may be thirty or less (Statzner, Gore & Resh personal communication). It will be necessary to provide a cost-effective, high precision sampling strategy in order to make macroinvertebrate studies useful for regulated flow management.

Although regulated flows have come to be defined as river and stream reaches downstream of impoundments (Ward & Stanford 1984), much of the original development of instream flow methodologies was developed for flows which were "regulated" as a consequence of irrigation withdrawal and dewatering for electricity generation in the Northern Great Plains of the United States. In these cases, instream flow models have been used to recommend minimum flows which must be reserved for maintenance of aquatic biota (Sweetman 1980). Considering that continuous downstream drift of most aquatic insects, combined with upstream flight of adults for oviposition, is a natural process during the life cycle of these organisms (Hynes 1970; Waters 1972), it may become necessary to determine the minimum stretch of river for which flows must be reserved to maintain this life cycle process. This has become an important consideration on the Dan River and upper Jordan River where waters are removed for various forms of domestic and agricultural uses and returned further downstream (Gershon Herbst, Hebrew University, personal communication).

The flexibility of instream flow models is demonstrated in their potential application to areas other than regulated flow situations. Gore (1985) and Gore and Johnson (1980) have demonstrated the use of instream flow techniques to assess the effects of reclamation structures on increasing habitat availability. Preference curves themselves offer a measure of eurytopy and could be used in habitat overlap and resource partitioning studies. Since weighted usable area functions can be calculated for an experimental reach, measures of hydraulic resource and diet item (macroinvertebrate) resource availability can be determined for these same types of studies.

With the exception of the initial studies of Gore (1978) and Gore and Judy (1981) and evaluations by Orth and Maughan (1983) and Morin et al. (1986), few studies have been conducted to verify or assess the use of instream flow models for macroinvertebrates. I suspect that this may be because of the apparent high cost to conduct these studies (large sample size and site-by-site consideration). Some of the "cost" related problems will be relieved with the current studies by Statzner, Gore, and Resh (previously mentioned).

I suggest that a greater number of macroinvertebrate instream flow studies must be performed, not only to verify the utility of proposed models, but to assess the effects of regulated flows upon interactions within communities and various species assemblages. For example, flow reductions which reduce numbers of only one population may result in significant changes in competitive interactions which could ultimately lead to alteration of community structure. In a similar manner, flow reductions which result in the loss of only a few macroinvertebrate species yet only a slight decline in total macroinvertebrate biomass may not be viewed as a significant loss. However, these losses may be more important when diet specificity of a managed fish species is examined. Few studies have examined impacts of regulated discharge upon loss of pre-regulation diet items to managed fish and resulting loss in overall condition or numbers of fish individuals.

CONCLUSIONS

Instream flow models which have been designed to predict density and biomass changes among benthic macroinvertebrates have been shown to be effective for regulated flow management. Macroinvertebrate flow needs, as they relate to the flow needs of managed fish populations, are particularly important since macroinvertebrates, being less mobile, may have higher discharge requirements under certain flow conditions. Models which have integrated depth, velocity, and substrate values into a single hydraulic index can be used in place of the composite suitability values of the IFG-IV programs and may increase the precision of the flow requirement estimates as well as reducing the variance problems sometimes encountered with the incremental technique. More research is needed to support the general application of macroinvertebrate instream flow studies for regulated flow management. The utility of the model for predicting density and biomass is well documented. It will be the decision of regulatory agencies and stream managers to apply these models and predictions for the maintenance of biotic stability in fluctuating flow environments.

ACKNOWLEDGEMENTS

I thank Dr. Bernhard Statzner, University of Karlsruhe and Dr. Vincent Resh, University of California, Berkeley, for discussions and comments during the preparation of the manuscript. Anonymous referees provided valuable comments to improve this manuscript. Preparation of the manuscript was supported, in part, by a research fellowship from the College of Arts and Sciences, University of Tulsa.

REFERENCES

Ambühl, H. 1959. Die Bedeutung der Strömung als ökologischer
    Faktor. Schweiz. Z. Hydrol. **21**: 133-264.
Bovee, K. D. 1978. Probability-of-use criteria for the family Salmonidae.
    Instream Flow Info. Paper No. 4, U. S. Fish. Wildl. Serv.,
    FWS/OBS-78/07.
Bovee, K. D. 1982. A guide to stream habitat analysis using the instream
    flow incremental methodology. Instream Flow Info. Paper No. 12,
    U. S. Fish Wildl. Serv., FWS/OBS-82/26.
Bovee, K. D. and Cochnauer, T. 1977. Development and evaluation of weighted
    criteria, probability-of-use curves for instream flow assessments:
    fisheries. Instream Flow Info Paper No. 3, U. S. Fish Wildl. Serv.,
    FWS/OBS-77/63.
Bovee, K. D., Gore, J. A. and Silverman, A. J. 1978. Field testing and
    adaptation of a methodology to measure "in-stream" values in the
    Tongue River, Northern Great Plains (NGP) region. U. S. Environ.
    Prot. Agency, EPA-908/4-78-004A.
Bovee, K. D. and Milhous, R. 1978. Hydraulic simulation in instream flow
    studies: theory and techniques. Instream Flow Info Paper No. 5,
    U. S. Fish Wildl. Serv., FWS/OBS-78/33.
Brusven, M. A., MacPhee, C. and Biggam, R. C. 1974. Effects of water
    fluctuations on benthic insects. In: Anatomy of a River.
    pp. 67-79. Pacific Northwest River Basins Rpt., Vancouver, Wash.
Chow, V. T. (Ed.) 1964. Handbook of Applied Hydrology. McGraw-Hill, New
    York.
Connor, E. F. and Simberloff, D. 1984. Neutral models of species
    co-occurrence patterns. In: Ecological Communities (Ed. by
    D. R. Strong, Jr., D. Simberloff, L. G. Abele and A. B. Thistle),
    pp. 316-337, Princeton Univ. Press, Princeton, N.J.
Décamps, H., Larrowy, G. and Trivellato, D. 1975. Approche hydrodynamique
    de la microdistribution d'invertébrés en eau courante.
    Ann. Limnol. **11**: 79-100.
Dodds, G. S. and Hisaw, F. L. 1924. Ecological studies of aquatic
    insects. I. Adaptations of mayfly nymphs to swift streams. Ecology
    5: 137-148.
Dodds, G. S. and Hisaw, F. L. 1925. Ecological studies on aquatic
    insects. III. Adaptations of caddisfly larvae to swift
    streams. Ecology **6**: 123-137.
Fremling, S. 1973. Changes of the ice regime in Swedish rivers due to the
    development of the hydroelectric power. Comm. Int. des Grands
    Barrages, Madrid.
Gilpin, M. E. and Diamond, J. M. 1984. Are species co-occurrences on
    islands non-random, and are null hypotheses useful in community
    ecology? In: Ecological Communities (Ed. by D. R. Strong, Jr.,
    D. Simberloff, L. G. Abele, and A. B. Thistle), pp. 297-315, Princeton
    Univ. Press, Princeton, N.J.
Gore, J. A. 1977. Reservoir manipulations and benthic macroinvertebrates in
    a prairie river. Hydrobiologia **55**: 113-123.
Gore, J. A. 1978. A technique for predicting the in-stream flow
    requirements of benthic macroinvertebrates. Freshwat. Biol. **8**:
    141-151.
Gore, J. A. 1983. Considerations of size related flow preferences among
    macroinvertebrates used in instream flow studies. In: Developments in
    Ecology and Environmental Quality, Vol. II (Ed. by H. I. Shuval),
    pp. 389-397. Balaban Int. Press, Jerusalem.
Gore, J. A. 1985. Mechanisms of colonization and habitat enhancement for
    benthic macroinvertebrates in restored river channels. In: The
    Restoration of Rivers and Streams (Ed. by J. A. Gore), pp. 81-101,
    Butterworths Publ., Boston.

Gore, J. A. and Johnson, L. S. 1980. Establishment of biotic and hydrologic stability in a reclaimed coal strip-mined river channel. Inst. Energy and Environ., Univ. of Wyoming, Laramie.

Gore, J. A. and Judy, Jr., R. D. 1981. Predictive models of benthic macroinvertebrate density for use in instream flow and regulated flow management. Can. J. Fish. Aquat. Sci. **38**: 1363-1370.

Grenney, W. J. and Kraszewski, A. K. 1981. Description and application of the stream simulation and assessment mode: Version IV (SSAM IV), U.S.D.I., Fish Wildl. Serv., FWS/OBS-81/46.

Hynes, H. B. N. 1970. The Ecology of Running Waters. Univ. Toronto Press, Toronto, Ont.

Johnson, L. S. 1980. Tracking movement and identification of instream flow needs of brown trout (*Salmo trutta*) by use of radio-isotopes. M.S. Thesis, Univ. Wyoming, Laramie.

Johnson, L. S., Wichers, D. L., Wesche, T. A. and Gore, J. A. 1982. Instream salmonid habitat exclusion by ice-cover. Water Res. Series No. 84, Wyoming WRRI, Univ. Wyoming, Laramie.

Kovalak, W. P. 1978. Relationships between size of stream insects and current velocity. Can. J. Zool. **56**: 178-186.

Logan, S. M. 1963. Winter observations on bottom organisms and trout in Bridger Creek, Montana. Trans. Am. Fish. Soc. **92**: 140-145.

Mathur, D., Bason, W. H., Purdy, Jr., E. J. and Silver, C. A. 1985. A critique of the instream flow incremental methodology. Can. J. Fish. Aquat. Sci. **42**: 825-831.

Michel, B. 1971. Winter regime of rivers and lakes. Cold Regions Res. Eng. Lab., Monogr. III-Bla. Hanover, New Hampshire.

Minshall, G. W. 1984. Aquatic insect-substratum relationships. In: The Ecology of Aquatic Insects (Ed. by V. H. Resh and D. M. Rosenberg), pp. 358-400. Praeger, N.Y.

Morin, A., Harper, P.-P., and Peters, R. H. 1986. Microhabitat preference curves of black fly larvae (Diptera: Simulidae): a comparison of three estimation methods. Can. J. Fish. Aquat. Sci. (in press).

Newbury, R. W. 1984. Hydrologic determinants of aquatic insect habitats. In: The Ecology of Aquatic Insects (Ed. by V. H. Resh and D. M. Rosenberg), pp. 323-357, Praeger, N.Y.

Nielsen, A. 1950. The torrential invertebrate fauna. Oikos **2**: 176-196.

Olsson, T. I. 1981. Overwintering of benthic macroinvertebrates in ice and frozen sediment in a North Swedish River. Hol. Ecol. **4**: 161-166.

Olsson, T. I. 1983. Seasonal variation in the lateral distribution of mayfly nymphs in a boreal river. Hol. Ecol. **6**: 333-339.

Orth, D. J. and Maughan, O. E. 1983. Microhabitat preferences of benthic fauna in a woodland stream. Hydrobiologia **106**: 157-168.

Patten, B. D. 1979. Summary report of module B-instream fishery ecosystems. Instream flow criteria and modeling workshop. Colorado State Univ., Exp. Station, IS No. 40.

Rabeni, C. F. 1985. Resource partitioning by stream dwelling crayfish: the influence of body size. Amer. Midl. Nat. **113**: 20-29.

Railsback, S. F., Herricks, E. E., and Alavian, V. 1981. Modeling aquatic insect habitat. Proc. Symp. Acquisition and Utilization of Aquatic Habitat Inventory Information, Portland, Oregon.

Resh, V. H. and Price, D. G. 1984. Sequential sampling: a cost-effective approach for monitoring benthic macroinvertebrates in environmental impact assessments. Env. Mgmt. **8**: 75-80.

Robinson, W. L. 1978. The Columbia: a river system under siege. Oregon Wildlife **33**: 3-12.

Sheldon, A. L. 1985. Cost and precision in a stream sampling program. Hydrobiologia (in press).

Stalnaker, C. B. 1979. The use of habitat structure preferenda for establishing flow regimes for maintenance of fish habitat. In: The Ecology of Regulated Streams (Ed. by J. V. Ward and J. A. Stanford), pp. 321-337, Plenum Press, N.Y.

Stalnaker, C. B. 1982. Instream flow assessments come of age in the decade of the 1970s. In: Research on Fish and Wildlife Habitat (Ed. by W. T. Mason Jr. and S. Iker), pp. 119-142, U.S. Environ. Prot. Agency, EPA-600/8-82-022.

Stanford, J. A. and Ward, J. V. 1979. Stream regulation in North America. In: The Ecology of Regulated Streams (Ed. by J. V. Ward and J. A. Stanford), pp. 215-236, Plenum Press, N.Y.

Statzner, B. 1979. Der Obere und Untere Schierenseebach (Schleswig-Holstein). Strukturen und Funktionen in zwei norddeutschen See-Auffluss-Systemen, unter besonderer Berücksichtigung der Makroinvertebraten. Thesis. Christian-Albrechts-Universitat, Kiel.

Statzner, B. 1981a. The relation between "hydraulic stress" and microdistribution of benthic macroinvertebrates in a lowland running water system, the Schierenseebrooks (North Germany). Arch. Hydrobiol. **91**: 192-218.

Statzner, B. 1981b. A method to estimate the populations size of benthic macroinvertebrates in streams. Oecologia (Berl) **51**: 157-161.

Statzner, B. and Holm, T. F. 1982. Morphological adaptations of benthic invertebrates to stream flow - an old question studied by means of a new technique (Laser Doppler Anemometry). Oecologia (Berl) **53**: 290-292.

Steinmann, P. 1907. Die Tierwelt der Gebirgsbache. Annls. Biol. Lacustr. **2**: 30-150.

Sweetman, D. A. 1980. Protecting instream flows in Montana: Yellowstone River reservations case study. Instream Flow Info Paper No. 10, U. S. Fish Wildl. Serv., FWS/OBS-79/36.

Thienemann, A. 1926. Die Binnengewasser Mitteleuropas, Die Binnengewasser **1**, Schweizerbart, Stuttgart.

Wallace, J. B. and Merritt, R. W. 1980. Filter-feeding ecology of aquatic insects. Ann. Rev. Ent. **25**: 103-132.

Ward, J. V. and Stanford, J. A. 1984. The regulated stream as a testing ground for ecological theory. In: Regulated Rivers (Ed. by A. Tillehammer and S. J. Saltveit), pp. 23-38, Universitetsforlaget AS, Oslo.

Waters, T. F. 1972. The drift of stream insects. Ann. Rev. Ent. **17**: 253-272.

Winget, R. N. 1985. Methods for determining successful reclamation of stream ecosystems. In: The Restoration of Rivers and Streams (Ed. by J. A. Gore), pp. 165-192, Butterworths Publ., Boston.

STONEFLIES AND RIVER REGULATION - A REVIEW

Svein Jakob Saltveit, John E. Brittain and
Albert Lillehammer

Zoological Museum, University of Oslo
Sars gate 1, 0562 Oslo 5, Norway

INTRODUCTION

Stoneflies (Plecoptera) are a widespread and abundant order of benthic
freshwater insects, especially in the running waters of temperate
regions. They occupy a central role in trophic relationships both in terms
of functional feeding groups and as fish food organisms. Although recorded
from a range of freshwater habitats, the greatest species diversity is to
be found in cool lotic habitats (Hynes 1976; Baumann 1979). They are
considered to be good indicators of water quality and environmental change
(Gaufin 1973; Baumann 1979).

Current is a major factor controlling the distribution and abundance
of most lotic organisms and this is also the case for stoneflies (Hynes
1970). River regulation invariably modifies both flow patterns and
discharge (Ward 1976a). This can lead not only to changes in current speed
and substrate, but also to changes in water temperature, oxygen content,
drift and trophic structure (Armitage 1984). Temperature is another major
factor controlling the distribution, abundance and life cycles of aquatic
insects (Ward & Stanford 1982) and several of the studies on the effect of
river regulation on stoneflies have cited temperature as a major
environmental factor. In the present review we summarize the known effects
of river regulation on stoneflies and put forward hypotheses to explain its
differing effect, both under varying regulation schemes and in various
regions of the world.

DISCHARGE AND FLOW PATTERNS

Discharge and flow patterns are invariably modified in regulated
rivers. The nature of the change depends on the type of regulation.
Discharge may be reduced, increased or made more or less stable. The
reduction in discharge may be such that periods of drought occur. Changes
in discharge may range from seasonal modifications to short-term diurnal
fluctuations.

Change in flow pattern and discharge alone may affect stonefly
communities in regulated rivers. However, discharge affects many other
environmental parameters such as temperature, current, substrate
composition and water quality. It is often difficult to separate cause and

effect in such instances without controlled field or laboratory
experiments. Nevertheless, stonefly species have certain preferential and
absolute limits with respect to current velocity and water depth.

Short-term Fluctuation

Short-term fluctuations in discharge are a common feature in many
regulated rivers throughout the world. In North America there have been
several studies of such systems. The stonefly genera *Acroneuria* and
Paragnetina were more than twice as abundant above an impoundment in Maine,
U.S.A., than below (Trotzky & Gregory 1974). However, *Alloperla*
(s.l.) (Chloroperlidae) was able to tolerate the wide range in discharge
and was in fact twice as abundant below the dam. A similar dominance of
Alloperla was found in a regulated Colorado river (Ward & Short 1978) and a
regulated Albertan stream (Radford & Hartland-Rowe 1971). Nymphs of
Alloperla (s.l.) are therefore apparently tolerant of large flow
fluctuations, their cylindrical body form enabling them to move into
substrate interstices. Nevertheless, rapid fluctuations in flow can also
affect even *Alloperla* nymphs and in the Snake River, Wyoming, Kroger (1973)
found that a reduction in discharge from 2.8 to 0.3 m^3 s^{-1} in less than 5
min eliminated both *Alloperla* and *Isoperla* nymphs. In the same study nymphs
of *Pteronarcella badia*, which migrated into newly created shallow areas
along the edge of the river prior to emergence, were frequently observed
stranded after a reduction in water flow from 132 to 62 m^3 s^{-1} during the
course of 3 days. Sudden flow reductions were also responsible for major
mortality of *Utacapnia* and *Capnia* nymphs in the Flathead River, Montana,
during 1973 (Stanford & Ward 1979). The sudden reduction permitted icing
conditions to develop in the shoreline areas where the mature nymphs of
these winter stoneflies had congregated just prior to emergence.

In addition to *Alloperla*, the carnivorous perlid *Neoperla clymene* is
reported to be tolerant of large flow fluctuations and occurs in the
regulated Brazos River in Texas (Vaught & Stewart 1974). A European
chloroperlid genus, *Chloroperla*, together with *Leuctra fusca*, was abundant
both in an unregulated and a regulated river in mid-Wales (Brooker & Morris
1980). The regulated river was subject to irregular discharges up to 15
m^3 s^{-1} compared to the usual discharge of about 1.8 m^3 s^{-1}. In two other
regulated Welsh rivers, artificial reservoir releases which increased
discharge between three- and six-fold caused a consistent, but
statistically insignificant, decrease in stonefly densities (Scullion &
Sinton 1983).

Reduced Discharge and Cessation of Flow

A reduction in discharge of 90% during winter and 10-70% during summer
in a Norwegian river, Søre Osa, gave rise to a drastic reduction in benthic
densities, including the stoneflies, *Isoperla* and *Dinocras cephalotes*
(Garnås 1985). *Protonemura meyeri* did, however, increase at one
locality. Surprisingly, possibly due to the lack of other food organisms in
the river, Plecoptera increased in importance in the diet of the fish in
the river.

In certain cases reduction in discharge leads to cessation of flow,
but only a few studies have been conducted in such rivers. However,
information also exists from studies in temporary streams. Generally,
stoneflies surviving drought are either able to go into a diapause as eggs
or as nymphs, or are capable of going into the hyporheic (Williams & Hynes
1976; Canton et al. 1984). *Leuctra fusca* and *L. digitata* were not affected
in a Norwegian regulated river, the Nøra, which dried up from October to
May-June (Josefsen 1953). In another Norwegian river, the Nea, a severe
reduction in discharge, coupled with periods with virtual drought, led to a

reduction in stonefly species. Only *L. fusca* and *Diura nanseni* remained,
L. fusca being dominant (Langeland & Haukebø 1979). *L. fusca* and
L. geniculata persisted in a regulated section of a French river, although
a large number of plecopteran species were eliminated (Berthélemy & Laur
1975). In northern Sweden, Ulfstand et al. (1971) did not find any
stoneflies in a river subject to widely fluctuating discharge including
times when the river bed was almost dry, although neighbouring rivers
supported a rich stonefly fauna.

A sustained reduction in water flow can also lead to faunal changes in
regulated rivers. In the Green River on the Utah-Colorado border, U.S.A.,
the perlid *Claassenia sabulosa* and *Nemoura* disappeared after impoundment
reduced flows (Pearson et al. 1968). Similarly a reduction in flow in the
French mountain river, the Aston, led to a reduction in *Protonemura*,
although countered by an increase in *Amphinemura* (Chauvet 1983).

Low winter flows were considered to be responsible for the elimination
of the winter stonefly *Capnia atra* downstream of a dam on the Norwegian
River Glomma (Brittain, Lillehammer & Bildeng 1984). Low winter discharge
resulted in a reduction in the availability of allochthonous material, much
of the autumnal input remaining along the banks of the river. Allochthonous
leaf material is an important food resource for many stoneflies, such as
Capnia atra. Carnivorous species, such as *Diura nanseni*, are less dependent
on the supply of such material. This species also occurred in the egg stage
during the first winter and was nearly fully grown at the beginning of the
second winter and in fact increased its numerical dominance from 60 to 80%
of total stonefly numbers, compared with the unregulated stretch of river.

Uniformly low flows over extended periods appeared to be the major
cause of the observed community changes below a dam on the Strawberry
River, Utah, U.S.A. (Williams & Winget 1979). The elimination of high
seasonal discharges permitted enhanced algal growth, beaver dams and
increased sedimentation. The latter reduced or changed habitat
heterogeneity, including reducing the availability of the hyporheic
zone. All the stoneflies (*Pteronarcella badia, Isoperla quinquepunctata,
I. fulva* and *Megarcys* sp.) were eliminated in the regulated section.

Increased Discharge

Examples of more or less stable increase in discharge are uncommon in
regulated rivers and are generally limited to irrigation schemes in warm,
dry countries, or to river stretches upstream of power stations. High
summer flows in an Australian river appeared to result in a reduction in
stonefly species because the resultant depression of water temperatures
caused life cycle disruption (Doeg 1984).

COMMUNITY STRUCTURE

Change in community structure may also arise in regulated rivers,
often as an indirect result of changes in discharge. Although the faunas of
two Welsh rivers were generally similar, *Perla bipunctata* were absent in
the regulated River Elan, probably due to low numbers of simuliids with
which this species is highly associated (Scullion et al. 1982). Plecoptera
increased in relative importance in two regulated Welsh rivers (Scullion
1983). This was due to a general decrease in the density and species
richness of total benthos in the regulated rivers. Regulation occluded
streambed surfaces with iron and manganese rich deposits, which the
stoneflies *Protonemura meyeri* and *Leuctra fusca* could tolerate (Inverarity
et al. 1983). However, the species richness in the pools in the regulated
Elan was less than in the riffles (Scullion et al. 1982). This was also the

case in the River Ekse, Norway, after the building of weirs to reduce the deleterious effects of low flows (Baekken et al. 1984). Stonefly species adapted to swift-flowing water, *Taeniopteryx nebulosa*, *Diura nanseni*, *Brachyptera risi* and *Protonemura meyeri*, were eliminated in the weir basins and replaced by the more lentic species, *Nemoura cinerea* and *Nemurella pictetii*. Stonefly production was 3.5 times higher in the riffle areas outside the weir basin.

DRIFT

It has long been recognized that increases in discharge or water velocity, including spates, lead to increased invertebrate drift. Several invertebrates appear more likely to enter the drift when the current velocity is changing (Ciborowski et al. 1977), although drift maxima have also been found to occur well after maximum flow. For example, in the regulated River Elan in Wales, stonefly numbers in the drift did not increase immediately after an artificial morning freshet, but during the following night (Scullion & Sinton 1983).

Conditions of low discharge and current velocity due to river regulation have also been reported to increase drift. Minshall and Winger (1968) showed an increase in the drift of *Nemoura cinctipes* when stream discharge was artificially reduced. This increase in drift was ascribed to factors related to depth and water velocity. In the regulated Tongue River, Montana, U.S.A., a reduction in discharge of about 30% produced a dramatic increase in invertebrate drift. A further 10% reduction in discharge gave rise to massive drift, suggesting that the minimum discharge sufficient to maintain this benthic community had been reached (Gore 1977). Increased drift was also considered to be responsible for the downstream extension of certain species, including the plecopteran *Capnia limata*.

The stonefly *Alloperla* was frequent in the drift and had similar drift rates in both an unregulated and a regulated Albertan stream (Radford & Hartland-Rowe 1971). However, reduction in flows caused increased drift, as probably did reservoir releases. In a British study comparing drift in the regulated River Tees with an unregulated tributary, Maize Beck, Armitage (1977) found that Plecoptera were always more abundant in the drift from the unregulated tributary. This was ascribed to the lower benthic densities in the Tees leading to less competition for food and space and to the greater abundance of moss and algae in the Tees, which as well as providing food and increased habitat space, may facilitate more rapid return to the substrate for drifting animals.

GAS SUPERSATURATION

Gas supersaturation is a well-known and serious problem in connection with stream regulation and power plants (Weitkamp & Katz 1980). However, little information is available on fish food organisms. The stonefly species *Doroneuria californica*, *Hesperoperla pacifica* and *Pteronarcys californica* survived more than 125% gas saturation for at least 10 days (Nebeker 1976). Their tolerance of gas supersaturation is greater than fish and is explained by their relatively open, simple respiratory system and by their rigid exoskeleton. However, both internal and external gas bubbles form, which increases buoyancy and therefore may increase mortality indirectly through increased predation. This was also the case for the stoneflies, *P. californica*, *Claassenia sabulosa* and *Isoperla* sp. (Montgomery & Fickeisen 1979).

SUBSTRATE, SEDIMENTATION AND SCOURING

The morphometric characteristics of the streambed influence the diversity, density and distribution of benthic animals. Changes in substrate composition through sedimentation or scouring may therefore alter species composition and community structure dramatically. Indirect changes may occur through changes in periphyton biomass and detrital processing.

Changes in substrate characteristics may occur as a result of the addition of sand and silt from dam construction, and usually result in a decrease in the stonefly fauna (Briggs 1948; Blyth et al. 1984; Cline & Ward 1984). Stoneflies seem especially sensitive to introduced sediments, due to sediment filling the substrate interstices and reducing the effective size of the surface cobbles (McClelland & Brusven 1980) and by increasing their propensity to drift (Rosenberg & Wiens 1978). The most sensitive stonefly to the adding of low and moderate amount of sediments in an experimental stream was *Pteronarcys californica* (McClelland & Brusven 1980). During the late instars this species attains a length of >5 cm and is often associated with loosely aggregated cobbles. At the highest level of sedimentation all stoneflies tested (*P. californica, Hesperoperla pacifica, Cultus* sp. and *Skwala* sp.) clearly preferred the unsedimented control region. Increase in fine sediments and reduction of detrital matter, caused by a change from high spring runoff regime to a moderate flow regime in a regulated Utah river, led to increase in algal scrapers, filter feeders and low flow tolerant species (Williams & Winget 1979). No stoneflies occurred after regulation. The elimination of *Pteonarcella badia* was due to the fact that this moss and detritus feeder is only tolerant of moderate silting.

Stoneflies are also susceptible to the scouring action of spates. Downstream of some river impoundment schemes, the erosive effects of highly variable discharge have resulted in the washout of fine particles and general reductions in invertebrate abundance (Radford & Hartland-Rowe 1971; Trotsky & Gregory 1974; Ward & Short 1978). Burrowing stoneflies, such as *Alloperla*, which normally inhabit the hyporheic zone have frequently been reported under such flow regimes. Radford and Hartland-Rowe (1971) found that *Paraleuctra* was also common on scoured substrates, while *Nemoura* sp., highly dependent on aggregation of leaf litter, were much more common in a nearby unregulated stream.

TEMPERATURE

Thermal changes are most pronounced just below deep release dams or power stations taking water from stratified reservoirs (Ward 1976b), although dams with surface release may change faunal composition in the receiving stream. Generally water released from the hypolimnion of deep reservoirs decreases river temperatures during summer and increases them during winter, compared to the unregulated conditions. Summer temperatures up to 20°C below that of unregulated rivers in the same region have been reported (Ward & Stanford 1979). The resulting decrease in both annual and diurnal temperature amplitudes, have in one case led to an almost constant temperature regime (Stanford & Ward 1979). Surface release usually leads to increase in summer temperatures.

Hypolimnetic Release

Most of the studies on hypolimnetic release below dams and the effect of temperature have been carried out in North America, where this is the most common regulation type (Ward & Stanford 1979). Stoneflies seem to be one of the groups most affected, being either absent or strongly reduced

both in density and species diversity below deep release dams (Ward &
Stanford 1979).

Spence and Hynes (1971) found a pronounced effect on stoneflies below
an impoundment in Canada. Upstream of the dam the stonefly species
*Acroneuria lycorias, Paragnetina media, Allocapnia pygmaea, Taeniopteryx
maura* and *Amphinemura delosa* were found, while no stoneflies occurred
downstream. The effect was comparable to that occurring after mild organic
enrichment and according to the authors the absence of stoneflies was
probably due to oxygen exhaustion in the surface microlayers. However,
reduced summer temperature and a suppressed diel temperature cycle may have
caused the extinction of stoneflies.

Pfitzer (1954), Elder and Gaufin (1973), Ward (1976a & b), Gore (1977)
and Donald and Mutch (1980), mention stonefly species lacking or reduced
below dams in North America. Colorado mountain rivers normally contain 5-10
stonefly species, but just below dams Ward and Short (1978) recorded
none. Furthermore, Zimmermann and Ward (1984) concluded from a survey of
Rocky Mountain rivers that among the stoneflies only *Alloperla* and *Isoperla*
were commonly collected below deep release reservoirs. Wais (1984) also
related the lesser stonefly diversity on the Argentine side of the Negro
and Valdivia basins to the construction of dams. In contrast in the
Brazos River, Texas, *Neoperla clymene* appeared only after the building of a
dam with deep release (Vaught & Stewart 1974).

Stanford and Ward (1984) found the faunal diversity of the Gunnison
River, Colorado, was drastically reduced by the summer cold hypolimnetic
releases and the large stonefly, *Pteronarcys californica*, was virtually
eliminated. In the Tongue river, Montana, U.S.A., a hypolimnetic release
reservoir led to a reduction in the abundance of the stoneflies
Strophopteryx fasciata and *Leuctra* sp. below a dam (Gore 1980). Ward (1974)
reported low species diversity below the Chessman reservoir in Colorado due
to the altered thermal condition. The effect reached up to 5.0 km from the
dam (Ward 1976b). *Isoperla quinquepunctata* was the only stonefly recorded
in the most affected reaches. Its density increased downstream. No
Euholognathan stoneflies were found; in contrast 13 species were recorded
in a virtually unregulated stretch about 30 km further down (Ward
1976b). In the Trinity River in Northern California, studies made just
after impoundment concluded that temperature due to hypolimnetic release
was one of several factors producing a poor stonefly fauna (Boles
1981). *Pteronarcys californica* was not present until 13 km below the dam.

Most of these American studies conclude that stoneflies are scarce
downstream of impoundments as a result of altered temperature
regimes. However, below an impoundment in the river Elan in Britain,
Scullion et al. (1982) found stoneflies to be slightly more abundant, even
though temperatures were modified by a deep release dam (Brooker 1981).

Below the Cow Green dam in River Tees in England, Armitage (1976)
found that *Perla bipunctata* and *Perlodes microcephala*, were absent below
the dam, while *Dinocras cephalotes* was recorded occasionally. Changes in
substrate were postulated as the main cause. However, the disappearance of
the winter growing species, *Brachyptera risi, Protonemura meyeri* and
P. praecox, may have been due to slightly altered temperatures.

According to Raddum (1978) a reduced summer temperature of about 3°C
and an increased winter temperature of 1-2°C in Aurlandselva, Western
Norway, gave rise to a greater abundance of stoneflies, but with a reduced
biomass. This was due to shift to smaller individuals caused by the drift
of large individuals out of the area.

In a regulated Swedish river the plecopteran fauna was reduced from 21 species to 16, and several species with a hiemal growth type (main nymphal growth during autumn and winter), such as *Capnia atra, Leuctra hippopus* and *Taeniopteryx nebulosa* were negatively influenced, while aestival species (main growth during spring and summer) such as *Leuctra fusca* and *Amphinemura standfussi* increased in relative abundance (Henricson & Müller 1979). In this case, other factors such as altered flow regime occurred together with the thermal changes.

Temperature seems to have the greatest effect in continental and warm temperate areas, where hypolimnion releases can produce very large changes. Similar effects as in the U.S.A. are also seen below dams in Australia and Spain. Malipatil and Blyth (1982) found only one species of Plecoptera, *Riekoperla,* sp. to be present 0.5 km below a dam on the Thomson River, Australia, with hypolimnion release, while several species occurred upstream. Higher summer flows and rapid fluctuations also occurred, but altered temperature conditions were considered to be the main factor. The same authors report little effect on stoneflies below an epilimnetic release reservoir. In the Mitta River, Australia, a depauperate fauna resulted from dam construction, both due to sedimentation and changes in temperature. The hypolimnetic discharge led to decline in *Dinotoperla* 2 km below the dam, while Gripopterygidae were eliminated (Doeg 1984). In Spain colder waters released from the bottom of irrigation reservoirs have a major impact on natural stream communities and no Plecoptera were found until 6 km below one such dam, although they were common above (Garcia de Jalon 1980).

Hypolimnetic release in the South Fork of the Flathead River, Montana, U.S.A., produced a thermal regime which was relatively constant around 7°C, compared to a thermal amplitude in the unregulated reaches of 0-18°C (Stanford & Ward 1984). Plecopteran species were reduced from 38 species to five. The temperature effect is reduced in the main river due to unregulated discharge from other tributaries. Even if the constant temperature provides the appropriate degree days for growth, thermal cues for emergence are lacking (Ward & Stanford 1979). Below this reservoir the depressed daily mean temperature due to an increase in late summer discharge was supposed to be the main reason for the stonefly *Claassenia sabulosa* not receiving its appropriate thermal cue and failing to emerge. A similar change may have affected the emergence of *Isocapnia* in spring 1974, although higher winter temperatures may have caused faster winter growth resulting in emergence at an inappropriate time.

Surface Release

Surface releases generally increase summer temperatures in the receiving stream. Probably the best documented studies on effects of surface release reservoir are from a shallow reservoir on the Madison River, Montana, where there was an average increase in the water temperature of 3.4°C from June to August 1978. This increase negatively influenced the population of two stoneflies, *Suwallia lineosa* and *S. pallidula*, which were not recorded below the dam (Fraley 1979). In contrast, *Isoperla quinquepunctata* was obviously favoured in the warmer section, as it did not occur in the upper unregulated part. Increased temperatures also produced earlier emergence in *Pteronarcys californica, Claassenia sabulosa, Hesperoperla pacifica* and *Isoperla fulva*. However, 4 km below the reservoir where a pattern of constant temperature existed, no emerging *P. californica* were taken, suggesting that its emergence is controlled by a maximum temperature which was not reached. The mean body weight of adults of *P. californica* was also 13% less in the lower river. Vannote and Sweeney (1980) found a decrease in the adult weight of insects reared in thermal regimes above or below optimum levels.

High summer temperatures below a surface release dam in a mountain river in Colorado was supposed to be the main reason for the absence of Plecoptera below a dam (Ward & Short 1978), while below another Colorado surface release dam at a higher elevation stoneflies actually increased in density, although the number of taxa did fall (Zimmermann & Ward 1984). Where temperature differences are small, surface release dams seem to have little effect on the benthic fauna (Malipatil & Blyth 1982; Kraft & Mundahl 1984).

CONCLUSIONS

In general stoneflies are negatively affected by river regulation. However, certain species tolerate the various environmental changes better than others. There are also considerable regional differences due to differences in climate, species richness and the regulation scheme.

In North America the construction of dams on rivers has severe effects on the fauna, and stoneflies are one of the groups most affected. Such dams alter a wide range of environmental parameters, including temperature, discharge and community structure and it is difficult to single out one main factor. However, in addition to temperature, changes in flow conditions seem to be the main factor responsible for the observed changes. Rapid fluctuations in discharge have the most pronounced effect. Most stoneflies are either stranded or enter the drift with sudden drawdowns. Nevertheless, species in the chloroperlid genus, *Alloperla* (s.l.), which have cylindrical bodies enabling them to burrow into the substrate, are commonly reported under such conditions. Chloroperlids are also reported to withstand periods of drought (Canton et al. 1984). *Alloperla* nymphs tolerate scouring, although severe flow fluctuations affect even this genus (Kroger 1973).

In Europe, in addition to *Chloroperla*, the genus *Leuctra* seems to be most tolerant of changes in discharge. *Leuctra fusca* and *L. digitata* have life cycle strategies which enable them to survive severe winter conditions, including freezing, in the egg stage. In addition to being burrowing species as is *Chloroperla*, they are univoltine, aestival species, with rapid summer growth (Lillehammer 1985). This enables them to utilize variable discharge.

Permanent reduction in flow decreases the productive area, but does not affect the species composition to the same degree as increased winter and reduced summer flows, which change the trophic structure (Lillehammer & Saltveit 1979; 1984). Nevertheless, low discharge and/or reduced current velocity can give rise to major changes in community structure, and stonefly species typical of more lentic habitats may increase in importance.

Below hypolimnetic release dams most species are reduced in numbers or eliminated, probably as a result of altered temperature conditions. *Isoperla quinquepunctata* is one of the few species colonizing such habitats. Its success is not easily explained by its life cycle. However, being eurythermal with wide environmental limits (Knight & Gaufin 1966), it may be excluded in competition with more specialized species in habitats with diverse faunas.

The plecopteran species eliminated below impoundments cover a wide range of life cycle types, making it difficult to make any generalizations in these terms. However, Ward and Stanford (1982) stressed that a high species diversity among aquatic insects is usually associated with wide annual temperature range. Moreover, Vannote and Sweeney (1980) have suggested that the course of biotic diversity along a stream closely parallels that of thermal variation. Thus a reduction in both diel and annual thermal ranges in connection with river regulation will almost certainly lead to a reduction in species diversity among stoneflies. However, the confluence of regulated and unregulated tributaries in the Flathead system, Montana, did increase the number of species in the genus *Utacapnia* by optimizing the range in annual temperature for this particular genus (Stanford & Ward 1983).

These effects may be greater in areas with high species diversity, such as certain parts of North America and less marked in areas with a low species diversity such as north-western Europe. In continental areas with a highly diverse fauna small environmental changes may have a much greater effect than in oceanic areas with a less diverse fauna due to a greater specialization and a decreased niche breadth in faunally rich areas.

Vannote and Sweeney (1980) and Brittain, Lillehammer and Saltveit (1984) provide evidence that suboptimal temperatures affect the size of eggs, nymphs and adults as well as fecundity in aquatic insects, which influences their abundance. Such temperature effects have been reported below a surface release dam in Montana, U.S.A. (Fraley 1979) and may at least partly explain some Norwegian data (Raddum 1978). Necessary temperature and other environmental cues may also not be reached after regulation, as shown by Lehmkuhl (1974) for the mayfly, *Ephoron album* and by Ward and Stanford (1979) for the stonefly *Claassenia sabulosa*.

Except for the laboratory studies on sedimentation and gas supersaturation, no studies have really attempted to single out causes and consequences of the environmental modifications resulting from river regulation, or to relate them to life cycle patterns or other biological characteristics of particular stonefly species. There is a need for critical experimental studies on the effects of river regulation on stoneflies and for more basic knowledge of their life cycle parameters.

ACKNOWLEDGEMENTS

We are grateful to Dr. J.V. Ward (University of Colorado) and Dr. J.A. Stanford (University of Montana) for constructive comments on our manuscript.

REFERENCES

Armitage, P. D. 1976. A quantitative study of the invertebrate fauna of the River Tees below Cow Green Reservoir. Freshwat. Biol., **6**: 229-240.
Armitage, P. D. 1977. Invertebrate drift in the regulated River Tees, and an unregulated tributary Maize Beck, below Cow Green dam. Freshwat. Biol. **7**: 167-183.
Armitage, P. D. 1984. Environmental changes induced by stream regulation and their effect on lotic macroinvertebrate communities. In: Regulated Rivers (Ed. by A. Lillehammer and S. J. Saltveit) pp. 139-165. Universitetsforlaget, Oslo.
Baekken, T., Fjellheim, A. and Larsen, R. 1984. Benthic animal production in a weir basin in Western Norway. In: Regulated Rivers (Ed. by A. Lillehammer and S. J. Saltveit) pp. 223-232. Universitetsforlaget, Oslo.
Baumann, R. W. 1979. Nearctic stonefly genera as indicators of ecological parameters (Plecoptera: Insecta). Great Basin Nat., **39**: 241-244.

Berthélemy, C. and Laur, C. 1975. Plécoptères et Coléoptères aquatiques du Lot (Massif Central Francais). Annls Limnol., **11**: 263-285.

Blyth, J. D., Doeg, T. J. and StClair, R. M. 1984. Response of the macroinvertebrate fauna of the Mitta Mitta River, Victoria, to the construction and operation of Dartmouth Dam. 1. Construction and initial filling period. Occ. Pap. Mus. Vict., **1**: 83-100.

Boles, G. L. 1981. Macroinvertebrate colonization of replacement substrate below a hypolimnial release reservoir. Hydrobiologia, **78**: 133-146.

Briggs, J. C. 1948. The quantitative effects of a dam on the bottom fauna of a small Californian stream. Trans. Am. Fish. Soc., **78**: 70-81.

Brittain, J. E., Lillehammer A. and Bildeng, R. 1984. The impact of a water transfer scheme on the benthic macroinvertebrates of a Norwegian river. In: Regulated Rivers (Ed. by A. Lillehammer and S. J. Saltveit) pp. 189-199. Universitetsforlaget, Oslo.

Brittain, J. E., Lillehammer, A. and Saltveit, S. J. 1984. The effect of temperature on intraspecific variation in egg biology and nymphal size in the stonefly *Capnia atra* (Plecoptera). J. Anim. Ecol., **53**: 161-169.

Brooker, M. P. 1981. The impact of impoundments on the downstream fisheries and general ecology of rivers. Adv. Appl. Ecol., **6**: 91-152.

Brooker, M. P. and Morris, D. L. 1980. A survey of the macroinvertebrate riffle fauna of the rivers Ystwyth and Rheindol, Wales. Freshwat. Biol., **10**: 459-474.

Canton, S. P., Cline, L. D., Short, R. A. and Ward, J. V. 1984. The macroinvertebrates and fish of a Colorado stream during a period of fluctuating discharge. Freshwat. Biol., **14**: 311-316.

Chauvet, E. 1983. Influence d'une réduction de débit sur un torrent de montagne: l'Aston (Ariège). Annls Limnol., **19**: 45-49.

Ciborowski, J. J., Pointing, P. J. and Corkum, L. D. 1977. The effect of current velocity and sediment on the drift of the mayfly *Ephemerella subvaria* McDunnough. Freshwat. Biol., **7**: 567-572.

Cline, L. D. and Ward, J. V. 1984. Biological and physiochemical changes downstream from construction of a subalpine reservoir, Colorado, U.S.A. In: Regulated Rivers (Ed. by A. Lillehammer and S. J. Saltveit) pp. 233-243. Universitetsforlaget, Oslo.

Doeg, T. J. 1984. Response of the macroinvertebrate fauna of the Mitta Mitta River, Victoria, to the construction and operation of Dartmouth Dam. 2. Irrigation release. Occ. Pap. Mus. Vict., **1**: 101-127.

Donald, D. B. and Mutch, R. A. 1980. The effect of hydroelectric dams and sewage on the distribution of stoneflies (Plecoptera) along the Bow River. Quaest. ent. **16**: 665-670.

Elder, J. A. and Gaufin, A. R. 1973. Notes on the occurrence and distribution of *Pteronarcys californica* Newport (Plecoptera) within streams. Gt. Basin Nat. **33**: 218-220.

Fraley, J. J. 1979. Effects of elevated stream temperatures below a shallow reservoir on a cold water macroinvertebrate fauna. In: The Ecology of Regulated Streams (Ed. by J. V. Ward and J. A. Stanford) pp. 257-272. Plenum Press, New York.

Garcia De Jalon Lastra, D. 1980. Efectos del embalse de "Pinilla" (Madrid) sobre las comunidades de macroinvertebrados benticos del Rio Lozoya. Boln. Estac. cent. Ecol., **9**: 47-52.

Garnås, E. 1985. Effekt av redusert vannføring på bunndyr og fisk fra 1982-1984 i Søre Osa, Hedmark. Rapp. DVF-Reguleringsundersøkelsene, 9-1985, 84 pp. In Norwegian.

Gaufin, A. R. 1973. Use of aquatic invertebrates in the assessment of water quality. Special technical publication 528 American Society for Testing and Materials. Philadelphia: 96-116.

Gore, J. A. 1977. Reservoir manipulations and benthic macroinvertebrates in a prairie river. Hydrobiologia, **55**: 113-123.

Gore, J. A. 1980. Ordinational analysis of benthic communities upstream and downstream of a prairie storage reservoir. Hydrobiologia, **69**: 33-44.

Henricson, J. and Müller, K. 1979. Stream regulation in Sweden with some
    examples from Central Europe. In: The Ecology of Regulated Streams
    (Ed. by J. V. Ward and J. A. Stanford) pp. 183-199. Plenum Press, New
    York.
Hynes, H. B. N. 1970. The Ecology of Running Waters. Liverpool University
    Press, U.K. 555 pp.
Hynes, H. B. N. 1976. Biology of stoneflies. Ann. Rev. Entomol.,
    21: 135-153.
Inverarity, R. J., Rosehill, G. D. and Brooker, M. P. 1983. The effect of
    impoundment on the downstream macroinvertebrate riffle fauna of the
    River Elan, Mid-Wales. Environ. Pollut. (A), 32: 245-267.
Josefsen, E. 1953. Reguleringsundersøkelser i Tisleia, Flya og
    Nøra. Unpubl. thesis, Univ. Oslo, 80 pp. In Norwegian.
Knight, A. W. and Gaufin, A. R. 1966. Altitudinal distribution of
    stoneflies (Plecoptera) in a Rocky Mountain drainage system. J. Kansas
    Ent. Soc., 39: 668-675.
Kraft, K. J. and Mundahl, N. D. 1984. Effect of intermittent flow
    regulation on temperature and macroinvertebrate distribution and
    abundance in a Michigan river. Freshwat. Invert. Biol., 3: 21-35.
Kroger, R. L. 1973. Biological effects of fluctuating water levels in the
    Snake River, Grand Teton National Park, Wyoming. Am. Midl. Nat,
    89: 478-481.
Langeland, A. and Haukebø, T. 1979. Trout, burbot and bottom invertebrates
    in the River Nea, Central Norway, before the building of
    weirs. Inf. Terskelprosjektet, NVE, Oslo, 9, 56 pp. (In Norwegian,
    English summary).
Lehmkuhl, D. M. 1974. Thermal regime alteration and vital environmental
    physiological signals in aquatic organisms. In: Thermal
    Ecology. (Ed. by J. W. Gibbons and R. R. Scharits). pp. 216-222. AEC
    Symposium series (CONF 730505).
Lillehammer, A. 1985. Temperature influence on egg incubation period and
    nymphal growth of the stoneflies *Leuctra digitata* and *L. fusca*
    (Plecoptera: Leuctridae). Entomol. Gener., 11: 59-67.
Lillehammer, A. and Saltveit, S. J. 1979. Stream regulation in
    Norway. In: The Ecology of Regulated Streams (Ed. by J. V. Ward and
    J. A. Stanford) pp. 201-213. Plenum Press, New York.
Lillehammer, A. and Saltveit, S. J. 1984. The effect of the regulation on
    the aquatic macroinvertebrate fauna of the River Suldalslågen, Western
    Norway. In: Regulated Rivers (Ed. by A. Lillehammer and
    S. J. Saltveit) pp. 201-210. Universitetsforlaget, Oslo.
Malipatil, M. B. and Blyth, J. D. 1982. A qualitative study of the
    macroinvertebrate fauna of the Thomson River and its major
    tributaries, Gippsland, Victoria. Rep. Nat. Mus. Vict., 1- 1982: 1-95.
McClelland, W. T. and Brusven, M. A. 1980. Effects of sedimentation on the
    behavior and distribution of riffle insects in a laboratory
    stream. Aquat. Ins., 2: 161-169.
Minshall, G. W. and Winger, P. V. 1968. The effect of reduction in stream
    flow on invertebrate drift. Ecology, 49: 580-582.
Montgomery, J. C. and Fickeisen, D. H. 1979. Tolerance and buoyancy of
    aquatic insect larvae exposed to gas supersaturated water.
    Environ. Ent., 8: 655-657.
Nebeker, A. V. 1976. Survival of *Daphnia*, and stoneflies in
    air-supersaturated water. J. Fish. Res. Board Can., 33: 1208-1212.
Pearson, W. D., Dramer, R. H. and Franklin, D. R. 1968. Macroinvertebrates
    in the Green River below Flaming Gorge Dam, 1964-65 and
    1967. Proc. Utah Acad. Sci., 45: 148-167.
Pfitzer, D. W. 1954. Investigations of waters below storage reservoirs in
    Tennessee. Trans. N. Am. Wildl. Conf. 19: 271-282.
Raddum, G. 1978. Reguleringsvirkning pa bunnfaunaen i Aurlandselven.
    Rapp. Lab. Ferskv. Økol. Innlandsfiske, Bergen, 25, 49 pp. In
    Norwegian.

Radford, D. S. and Hartland-Rowe, R. 1971. A preliminary investigation of bottom fauna and invertebrate drift in an unregulated and a regulated stream in Alberta. J. Appl. Ecol. **8**: 883-903.

Rosenberg, D. M. and Wiens, A. P. 1978. Effects of sediment addition on macrobenthic invertebrates in a northern Canadian river. Wat. Res. **12**: 753-763.

Scullion, J. 1983. Effects of impoundments on downstream bed materials of two upland rivers in mid-Wales and some ecological implications of such effects. Arch. Hydrobiol., **96**: 329-344.

Scullion, J., Parish, C. A., Morgan, N. and Edwards, R. W. 1982. Comparison of benthic macroinvertebrate fauna and substratum composition in riffles and pools in the impounded River Elan and the unregulated River Wye, mid-Wales. Freshwat. Biol., **12**: 579-595.

Scullion, J. and Sinton, A. 1983. Effects of artificial freshets on substratum composition, benthic invertebrate fauna and invertebrate drift in two impounded rivers in mid-Wales. Hydrobiologia, **107**: 261-269.

Spence, J. A. and Hynes, H. B. N. 1971. Differences in benthos upstream and downstream of an impoundment. J. Fish. Res. Bd. Can., **28**: 35-43.

Stanford, J. A. and Ward, J. V. 1979. Stream regulation in North America. In: The Ecology of Regulated Streams (Ed. by J. V. Ward and J. A. Stanford) pp. 215-236. Plenum Press, New York.

Stanford, J. A. and Ward, J. V. 1983. Insect species diversity as a function of environmental variability and disturbance in stream systems. In: Stream Ecology - Application and Testing of General Ecological Theory (Ed. by J. R. Barnes and G. W. Minshall) pp. 265-278. Plenum Press, New York.

Stanford, J. A. and Ward, J. V. 1984. The effects of regulation on the limnology of the Gunnison river: a North American case history. In: Regulated Rivers (Ed. by A. Lillehammer and S. J. Saltveit) pp. 467-480. Universitetsforlaget, Oslo.

Trotzky, H. M. and Gregory, R. W. 1974. The effects of water flow manipulation below a hydroelectric power dam on the bottom fauna of the upper Kennebec River, Maine. Trans. Am. Fish. Soc., **103**: 318-324.

Ulfstrand, S., Svensson, B., Enckell, P. H., Hagerman, L. and Otto, C. 1971. Benthic insect communities of streams in Stora Sjöfallet National Park, Swedish Lapland. Ent. scand., **2**: 309-336.

Vannote, R. L. and Sweeney, B. S. 1980. Geographic analysis of thermal equilibria: A conceptual model for evaluating the effect of natural and modified thermal regimes on aquatic insect communities. Am. Nat., **115**: 667-695.

Vaught, G. L. and Stewart, K. W. 1974. The life history of the stonefly *Neoperla clymene* (Newman) (Plecoptera: Perlidae). Annls Ent. Soc. Am., **67**: 167-178.

Wais, I. R. 1984. Two Patagonian basins - Negro (Argentina) and Valdivia (Chile) - as habitats for Plecoptera. Annls Limnol., **20**: 115-122.

Ward, J. V. 1974. A temperature-stressed stream ecosystem below a hypolimnial release mountain reservoir. Arch. Hydrobiol., **74**: 247-275.

Ward, J. V. 1976a. Effects of flow patterns below large dams on stream benthos: A review. In: Instream Flow Needs Symposium. (Ed. by J. F. Orsborn and C. H. Allman) **2**: 235-253. Am. Fish. Soc., Bethesda, Maryland.

Ward, J. V. 1976b. Comparative limnology of differentially regulated sections of a Colorado mountain river. Arch. Hydrobiol., **78**: 319-342.

Ward, J. V. and Short, R. A. 1978. Macroinvertebrate community structure of four special lotic habitats in Colorado, U.S.A. Verh. int. ver. Limnol., **20**: 1382-1387.

Ward, J. V. and Stanford, J. A. 1979. Ecological factors controlling stream zoobenthos with emphasis on thermal modification of regulated streams. In: The Ecology of Regulated Streams (Ed. by J. V. Ward and J. A. Stanford) pp. 35-55. Plenum Press, New York.

Ward, J. V. and Stanford, J. A. 1982. Thermal responses in the evolutionary
    ecology of aquatic insects. Ann. Rev. Entomol., **27**: 97-117.
Ward, J. V. and Stanford, J. A. 1983. The serial discontinuity concept of
    lotic ecosystems. In: Dynamics of Lotic Ecosystems. (Ed. by T. D.
    Fontaine and S. M. Bartell) pp. 29-42. Ann Arbor Science.
Weitkamp, D.E. and Katz, M. 1980. A review of dissolved gas supersaturation
    literature. Trans. Am. Fish. Soc., **109**: 659-702.
Williams, D. D. and Hynes, H. B. N. 1976. The ecology of temporary streams,
    1. The fauna of two Canadian streams. Int. Rev. ges. Hydrobiol.,
    61: 761-787.
Williams, R. D. and Winget, R. N. 1979. Macroinvertebrate response to flow
    manipulation in the Strawberry river, Utah (U.S.A.). In: The Ecology
    of Regulated Streams (Ed. by J. V. Ward and J. A. Stanford)
    pp. 365-376. Plenum Press, New York.
Zimmermann, H. J. and Ward, J. V. 1984. A survey of regulated streams in
    the Rocky Mountains of Colorado, U.S.A. In: Regulated Rivers
    (Ed. by A. Lillehammer and S. J. Saltveit) pp. 251-259.
    Universitetsforlaget, Oslo.

THE CLASSIFICATION OF TAILWATER SITES RECEIVING RESIDUAL FLOWS FROM

UPLAND RESERVOIRS IN GREAT BRITAIN, USING MACROINVERTEBRATE DATA

Patrick Armitage

Freshwater Biological Association
River Laboratory
East Stoke, Wareham, Dorset BH20 6BB, U.K.

INTRODUCTION

The amount of water left in rivers and streams after abstraction for public supply is referred to as the residual or compensation flow. In the past the quantitites of water released from reservoirs have been the subject of arbitrary rulings (Sheail 1984). Throughout Great Britain there are a large number of reservoirs where decisions as to the amount of compensation flow to be released were made many years before the present archive of river flow data was available. These data are now being used by the Institute of Hydrology as the basis for preparing guidelines for resetting reservoir operating policy and compensation releases in order to minimise disturbance to the flow regime downstream and wherever possible to conserve water for supply purposes.

Before such guidelines can be prepared it is important that information should be obtained on the biological consequences of particular compensation flows and in 1983 the macroinvertebrate fauna below 29 reservoirs was examined. The chief aims of the biological study were:

1.   to relate below-reservoir faunas to conditions of flow and other physical and chemical characteristics of the sites,

2.   to determine whether the faunal communities below reservoirs are distinctive by classifying the regulated-stream data with macroinvertebrate data from unregulated sites sampled for the Freshwater Biological Association's Project 103, "The analysis of natural river communities in Great Britain" (Wright et al. 1984). Henceforth in this paper this project will be referred to as FBA 103.

All analyses of faunal data were carried out at family level as this allowed a greater number of sites to be examined within the time available.

STUDY AREA AND METHODS

Hydro-electric and regulating reservoirs with widely fluctuating discharges were excluded from the pool of sites available because the main interest was in reservoirs whose discharge was released as compensation

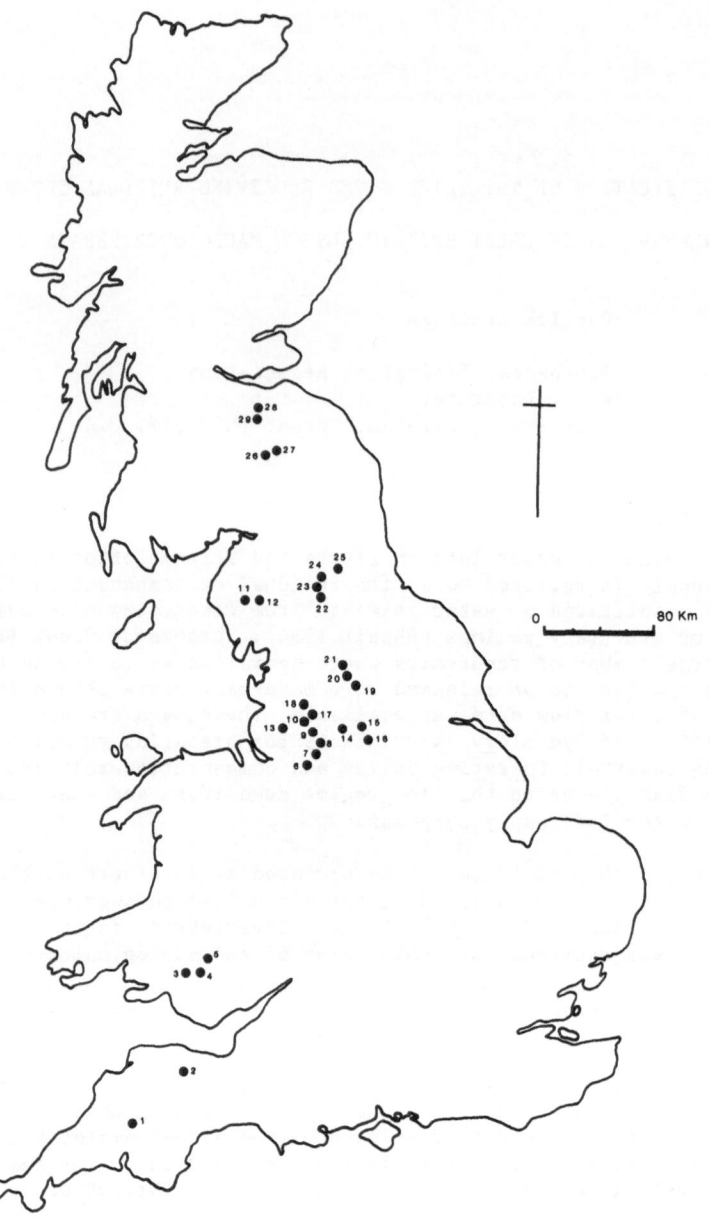

Figure 1.   Location of the sites. 1 Meldon, 2 Wimbleball, 3 Llwyn-on,
4 Taf Fechan, 5 Talybont, 6 Teggsnose, 7 Fernilee, 8 Kinder (1),
Kinder (2), Sett, 9 Longdendale, 10 Castleshaw, 11 Thirlmere,
12 Haweswater, 13 Belmont, 14 Winscar, 15 Scout Dike, 16 More
Hall, 17 Boothwood, 18 Widdop, 19 Lindley Wood 1, Lindley Wood
2, 20 Thruscross, 21 Gouthwaite, 22 Hury, 23 Grassholme, 24
Burnhope, Wearhead, 25 Derwent, 26 Fruid, 27 Talla,
28 Baddinsgill, 29 Westwater.

flow. Emphasis was placed on sampling a wide range of compensation flows from upland reservoirs throughout the country (Figure 1). None of these reservoirs released deoxygenated hypolimnetic water downstream. Ease of access, absence of pollution and well-monitored discharge were factors important in the final selection. In addition to the 29 sites below reservoirs two unregulated streams which were adjacent to regulated sites were sampled for reference. The full site list is presented in Table 1 together with selected environmental data.

Sampling Frequency and Methods

Sampling and data collection followed procedures adopted for FBA 103 (Wright et al. 1984). Samples were taken in spring (April 10-16), summer (June 28-July 4) and autumn (September 19-25) downstream of the reservoir outflows. The macroinvertebrate fauna was sampled by disturbing the substratum upstream of a pond net (mesh size 900 μm) for 3 minutes across all major biotopes (Furse et al. 1981). Animals were identified to family level and a log scale was used to categorise the abundance of animals in each family (10 animals = 1, 100 = 2, 1000 = 3, 10,000 = 4, >10,000 = 5). Physical data were gathered at the site or from maps and chemical data were obtained from the relevant Water Authority or River Purification Board. Discharge data were provided by the Institute of Hydrology.

Data Analyses

Emphasis was placed on ordination and classification techniques so that the relationship between unregulated sites sampled for FBA 103 and the regulated sites sampled for this study could be demonstrated.

Ordination was carried out using detrended correspondence analysis (Hill & Gauch 1980) and the sites were classified using two-way indicator species analysis (TWINSPAN) (Hill 1979) which classifies both samples and species and constructs ordered two-way tables to show the relationship between them as clearly as possible. This is a divisive technique and the division of samples into progressively smaller groupings may be continued for as long as seems profitable. The program also constructs a key to the sample classification by identifying one or more differential taxa which are particularly diagnostic of each division in the classification. The key can therefore be used to classify new sites without the need to reclassify all sites. Detailed descriptions and discussion of the merits and application of these methods are found in Wright et al. (1984).

In addition to these methods the BMWP biotic score based on macroinvertebrates and derived by summing the individual scores of all families present (Armitage et al. 1983) was calculated for regulated sites and compared with values from unregulated sites. The score and average score per taxon provide a measure of the pollution status of a site where in general high values indicate `clean´ conditions and low values indicate a possible reduction in water quality.

RESULTS

The Fauna

A total of 70 families was recorded from the 33 sites when the samples taken in the three seasons were combined. The mean numbers of families per site in spring, summer and autumn samples were 24, 25 and 23, respectively. Seasonal variation in the number of families recorded at the 33 sites is shown in Table 2 which also presents biotic score values for combined seasons data. Thirty-six of the total 70 families occurred in more

Table 1. Year of completion of earliest upstream reservoir, distance
of site from dam; compensation flow as millions of litres per
day; substratum (phi values) mean of 3 seasons estimates,
alkalinity (mg l^{-1} CaCO$_3$) mean of all determinations available
for one year. [(a) upstream of Sett confluence , (b) downstream
of Sett confluence with Kinder Stream; (c) and (d) upstream and
downstream of fish hatchery; (e) unregulated site on R. Wear
upstream of confluence with Burnhope Burn.]

Site Name	Age	Distance from dam (km)	Compensation flow as MLD	Substratum	Alkalinity
SOUTH WEST					
Meldon	1972	0.11	7.7	-6.73	9.67
Wimbleball	1977	0.60	9.1	-5.65	27.33
WELSH					
Llwyn-on	1927	0.37	23.2	-7.36	45.73
Taf Fechan	1927	0.41	21.8	-6.47	76.86
Talybont	1938	0.23	13.6-25.0	-6.14	43.50
NORTH WEST					
Teggsnose	1850	0.22	2.4	-1.74	25.00
Fernilee	1927	0.25	13.6	-5.17	17.43
Kinder (1)(a)	1912	0.60	1.1	-4.73	11.00
Kinder (2)(b)	1912	1.62	-	-4.23	11.00
Sett	-	-	-	-4.60	-
Longdendale	1877	0.25	22.7-68.2	-4.47	15.71
Castleshaw	1891	0.35	0.34- 2.3	-4.85	36.11
Thirlmere	1894	0.41	13.6	-6.15	6.50
Haweswater	1941	0.75	21.8	-6.51	14.75
Belmont	1826	2.25	15.6	-5.84	30.00
YORKSHIRE					
Winscar	1975	0.50	9.1-11.8	-5.20	62.98
Scout Dike	1924	0.30	2.7- 3.9	-4.19	35.09
More Hall	1930	0.30	9.1-12.0	-5.83	57.51
Boothwood	1971	0.25	9.7	-6.33	3.19
Widdop	1878	1.75	20.0	-5.17	0.50
Lindley Wood (1)(c)	1875	0.60	<18.2	-4.19	66.67
Lindley Wood (2)(d)	1875	0.61	<18.2	-3.94	66.67
Thruscross	1967	0.30	spill only	-6.32	5.67
Gouthwaite	1901	0.55	22.7-70.9	-5.20	29.86
NORTHUMBRIAN					
Hury	1894	0.20	15.2	-5.39	27.00
Grassholme	1915	1.90	28.4	-6.43	33.00
Burnhope	1936	1.10	9.1	-3.23	23.02
Wearhead (e)	-	-	-	-6.50	-
Derwent	1916	0.40	22.7-25.0	-5.76	17.83
TWEED					
Fruid	1968	0.25	9.0-13.0	-6.09	33.75
Talla	1905	0.25	0-16.6	-6.48	33.75
Baddinsgill	1930	0.25	2.3- 4.0	-5.85	62.23
Westwater	1967	0.50	1.6- 2.9	-5.36	70.49

Table 2. Seasonal variations in the number of families and combined
          seasons biotic score values recorded at 33 sites.
          (Sp = spring; Su = summer; Au = autumn; Co = combined seasons;
          S = BMWP score; A = average score per taxon).

| Sites | Sp | Total Families | | | Score (Co) | |
		Su	Au	Co	S	A
Meldon	7	14	9	16	62	6.20
Wimbleball	31	21	31	39	195	6.96
Llwyn-on	18	18	23	29	126	6.30
Taf Fechan	17	23	18	31	133	6.33
Talybont	26	24	26	35	154	6.70
Teggsnose	31	30	25	41	142	5.46
Fernilee	17	17	18	28	124	6.20
Kinder 1	24	24	28	36	166	6.64
Kinder 2	29	28	28	40	192	6.86
Sett	25	21	21	36	172	6.88
Longdendale	13	22	20	30	94	5.53
Castleshaw	19	24	24	32	112	5.33
Thirlmere	22	27	29	36	145	6.30
Haweswater	29	27	26	39	183	6.31
Belmont	24	23	19	33	126	6.30
Winscar	19	22	19	28	98	6.13
Scout Dike	22	21	22	33	115	5.48
More Hall	26	26	24	34	138	6.00
Boothwood	13	12	16	18	70	6.36
Widdop	14	18	22	29	118	5.90
Lindley Wood 1	39	34	31	47	197	5.97
Lindley Wood 2	33	38	33	47	197	6.16
Thruscross	19	27	17	29	129	6.45
Gouthwaite	36	35	28	49	212	6.42
Hury	34	35	33	44	198	6.83
Grassholme	28	33	27	39	211	6.31
Burnhope	17	18	19	35	147	6.39
Wearhead	17	14	11	24	133	7.00
Derwent	28	27	26	36	162	6.23
Fruid	23	20	21	37	164	6.31
Talla	26	26	29	38	194	6.93
Baddinsgill	25	27	23	35	154	6.42
Westwater	31	36	26	43	212	6.63
Total families and mean score values	63	61	61	70	151	6.32

than 50% of the sites. Of these 12 were abundant and included stoneflies (Nemouridae, Leuctridae), mayflies (Baetidae), caseless caddis (Polycentropodidae, Hydropsychidae and Rhyacophilidae), sub-families/tribes of Chironomidae (Orthocladiinae, Tanypodinae, Tanytarsini), amphipods (Gammaridae) and oligochaetes (Naididae, Lumbriculidae). In contrast, 28 families occurred at 9 or less sites and were never abundant.

The BMWP score system was applied to the 33 residual flow sites (Table 2). Armitage et al. (1983) have shown that there is considerable variation in achievable score and average score per taxon (ASPT) in different categories of unpolluted sites. Scores >150 and ASPT >6.00 are to be expected in clean upland streams of the type sampled in this survey. However, four sites, Meldon, Longdendale, Winscar and Boothwood had scores below 100 and six sites, Teggsnose, Longdendale, Scout Dike, Castleshaw, Widdop and Lindley Wood (1) had ASPT values below 6.00. The relatively low score values at Meldon, Boothwood and Winscar suggest that regulation causes a reduction in the total number of families without a decrease in high scoring taxa which results in low scores and high ASPT. In contrast at the other sites with relatively high scores and low ASPT regulation is creating conditions which lead to an increase in low scoring taxa which is sometimes accompanied by a decrease in high scoring taxa. Thus although interpretation of the meaning of the biotic score can be difficult its ability to summarise faunal conditions, particularly in large data sets, is useful in focussing attention on faunistically unusual sites.

The number of families present at a site showed statistically significant correlations with reservoir age, distance from source of stream, altitude, water velocity and pH. The r values and associated probability levels for these and other variables are presented in Table 3. None of the measures of discharge showed a significant correlation with either family richness or abundance but water velocity was negatively correlated. In general, high altitude low pH outflows from reservoirs situated close to the stream´s source supported low family diversity and abundance and family-rich sites occurred below older reservoirs. Five out of six of the richest sites are below reservoirs greater than 70 years old. In contrast, three out of the four poorest sites are below reservoirs which are less than 12 years old.

Ordination

The ordination uses faunal composition and relative abundance to arrange the sites mathematically along axes of variation. The eigenvalues of each axis may be regarded as a measure of the relative importance of individual axes. They can be used to derive the amount of variance explained by each axis. In the present analysis most variation (75.6%) is accounted for by the first two axes which are plotted in Figure 2. By examining the relationship between environmental variables and axis scores (Table 3) it is possible to interpret the figure as follows. Along axis 1 there is a gradation of sites from left to right showing a decrease in altitude and water velocity and increase in pH, TON and ALK. Along axis 2, from bottom to top, sites increase in distance from source, channel width and pH, and decrease in slope and velocity. Positive correlations were also noted between axis 2 scores and family richness (r = 0.756, n = 33, p <0.001) and abundance (r = 0.596, n = 33, P <0.001).

Classification

The relationship between the 268 mainly unregulated sites sampled for the FBA 103 project (Wright et al. 1984) and the 33 ´residual-flow´ sites sampled for this study was investigated by amalgamating the two combined

Table 3. Correlation coefficients between faunal parameters and environmental variables. Log transformation was applied to non-normally distributed variables before the calculation of the the coefficients, these variables are prefixed by L. (LAGE = age of reservoir, LALT = site altitude, LKM = distance of site from true source of stream, MEDVEL = median surface velocity at site , LMDEPTH = mean depth, LCF = compensation flow, LCFAF = compensation flow as a percentage of pre-regulation Average Daily Flow, MSUBST = mean of seasonal estimates of substratum, MINMAC = minimum macrophyte cover in 3 seasons; pH, Total Oxidised Nitrogen (mg N 1^{-1}), Chloride (mg Cl 1^{-1}), Orthophosphate (mg P 1^{-1}) and Total Alkalinity (mg $CaCO_3$ 1^{-1}) mean of all determinations available for one year) [P <0.01** <0.05*]. (FAM = number of families, ABUND = total numbers of animals per site based on mean of 3 seasons, BMWP score and ASPT, AXIS1 and AXIS2 ordination scores).

Variable (N)	FAM	ABUND	BMWP	ASPT	AXIS1	AXIS2
LAGE	0.409*	0.376*	0.212	-0.301	0.289	0.213
LALT	-0.432*	-0.454**	-0.209	0.320	-0.464**	-0.119
LKM	0.373*	0.271	0.410*	0.136	0.078	0.530**
MEDVEL	-0.471**	-0.307	-0.327	0.127	-0.388*	-0.298
LMDEPTH	0.032	-0.128	0.164	0.383*	-0.220	0.281
LCF	-0.061	-0.110	-0.099	-0.037	0.069	0.076
LCFAF	-0.131	-0.087	-0.252	-0.256	0.195	-0.164
MSUBST	0.313	0.186	0.092	-0.323	0.290	0.029
MINMAC	0.082	0.212	0.117	0.076	-0.077	-0.031
pH	0.462**	0.467**	0.345	-0.037	0.359*	0.428*
LTON	0.125	0.203	-0.045	-0.349	0.361*	-0.011
LCL	-0.170	-0.052	-0.378*	-0.514**	0.273	-0.426*
LORPH	0.255	0.336	0.088	-0.185	0.258	-0.017
ALK	0.269	0.346	0.157	-0.182	0.353*	0.216

seasons data sets and classifying the 301 sites at family level using TWINSPAN. Figure 3 presents a dendrogram showing how TWINSPAN groups are developed. Also shown are the number of sites per group and those groups which contain `residual-flow´ sites. Table 4 gives data on the mean of 9 environmental variables for unregulated and regulated sites within TWINSPAN groups containing `residual-flow´ sites.

The faunal similarity is emphasised here where 24 of the 33 sites occupy adjacent groups 56, 57 and 58. Only two sites, Scout Dike and

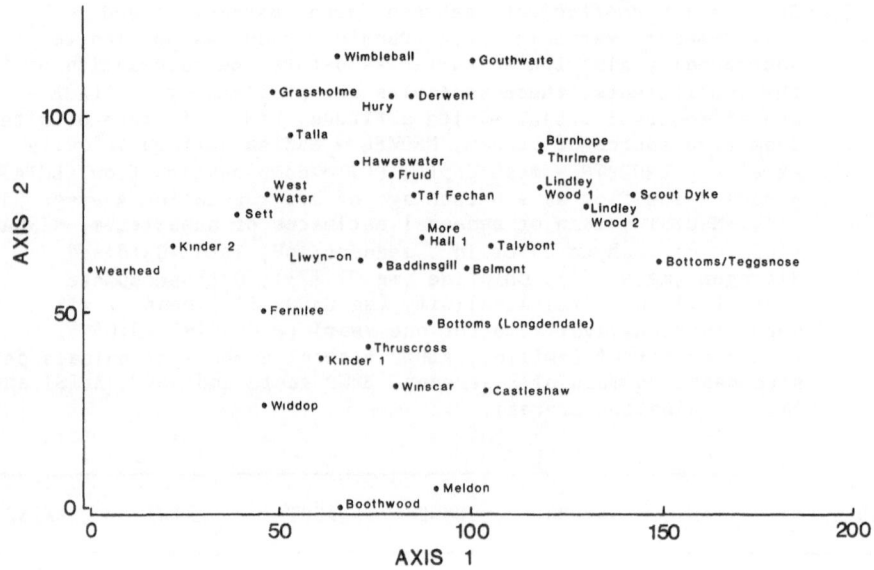

Figure 2. Ordination plot (Axis 1 v Axis 2) of 33 residual flow sites.

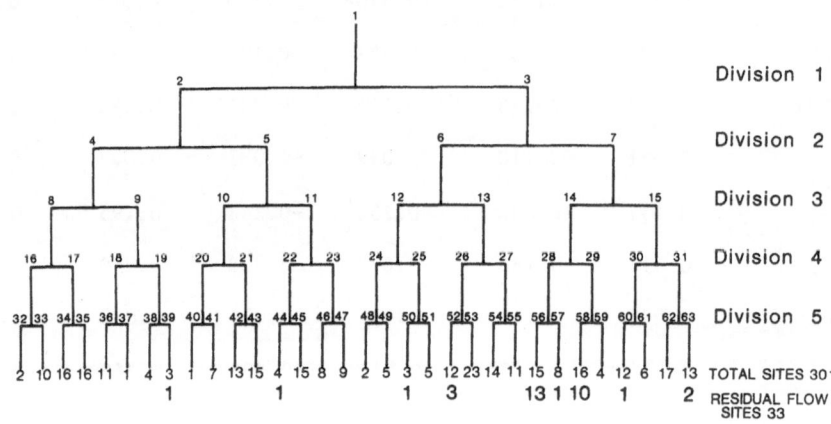

Figure 3. TWINSPAN classification of combined River Communities Project sites (268) and residual flow sites (33). Group numbers are indicated at each division (1-63) together with the number of sites in each of the 32 groups produced at the fifth division.

Teggsnose, were separated from the remainder at the first division. Scout Dike differed from the unregulated sites in its group in having a steeper slope, being nearer to the true source of the river and having a lower channel depth and discharge. Teggsnose was again nearer to the source and had a steeper slope than the other sites in its group. Alkalinity was lower at Teggsnose but discharge was similar in all sites.

Lindley Wood (2) in group 50 is further from the source, has a higher macrophyte cover and lower alkalinity than unregulated sites in its group. Lindley Wood (1), Thirlmere and Gouthwaite (group 52), combine with sites situated low down on spatey streams or near the source of small

Table 4. Mean values for 9 environmental variables for regulated (R) and unregulated (U) sites within TWINSPAN groups generated from a combined data set including 268 FBA[103] sites and the 33 residual flow sites. The numbers of R and U sites in each group are indicated by subscripts. (Discharge categories $1 \leq 0.3$ cumecs; $2 \leq 0.62$; $3 \leq 1.25$; $4 \leq 2.5$; $5 \leq 5.0$; $6 \leq 10$; $7 \leq 20.0$; $8 \leq 40$; substratum S = sand, G = gravel, P = pebbles, C = cobbles).

Sites grp^{-1}	39		44		50			52		56	57		58		60		63	
	R_1	U_2	R_1	U_3	R_1	U_2	R_3	U_9	R_{13}	U_1	R_1	U_7	R_9	U_6	R_1	U_{11}	R_1	U_{11}
Distance from source (km)	7.6	45.7	2.3	8.9	20.9	3.2	15.8	17.7	8.7	66.0	4.0	30.1	8.5	6.5	7.1	16.2	2.8	10.9
Slope (m km^{-1})	14.1	1.7	36.3	2.0	8.9	10.0	9.5	1.5	16.6	0.9	17.0	3.5	21.8	6.0	21.9	4.4	34.7	9.8
Altitude (m)	204	47	182	28	68	74	112	39	234	200	282	32	151	78	209	105	214	295
Discharge category	1	6	1	2	1	1	1	4	1	8	1	6	1	3	1	5	1	5
Channel width (m)	5.0	39.8	4.0	4.9	7.9	3.7	9.8	9.8	7.4	44.7	3.0	19.5	5.7	6.3	6.0	11.5	4.4	13.8
Channel depth (cm)	13	107	13	28	23	12	23	25	29	63	18	30	21	19	26	27	16	25
Water velocity (cm s^{-1})	<50	<25	<25	<100	<50	<50	<50	<50	<100	<100	<50	<100	<50	<50	<50	<100	>100	<100
Substratum	P	S	G	G	P	P	P	G	C	C	P	P	P	P	P	C	C	C
Macrophyte cover %	2	10	15	36	38	2	31	38	37	1	53	12	29	9	4	4	14	1

streams with stable flow regimes and which include the Cumbrian Derwent at Ouse Bridge 0.1 km below Bassenthwaite Lake.

Group 56 contains 13 `residual-flow´ sites and two others, one on the R. Tees about 300 m below Cow Green Reservoir and the other on the R. Spey at Boat of Garten which may be affected by the lake-like section of the river at Kincraig. In group 57, Westwater, a regulated site situated close to the source of the stream, is associated with sites which are generally low down on spatey streams or have a lake along the course of the stream.

Group 58 contains 10 `residual-flow´ sites including the unregulated control on the R. Sett adjacent to Kinder 1. The discharge is generally lower in the regulated sites but most other physical and chemical variables are similar to those of the unregulated sites.

Group 60 contains one regulated site, Kinder 2, which has similar physical characteristics to other sites in the group and which includes 2 sites below lakes.

Group 63 includes the unregulated control on the R. Wear at Wearhead and the regulated site Boothwood. The latter site has the lowest discharge and highest macrophyte cover in the group.

Table 5 summarises the faunal characteristics of residual flow sites within the 9 TWINSPAN groups. Diptera form a high proportion of the fauna in all groups and it is amongst other orders that trends are discernible. For example groups on the left of the classification generally have higher proportions of Oligochaeta and Crustacea than those on the right. Mollusca are abundant only in group 44. Ephemeroptera and Plecoptera form the highest proportion of the fauna in groups 56-60 and 58-63 respectively. The relative numbers of Trichoptera are greatest in groups 52-60. Total mean numbers per sample tend to be highest in groups 39-52.

Prediction

Prediction of the fauna to be expected at a site in the absence of regulation would provide a measure of deviation from the `norm´ or an estimate of the effects of regulation. An example of this is presented below for the two sites, Scout Dike and Teggsnose, apparently most affected by regulation. The method employs multiple discriminant analysis (MDA) (Klecka 1975) which uses linear combinations of environmental variables to make predictions of probabilities of group membership. The MDA equations are derived from the FBA 103 268-site classification (Furse et al. 1984) which is based on combined seasons quantitative family data. The equations utilize information on 27 environmental variables including measures of water chemistry, substratum characteristics, channel morphology, geographic location and estimates of discharge in the absence of regulation. Details of the variables are given in Wright et al. (1984).

When the MDA equations are applied using the environmental data from Scout Dike and Teggsnose the sites are predicted to occur in groups 63 and 56 respectively, of the 268-site classification. The faunal composition of these groups can be determined and compared with the actual fauna of the two sites.

The method can be criticised on the grounds that factors which are likely to be affected by regulation are being used to predict its effects. However, channel shape, substratum characteristics, and water chemistry will remain the same or nearly so after regulation within the limits of measurement used in this study. Geographic features will be

Table 5. Faunal composition (%) and mean total numbers per sample of regulated residual flow sites within TWINSPAN groups, based on combined seasons data. (n = number of regulated sites).

Fauna		TWINSPAN Groups n	39 1	44 1	50 1	52 3	56 13	57 1	58 9	60 1	63 1
Mollusca	%		2	28	5	6	4	2	2	2	-
Oligochaeta	%		11	22	36	7	6	3	5	4	18
Hirudinea	%		<1	<1	2	1	<1	<1	<1	-	-
Crustacea	%		22	18	4	4	3	12	4	<1	-
Ephemeroptera	%		5	1	9	5	19	29	14	39	6
Plecoptera	%		<1	<1	2	11	6	4	28	11	21
Coleoptera	%		<1	<1	<1	1	2	25	1	6	-
Trichoptera	%		6	2	8	31	15	9	14	21	7
Diptera	%		52	27	31	31	42	15	30	15	48
Others	%		<1	<1	2	2	2	1	1	1	-
Total nos.			2225	5529	5402	1117	755	3979	413	862	375

unaffected and discharge figures used in the equation are pre-regulation values.

Observed and predicted faunas at the two sites are shown in Figure 4. It is clear that the observed is markedly different from the predicted fauna due mainly to increased densities of Oligochaeta, Crustacea, Trichoptera, Chironomidae and Mollusca and reduced abundance of Plecoptera and Coleoptera. This is an indication that the sites are indeed greatly affected by regulation. A measure of the difference between observed and predicted would provide an index of the impact of a regulatory flow regime.

DISCUSSION AND CONCLUSION

This survey has demonstrated that although a range of communities occurred below reservoirs releasing water as compensation flow, most sites fell into 2 closely related groups. A positive relationship was observed between age of reservoir and number of families recorded. This might be expected since the fauna would have had longer to colonise and adapt below older reservoirs but there is the complicating factor that older reservoirs were generally built at lower altitudes than are modern ones and altitude has been shown to be negatively correlated with family richness. In some groups of sites regulation altered the complement of families to a community type normally associated with conditions further downstream. It had in effect `aged´ the stream. This may in part be attributable to the reduction of extreme flow fluctuations which are commonly found in steeply-sloping hill streams. The temperature regime may also have been affected below the dam (Ward 1976) but temperature data were not available to confirm this. Another feature observed but not quantified was the thin layer of fine sediment which coated the coarse substratum in the majority of sites. The releases from the reservoirs were insufficient to mobilise this material whose settlement favoured the development of large populations of Oligochaeta, Chironomidae and deposit feeding molluscs at certain sites. The diversity and abundance of macroinvertebrates at most sites were not apparently reduced by the deposition. However an increase of fine sediment which fills interstitial spaces is likely to increase the proportion of deposit feeders at the expense of grazers and reduce faunal diversity.

In the two unregulated control sites the number of families recorded was the same or lower than in the adjacent regulated streams. This agrees with Ward and Stanford (1979) who found that most major taxa (with the exception of Plecoptera, Ephemeroptera and Coleoptera) are relatively more abundant downstream of reservoirs than at nearby reference sites.

The classification identified certain faunistically unusual sites. MDA equations failed to predict correct group membership of these sites with the available environmental data. This suggests that some regulated sites possess important characteristics which are not used in the predictive equations. Chief amongst these is the post-regulation discharge. Temperature and sediment are two additional features, already mentioned above, which were not included in the equation. Despite these deficiencies

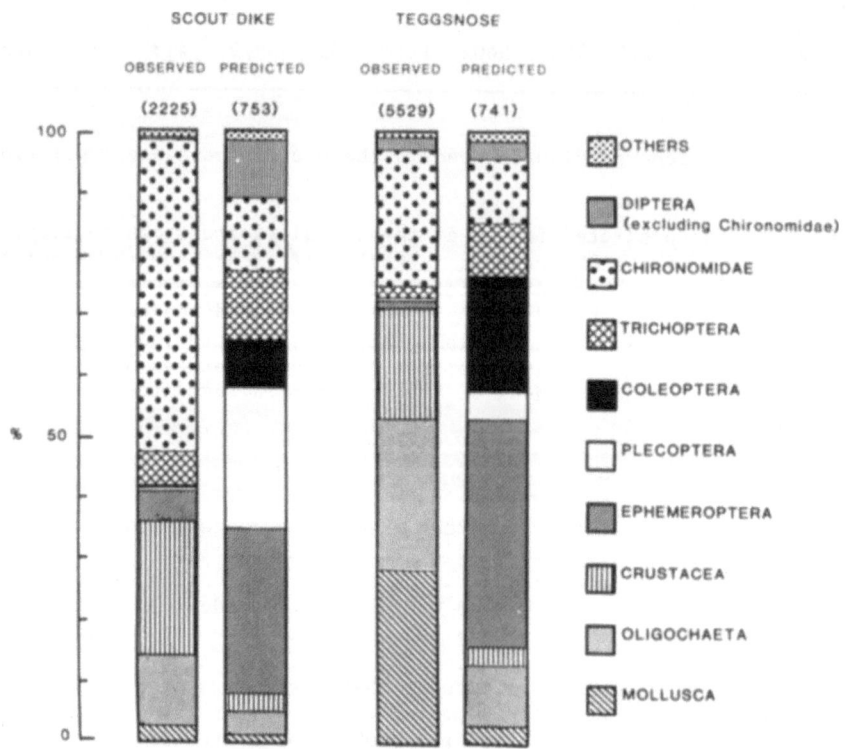

Figure 4.   The percentage composition of observed and predicted faunas at two regulated sites.

and largely because of them it has been possible to use the the prediction of group membership based on environmental data as a means of determining how much the regulated stream fauna differs from the pre-regulation state. With more testing this method will enable accurate assessments of environmental impact to be made by comparing observed with predicted faunas.

The extreme detrimental effects of flow regulation, such as severely reduced diversity and abundance, often reported in the literature (Ward & Stanford 1979; Armitage 1984) were not noticeable within the range of sites examined during this survey and biotic score values were generally high. It is possible that the effects of constant and reduced discharge may have been alleviated by winter spillage and occasional freshets. Although no major adverse effects were observed using family level identifications it should not be assumed that such effects might not be demonstrated if the data had been analysed at species level. Low species diversity is a feature of some regulated sites (Armitage 1978).

Despite the lack of detrimental effects apparent at family level there is no doubt that changes occur in faunal communities in response to regulation and reservoir operating policies will have to take this into account when recommending compensation releases. Increased demand for reservoir water which would reduce winter spillage and the surplus required for freshets is likely to reduce faunal diversity below dams.

In conclusion this survey has revealed two interesting areas for future study. The first concerns the statistically significant but low correlation of faunal diversity with age of site. Further data are needed from sites with similar release patterns to test this observation. Secondly the apparent `ageing´ effect of reservoirs on some downstream sites provides evidence of the major disruption of continuum processes (Vannote et al. 1980; Ward & Stanford 1983) and this would repay a more detailed investigation. In addition, a predictive method for assessing environmental impact has been demonstrated. The method is based on results obtained by the Freshwater Biological Association´s River Communities Project (Wright et al. 1984) and is provisional. Further studies are in progress to improve the system, by using a smaller number of variables which should be integrating and independent, and predicting actual families rather than group membership. By increasing the data base for the model it should eventually be possible to predict the effects of a particular regulation scheme on a pristine site. The model could be used for testing the likely impact of changes in flow regime and determining optimum management involving a balance between maximum resource usage and minimum effect on faunal communities.

ACKNOWLEDGEMENTS

Assistance with sampling and environmental data collection was provided by Mr. R.J.M. Gunn and Mr. M.P. Williams. Samples were sorted and identified by Mr. J.H. Blackburn. Mrs. M. Scott assisted with data processing and graphics. All computing analyses were carried out by Dr. D. Moss of ITE Bangor. Dr. J.F. Wright, Mr. M.T. Furse and Dr. K. Walker provided helpful advice and discussion. Chemical data were supplied by the following Water Authorities and River Purification Board, South West, Welsh, North West, Yorkshire, Northumbrian and Tweed. The Institute of Hydrology assisted in site selection and collation of discharge data. I am most grateful to all these individuals and organisations for their help. The work was funded by a grant from the Department of the Environment.

REFERENCES

Armitage, P. D. 1978. Downstream changes in the composition, numbers and
    biomass of bottom fauna in the Tees below Cow Green reservoir and in
    an unregulated tributary Maize Beck, in the first five years after
    impoundment. Hydrobiologia **58**: 145-156.
Armitage, P. D. 1984. Environmental changes induced by stream regulation
    and their effect on lotic macroinvertebrate communities. In:
    Regulated Rivers (Ed. by A. Lillehammer & S. J. Saltveit).
    pp. 139-165. University Press, Oslo.
Armitage, P. D., Moss, D., Wright, J. F. and Furse, M. T. 1983. The
    performance of a new biological water quality score system based on
    macroinvertebrates, over a wide range of unpolluted running water
    sites. Wat. Res. **17**: 333-347.
Furse, M. T., Wright, J. F., Armitage, P. D. and Moss, D. 1981. An
    appraisal of pond-net samples for biological monitoring of lotic
    macro-invertebrates. Wat. Res. **15**: 679-689.
Furse, M. T., Moss, D., Wright, J. F. and Armitage, P. D. 1984. The
    influence of seasonal and taxonomic factors on the ordination and
    classification of running-water sites in Great Britain and on the
    prediction of their macro-invertebrate communities. Freshwat.
    Biol. **14**: 257-280.
Hill, M. O. 1979. TWINSPAN - A FORTRAN Program for Arranging Multivariate
    Data in an Ordered Two-way Table by Classification of the Individuals
    and Attributes. Ecology and Systematics; Cornell University, Ithaca,
    New York 14850.
Hill, M. O. and Gauch, H. G. 1980. Detrended correspondence analysis: an
    improved ordination technique. Vegetatio **42**: 47-58.
Klecka, W. R. 1975. Discriminant analysis. In: SPSS. Statistical Package
    for Social Sciences. (Ed. by N. H. Nie, C. H. Hull, J. G. Jenkins,
    K. Steinbrenner & D. H. Bent). pp. 434-467. McGraw-Hill, New York.
Sheail, J. 1984. Constraints on water-resource development in England and
    Wales: the concept and management of compensation
    flows. J. Environmental Management **19**: 351-361.
Vannote, R. L., Minshall, G. W., Cummins, K. W., Sedell, J. R. and Cushing,
    C. E. 1980. The river continuum concept. Can. J. Fish
    Aquat. Sci. **37**: 130-137.
Ward, J. V. 1976. Effects of thermal constancy and seasonal termperature
    displacement on community structure of stream macroinvertebrates.
    In: Thermal Ecology II. (Ed. by G. W. Esch & R. W. McFarlane).
    pp. 302-307. ERDA Symp. Series [CONF.-750425], National Technical
    Information Series, Springfield, Illinois, U.S.A.
Ward, J. V. and Stanford, J. A. 1979. Ecological factors controlling stream
    zoobenthos with emphasis on thermal modification of regulated
    streams. In: The Ecology of Regulated Streams. (Ed. by J. V. Ward and
    J. A. Stanford). pp. 35-55. Plenum Press, New York and London.
Ward, J. V. and Stanford, J. A. 1983. The serial discontinuity concept of
    lotic ecosystems. In: Dynamics of Lotic Ecosystems (Ed. by T. D.
    Fontaine and S. M. Bartell). pp. 29-42. Ann Arbor Science, Michigan.
Wright, J. F., Moss, D., Armitage, P. D. & Furse, M. T. 1984. A preliminary
    classification of running-water sites in Great Britain based on
    macroinvertebrate species and the prediction of community type using
    environmental data. Freshwat. Biol. **14**: 221-256.

IMPACT OF LARGE DISCHARGE FLUCTUATIONS ON THE

MACROINVERTEBRATE POPULATIONS DOWNSTREAM OF A DAM

Odile Gaschignard and Alain Berly

Département de Biologie Animale et Ecologie
U.A. C.N.R.S. 367 "Ecologie des eaux douces"
Université Claude-Bernard, LYON I
69622 Villeurbanne-Cédex, France

INTRODUCTION

The impacts of river regulation on downstream habitats have received frequent study, and good reviews are given in Ward (1982), Armitage (1984) and Décamps (1984). The impacts of hydro-developments that include a reservoir are well known, however, little attention has been given to projects that lack a reservoir; situations in which only the flow rate has been modified. This paper deals with a comparison of the macroinvertebrate fauna in the up- and downstream sections of the Rhône River, close to Lyon (France), where only the flow rate has been modified.

STUDY AREA AND METHODS

Most of the hydroelectric facilities built by the "Compagnie Nationale du Rhône" consist of a retention canal or reservoir, headrace and tailrace with a by-passed section of the Rhône acting as an outlet for surplus flow. Two sampling sites have been chosen on the Rhône, upstream of Lyon, adjacent to one such hydroelectric development (Figure 1).

The first site, the Rhône River upstream of the Jons Dam, is located on the retention canal. This canal is not a reservoir because the bed is not enlarged and the current velocity is not modified. At this location, the hydrology of the Rhône is complex, and floods are possible during all seasons, depending on the year. The Rhône River at the Jons Dam is about 120 m wide and 2-5 m deep. The current speed, measured near the substrate, fluctuates from 10-110 cm s^{-1} (Figure 2).

The second site is on a by-passed section of the Rhône, called Miribel Canal, which is 80 m wide. This arm receives a minimum flow rate of 30 m^{3}s^{-1}, which may increase suddenly to more than 1000 m^{3}s^{-1}. The flow rate of Miribel Canal and the Rhône at the Jons Dam during the sampling period are presented in Figure 3. All macroinvertebrate samples were taken at low flow rates, so the depth was minimal (less than one meter) and the velocity was low (less than 80 cm s^{-1}; Figure 2). During floods, depths and current speeds may be the same in this arm as in the Rhône at Jons.

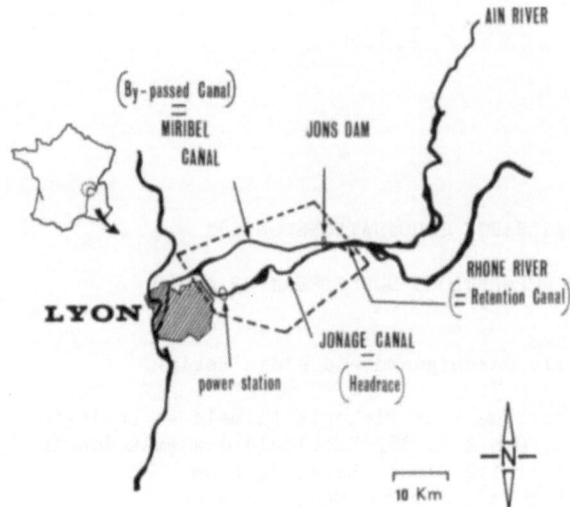

Figure 1.   Map of the study area.

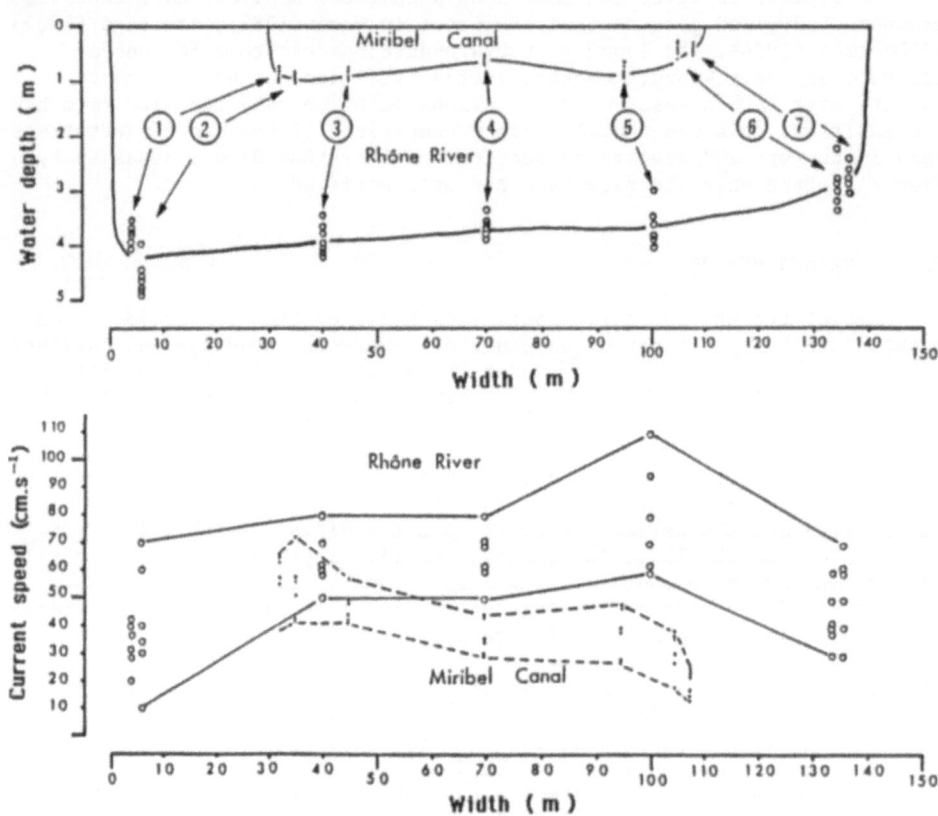

Figure 2.   Water depth and current speed across the channel at the two
sampling stations.

Between October 1981 and September 1982, six series of transect samples were taken from the Rhône River and Miribel Canal (7 sampling points at each site). A total of 84 samples were taken with a Rallier-du-Baty Style dredge (Boillot 1964), but of reduced size (weight 35 kg). A protective plastic bag encloses a fine net (0.5 mm mesh), which retains both the substrate materials and organisms. All samples were preserved in 4% formalin and handsorted in the laboratory.

The Rhône River east of Lyon flows on recent alluvial deposits, and at Jons, the substrate is dominated by pebbles (16-32 mm). Sand accounts for 2% of the total weight of the sediment collected. In Miribel Canal, small particles such as sand are rare because the Jons Dam limits the passage of lighter sediments, except during flooding (Gaschignard & el Hamdi 1984). Compared to the Rhône River, the sediment of Miribel Canal is generally smaller and less complex when evaluated by the number of particle sizes in each sample. The dam creates a more homogenous substrate by with-holding the smaller particles.

The mean densities of small organic particles (plant debris and algae of diam. 0.3 - 2 mm) are 28.8 mg 1^{-1} of sediment in the Rhône River and 8.7 mg 1^{-1} in Miribel Canal. For the larger particles (>2 mm), 29.5 mg 1^{-1} have been taken in the retention canal and only 3.1 mg 1^{-1} in the by-passed section; thus, Miribel Canal receives a relatively small amount of organic particles. Smaller organic particles are plentiful in Miribel Canal only during May when there is an abundance of epilithic algae. At Jons, the algae are well developed only in the shallower sections of the river.

RESULTS

Ninety taxa of invertebrates have been collected, most of which are known to be common in the Rhône River (Perrin & Roux 1978; Cellot 1982). Eleven of these ninety were absent in the Rhône at Jons and 22 were absent in Miribel Canal. Thus, faunal richness is lower in the by-passed section than in the Rhône, and Miribel Canal is much more sensitive to floods than is the Rhône. The mean densities followed the same pattern in both arms of the river.

The Rhône at Jons had high densities of numerous taxa not present in Miribel Canal (Table 1). They often consisted of lentic species (e.g., *Helobdella stagnalis* (L.), *Valvata cristata* (Mull.), *Sphaerium* sp.). Some Trichoptera, *Setodes punctatus* (Fbr.) and *Ceraclea* spp., were never found in Miribel Canal.

The most remarkable of the taxa collected in Miribel Canal and absent in the main part of the Rhône, are the phreatic invertebrates *Niphargus* sp. and *Niphargopsis* sp. These amphipods were collected at the top part of the sediment only, shortly after a reduction in the flow rate. The movement of groundwater, which imported these phreatic invertebrates in the by-passed section, is freer than in the Rhône because the sediment of Miribel Canal is more unstable and free of fine deposits. Moreover, exchanges between the epigean zone and groundwater were facilitated after the dam was put in operation because the riverbed below the dam has been deepened by erosion.

In general, the mean densities are greater in the Rhône River than in Miribel Canal; however, it is interesting to compare the densities of each taxon at the two stations. Figure 4 presents the ratio of the densities in the two arms for the 57 invertebrate taxa noted at the two sites (Appendix 1).

Table 1. Invertebrate Taxa Collected Only in the Rhône River or Only in
Miribel Canal. Accidental Invertebrates: only present in one
sample and only a few individuals; Rare Invertebrates: collected
several times but always in small numbers; Usual Invertebrates:
noted often and in large numbers (no animal from Miribel Canal
could be placed in this latter group).

Accidental Taxa	Rare Taxa	Usual Taxa

Rhône River

Heptagenia coerulans Rost.	*Helobdella stagnalis* (L.)	*Valvata cristata* Müll.
Protonemura sp.	*Potamopyrgus jenkinsi*	*Sphaerium sp.*
Perla bipunctata Pict.	Smith	*Dreissena polymorpha*
Dryops sp.	*Athripsodes spp.*	Pallas
Haliplus lineatocollis Mrsh.	(early instars)	*Setodes punctatus* Fbr.
Riolus subviolaceus P. Müll.	*Lepidostoma hirtum* Fbr.	*Ceraclea spp.*
Lype reducta Hag.	*Macronychus quadrituberculatus*	*(early instars)*
Oecetis notata Ramb.	P. Müll.	*Ceraclea spp.*
	Stenophylacini	Chironomini
		Robackia spp.

Miribel Canal

Niphargus sp.	*Planorbis sp.*	
Niphargopsis sp.	Perlodidae	
Oligoneuriella sp.	*Sialis sp.*	
Velia sp.	Empididae	
Riolus cupreus P. Müll.		
Hydropsyche angustipennis		
Curtis		
Rhyacophila sp.		

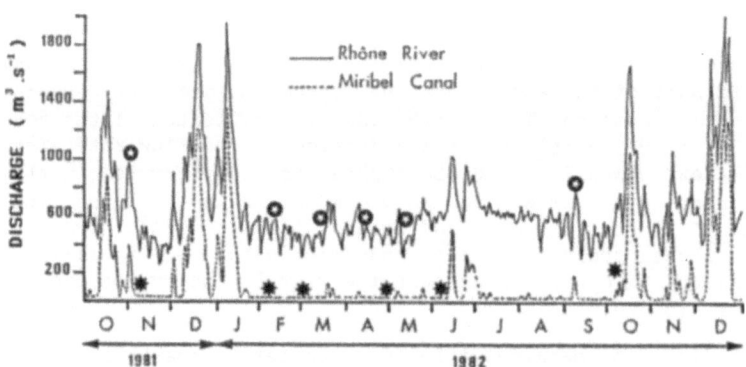

Figure 3. Flowrate of the Rhône River and Miribel Canal during the
sampling period. (✳ : Sampling dates for macroinvertebrates
in the Miribel Canal; ◉ : Sampling dates for macroinvertebrates
in the Rhône River).

Only 13 invertebrate taxa have approximately the same mean densities at both sampling stations. The course over time of these populations, whose densities are greater than 1 ind. (10 1)$^{-1}$ of sediment, is given in Figure 5. *Dugesia tigrina* (Girard) and *Psychomyia pusilla* (Fbr.) were collected principally during September 1982. These two species are sensitive to floods (Gaschignard 1984) and the relative hydraulic calm observed during 1982 allowed their populations to increase. The drastic reduction of

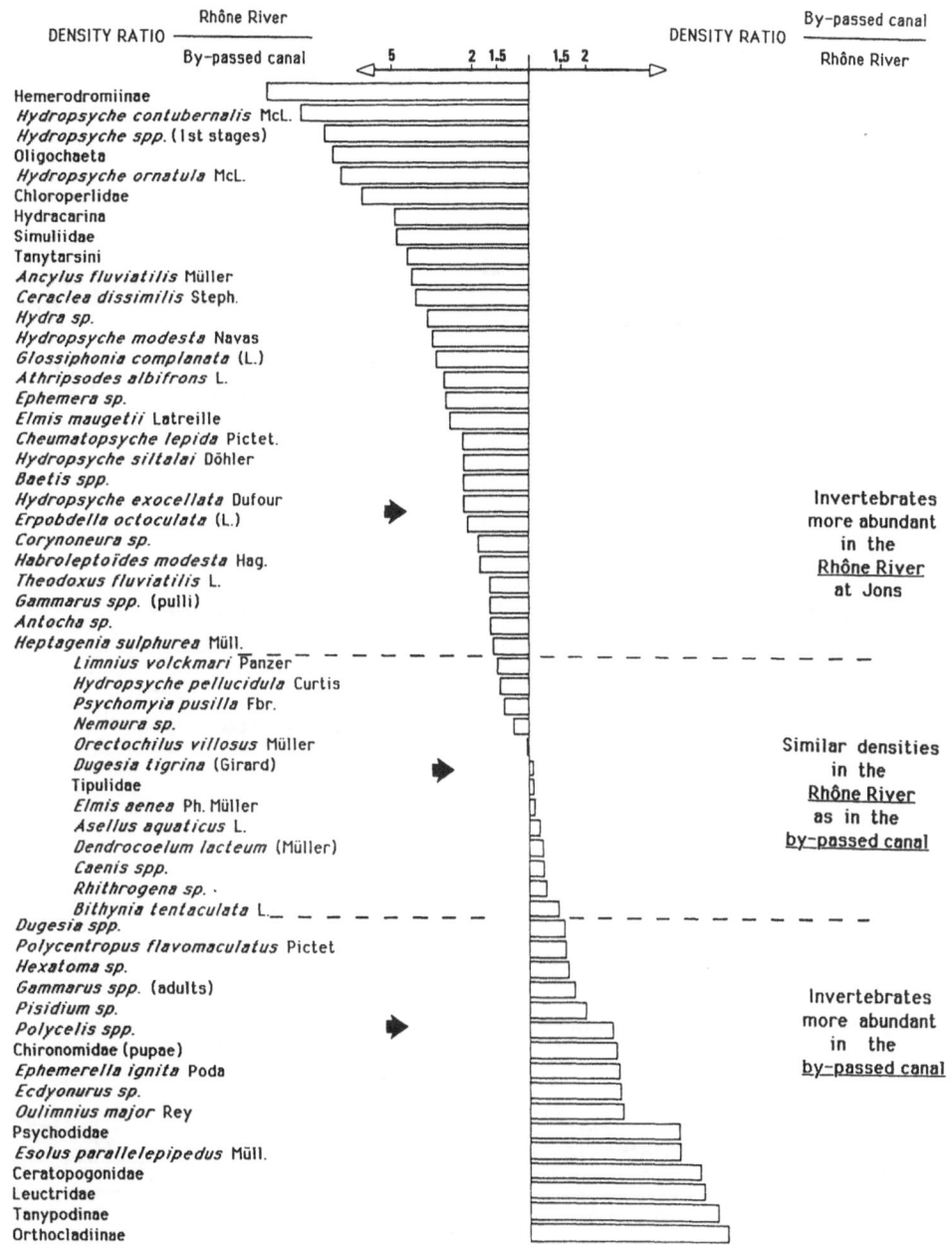

Figure 4. Classification of invertebrate taxa based on their density ratio at the two stations.

Dugesia tigrina (Girard), notably in October, and *Psychomyia pusilla* (Fbr.), notably in winter, because of flooding is more visible in the by-passed section than in the Rhône. The latter species needed three months to re-establish itself in Miribel Canal after a flood that occurred in January 1982.

The population of *Rhithrogena* sp. followed the same temporal pattern in both arms but was present in greater density in the by-passed section during March, April and September.

The temporal patterns of *Asellus aquaticus* (L.) and *Dendrocoelum lacteum* (Müller) are different at the two locations: these species are absent in the by-passed section during winter, although they are abundant after May. These invertebrates are lentic and they can only establish

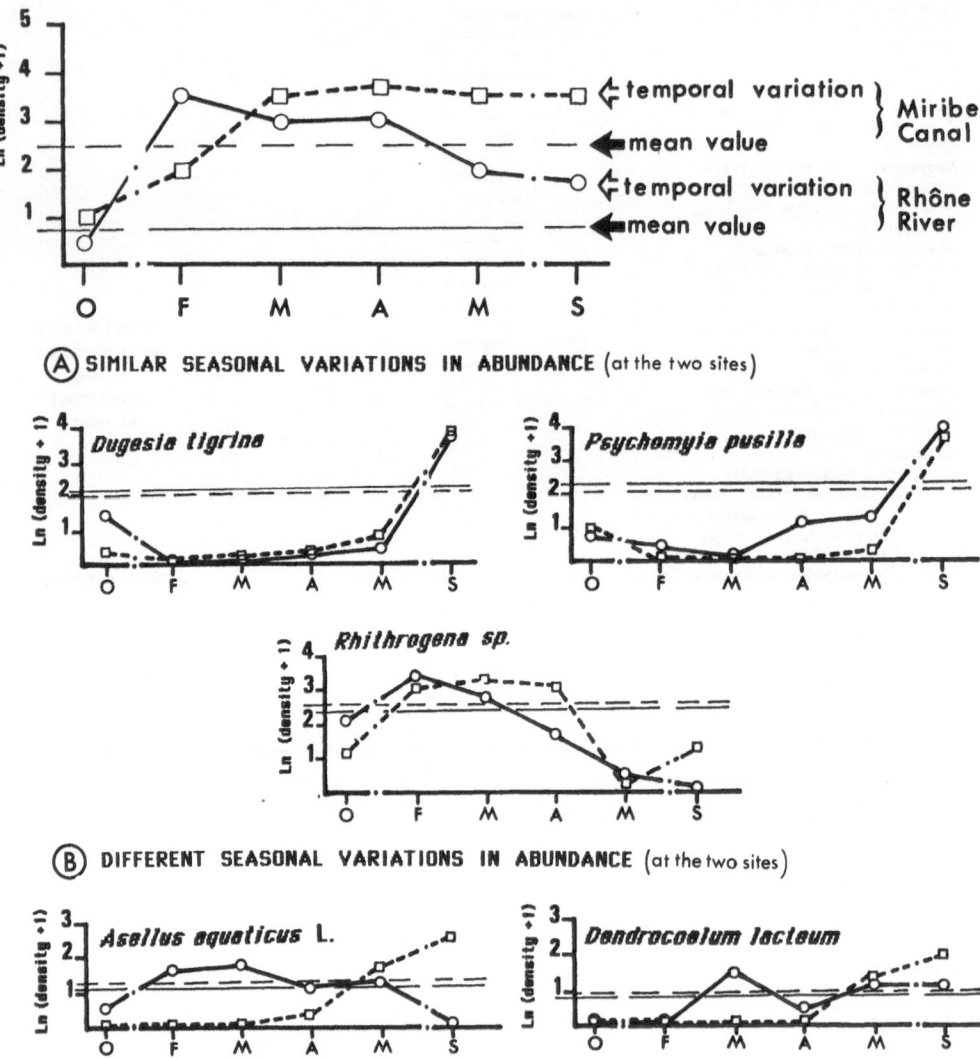

Figure 5. Temporal variation of invertebrate taxa with similar abundances at both stations and having densities greater than 1 ind. $(10 \ 1)^{-1}$ of sediment.

themselves downstream of the dam a long time after a flood. The re-establishment of this detrivorous species in Miribel Canal may also be inhibited by the reduced amount of detritus, which is washed out during floods.

The mean densities (greater than 1 ind. $(10\ 1)^{-1}$) of 18 invertebrate taxa were higher in the Rhône River than in the by-passed section. Their temporal patterns are presented in Figure 6. Tanytarsini, Hydracarina and *Athripsodes albifrons* (L.) are abundant in May and their temporal patterns are the same at the two sampling sites. The highest densities observed in the Rhône River were for *Hydra* sp. and *Caenislea dissimilis* (Steph.) in May, and for *Baetis* spp. in September. Most of these invertebrates are detrivores or filter feeders; their lower densities in Miribel Canal may be due to the lower amounts of organic matter present, in the sediment.

Many Hydropsychidae (*Hydropsyche modesta* (Navas), *H. contubernalis* (Mcl.), *H. exocellata* (Dufour) and their early instars, which can not be identified to species, had larger populations in autumn, particularly in 1982. The importance of hydraulic stability (Figure 3) can be seen here from the high densities in September 1982 compared to October 1981. The relative abundance of organic matter in the retention canal at Jons may help the establishment of Hydropsychidae in the Rhône River. *Glossiphonia complanata* (L), is also more abundant in the Rhône River. This leech is characteristic of slow to moderate fluvial systems and is expected to be less abundant from the dynamic and unstable conditions of the Miribel Canal (Jones & Peters 1977).

Chloroperlidae, *Erpobdella octoculata* (L.) and *Hydropsyche siltalai* (Dohler) are more often collected between February and May. Probably vulnerable to sharp variations in flow rate (Jones & Peters 1977), they are less abundant in the by-passed section. Oligochaeta and *Gammarus* spp. pulli (pulli: individuals of less than 5 mm length) are numerous in the Rhône River during May and are sparse in Miribel Canal during March; the causes of this peculiar distribution are not clearly known. Some Hemerodromiinae were sampled in Miribel Canal during September 1981, but they were never found again.

In conclusion, the macroinvertebrate fauna in the Rhône River at Jons is distinct from Miribel Canal because of its high densities of Trichoptera, Ephemeroptera and Plecoptera, and high abundance of these taxa in their early instars.

Ten taxa have higher mean densities in Miribel Canal than in the Rhône at Jons (Figure 7). Orthocladiinae, Chironomidae pupae, Tanypodinae, Leuctridae and *Ephemerella ignita* (Poda) are abundant in May because of abundant algal cover, whose development on the substrate of this arm is favoured by the low flow rate. The temporal pattern of *Dugesia* spp. is similar to that of *Dugesia tigrina* (Figure 5), but *Dugesia* spp. are more abundant in Miribel Canal whereas *Dugesia tigrina* is more abundant in the Rhône.

Gammarus spp. adults, *Esolus parallelepipedus* (Müll.), *Hexatoma* spp. and *Polycelis* spp. have different temporal patterns at the two sites. The difference is small for *Gammarus* spp. but more important for *Esolus parallelepipedus* (Müll.), a phreatic Coleoptera whose abundance at the sediment surface in the by-passed section is very likely related to the importance of the groundwater movements in that zone. The temporal pattern of *Polycelis* spp. is similar to that of *Asellus aquaticus* (L.) (Figure 5).

The macrofauna of Miribel Canal includes high densities of algivorous Diptera (e.g. Orthocladiinae) in May and is also different from the Rhône

at Jons in its high abundance of interstitial benthic invertebrates
(e.g. *Niphargus* sp., *Niphargopsis* sp., *Esolus parallelepipedus* Müll.).

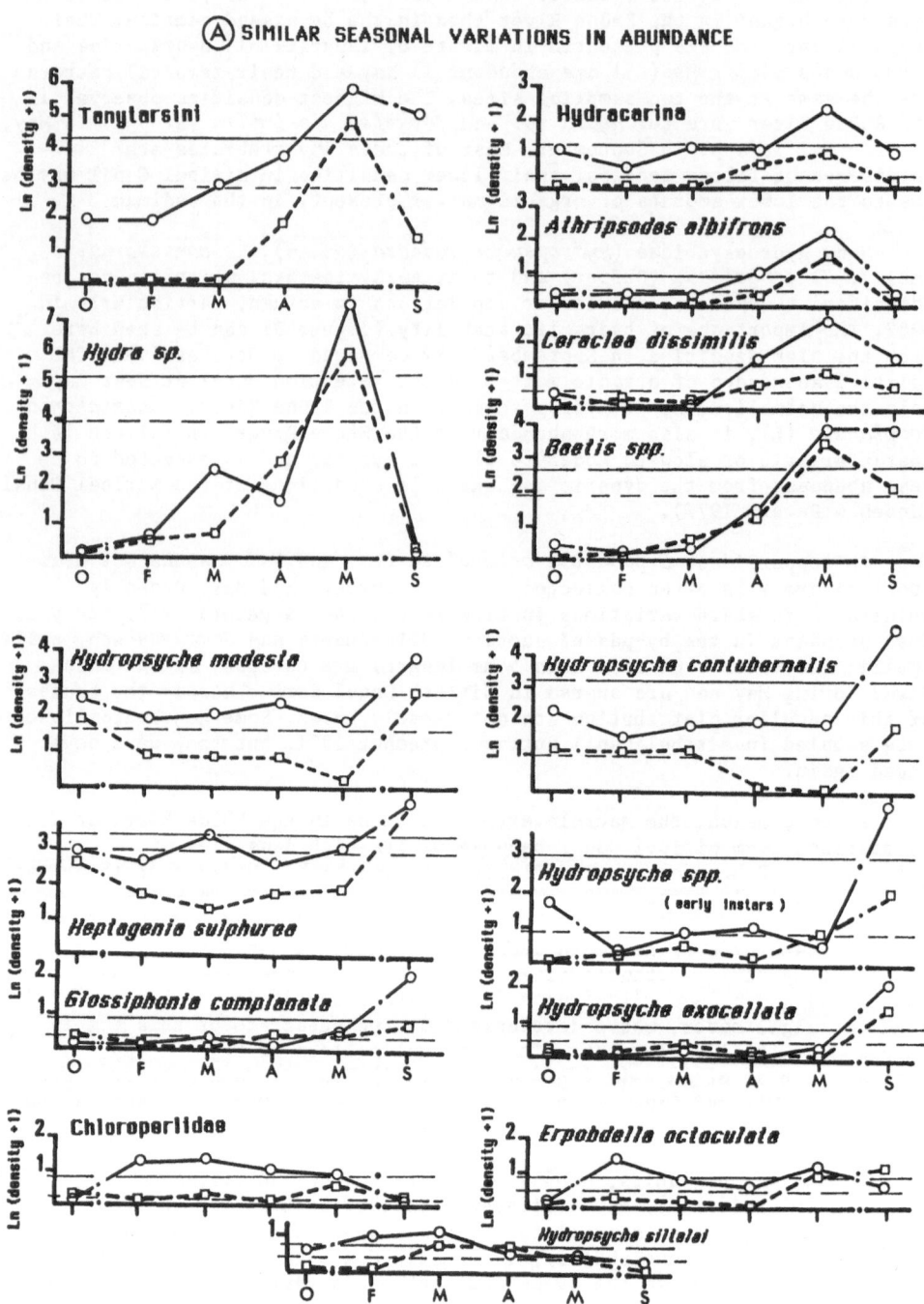

(A) SIMILAR SEASONAL VARIATIONS IN ABUNDANCE

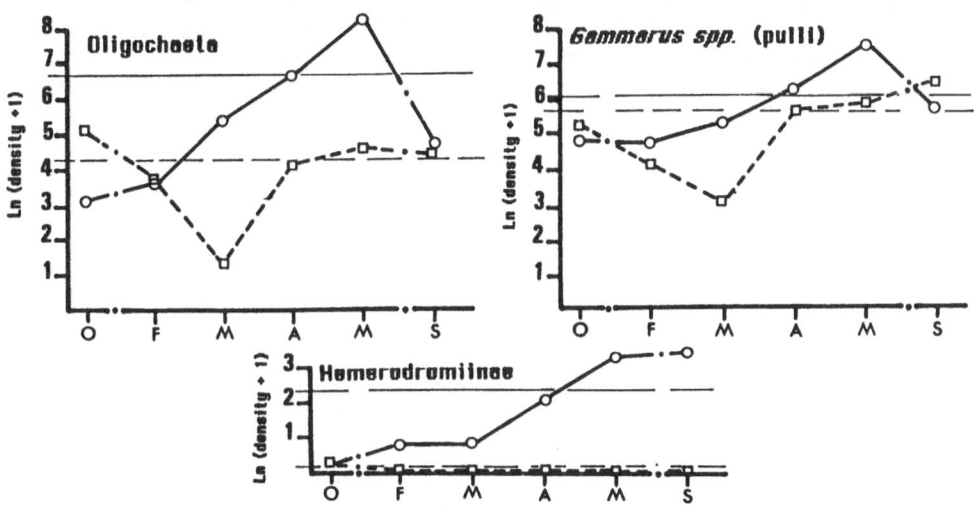

Figure 6. Temporal variation of taxa more abundant in the Rhône River than in Miribel Canal and having densities greater than 1 ind. $(10\ 1)^{-1}$ of sediment.

DISCUSSION

The presence of Jons Dam has modified the benthic habitat in the downstream section of the by-passed river through changes in hydraulic characteristics, sediments and organic matter content (Figure 8). These changes have accentuated the instability of the downstream environment. Abrupt hydraulic fluctuations in the by-passed section are responsible for the increased mobility of the substrate and increased hydraulic stress (Statzner 1981), which may cause the catastrophic drift of normally non-mobile invertebrates such as *Hydra* sp., *Sphaerium* sp., *Psychomyia pusilla* (Fbr.), *Ceraclea dissimilis* (Steph.), *Athripsodes albifrons* (L.), *Setodes punctatus* (Fbr.), etc. Such an increase is particularly drastic for filter feeders (Ward 1976) and may cause the rarefaction of Hydropsychidae and bivalve molluscs.

An increase of substrate mobility also causes difficulties for shelter-building invertebrates (*Psychomyia pusilla* (Fbr.), *Antocha* sp. and fixed molluscs (*Ancylus fluviatilis* (Muller), *Theodoxus fluviatilis* (L.), and *Dreissena polymorpha* (Pallas)). The relative abundance of such invertebrates is a useful index of substrate stability (Petran & Kothe 1978).

The differences in benthic fauna observed in these two parts of the Rhône River can be compared to the results described by Jones and Peters (1977) for streams with different hydraulic patterns: *Elmis* sp., *Baetis* sp., *Sphaerium* sp., *Glossiphonia complanata* (L.), *Erpobdella octoculata* (L.), *Helobdella stagnalis* (L.) and *Asellus* sp. are characteristic of uniform hydraulic conditions and are more numerous in the retention canal; contrastingly, *Ecdyonurus* sp., *Rhithrogena* sp. and *Leuctra* sp., invertebrates characteristic of irregular hydraulic conditions, are more abundant in the by-passed section.

The presence of the Jons Dam has induced a deepening of the downstream riverbed that facilitates, in conjunction with the flow rate fluctuations, the vertical movement of surface and ground waters. These water movements prevent fine deposits from plugging the sediment and thus facilitate exchanges between benthic and interstitial invertebrates such as *Niphargus* sp., *Niphargopsis* sp. and *Esolus parallelepipedus* (Müll.) (Gaschignard 1986). The low densities of *Gammarus* spp. observed in the top part of the sediment may indicate a deeper penetration of these young animals in the substrate at Miribel than at Jons.

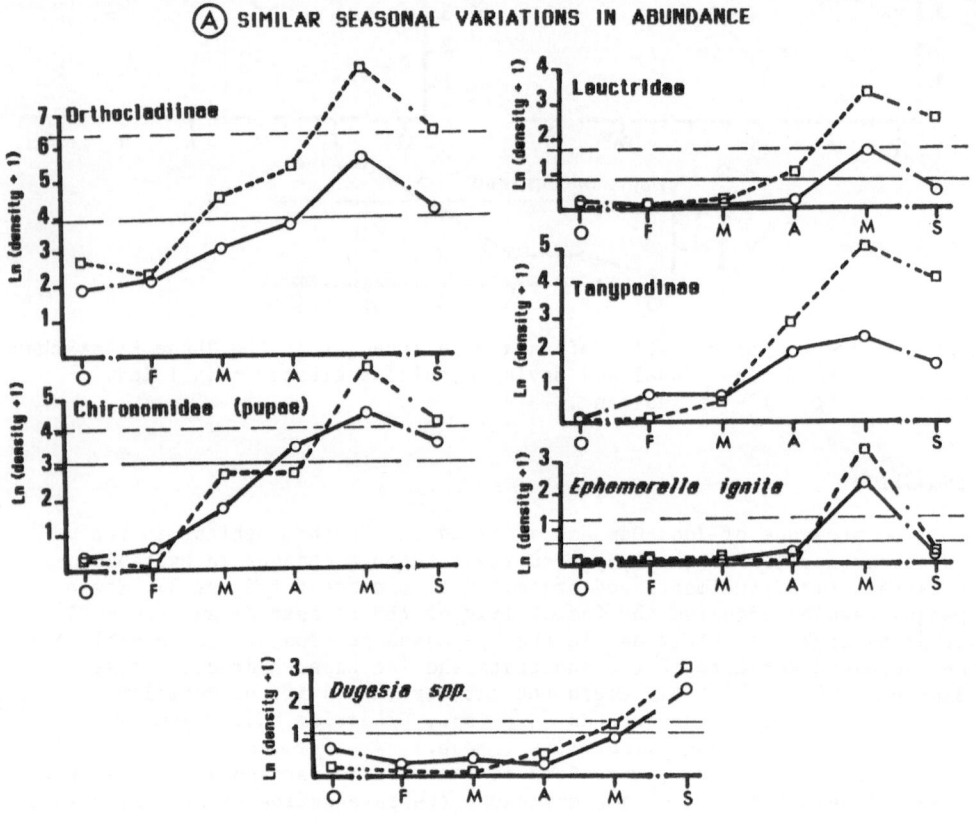

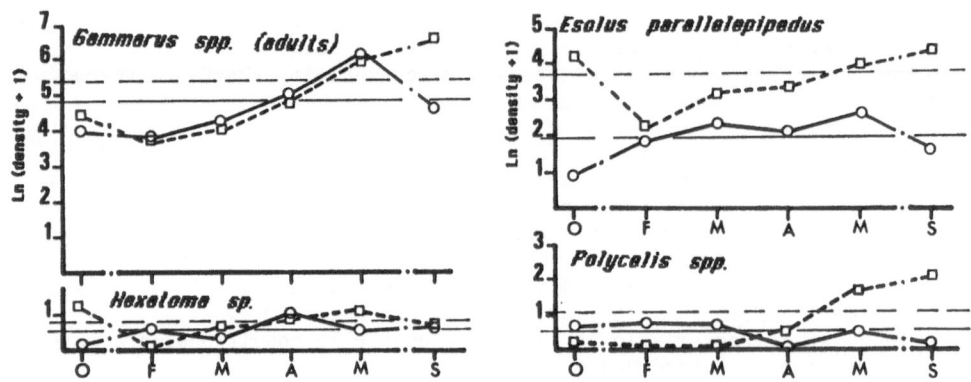

Figure 7. Temporal variations of taxa more abundant in Miribel Canal than in the Rhône River and with density greater than 1 ind $(10\ 1)^{-1}$ of sediment.

The Miribel Canal represents a much more uniform habitat than does the Rhône at Jons, which still contains small areas where protection from fast currents and deposition of fine organic and mineral substrate promotes the growth of macrophytes. Some invertebrates, sometimes collected in the Rhône at Jons (*Helobdella stagnalis* (L.), *Macronychus quadrituberculatus* (P. Mull.), Stenophylacini, *Lepidostoma hirtum* (Fbr.), *Dryops* sp., *Oecetis notata* Ramb), probably come from these lentic areas or from those dead arms still in communication with the upstream channel. The banks of Miribel Canal have no such lentic habitats and have no associated dead arms.

The rarefaction of coarse sediment, as mentioned earlier, is important because such substrate often shelters a rich invertebrate fauna (Resh 1977; Williams & Mundie 1978; Hart 1978). Greater sedimentary heterogeneity is beneficial to lithophilic invertebrates (*Baetis* spp., *Erpobdella octoculata* (L.), *Ancylus fluviatilis* (Müller) and *Theodoxus fluviatilis* (L.)) as well as to burrowing species (Hemerodromiinae, some Oligochaeta, *Sphaerium* sp., *Robackia* sp. and *Ephemera* sp.). Moreover, the structural complexity of the sediment determines the nature of the interstitial spaces and thus the importance of benthic organic matter (Otto 1981; Pringle 1982). The more complex structure of the sediment in the Rhône River, compared to the by-passed section, may be beneficial to detrivorous invertebrates such as Chironomini, Oligochaeta, *Athripsodes albifrons* (L.), *Ceraclea dissimilis* (Steph.) and *Setodes punctatus* (Fbr.).

CONCLUSIONS

The modifications of the hydrosystem due to the Jons Dam have induced an environmental situation characterized by increased instability and

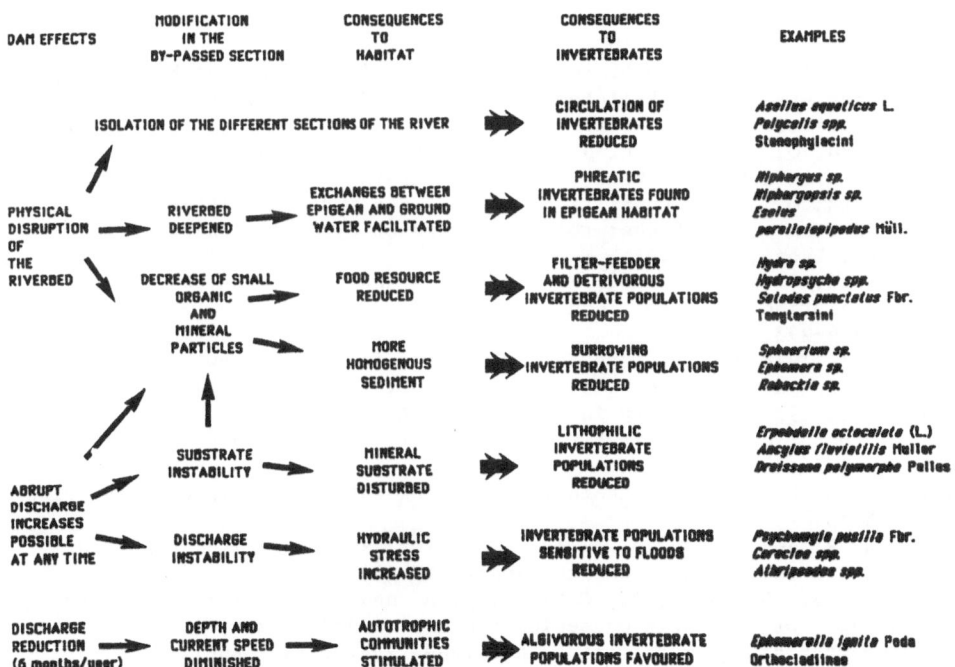

Figure 8.   Habitat modifications and consequences to benthic invertebrate populations in Miribel Canal compared to those in the Rhône River.

homogeneity in the by-passed section. The effects of increased instability on benthic macroinvertebrates have been often reported (Hynes 1970; Ward 1976; Gore 1978; Williams & Mundie 1978; Wise 1980; Barton 1980; Milner et al. 1981) and are evident in the sensitivity of Miribel Canal to large fluctuations in flow rate, such as during floods.

In Odum´s terms (1975), the Jons Dam can be considered as a supplementary "disturbance cause" to the benthic environment, which has reduced the diversity and richness, according to the "medium disturbance hypothesis" (Connell 1978; Ward & Stanford 1983), of the macroinvertebrate populations downstream of the Jons Dam.

ACKNOWLEDGEMENTS

The authors wish to thank M. Bournaud, P. Marmonier, H. Tachet and P. Usseglio-Polatera for reviewing the manuscript, G. H. Copp for editing the final draft and J. Tachet for typing it.

REFERENCES

Armitage, P. D. 1984. Environmental changes induced by stream regulation and their effects on lotic macroinvertebrate communities. In: Regulated Rivers. (Ed. by A. Lillehammer and S. J. Saltveit). pp. 139-165. Universitetforlaget AS Norway.

Barton, D. R. 1980. Benthic macroinvertebrate communities of the Athabasca River near Ft. MacKay, Alberta. Hydrobiologia, **74**: 151-160.

Boillot, G. 1964. Géologie de la Manche occidentale. Fonds rocheux, dépôts quaternaires, sédiments actuels. Annls. Inst. Oceanogr. **42**: 1-219.

Cellot, B. 1982. Cycle annuel et zonation de la dérive des macroinvertébrés du Rhône en amont de Lyon. Thèse Doct. Spéc., Lyon, 167 pp.

Connel, J. H. 1978. Diversity in tropical rain forests and coral reefs. Science, **199**: 1302-1310.

Décamps, H. 1984. Biology of regulated rivers in France. In: Regulated Rivers. (Ed. by A. Lillehammer and S. J. Saltveit). pp. 495-514. Universitetforlaget AS Norway.

Gaschignard, O. 1984. Impact d´une crue sur les macroinvertébrés benthiques d´un bras du Rhône. Verh. Internat. Verein. Limnol., **22**: 1997-2001.

Gaschignard, O. 1986. Répartition spatiale des macroinvertébrés benthiques d´un bras vif du Rhône; rôle des crues et dynamique saisonnière. Thèse Doct. Lyon, 250 pp.

Gaschignard, O. and el Hamdi, M. 1984. La granulométrie du sédiment et l´évolution du benthos dans un bras vif du Rhône. Rev. Fr. Sci. Eau, **3**: 279-293.

Gore, J. A. 1978. A technique for predicting in stream flow requirements of benthic macroinvertebrates. Freshwat. Biol., **8**: 141-151.

Hart, D. D. 1978. Diversity in stream insects: regulation by rock size and microspatial complexity. Verh. Internat. Verein. Limnol., **20**: 1376-1381.

Hynes, H. B. N. 1970. The Ecology of Running Waters. Liverpool Univ. Press, 555 pp.

Jones, H. R. and Peters, J. C. 1977. Physical and biological typing of unpolluted rivers, Technical Report **TR 41**: 1-48.

Milner, N. J., Scullion, J., Carling, P. A. and Crisp, D. T. 1981. The effects of discharge on sediment dynamics and consequent effects on invertebrates and salmonids in upland rivers. Adv. Appl. Biol., **6**: 153-220.

Odum, P. 1975. Ecologie. H. R. V. Ltée Ed. Montréal: 254 pp.

Otto, C. 1981. Food related adaptations in stream living caddisfly larvae feeding on leaves. Oikos, **37**: 117-122.

Perrin, J. F. and Roux, A. L. 1978. Structure et fonctionnement des écosystèmes du Haut-Rhône français - VI: La macrofaune benthique du fleuve. Verh. Internat. Verein. Limnol., **20**: 1994-1502.

Petran, M. and Kothe, P. 1978. Influence of bedload transport on the macro-benthos of running waters. Verh. Internat. Verein. Limnol., **20**: 1867-1872.

Pringle, S. 1982. Factors affecting the microdistribution of different sizes of the Amphipod *Gammarus pulex*. Oikos, **38**: 369-373.

Resh, V. H. 1977. Habitat and substrate influences on population and production dynamics of stream caddisfly *Ceraclea ancylus* (Leptoceridae). Freshwat. Biol., **7**: 261-277.

Statzner, B. 1981. The relation between "hydraulic stress" and microdistribution of benthic macroinvertebrates in a lowland running water system: the Schierenseebrooks (North Germany). Arch. Hydrobiol., **91**: 192-218.

Ward, J. V. 1976. Effects of flow patterns below large dams on stream benthos: a review. In: Instream Flow Needs Symposium. (Ed. by J. F. Orsborn and C. H. Allman), Amer. Fish. Soc. **2**: 235-253.

Ward, J. V. 1982. Ecological aspects of stream regulation: responses in downstream lotic reaches. Water pollution and management reviews (New Delhi), **2**: 1-26.

Ward, J. V. and Stanford, J. A. 1983. The intermediate disturbance hypothesis: an explanation for biotic diversity patterns in lotic ecosystems. Dynamics and control of system components, **17**: 347-356.

Williams, D. D. and Mundie, J. H. 1978. Substrate size selection by stream invertebrates and the influence of sand. Limnol. Oceanogr., **23**: 1030-1033.

Wise, E. J. 1980. Seasonal distribution and life histories of Ephemeroptera in a Northumbria River. Freshwat. Biol., **10**: 101-111.

Appendix 1. Mean Densities (No. (10 l)$^{-1}$ of sediment) of the 57 invertebrate taxa collected at both stations during each month (S: standard deviation).

Rhône River

Month Date	OCTOBER 30/10		FEBRUARY 18/02		MARCH 12/03		APRIL 14/04		MAY 13/05		SEPTEMBER 09/09		TOTAL MEAN DENSITY
	x̄	s	x̄	s	x̄	s	x̄	s	x̄	s	x̄	s	
Hydra spp.	0.2	0.7	0.9	1.9	12.1	8.1	4.6	4.4	2098.3	3143.3	0.2	0.8	236.73
Dendrocoelum lacteum					3.3	6.9	0.5	0.7	2.0	1.8	1.8	5.6	1.20
Dugesia tigrina	3.2	10.2					0.5	1.1	0.7	1.3	41.2	5.0	6.92
Dugesia spp.	1.2	4.3	0.4	1.0	0.6	0.6	0.5	2.0	2.0	2.1	11.5	33.7	2.37
Polycelis spp.	1.0	1.2	1.1	1.1	1.0	1.5			0.7	1.4	0.2	0.8	0.48
Oligochaeta	21.8	46.6	39.2	28.2	225.3	315.8	786.1	1282.5	5725.6	4540.8	106.4	79.9	830.25
Glossiphonia complanata	0.2	0.7			0.4	0.5			0.7	2.3	7.6	20.6	1.41
Erpobdella octoculata	0.2	0.3	3.1	3.8	1.2	0.9	0.9	1.4	2.4	3.1	1.0	2.5	1.33
Ancylus fluviatilis	0.2	0.3							0.3	0.9	0.6	0.6	0.14
Bithynis tentaculata													0.11
Theodoxus fluviatilis							0.7	0.6			0.8	1.8	0.13
Pisidium spp.											0.6	0.8	0.10
Asellus aquaticus	0.8	1.5	4.1	7.3	4.9	4.8	2.1	5.8	2.8	3.9	300.2	244.7	2.25
Gammarus spp. pulli	140.1	117.5	122.2	62.0	200.2	318.8	536.5	551.8	1820.5	2088.6	93.6	171.6	396.51
Gammarus spp. adults	50.2	51.9	43.6	29.0	67.7	39.8	146.7	134.3	432.3	257.0	1.8	2.9	107.11
Hydracarina	2.0	4.2	0.9	1.1	2.3	3.1	2.1	2.3	9.1	15.2			2.21
Heptagenia sulphurea	19.0	24.4	13.3	8.3	29.4	22.8	12.9	10.3	20.2	16.4	83.5	175.9	25.70
Rhithrogena spp.	7.5	7.7	29.0	38.6	15.8	15.9	4.6	4.9	0.7	1.7	0.2	0.4	8.73
Caenis spp.	0.4	1.9	0.9	1.5	0.6	1.9	0.2	0.6	11.1	16.5	0.8	3.7	1.67
Baetis spp.	0.6	0.9	0.2	0.4	0.4	1.0	3.5	6.1	47.1	39.1	47.1	133.0	13.65
Ecdyonurus spp.	0.2	0.9	0.4	1.1	0.2	1.0	0.2	0.4	0.3	0.9	0.4	1.3	0.25
Habroleptoides modesta	0.6	0.9	0.2	0.7									0.04
Ephemerella ignita							0.2	0.4	9.4	13.3	0.2	0.8	1.12
Ephemera sp.									0.7	1.2			0.11
Nemoura sp.					0.2	1.0	0.7	1.7			0.2	0.8	0.15
Chloroperla sp.			2.4	4.1	2.5	2.2	1.6	2.0	1.3	1.9	0.6	1.3	1.27
Leuctridae	0.2	0.4					0.2	0.6	4.4	5.1	0.8	1.6	0.62
Oreotochilus villosus	0.2	0.5					0.2	0.4			4.4	6.3	0.17
Esolus parallelepipedus	1.4	3.4	5.7	5.6	9.9	5.0	7.2	11.9	13.1	8.2	0.4	0.5	6.12
Elmis aenea			0.7	1.1	0.8	1.0	0.2	0.4	1.3	1.3	0.2	0.4	0.52
Elmis maugetii			0.2	1.0					0.3	0.9			0.11
Limnius volokmari			0.2	0.7			0.2	1.7	0.3	0.9	0.4	0.8	0.18

Appendix 1. Mean Densities (No. $(10\ l)^{-1}$ of sediment) of the 57 invertebrate taxa collected at both stations during each month (S: standard deviation) (continued).

Rhône River

Month / Date	OCTOBER 30/10 x̄	s	FEBRUARY 18/02 x̄	s	MARCH 12/03 x̄	s	APRIL 14/04 x̄	s	MAY 13/05 x̄	s	SEPTEMBER 09/09 x̄	s	TOTAL MEAN DENSITY
Oulimnius major			0.2	0.8									0.04
Polycentropus sp.					0.8	1.1					48.9	117.7	0.15
Athripsodes albifrons	0.4	1.4	0.4	1.1	0.4	1.0	1.6	6.3	8.4	10.9	3.8	4.4	1.42
Psychomiya pusilla	1.2	1.4	0.7	1.3	0.2	0.3	2.1	3.2	2.7	5.7	115.1	52.7	8.81
Ceracles dissimilis	0.6	0.9					3.2	3.1	14.5	16.3	133.6	70.6	2.76
Hydropsyche spp.	4.6	4.3	0.4	1.0	1.4	3.9	1.8	3.2	0.7	1.3	8.5	4.7	19.61
Hydropsyche contubernalis	9.7	11.1	3.9	2.6	6.2	9.2	7.2	6.2	6.7	7.8	51.3	22.8	25.64
Hydropsyche exocellata					0.2	0.3			0.3	0.8			1.46
Hydropsyche modesta	3.1	10.3	6.8	4.4	8.4	8.6	12.2	18.4	6.4	6.0	3.4	7.2	13.82
Hydropsyche ornatula	0.4	0.6	0.7	1.1			0.2	0.8	0.3	0.4	1.2	1.7	0.75
Hydropsyche pellucidula	0.4	0.9			0.2	0.3			0.3	0.8	0.4	0.8	0.27
Hydropsyche siltalai	0.8	1.8	1.7	2.6	2.3	2.2	0.7	0.9	0.7	1.1			0.97
Cheumatopsyche lepida			0.4	1.7			0.2	0.5	0.7	1.7			0.19
Psychodidae	0.2	1.9			0.2	0.7	0.2	1.0	0.3				0.11
Orthocladiinae	6.0	2.9	8.7	5.4	22.8	26.4	45.3	68.1	322.9	271.4	72.8	80.2	60.91
Tanytarsini	6.0	5.4	5.9	5.7	17.7	13.1	46.2	47.4	353.9	256.4	157.3	94.7	76.96
Tanypodinae	0.4	0.5	1.1	1.2	1.0	1.1	6.9	3.9	3.9	10.1	4.2	7.1	3.32
Chironomidae (pupae)	0.2	0.5	0.7	1.2	4.7	3.9	34.6	55.1	98.3		38.2	34.4	23.82
Hexatoma sp.			0.9	2.2	0.4	0.4	1.8	9.8	0.7	1.2	1.0	2.2	0.76
Antocha sp.	0.2	0.7	0.7	1.7			0.7	1.6	0.3	0.8			0.26
Hemerodromiinae	0.8	0.8	1.3	1.4	1.2	1.2	7.6	15.7	29.6	38.1	33.4	48.1	10.47
Simuliidae							0.2	1.7			1.4	2.4	0.27
Corynoneura sp.	0.4		0.4	0.7									0.08
Tipulidae	0.2		0.2	0.7									0.04
Ceratopogonidae	0.2	0.3											0.03
TOTAL MEAN DENSITY	309.0		326.0		688.0		1745.0		11114.0		1673.0		1976.46
TOTAL NUMBER OF SPECIES	46		44		46		44		54		60		

Appendix 1. Mean Densities (No. $(10\ 1)^{-1}$ of sediment) of the 57 invertebrate taxa collected at both stations during each month (S: standard deviation) (continued).

Miribel Canal

Month Date	OCTOBER 10/11		FEBRUARY 07/02		MARCH 01/03		APRIL 29/03		MAY 07/06		SEPTEMBER 10/10		TOTAL MEAN DENSITY
	x̄	S	x̄	S	x̄	S	x̄	S	x̄	S	x̄	S	
Hydra sp?.			0.7	1.3	1.1	1.3	17.0	12.2	452.7	1152.0	0.2	0.9	60.96
Dendrocoelum lacteum									2.6	3.0	6.2	6.5	1.42
Dugesia tigrina	0.5	1.4			0.2	0.8	0.4	0.7	1.3	1.7	43.2	63.6	7.98
Dugesia spp.	0.3	0.9	0.2	0.5			0.9	1.2	3.2	5.2	19.5	30.9	4.09
Polycelis spp.	0.3	0.9					0.6	0.7	4.5	8.4	6.9	11.6	1.94
Oligochaeta	170.1	163.2	42.7	136.0	2.8	5.1	63.4	190.6	103.2	150.4	85.1	106.7	73.64
Glossiphonia complanata	0.5	1.4					0.4	0.9	0.6	1.0	1.1	2.8	0.45
Erpobdella octoculata			0.2	0.6	0.2	0.5			1.6	4.2	2.1	3.1	0.65
Ancylus fluviatilis									0.3	0.6			0.04
Bithynia tentaculata	0.3	1.4	0.2	0.4			0.4	1.3			0.2	0.4	0.16
Theodoxus fluviatilis	1.1	1.7					0.2	0.6					0.08
Psidium spp.							0.2	0.4					0.20
Asellus aquaticus							0.4	0.4	4.5	8.4	12.0	13.7	2.75
Gammarus spp. pulli	196.3	244.4	65.8	66.0	20.8	18.6	297.2	265.6	360.7	441.6	640.0	386.6	259.72
Gammarus spp. adults	80.3	75.6	38.1	37.9	54.3	68.1	119.6	80.0	366.5	338.1	635.9	348.2	209.92
HYDRACARINA							1.1	1.6	1.9	2.5			0.49
Heptagenia sulphurea	12.5	28.2	4.3	3.1	2.6	3.1	4.7	8.7	5.8	9.7	80.2	34.7	18.87
Rhithrogena spp.	2.1	2.5	19.8	14.3	25.5	24.4	19.4	30.8	0.3	0.6	2.8	3.5	12.63
Caenis spp.	0.3	0.4	0.2	0.4	0.4	0.8	0.4	0.4	11.8	12.0	2.1	3.4	2.11
Baetis spp.			0.2	0.8	0.9	2.1	2.8	2.2	39.7	24.3	7.8	5.7	5.99
Ecdyonurus spp.	0.3	0.9			0.4	0.7	1.1	2.9	0.6	3.0	2.8	4.9	0.89
Habroleptoides modesta					0.2						0.2	0.9	0.08
Ephemerella ignita									26.5	5.6	0.2	0.9	3.40
Ephemera sp.											0.2	0.9	0.04
Nemoura sp.	0.5	0.9			0.2	0.5			0.3	0.6			0.12
Chloroperla sp.	0.3	0.8			0.2	0.5			0.6	2.5			0.16
Leuctridae							1.9	3.8	27.5	32.9	11.7	10.2	5.95
Orestochilus villosus											1.1	1.7	0.20
Esolus parallelepipedus	65.9	89.1	8.7	7.4	22.9	34.0	28.7	26.9	53.0	66.2	79.5	101.0	41.94
Elmis aenea			0.5	1.1	0.2	0.5	0.2	0.3	3.2	2.9	0.2	0.4	0.57
Elmis maugetii													0.04
Limnius volckmari							0.4	0.7			0.2	0.4	0.12

Appendix 1. Mean Densities (No. $(10\ l)^{-1}$ of sediment) of the 57 invertebrate taxa collected at both stations during each month (S: standard deviation) (concluded).

Miribel Canal

Month Date	OCTOBER 10/11 $\bar{x}$	s	FEBRUARY 07/02 $\bar{x}$	s	MARCH 01/03 $\bar{x}$	s	APRIL 29/03 $\bar{x}$	s	MAY 07/06 $\bar{x}$	s	SEPTEMBER 10/10 $\bar{x}$	s	TOTAL MEAN DENSITY
Oulimnius major											0.5	1.0	0.12
Polycentropus sp.							0.2	0.4	0.3	0.6	0.7	1.3	0.24
Athripsodes albifrons							0.4	1.0	3.5	4.5			0.53
Psychomyia pusilla	1.6	2.3							0.3	1.3	37.0	57.2	8.80
Ceraclea dissimilis			0.2	0.5	0.2	0.5	1.1	2.5	1.9	2.5	0.9	1.9	0.69
Hydropsyche spp.			0.2	0.6	0.6	1.6	0.2	1.3	1.3	2.9	6.7	6.8	1.54
Hydropsyche contubernalis	2.4	5.4	1.0	2.6	1.1	1.6	0.2	0.4			5.7	37.2	1.78
Hydropsyche exocellata					0.4	0.9					3.2	3.4	3.60
Hydropsyche modesta	6.4	14.3	1.7	3.6	1.5	2.6	1.5	2.8	0.3	0.8	17.2	19.8	4.90
Hydropsyche ornatula							0.2	0.6	0.3	0.6			0.08
Hydropsyche pellucidula			0.2	0.5			0.2	0.9	0.3	0.8	0.7	1.9	0.24
Hydropsyche siltalai					1.1	1.1	1.1	1.6	0.6	3.0			0.49
Cheumatopsyche lepida	6.4	9.3					0.4	0.9					0.08
Psychodidae	5.1	5.9	9.9	12.8									0.97
Orthocladiinae					101.1	52.1	245.1	198.9	4252.7	5364.0	676.3	1173.2	725.99
Tanytarsini							5.3	5.0	122.0	138.3	3.0	4.6	17.00
Tanypodinae					0.9	1.0	17.4	20.8	159.1	110.4	61.8	163.6	34.53
Chironomidae (pupae)	0.3	5.3			14.1	17.0	15.1	8.1	408.3	758.1	69.0	183.5	69.43
Hexatoma sp.	2.7	5.3			0.9	1.5	1.7	3.6	1.9	1.7	1.1	3.7	1.34
Antocha sp.					0.2	0.8					0.7	1.9	0.16
Hemerodromiinae	0.3	0.4											0.04
Simuliidae					0.4	0.7			0.3	0.6			0.08
Corynoneura sp.									0.3	1.3			0.04
Tipulidae									0.3	1.5			0.04
Ceratopogonidae					0.2	0.5	0.2	0.4	0.3		1.1	2.1	0.32
TOTAL MEAN DENSITY	557.0		196.0		256.0		854.0		6420.0		2531.0		1587.17
TOTAL NUMBER OF SPECIES	26.0		23.0		31.0		41.0		45.0		47.0		

SECTION III

PHYSICAL PROCESSES

THERMAL "RESETTING" OF STREAMS BY RESERVOIR RELEASES
WITH SPECIAL REFERENCE TO EFFECTS ON SALMONID FISHES

D.T. Crisp

Freshwater Biological Association, The Ferry House
Far Sawrey, Ambleside, Cumbria   LA22 OLP   U.K.

INTRODUCTION

A general review of effects of impoundments, including temperature effects, (Brooker 1981), brief reviews of the thermal regimes in natural and regulated streams (Ward & Stanford 1979; Ward 1982) and a review of thermal regimes in rivers (Smith 1972) have been published in recent years. The nature and size of thermal changes associated with impoundment are variable and we do not yet know enough to reach firm generalizations. Therefore, this paper makes a brief general survey and illustrates this by a small number of specific examples. This leads to an examination of the effects upon fish populations, with particular reference to salmonid fishes, especially the brown trout (*Salmo trutta* L.).

The complexities of the temperature regimes of natural streams are increased by impoundment. In making general comparisons between or within streams it is, therefore, necessary to summarize the results in the form of a limited number of quantifiable characteristics. This avoids the comparisons becoming too unwieldy. In the present paper some of the complexities are indicated in the text, but in the comparative tables the regimes have been summarized by use of the following characteristics:

a.   Annual mean temperature.

b.   Amplitude of monthly mean temperatures (difference between highest and lowest monthly means).

c.   Mean diel range (annual mean of daily ranges; where daily range = daily maximum - daily minimum).

d.   Highest monthly mean and its month of occurrence.

e.   Lowest monthly mean and its month of occurrence.

Where available, the positions and altitudes of the temperature recording stations and the annual mean discharges of the streams have been indicated. Different authors use a variety of methods to summarize their data and this presents difficulties in comparing published results. Therefore, most of the detailed comparisons are based on data readily available to this author and there is undue emphasis on U.K. rivers. As far as possible, this imbalance has been redressed in the text.

TEMPERATURE REGIMES IN NATURAL STREAMS

Source Temperatures

   Stream sources that are groundwater springs have very equable
temperatures (Ward & Dufford 1979); for example, 11.0 + 1.0°C (95%
C.L.) in southern England (Crisp et al. 1982 & Table 1)⁻. In surface-fed
streams, source temperatures will show larger seasonal and diel
fluctuations. Source temperatures thus reflect the nature of the source.

Changes with Distance Downstream of the Source

   In general, with increasing distance downstream of the source, water
temperature in headstreams is progressively modified towards air
temperature and the magnitude of diel and annual fluctuations increases.
The rate of equilibration is modified by aspect, vegetation cover, channel
geometry and discharge (Macan 1958; Edington 1965; Brown 1971).
Individual headstreams carry relatively small volumes of water and
short-term fluctuations are imposed on the underlying diel cycle as a
result of changes in cloud cover, rainfall, sunshine, snowmelt and local
variations in vegetation cover and aspect (Macan 1958; Kamler 1965; Gray
& Edington 1969 ; Crisp & Le Cren 1970; Smith & Lavis 1975).

   As the volume of water, hence the thermal capacity of the streams,
increases, the influence of short-term local effects generally becomes less
important and the temperature regimes of large streams and rivers often
show greater constancy than those of small headstreams. In general, water
temperature increases with distance downstream (Smith 1968; Boon & Shires
1976), presumably as a result of air temperature increasing with decreased
altitude.

   A simple summary of temperature regimes in different portions of the
upper parts of two very different river systems in the U.K. (Table 1)
demonstrates some of the above points. A typical chalk spring source shows
a remarkably equable temperature regime with very small annual and diel
fluctuations. After discharge into a small chalk stream there is a
substantial increase in diel and annual fluctuation and a small rise in
annual mean temperature. Within the mainstem chalk river, annual mean is
similar to that of the chalk stream, annual temperature amplitude is
further increased and diel fluctuations are reduced. The latter probably
reflects the increased thermal capacity of the river. Within the River
Tees, annual mean temperature shows a broad tendency to increase with
distance downstream from a small headstream through four stations on the
main river, though this trend is not entirely consistent, probably because
the data refer to several different years. The amplitude of annual
fluctuation shows no clear systematic trend but is large at all five
stations, whilst mean diel range increases between the headstream and the
upper parts of the main river. In the nearby R. North Tyne the largest
diel fluctuations were observed in the middle reaches of the river (mean
discharge c. 6 m^3 s^{-1}), the smallest in the lower reaches (mean discharge
c. 15 m^3 s^{-1}) and intermediate values in the upper reaches of the main
river (<2.5 m^3 s^{-1}) (Boon & Shires 1976). Smith (1972) argues that this
pattern is likely within most rivers with upland headwaters and limited
groundwater inputs and he cites supporting evidence from the Austrian Alps
(Eckel 1953).

Table 1. Summary of temperature regimes in different parts of two river systems. The R. Frome is a chalk river in Dorst. Its main sources are springs similar to the one included in the table. Bere Stream is a typical Frome tributary. The River Tees is a predominatly surface-fed river in the northern Pennines and Rough Sike is a small tributary of the River Tees. R. Frome data from Crisp et al. (1982); R. Tees data from Smith (1968), Crisp et al. (1975) & Crisp (1977).

River system	Station	Nat. Grid. Ref.	Altitude (m.O.D.)	Approximate mean discharge (m^3s^{-1})	Annual mean temp. (°C)	Amplitude of monthly means (°C)	Mean diel range (°C)
R. Frome, S. England	Spring	SY/852935	30.5	0.0004	10.7	0.4	0.3
	Bere Stream	SY/858923	30.0	0.85	11.6	7.4	3.4
	R. Frome	SY/868868	15.2	4.0	11.4	10.9	1.7
R. Tees, N. England	Rough Sike (headstream)	NY/756328	565.0	0.04	6.0	12.8	2.4
	R. Tees at Tees Bridge	NY/760338	533.0	0.75	6.7	13.7	4.1
	R. Tees at Cow Green	NY/813289	466.0	2.3	7.8	12.4	4.2
	R. Tees at Dent Bank	NY/922260	226.0	7.0	7.3	12.8	-
	R. Tees at Broken Scar	NY/259138	37.0	15.0	8.5	14.3	-

Prediction of Water Temperature in Streams and Rivers

Brown (1969) developed a model to predict temperatures in small streams, from energy budgets. The model is relatively complex and involves such variables as barometric pressure, wind speed, vapour pressure and radiation. It proved useful in predicting hourly temperature values in a non-forested section of stream and was subsequently used (Brown 1970) to predict the effect of clear cutting upon water temperatures. However, this type of approach requires rather complex sets of measurements and its general use may be limited by this constraint.

Smith & Lavis (1975) examined a small upland stream in northern England and found good correlations (r^2 = 0.70 to 0.92) between daily maximum and minimum water and air temperatures close to the stream at low to moderate discharges. At high discharges the fit was less satisfactory (r^2 = 0.19 to 0.61). There were also the marked effects of snowmelt and rainfall in causing short-term temperature fluctuations.

Crisp & Howson (1982) compared weekly mean water temperatures in nine streams in northern England with weekly mean air temperatures from several standard meteorological stations. The altitudes of the water temperature stations varied from 70 to 533 m.O.D., the mean annual discharges of the streams from c. 0.04 to c. 1.50 m^3 s^{-1}, and the distances between the stream stations and meteorological stations by 0.9 to 54.6 km. The regressions of water temperature upon air temperature (if weeks with sub-zero air temperatures were excluded at the high altitude stations) generally accounted for 90% of the variance of water temperature and never for less than 86%. The introduction of rainfall or discharge in multiple regressions did not improve the goodness of fit.

Perhaps the most important point arising is that, in a variety of water courses ranging from small streams to small rivers, air temperature appears to be a valuable predictor of water temperature, at least in terms of weekly means. This may also be true of daily maximum and minimum water temperatures, at least in small streams (Smith & Lavis 1975). This does not imply that short-term, local water temperature fluctuations caused by insolation, rainfall, snowmelt, cloud cover or changing discharge do not occur or have no significance, but it does indicate that such effects are not very apparent unless the data are analysed on a rather fine timescale. For many practical purposes, knowledge of these minor temporal variations may not be of great importance.

For large rivers, the ability to predict water temperatures by means of empirical, physical models may be rather simpler (Ward 1963; Collings 1969; Edinger et al. 1968; Morse 1970), though the array of variables to be measured (e.g. Morse 1970) may limit the application of some of the more complex approaches.

THERMAL REGIMES DOWNSTREAM OF IMPOUNDMENTS

General Considerations

The temperature regime of water discharged from a reservoir will be influenced by the local climate, the size, depth and exposure of the reservoir, the pattern of thermal stratification (if any) and the depth(s) and pattern of draw off. Ward (1982) summarized the main features of stream temperature regime which may be modified by impoundments with dimictic patterns of stratification (i.e. with both summer and winter stratification and mixing in autumn and spring).

166

Most reservoirs can be classified, according to their function, as being for direct supply or irrigation, river regulation, hydro-electric power generation or transfer; though some have multiple functions. Direct supply reservoirs provide water for piped domestic or industrial consumption. A large proportion of their water may be drawn off for supply and a residual flow returns directly to the river. The reduced flow in the river downstream reduces thermal capacity and leads to rapid equilibration of the temperature of the released water to that of the air. River regulating reservoirs release all of their water to the river but they modify the temporal pattern of discharge (usually at a seasonal level) so as to maintain the river at suitable levels for downstream abstraction or some other purpose (e.g. flood control). Hydro-electric schemes, in common with regulating reservoirs, modify the temporal pattern of discharge. There can be large, short-term (days or hours) fluctuations in response to changing power demands. During periods of low discharge (no power generation) the lowered thermal capacity of the river can result in rapid equilibration of water temperatures with air temperatures (Pfitzer 1967).

Inter-river transfer works may operate in conjunction with schemes for impoundment and regulation. In addition to possible effects upon temperatures in the recipient river as a result of transfer of water of differing temperature regime (arising from reservoir storage and/or passage through transfer channels or pipelines), the reduced flow in the donor river and increased flow in the recipient river will modify the thermal capacities of both.

All of these reservoir types discharge water whose temperature regime is modified, relative to that of the natural river, as a result of storage.

Temperature of the Released Water

Ward (1982) listed the nature of the changes in stream temperature regime associated with shallow- and deep-release reservoirs. It is not the intention here to reiterate Ward's summary, but rather to examine the nature and magnitude of the observed changes at four reservoirs from which adequate data are available in published form or were readily accessible to the author by other means.

The most obvious changes are (Table 2):

1.   Annual Mean Temperature
      A number of authors (Jaske & Goebel 1967; Lavis & Smith 1972; Ward 1976) have suggested that impoundment has relatively little effect on the annual mean water temperature. The data in Table 2 show a consistent pattern of reduction in annual mean following impoundment, though the change is generally small and might be fortuitous.

2.   Annual Amplitude of Monthly Means
      All four examples show a reduction in annual amplitude as a consequence of impoundment. The smallest reduction (0.3°C in the period considered but as large as 1.0°C in some individual years) was seen at Cow Green, a relatively shallow and exposed reservoir where stratification is of rare and brief occurrence. In the other three reservoirs, all of which are known or assumed to stratify annually, reductions of 3.2 to 4.6°C in amplitude were observed and the size of the reduction appears more likely to be related to reservoir depth than to area or capacity.

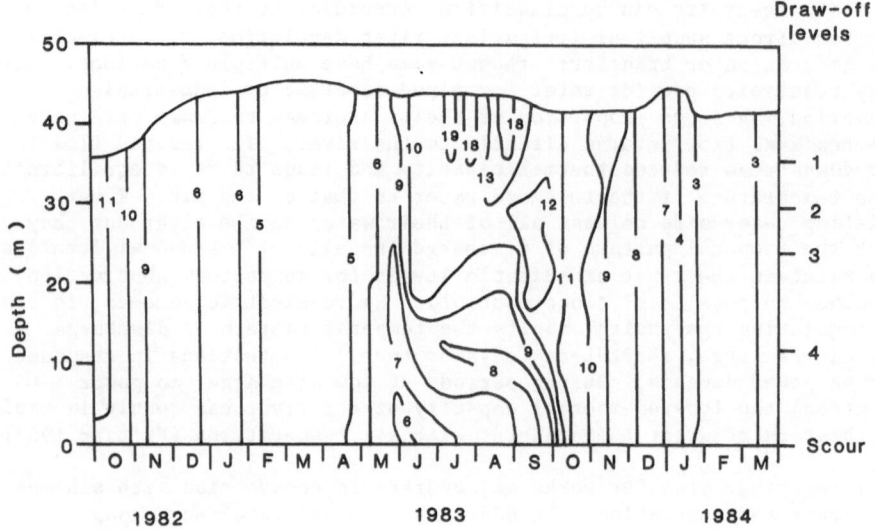

Figure 1. Depth-time diagram of water temperatures (°C) taken at Kielder
valve tower. The five draw-off levels are indicated.

3.   Mean Diel Range

Data are available for three reservoirs and at all of them the
annual mean of diel fluctuations was substantially reduced. At
Kielder the annual mean of diel ranges was reduced from 2.6 to
1.1°C. However, the latter value conceals the fact that during the
period of stratification the mean was 2.1°C and during the rest of the
year it was only 0.8°C. This may partly reflect the fact that large
changes in the temperature of discharged water at Kielder can occur
rapidly during the period of stratification as a result of switches
from one draw-off level to another (Figure 1).

4.   Summer Peak Temperature

In all of the examples, the highest monthly mean in the natural
river occurred in July or August, usually the former. Below the
three reservoirs which stratify, the highest monthly mean was delayed
until September and was depressed by 3.2 to 4.6°C. At Cow Green the
summer temperature peak was delayed by only one month and was
depressed by 1.3°C. The data, therefore suggest that two separate
effects are at work. First, a simple storage effect (as at Cow Green)
which results in a delay of the annual temperature cycle by some weeks
and in a small depression of the summer peak temperature. Second, and
added to the storage effect in those reservoirs which stratify, is a
delay of the summer peak until the stratification breaks down in the
autumn and a reduction of the summer peak, often by several degrees.
This is not an inherent feature of the reservoir but a reflection of
the choice of draw-off depths.

5.   Winter Low Temperature

The lowest monthly mean in the natural rivers occurred in January
or February. In the regulated rivers a delay of about one month
occurred in those fed by the three deepest reservoirs. In contrast,
at the shallowest reservoir there was no evidence of a major shift in
the timing of the winter low.

Table 2. Summary of temperature regimes (a) In natural rivers. (b) Immediately downstream of impoundments. Sources of data are: 1. Penaz et al. (1968), 2. Crisp (1977), 3. Brooker (1981), Edwards (1984) and unpublished, 4. Unpublished data (Northumbrian Water Authority and Freshwater Biological Association). * = larger values (mean = 2.1°C) during period of stratification, lower values (mean = 0.8°C) during rest of year.

Reservoir(s)	Area (ha)	Capacity (m³x10⁶)	Max depth (m)	Stratifies?	Annual mean (°C)	Amplitude of monthly means (°C)	Mean diel range (°C)	Summer peak	Winter low
1. Vir R. Valley dam (Czechoslovakia)	212	53	c.65	?	a 7.0 b 6.4	14.6 8.3	– –	14.9 July 10.3 Sept	0.1 Jan 2.0 Feb
2. Cow Green (U.K.)	312	41	23	Rarely and briefly	a 7.8 b 6.8	12.4 12.1	4.2 0.6	15.0 July 13.7 August	2.6 Feb 1.6 Feb
3. Elan Valley (U.K.)	340	51	40	June–Sept (11°C range)	a 8.9 b 7.4	12.2 9.4	5.6 1.8	15.4 July 12.2 Sept	3.2 Jan 2.8 Feb
4. Kielder Water (U.K.)	1087	201	52	June–Sept (12°C range)	a 8.2 b 7.8	15.4 9.3	2.6 *1.1	16.3 July/Aug 12.6 Sept	0.9 Feb 3.3 March

This sample of four European reservoirs cannot give as adequate a picture as would an analysis of data from a much wider range of reservoirs, including some much larger ones from major continental river systems such as the Mississippi or the Nile. Not only are there reservoirs of much larger size but also some where impoundment and hypolimnial releases can cause much greater changes in temperature regime, e.g. on the Flathead River, Montana, where an annual amplitude of c. 18°C has been reduced to almost zero (Ward & Stanford 1979).

Temperature Change with Distance Downstream

As the water released from impoundments proceeds downstream its temperature will approach equilibrium with ambient air temperatures. The rate of equilibration will be related to the difference between the water and air temperatures, to the discharge of the released water and to the form of the river channel. Discontinuities in the pattern of equilibration will occur with the entry of unregulated tributaries and their magnitude will reflect differences in temperature and discharge between the regulated river and the unregulated tributaries.

There is relatively little published information on the rate of equilibration of temperatures downstream of impoundments. A summary of selected examples, mainly from N. America, (Table 3) shows that the temperature effects of very large impoundments, especially where releases are from the hypolimnion, can persist for substantial distances downstream. However, Edwards and Crisp (1982) and Edwards (1984) suggested that for two British impoundments of more modest size the downstream temperature effects were likely to be relatively local (a few km) and the results of Penaz et al. (1968) (Table 3) for a European reservoir of similar size (see Table 2) support this suggestion.

IMPACTS OF MODIFIED THERMAL REGIME UPON FISH

General Comments

Both the physiology and the behaviour of fish are affected by temperature and the physiological relationships with temperature are rarely linear. Quite small changes in the timing and pattern of water temperature fluctuations can have disproportionate effects on the vital processes of fish.

There are many accounts of the effect of changed thermal regime in modifying the abundance and/or distribution of fish species, (Penaz et al. 1968; Spence & Hynes 1971; Holden & Stalnaker 1975; Edwards 1978). The general pattern is for deep releases of cold reservoir water to cause a decrease in numbers of warm-water species, though in some instances this loss is partially compensated by an improvement in conditions for cold water groups such as salmonids (Parsons 1955; Pfitzer 1967). However, there is much less information about the detail of the mechanisms whereby the fish populations are affected. This largely reflects our lack of detailed knowledge of the temperature relationships of individual species. The group for which most is known is the salmonids and these will be considered in some detail below.

Temperature Relationships of Salmonid Fishes

Salmonid eggs can tolerate a temperature range of about 0 to 16°C (Peterson et al. 1977), though some variation between species and possibly also some acclimation effects might be expected. In some species, at

Table 3. Brief summary of some published information on the downstream distances at which the temperature effects of reservoirs have been detected. Capacities are shown, when available in the literature.

Reservoir	Source of Information	Capacity ($m^3 \times 10^6$)	Downstream effects on water temperature
Fort Randall Resr. (S. Dakota, U.S.A.)	Neel (1963)	7524	Effect detectable at least 130 km downstream.
Cannonsville Resr. (R. Delaware, U.S.A.)	Williams (1968)	-	Temperature reductions of up to 14 C at 13 km and up to 4°C at 70 km.
Pepacton Resr. (R. Delaware, U.S.A.)	Williams (1968)	-	Temperature reductions of up to 11 C at 50 km and up to 2.8°C at 95 km.
Vir River Valley Resr. (Czechoslovakia)	Penaz et al. (1968)	53	Effects substantial 7 km downstream, barely detectable at 30 km.
Diefenbaker Lake (Saskatchewan, Canada)	Lehmukuhl (1972)	-	Effects clearly detectable to about 30 km downstream, little effect at 112 km.
Glen Canyon Dam (Colorado River, U.S.A.)	Holden & Stalnaker (1975)	-	Water too cold for most native fish species for over 400 km.
Cheesman Lake (Colorado, U.S.A.)	Ward (1976)	98	Marked effects below release point. No discernible effect 32 km downstream.
Lake Hume (Murray R., Australia)	Walker et al. (1979)	3070	Effects still apparent up to 200 km downstream.
Morris Sheppard Dam (R. Brazos, U.S.A.)	Zimmerman & Richmond (1981)	-	At baseflow equilibration is rapid. During releases temperature modified by 4-8°C within 30 km and <2°C beyond 50 km.

171

least, eggs may survive at 0°C or lower (McNeil 1966) but freezing of the intragravel water may be a significant cause of egg mortality for pink salmon (*Oncorhynchus gorbuscha* (Walbaum)) (McNeil 1966) and for Atlantic salmon (*Salmo salar* L.) and sea trout (*Salmo trutta* L.) (Somme 1960). Mortality in Atlantic salmon eggs immediately after fertilization is c. 20% at tempertaures below 4°C and less than 5% at temperatures above 4°C (Peterson et al. 1977) and there is also an increase in mortality towards the upper end of the range of tolerated temperatures (Gunnes 1979).

The rate of development of the eggs and alevins of salmonids is primarily influenced by temperature, though other factors such as oxygen supply and light have been shown to have some effect. The relationship between temperature and hatching time is curvilinear and is poorly approximated by a linear (degree-day) relationship, except over very limited temperature ranges. Several curvilinear models have been used (e.g.s. Embody 1934; Alderdice & Velsen 1978; Crisp 1981) and these usually, but not always, give reasonable accurate and useful predictions. Peterson et al. (1977) noted that sudden decreases in temperature at the eyed stage and at hatching caused slower development and increased mortality, and Kazakov (1971) noted similar effects of temperature reductions prior to blastopore closure. Gray (1928) suggested that embryonic development has a much lower temperature threshold than hatching and this has implications where reservoir releases delay the warming up of streams at times when hatching is imminent.

As spawning time amongst salmonids is largely determined by day length (Bye 1984), it is unlikely, at least in the short-term, that thermal "resetting" will have a marked effect on the date of oviposition for any given species at any given site. The main effect of temperature on the reproductive process will, therefore, be through its influence upon embryonic development. It is worth noting, however, that fish of other groups do not, necessarily, respond in the same manner as salmonids. Sexual development and spawning in cyprinids is influenced by both temperature and photoperiod, and temperature is predominant (Bye 1984). Circumstantial evidence suggests that the spawning period of the sculpin (*Cottus gobio* L.) downstream of Cow Green reservoir may have been delayed as a result of delay in the rise in water temperature in spring (Crisp et al. 1983).

Information on the interrelationships of temperature, assimilation rate, and growth is available for rainbow trout (*S. gairdneri* Richardson) (Wurtsbough & Davis 1977); young sockeye salmon (*O. nerka* (Walbaum)) (Brett et al. 1969; Brett & Shelbourne 1975) and brown trout (Elliott 1972; 1975a; 1975b; 1975c; 1975d). Edwards et al. (1979) used temperature records from 25 streams and rivers, together with the equations of Elliott (1975c), to show that the observed mean growth rates of brown trout were between 60 and 90% of those predicted on maximum rations. These equations therefore are useful in predicting growth in the field under different temperature regimes. The results also suggest that natural food supply may be a rather less important constraint on growth than has often been supposed. This approach has also been used in predicting the effects upon trout growth of the temperature changes caused by impoundment (Crisp 1977; Edwards 1984).

There are, therefore, models available to predict the hatching time of brown trout and Atlantic salmon eggs and the growth of brown trout, from water temperature. As they are mathematical simplifications of complex natural processes, it is unlikely that such models will work perfectly, especially as there are few data for low (0-3°C) temperatures. Nevertheless, they can provide a useful basis for prediction of some of the effects of changed temperature regime.

172

Temperature may be an important trigger to local and extensive movements and migrations of juveniles or resident adults and it may influence other forms of behaviour (McCleave 1978; Gibson 1973). Two important "trigger" temperatures for Atlantic salmon are indicated in the literature. At temperatures below 7°C the parr do not feed well (Allen 1940; Saunders & Henderson 1969) and show little growth (Siginevich 1967; Symons 1979), whilst seaward movement of smolts is reduced or inhibited below a certain temperature (Solomon 1978). This latter temperature varies according to the previous temperatures experienced by the fish, but appears to be around 11°C. The dates on which temperatures of 7°C and 11°C are first attained can thus be taken as approximate guides to the times when young salmon first begin to feed and when smolts begin to move downstream.

These various relationships are used below to explore the possible effects of three U.K. upland reservoirs.

Some Predicted Impacts of Thermal Resetting in the Tailwaters of Three United Kingdom Upland Impoundments Upon Salmonid Fishes

1. Data Used.
    Three reservoirs are considered. Cow Green regulating reservoir is of moderate size, in an exposed position and rarely stratifies (Table 2). Information on the growth of trout downstream of the reservoir is available (Crisp et al. 1983). Kielder Water is, by British standards, large. It stratifies and some hypolimnetic draw-off occurs (Table 2, Figure 1). Water temperature data are available but the monthly mean temperatures for the natural river are somewhat approximate. Temperature data from the Elan Valley reservoirs and from a comparable, but unregulated, station in the R. Wye have been published (see Table 2). The present account refers to an independent analysis of the raw temperature data.

    Monthly mean temperatures have been used in the calculations of growth rate and maintenance ration. At temperatures below 3.8°C both growth and maintenance ration have been assumed to be negligible.

2. Growth and Maintenance Ration of Brown Trout
    The predicted annual maintenance rations and the mean instantaneous growth rates of a "standard trout" (18.6 g on 1 January) are shown in Table 4. At all three sites the regulated temperature regime reduced the required annual maintenance ration by 16 to 22% and gave a more even distribution of the requirement between months (Figure 2).

    The prediction of little or no change in mean instantaneous growth rate at Cow Green is substantiated by observed growth rates (Crisp et al. 1983) and reflects the fact that the timing of the annual temperature cycle has been modified but that the summer peak temperature has been depressed by only a small amount as a consequence of impoundment. In contrast, at the other two sites, in addition to delay of the annual temperature cycle, the annual temperature peak has been depressed by several degrees and brought closer to the optimum value (c. 13 C) for trout growth. At Kielder this leads to an 18% increase in predicted mean growth rate, whereas in the R. Elan it leads to a reduction of c. 12%. This reduction appears to reflect mainly a depression of the temperature of the released water in May and June (Figure 2) to values well below the optimum for trout growth. The depressed temperatures of May and June are the main contributors to the appreciable difference (Table 4) between annual mean

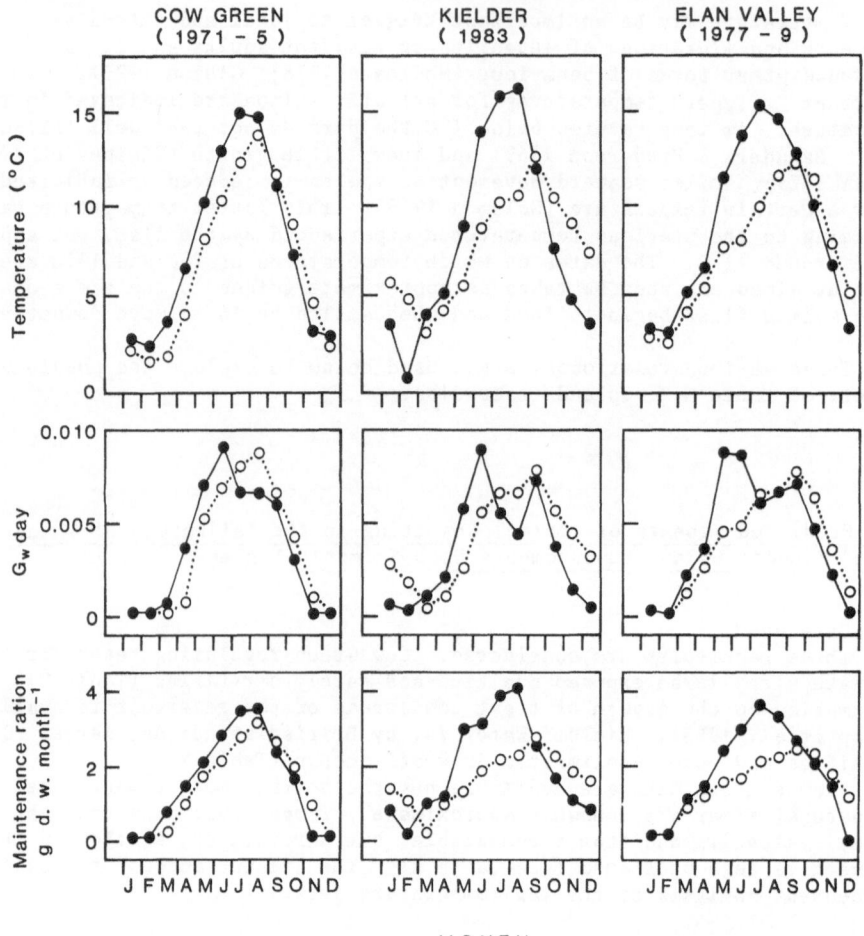

Figure 2. Monthly mean water temperatures for natural tributaries
(●————————●) and regulated river (○----------○) at Cow Green,
Kielder and the Elan Valley (see Table 2), together with
corresponding values of instantaneous growth rate (G_w) on maximum
ration for a brown trout with a weight of 18.6 g (length c. 11.75
cm) on 1 January and the calculated maintenance ration for a
trout of constant weight (18.6 g).

temperatures in the regulated and natural rivers. The prediction of a
reduced rate of growth in the regulated R. Elan supports the conclusion
from a similar but independent assessment (Edwards 1984). At all three
sites the regulated temperature regime delayed and smoothed the annual
growth cycle. However, the efect on mean annual growth rate varied
considerably between sites, dependent upon the fine detail of the
temperature pattern, and could not be predicted easily by direct inspection
of the temperature data.

3.   Hatching Time of Salmonid Eggs.
     Equations from Crisp (1981) have been used, in conjunction with
observed monthly mean water temperatures, to predict the date of 50% hatch
for brown trout and Atlantic salmon eggs under natural and regulated
conditions (Table 5) for two of the sites.

Table 4. Annual mean water temperatures for the regulated river and a comparable natural tributary at Cow Green reservoir, the Elan Valley reservoirs and at Keilder Water, together with mean instantaneous growth rates for brown trout which had a weight of 18.6 g on January (calculated from Elliott 1975 c) and annual maintenance ration for a fish with constant weight of 18.6 g. (calculated from Elliott 1975d). The monthly mean values are shown in Figure 2. Details of the impoundments are given in Table 2.

	Mean temperature (°C)	Mean instantaneous growth rate (G_W) day^{-1} on maximum ration	Annual maintenance ration (g dry weight)
River Tees at Cow Green reservoir.			
Natural	7.8	0.0035	18.0
Regulated	6.8	0.0035	15.5
R. North Tyne at Kielder Water			
Natural	8.2	0.0033	22.8
Regulated	7.8	0.0039	18.8
R. Elan Elan Valley Reservoirs			
Natural	8.9	0.0041	21.2
Regulated	7.4	0.0036	17.3

Table 5. Predicted dates of 50% hatch under natural and regulated temperature regimes for brown trout and Altantic salmon eggs, assuming fertilization on 1 November or 15 November.

Site	Species	Temperature regime	Date of fertilization	Predicted date of 50% hatch	Days elapsed	Difference (days)
Cow Green	trout	Natural	1 November	9 March	129 }	
		Regulated	1 November	12 March	132 }	+ 3
	trout	Natural	15 November	13 March	118 }	
		Regulated	15 November	14 January	60 }	-58
Kielder	salmon	Natural	15 November	23 March	128 }	
		Regulated	15 November	22 January	68 }	-60

Table 6. The first date on which daily mean temperature exceeded 7°C and 11°C in the regulated R. North Tyne immediately downstream of Kielder dam and in a natural tributary during 1983.

	Natural temperature regime	Regulated temperature regime	Difference (days)
Date on which mean first exceeded 7 C	16 April	27 May	+41
Date on which mean first exceeded 11 C	16 May	28 June	+43

In the region around Cow Green, most brown trout spawn in late October and early November and a fertilization date of 1 November is assumed. The predictions suggest that "thermal resetting" would have little effect on the hatching date of trout eggs incubated in the tailwaters. At Kielder 15 November is probably a more realistic oviposition date and the date of 50% hatch would then be advanced by c.60 days for both trout and salmon as a consequence of impoundment. There are no very satisfactory models to predict from temperature the time required from 50% hatch to emergence of the fry from the gravel ("swim-up") but the few data available suggest that, for most salmonids, at any constant temperature, the time required from fertilization to 50% swim-up is between 1.5 and 2 times the period taken from fertilization to 50% hatch.

4.   Dates When Salmon Parr Begin to Feed and When Smolts Begin to Move Downstream.

As a consequence of impoundment the dates in 1983 on which temperatures of 7°C and 11°C were attained in the North Tyne below Kielder dam were both delayed by about 40 days, relative to an unregulated tributary (Table 6).

In the unregulated river the temperature conditions for the parr to feed (c. 7°C) would occur soon after the eggs hatched (c.f. Tables 5 & 6) and, as the time of swim-up would be even later, the parr would be able to feed as soon as they emerged from the gravel. Under regulated temperatures, hatching would be advanced by c. 60 days and the date on which feeding could begin would be delayed by c. 40 days. For the fish to be able to feed as soon as they emerged from the gravel, development from 50% hatch to 50% swim-up would need to take c. 100 days. Otherwise, there would be a period during which the parr were unable to feed because the temperature would be too low and, if this period were at all lengthy, it could cause mortality. Tentative estimates suggest that swim-up would be between mid-March and mid-May and that there would be a period, perhaps of several weeks, between emergence from the gravel and the commencement of feeding.

The effect of the 40 day delay in the date at which salmon smolts begin to move seawards, could have important consequences. Departure in late June would result in their arrival in the estuary at a time when the discharge in the lower reaches of the river was low and the temperature and pollution load were relatively high and would increase the hazards to which the smolts were exposed during the transition from river to sea.

DISCUSSION & CONCLUSIONS

Even for salmonids, we still do not know enough about temperature relations, the effects of acclimation and possible differences between stocks to model and predict the effects of changes in temperature regime with certainty and accuracy. Nevertheless, the predictions do allow us to identify potential problems. Some of the predicted effects are of sufficient magnitude and of sufficient ecological importance to merit attention in fishery management. They also emphasize the importance of the choice of draw-off depth and indicate the need, for reservoirs with multiple draw-off depths, to develop operating regimes which meet the essential requirements of engineers whilst minimizing the harmful and/ or maximizing the beneficial effects upon fisheries. During the period of stratification it would be possible to balance the output from epilimnic and hypolimnic draw-offs so as to attain any desired output temperature which fell between the temperatures of the epilimnion and the hypolimnion. In this way, the output temperatures could be brought close to those of natural tributaries. However, during the winter and early spring when reservoir water temperatures are relatively uniform throughout the depth range, this type of manipulation would not be possible and the effects of resetting the annual temperature cycle could not be entirely avoided. As we have already seen, temperatures during this particular period can have a large influence upon young stages of salmonid fishes. It would be possible, theoretically, to raise the temperature of winter releases by conserving heat in the reservoir by taking summer draw-off entirely from the hypolimnion. However, the amount of benefit obtained would need careful quantification for each individual reservoir and would have to be balanced against any harmful effects arising from low summer release temperatures.

The predictions have been concerned with multiple draw-off reservoirs within the United Kingdom and have dealt with impacts in the area of maximum effect (i.e. immediately downstream of the release point). The magnitude of the impact is likely to reduce with distance downstream and may not be a serious problem at these reservoirs, where the reduction is likely to be fairly rapid. Evidence from the literature suggests that where much larger reservoirs are concerned, especially those with a fixed, hypolimnetic draw-off point, the downstream temperature effects (hence their impact upon fish) may be greater and very much more extensive.

Future research needs are:

1. More and better information on the rate at which the temperature of reservoir release water approaches equilibrium with air temperature, relative to distance downstream and river discharge.

2. More detailed and accurate information on the temperature relationships of the important fish species.

3. Critical field evaluation of the predicted effects of temperature regime upon the vital processes of fish species.

The design and operation of reservoirs require:

1. The need for river regulation systems to be designed with their effects on river ecology and fisheries in mind.

2. The operating strategies, particularly the choice of draw-off depths, should take account of the temperature requirements of downstream fish populations.

178

ACKNOWLEDGEMENTS

Collection of temperature data from the R. North Tyne was supported financially by the Department of the Environment and "in kind" by the Northumbrian Water Authority. The Northumbrian Water Authority also gave facilities to study temperature effects at Cow Green. Data on the Elan Valley reservoirs were provided by Dr. M. P. Brooker (Welsh Water Authority) and Prof. R. W. Edwards (University of Wales Institute of Technology). Access to the various study sites was granted by a number of landowners. Most of the fieldwork and data processing was carried out by Mrs. S. Robson, Miss S. J. Bidmead, Mrs. D. C. Crisp and Mr. P. R. Cubby. The typescript was made by Mrs. D. Jones and the figures were drawn by Mrs. S. Robson. Helpful comments on the draft were made by Mr. E.D. Le Cren. The author is deeply indebted for all this help, and for other help which is not specifically mentioned here.

REFERENCES

Alderdice, D. F. and Velsen, F. P. J. 1978. Relation between temperature and incubation time for eggs of Chinook salmon (*Oncorhynchus tshawytscha*). J. Fish. Res. Board Can. **35**: 69-75.

Allen, K. R. 1940. Studies on the biology of the early stages of the salmon (*Salmo salar*). I. Growth in the River Eden. J. Anim. Ecol. **9**: 1-23.

Boon, P. J. and Shiers, S. W. 1976. Temperature studies on a river system in north-east England. Freshwat. Biol. **6**: 23-32.

Brett, J. R., Shelbourne, J. E. and Shoop, C. T. 1969. Growth rate and body composition of fingerling sockeye salmon, *Oncorhynchus nerka*, in relation to temperature and ration size. J. Fish. Res. Board Can. **26**: 2363-2394.

Brett, J. R. and Shelbourne, J. E. 1975. Growth rate of young sockeye salmon, *Oncorhynchus nerka*, in relation to fish size and ration level. J. Fish. Res. Board Can. **32**: 2103-2110.

Brooker, M. P. 1981. The impact of impoundments on the downstream fisheries and general ecology of rivers. Advances in Applied Biology **6**: 91-152.

Brown, G. W. 1969. Predicting temperatures of small streams. Water Resources Research **5**: 68-75.

Brown, G. W. 1970. Predicting the effect of clear cutting on stream temperature. Journal of Soil & Water Conservation **25**: 11-13.

Brown, G. W. 1971. Water temperature in small streams as influenced by environmental factors and logging. In: Forest Land Uses & Stream Environment. (Ed. by T. T. Krygier and J. D. Hall). pp. 175-181. Oregon State University, Corvallis.

Bye, V. J. 1984. The role of environmental factors in the timing of reproductive cycles. In: Fish reproduction strategies and tactics. (Ed. by G. W. Potts and R. J. Wootton). pp. 187-205. Academic Press, London.

Collings, M. R. 1969. Temperature analysis of a stream. U.S. Geol. Surv. Prof. Paper. 650-B. B 174 - B 179.

Crisp, D. T. 1977. Some physical and chemical effects of the Cow Green (Upper Teesdale) impoundment. Freshwat. Biol. **7**: 109-120.

Crisp, D. T. 1981. A desk study of the relationship between temperature and hatching time for the eggs of five species of salmonid fishes. Freshwat. Biol. **11**: 361-368.

Crisp, D. T. and Le Cren, E. D. 1970. The temperature of three different streams in northwest England. Hydrobiologia **35**: 305-323.

Crisp, D. T., Mann, R. H. K. and McCormack, Jean C. 1975. The populations of fish in the River Tees system on the Moor House National Nature Reserve, Westmorland. J. Fish. Biol. **7**: 573-593.

Crisp, D. T. and Howson, G. 1982. Effect of air temperatures upon mean water temperature in streams in the north Pennines and English Lake District. Freshwat. Biol. **12**: 359-367.

Crisp, D. T., Matthews, A. M. and Westlake, D. W. 1982. The temperatures of nine flowing waters in southern England. Hydrobiologia **89**: 193-204.

Crisp, D. T., Mann, R. H. K. and Cubby, P. R. 1983. Effects of regulation of the River Tees upon fish populations below Cow Green Reservoir. J. Appl. Ecol. **20**: 371-386.

Eckel, O. 1953. Zur Thermik de Fliessgewässer: Uber die Anderung der Wassertemperatur entlang des Flusslaufs. Wett. u. Leben **2**: 41-47.

Edinger, J. E., Duttweiler, D. W. and Geyen, J. C. 1968. The response of water temperature to meteorological conditions. Water Resource Research **4**: 1137-1143.

Edington, J. M. 1965. Some observations on stream temperatures. Oikos **15**: 265-273.

Edwards, R. J. 1978. The effect of hypolimnion reservoir releases on fish distribution and species diversity. Trans. Am. Fish. Soc. **107**: 71-77.

Edwards, R. W. 1984. Predicting the enviromental impact of a major reservoir development. In: Planning & Ecology. (Ed. by R. D. Roberts and T. M. Roberts). pp. 55-79. Chapman & Hall, London.

Edwards, R. W., Densem, J. W. and Russell, P. A. 1979. An assessment of the importance of temperature as a factor controlling the growth rate of brown trout in streams. J. Anim. Ecol. **48**: 501-507.

Edwards, R. W. and Crisp, D. T. 1982. Ecological implications of river regulation in the United Kingdom. In: Gravel-bed Rivers Fluvial Processes, Engineering & Management. (Ed by R. D. Hey, J. C. Bathurst and C. R. Thorne). pp. 843-865. Wiley, New York.

Elliott, J. M. 1972. Rates of gastric evacuation in brown trout, *Salmo trutta* L. Freshwat. Biol. **2**: 1-18.

Elliott, J. M. 1975a. Weight of food and time required to satiate brown trout, *Salmo trutta* L. Freshwat. Biol. **5**: 51-64.

Elliott, J. M. 1975b. Number of meals in a day, maximum weight of food consumed in a day and maximum rate of feeding for brown trout, *Salmo trutta* L. Freshwat. Biol. **5**: 287-303.

Elliott, J. M. 1975c. The growth rate of brown trout, *Salmo trutta* L., fed on maximum rations. J. Anim. Ecol. **44**: 805-821.

Elliott, J. M. 1975d. The growth rate of brown trout *Salmo trutta* L. fed on reduced rations. J. Anim. Ecol. **44**: 823-842

Embody, G. C. 1934. Relations of temperature to the incubation periods of eggs of four species of trout. Trans. Am. Fish. Soc. **64**: 281-292.

Gibson, R, J, 1973. Interactions of juvenile Atlantic salmon (*Salmo salar* L.) and brook trout (*Salvelinus fontinalis* (Mitchill)). Int. Atl. Salmon Found. Spec. Publ. Ser. **4**: 181-202.

Gray, J. 1928. The growth of fish: III. The effects of temperature on the development of eggs of *Salmo fario*. British Journal of Experimental Biology **6**: 110-124.

Gray, J. R. A. and Edington, J. M. 1969. Effect of woodland clearance on stream temperatures. J. Fish Res. Bd. Can. **26**: 299-403.

Gunnes, K. 1979. Survival and development of Atlantic salmon eggs and fry at three different temperatures. Aquaculture **16**: 211-218.

Holden, P. B. and Stalnaker, C. B. 1975. Distribution and abundance of mainstream fishes in the middle and upper Colorado River basins, 1967-1973. Trans. Am. Fish. Soc. **104**: 217-2311.

Jaske, J. T. and Goebel, J. B. 1967. Effects of dam construction on temperature of Columbia River. J. Amer. Wat. Wks. Assoc. **59**: 935-942.

Kamler, E. 1965. Thermal conditions in mountain waters and their influence on the distribution of Plecoptera and Ephemeroptera larvae. Ekol. Pol. Ser. A. **13**: 377-414.

Kazakov, J. 1971. Experimental verification of the effect of the temperature regime of egg incubation on the condition of embryos and larvae of the Neva River population of Atlantic salmon (*Salmo salar* L.). J. Icthyol. **11**: 123-128.

Lavis, M. E. and Smith, K. 1972. Reservoir storage and the thermal regime of rivers, with special reference to the River Lune, Yorkshire. Sci. Total Environ. **1**: 81-90.

Lehmkuhl, D. M. 1972. Change in thermal regime as a cause of reduction of benthic fauna downstream of a reservoir. J. Fish. Res. Board Can. **29**: 1329-1332.

Macan, T. T. 1958. The temperature of a small stony stream. Hydrobiologia **12**: 89-106.

McCleave, J. D. 1978. Rhythmic aspects of estuarine migration of hatchery-reared Atlantic salmon (*Salmo salar*) smolts. J. Fish. Biol. **12**: 559-570.

McNeil, J. W. 1966. Effect of spawning bed environment on reproduction of pink and chum salmon. Bull, U.S. Fish. Wildl. Serv. Fish. **65**: 495-523.

Morse, W. L. 1970. Stream temperature prediction model. Water Resource Research **6**: 290-302.

Neel, J. K. 1963. Impact of reservoirs. In: Limnology in North America. (Ed. by D.G. Frey). pp. 573-593. University of Wisconsin Press, Wisconsin.

Parsons, J. W. 1955. The trout fishery of the tailwater below Dale Hollow Reservoir. Trans. Am. Fish. Soc. **85**: 75-92.

Penaz, M., Kubicek, F., Marvan, P. and Zelinka, M. 1968. Influence of the Vir River Valley reservoir on the hydrobiological and ichthyological conditions in the River Svratka. Acta. Sci. Nat. Bruno. **2**: 1-60.

Peterson, R. H., Spinney, H. C. E. and Sreedharan, A. 1977. Development of Atlantic salmon (*Salmo salar*) eggs and alevins under varied temperature regimes. J. Fish. Res. Board Can. **34**: 31-43.

Pfitzer, D. W. 1967. Evaluation of tailwater fishery resources resulting from high dams. In. Reservoir Fishery Resources Symposium. pp. 477-488. American Fisheries Society, Washington, D.C.

Saunders, R. L. and Henderson, E. B. 1969. Survival and growth of Atlantic salmon fry in relation to salinity and diet. J. Fish Res. Board Can. Tech. Pap. **148**: 1-7.

Siginevich, G. P. 1967. Nature of the relationship between increase in size of Baltic salmon fry and the water temperature. Gidrob. Zhurnal, **3**: 43-48. Fish. Res. Board Can. Transl. Serv. No. 952, 1-14.

Smith, K. 1968. Some thermal characteristics of two rivers in the Pennine area of northern England. Journal of Hydrology **6**: 54-65.

Smith K. 1972. River water temperatures .... an environmental review. Scottish Geographical Magazine **88**: 211-220.

Smith, K. and Lavis M. E. 1975. Environmental influences on the temperature of a small upland stream. Oikos **26**: 228-236.

Solomon, D. J. 1978. Some observations on salmon smolt migration in a chalk stream. J. Fish. Biol. **12**: 571-574.

Somme, S. 1960. The effects of impoundment on salmon and sea trout rivers. Int. Union for Conserv. Nat. Res. 7th Tech. Meeting, 1958. 77-80.

Spence, J. A. and Hynes, H. B. N. 1971. Differences in fish populations upstream and downstream of a mainstream impoundment. J. Fish. Res. Board Can. **28**: 45-46.

Symons, P. E. K. 1979. Estimated escapement of Atlantic salmon (*Salmo salar*) for maximum smolt production in rivers of different productivity. J. Fish. Res. Board Can. **36**: 132-140.

Walker, K. F., Hillman, T. J. and Williams, D. W. 1979. The effects of impoundment on rivers: an Australian case study. Verh. Int. Verein. Limnol. **20**: 1695-1701.

Ward, J. C. 1963. Annual variation of stream water temperature. Proceedings of the American Society of Civil Engineers (J. San. Engng. Div.) **89**: 1-16.

Ward, J. V. 1976. Comparative limnology of differentially regulated sections of a Colorado mountain river. Arch. Hydrobiol. **78**: 319-342.

Ward, J. V. 1982. Ecological aspects of stream regulation: Responses in downstream lotic reaches. Water Pollution and Management Reviews (New Delhi) **2**: 1-26.

Ward, J. V. and Dufford, R. G. 1979. Longitudinal and seasonal distribution of macroinvertebrates and eplithic algae in a Colorado springbrook-pond system. Arch. Hydrobiol. **86**: 284-321.

Ward, J. V. and Stanford, J. A. 1979. Ecological factors controlling stream zoobenthos with emphasis on thermal modification of regulated streams. In: The Ecology of Regulated Streams. (Ed. by J. Ward and J. A. Stanford). pp. 35-55. Plenum, New York.

Williams, O. O. 1968. Reservoir effect on downstream water temperatures in the upper Delaware river basin. U.S. Geol. Surv. Prof. Pap. 600-B. 195-199.

Wurtsbaugh, W. A. and Davis, G. E. 1977. Effects of temperature and ration level on the growth and food conversion efficiency of *Salmo gairdneri* Richardson. J. Fish. Biol. **11**: 87-98.

Zimmerman, E. G. and Richmond, M. C. 1981. Increased heterozygosity at the Mdh-B locus in fish inhabiting a rapidly fluctuating thermal environment. Trans. Am. Fish. Soc. **110**: 410-466.

INSTREAM TEMPERATURE MODELING AND FISH IMPACT ASSESSMENT

FOR A PROPOSED LARGE SCALE ALASKA HYDROELECTRIC PROJECT

William J. Wilson, Michael D. Kelly and Paul R. Meyer

Arctic Environmental Information and Data Center
University of Alaska
707 A Street
Anchorage, Alaska 99501 USA

INTRODUCTION

The State of Alaska is proposing to construct a two dam, 1620 megawatt hydroelectric project (U.S. Federal Energy Regulatory Commission N. 7114) on the Susitna River approximately 190 km NNE of Anchorage. A study is underway to determine the effects this project may have on the indigenous aquatic resources of the Susitna drainage, and in this paper we report on the studies of the expected alteration of the instream temperature regime of the Susitna River (Meyer et al. 1984). Twenty species of fish are known to inhabit the Susitna basin. This study focuses on the most numerous and economically valuable Pacific salmon species, approximately two million of which annually enter this river to spawn.

The Susitna River flows 520 km from its source at the glaciers on the southern slopes of the Alaska range to its mouth at Cook Inlet near Anchorage (Figure 1). It is seasonally turbid from the glacier melt contribution with summer turbidities of 74 to 730 NTU, and winter turbidities <1 NTU (R&M Consultants, Inc. & Larry A. Peterson & Associates 1981). The river drains a basin of approximately 50,800 km^2, the sixth largest river basin in the state. Like all northern rivers, the Susitna exhibits strong seasonal variation in flow, high during the spring and summer due to snow melt and summer rains, and low during the winter. With the project in place, high summer flows would be captured for winter release when the demand for power generation is greatest.

The project would be constructed in two stages. The first stage, Watana dam and reservoir, would be located at river kilometer (RK) 296 (296 km upriver from the mouth) and is scheduled for completion in 1996. The last year Watana dam would be operated alone is 2001. The second stage, Devil Canyon dam, would be located downstream at RK 243 and is scheduled to be operational in 2002. The development scenarios discussed in this paper are Watana in the year 2001 and Devil Canyon plus Watana in 2002.

The Susitna River has a mean annual flow of 275 m^3 s^{-1} measured at an index station in the study reach. Mean monthly flows for the summer months (June through August) range from 590–740 m^3 s^{-1}, with peak flows normally occurring during June. Flows begin receding in September, reaching winter lows of 25–30 m^3 s^{-1}.

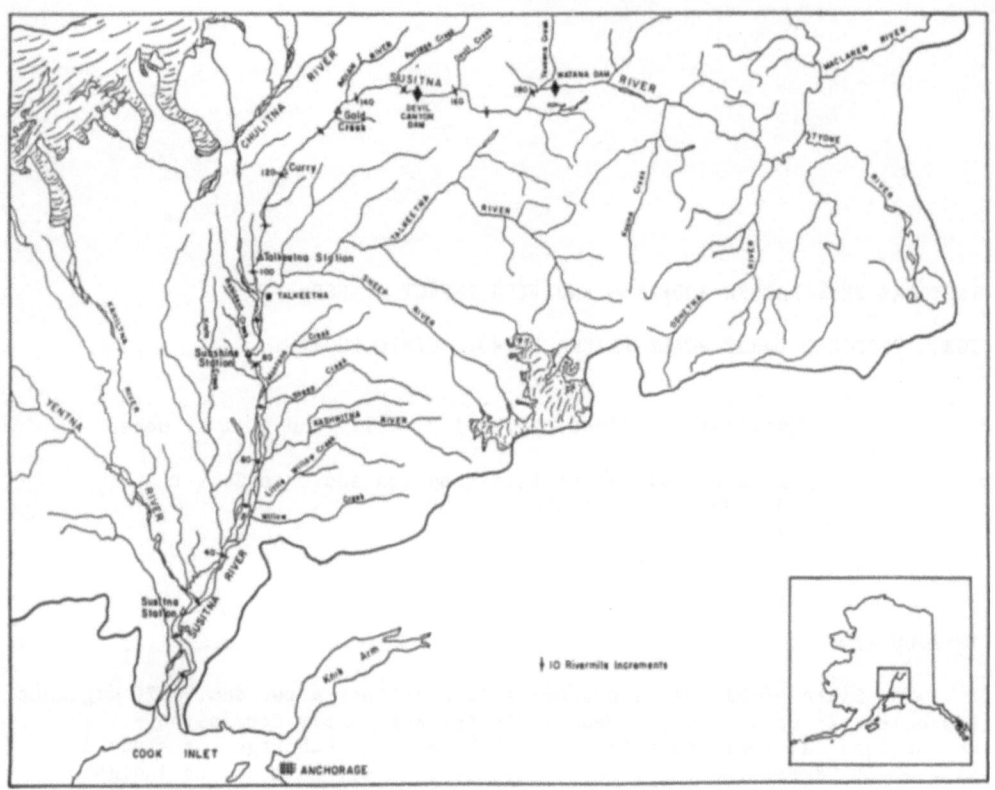

Figure 1. Susitna River Basin, Alaska, and locations of proposed hydroelectric facilities.

Under the regulation of the project, flow variation would be dampened considerably. With a Watana-only configuration, mean monthly flows would range from 210-340 m^3 s^{-1}, with peak flows released in August to facilitate access for salmon spawning, and during winter high-demand periods. With the addition of the second dam, mean monthly flows would range from 200-320 m^3 s^{-1}, with higher flows more uniform throughout the winter, and slightly lower summer flows.

Reservoirs store heat as well as storing water. The temperature of water released from the reservoir is expected to be cooler than natural during the summer, and warmer than natural during the winter. Since both reservoirs are expected to thermally stratify, multilevel intake structures have been incorporated into the dam design which would allow some degree of control on the release temperature.

Warmer-than-natural releases during the winter would alter the normal ice processes below the dams, delaying the formation of an ice cover and relocating the upstream end of the ice front. Cooler releases in the summer likewise would alter river temperature for a considerable distance downstream. To quantify this temperature change, an instream temperature model was used.

The model simulated effects of the hydroelectric development in an 80 km reach below the Devil Canyon dam. This is the only habitat available to

salmon in the upper part of the Susitna River, as the Devil Canyon dam site
blocks salmon passage further up river. Two large tributaries converge
with the Susitna downstream from this study reach, the resultant flow more
than double the flow upstream from this point. The dampening effect of
these tributaries, both with respect to flow and temperature, creates a
distinct lower boundary to the study reach. In 1984, the study reach
received an escapement of 26,060 chum (*Oncorhynchus keta*), 2,325 sockeye
(*O. nerka*), 29,300 pink (*O. gorbuscha*), 2,900 coho (*O. kisutch*) and
13,800 chinook (*O. tshawytscha*) salmon (Barrett et al. 1985). The modeling
system was run for a variety of power demands and hydrologic and
meteorologic conditions. Downstream temperature results from these
simulations were examined with respect to effects on salmon. This paper
discusses the process of instream temperature modeling and our subjective
assessment of effects of predicted with-project temperature regimes on
salmon.

METHODS

Assessment of temperature impacts on salmon involved a three stage
process. First, natural and with-project temperature regimes were
predicted through use of a stream temperature simulation model for a study
reach of mainstem river which extends approximately 80 km (RK 240 to RK
160) below the proposed dams. Next, fish temperature tolerance criteria
were developed based on literature, laboratory, and field studies.
Finally, these criteria were compared with the temperature model output and
an assessment of the effects was made.

The Stream Temperature Model

The Stream Network Temperature Simulation Model, SNTEMP, was
originally developed by the U.S. Fish and Wildlife Service's Instream Flow
and Aquatic Systems Group in Fort Collins, Colorado (Theurer et al. 1983).
The model requires hydrology, meteorology and stream geometry data as input
and computes heat flux relationships and heat transport through the
system. The model is one-dimensional, producing cross-section averaged
mean weekly temperatures at any mainstem location in the study reach.
A number of modifications were made to the model to better simulate
northern conditions.

1. A monthly variable shade factor was incorporated to account for the
    stream shading from topographical features, a serious concern in
    northern latitudes where solar angles are very small.

2. The model was modified to accept non-constant lapse rates for air
    temperature and humidity. This is of special value during the winter
    when temperature inversions often occur.

3. An influent groundwater temperature submodel was developed and
    incorporated into SNTEMP. This routine considers the effects of the
    depth to groundwater and the cyclical temperature pattern resulting
    from variations in elevation and time.

4. Regression models were developed to fill discontinuous temperature
    records, a common problem in Alaska.

Four summers and five winters were selected from the meteorological
record as representative periods of normal and extreme hydrology and
meteorology. Simulations were run under these conditions for natural
(i.e., without dams), single-dam (Watana) and two-dam (Watana plus Devil
Canyon) project configurations. In this way, the range of downstream

temperatures found naturally and predicted to occur with the project in place was identified.

Development of Temperature Criteria for Fish

To assess the effects of with-project instream temperatures on salmon, we first reviewed available information on the response of the five salmon species to different thermal conditions. Ideally, information used in an effects analysis should be specific to the water body in question and to its particular community of organisms. Little specific information exists on the effects of temperature changes on Susitna River fish stocks, necessitating the use of information from other areas and latitudes. Professional judgement was used to ascertain the applicability of each piece of information to the Susitna Basin. Generally, information proximal to the Susitna River was judged to be more pertinent than data from other areas of Alaska, which in turn was usually more useful than information from more southerly latitudes. Once the information was assembled, it was synthesized to produce thermal tolerance ranges. These criteria were the temperature ranges believed to be capable of supporting adult spawning migrations, spawning, incubation, rearing, and smolt migrations.

Assessment of Temperature Effects

Graphic techniques were used to demonstrate the relationships between simulated natural or with-project temperature regimes and the salmon thermal tolerance criteria. Illustrations were prepared showing the thermal tolerance "envelope" over a one-year time period for each salmon species. Overlays of natural and with-project temperatures were superimposed on the species-specific temperature tolerance graphics; separate illustrations were prepared for each of two representative mainstem river locations. This procedure was followed for each of the meteorological simulations.

We assumed that only in cases where the simulated temperature regimes fall outside the temperature tolerance ranges is an obvious adverse impact established. However, in cases where with-project temperatures do not exceed tolerances but yet appear to be substantially different from natural, a further subjective analysis and prediction of effects was conducted.

RESULTS AND DISCUSSION

Effect of Project on Mainstem Temperatures

Operation of either a single- or two-dam project would reduce mean summer river temperatures below the dam by as much as 2°C. The two-dam project would result in a greater change, primarily because the second dam would be located 53 km further downstream, reducing the length of river in which release waters would warm towards ambient air temperature.

Warmer winter release temperatures would delay the formation of an ice cover in the study reach 2 to 6 weeks with one dam and 4 to 7 weeks with both dams in place. The ice front would be located 16 to 47 km further downstream than under normal conditions (R&M Consultants, Inc., et al. 1985). A synopsis of natural and with-project mean seasonal temperatures for four summers and five winters is shown in Table 1.

One of the most notable effects of project operations on temperature would be the change in the timing of seasonal warming and cooling. River temperatures would warm later in the summer than they do naturally and cool

186

Table 1. Simulated mean seasonal temperatures (°C) at RK 209 for four summer and five winter scenarios.

SUMMER   (Water weeks 31-52; April 29 - September 30)

Year	Air Temperature	Available Runoff	Natural Temperature	1-Dam Project Temperature	2-Dam Project Temperature
1971	cold	wet	7.8	6.8	6.2
1974	warm	dry	8.7	7.5	7.2
1981	average	wet	8.6	7.9	6.8
1982	average	average	8.8	7.7	7.0

WINTER   (Water weeks 5-30; October 29 - April 28)

Year	Air Temperature	Available Runoff	Natural Temperature	1-Dam Project Temperature	2-Dam Project Temperature
1971-72	cold	wet	0.0	0.1	0.2
1974-75	average	dry	0.0	0.4	0.6
1976-77	warm	dry	0.0	Not simulated	0.4
1981-82	average	wet	0.0	1.0	1.7
1982-83	average	average	0.0	1.0	1.2

later in the fall than normal (Figures 2 & 3). Figure 4 compares natural and two-dam project temperatures at RK 209 for 1981 and illustrates this delay in the normal temperature pattern.

Temperature Criteria for Salmon

Thermal tolerance ranges were established during the course of this study for the five Pacific salmon species found in the Susitna drainage. These ranges were based on literature reports of fish distribution, laboratory studies, and field studies (Table 2). Observed Susitna drainage temperature data were utilized in conjunction with the literature reports to establish tolerance criteria for each life phase (Table 3). In cases where life phases overlap, that life phase most sensitive to temperature was chosen when preparing the tolerance criteria graphic overlays. The criteria, then, establish the narrowest temperature tolerance window for evaluation. Within these ranges Susitna salmon stocks were assumed to live and function free from the lethal effects of temperature.

Embryo incubation rates rise with increasing intragravel water temperature. Accumulated temperature units, or degree-days to hatching and emergence, were obtained from literature reports (ADF&G 1981b; 1983; Raymond 1981; Wangaard & Burger 1983) and used as criteria for incubation. Data from laboratory studies of salmon embryo development under different temperature regimes using Susitna chum salmon stocks (Wangaard & Burger 1983) were compared with other chum salmon embryo

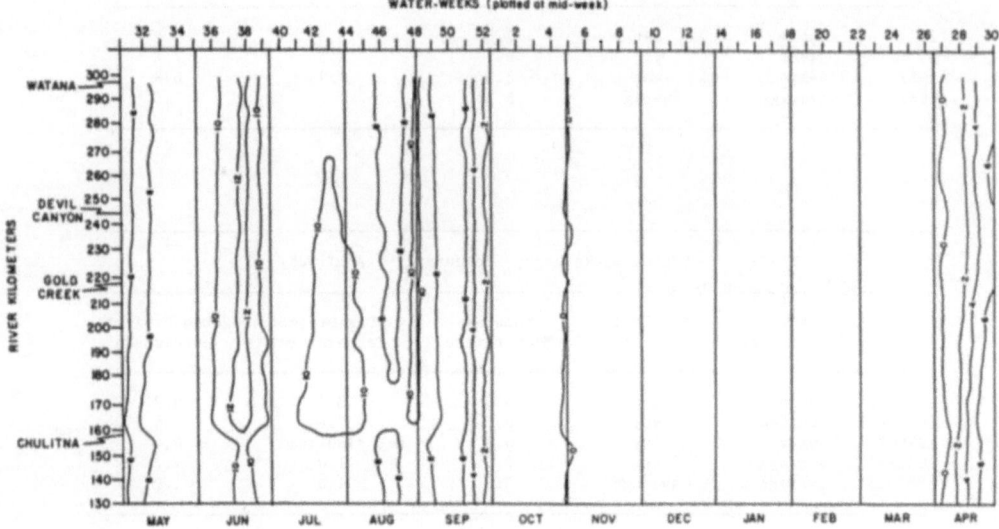

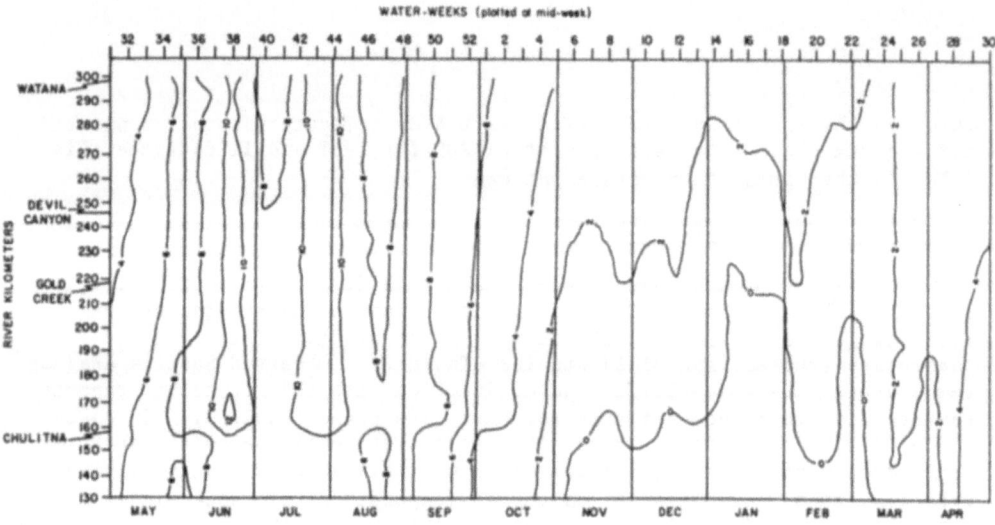

Figure 2. Isotherm plots of simulated instream temperature (°C) for natural and one-dam (Watana) conditions, May 1981 – April 1982.

NATURAL

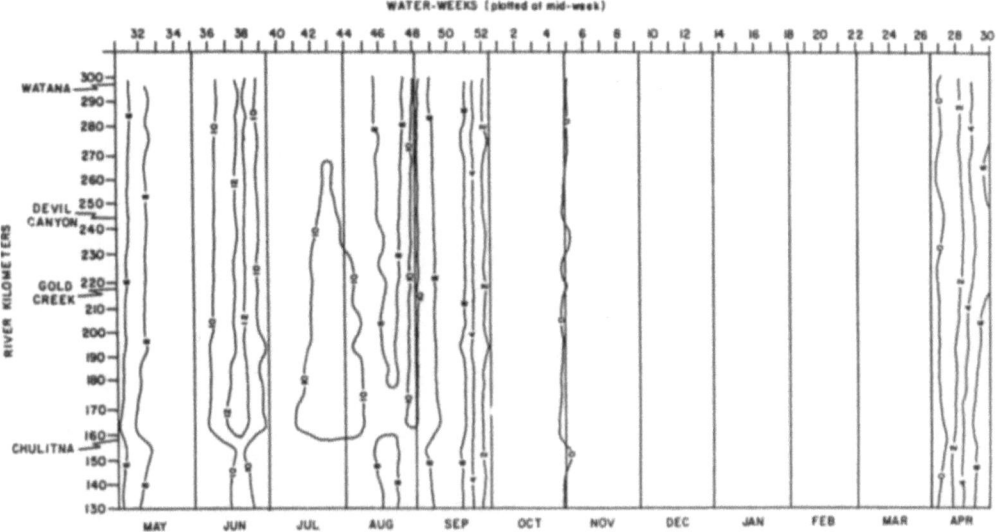

DEVIL CANYON 2002

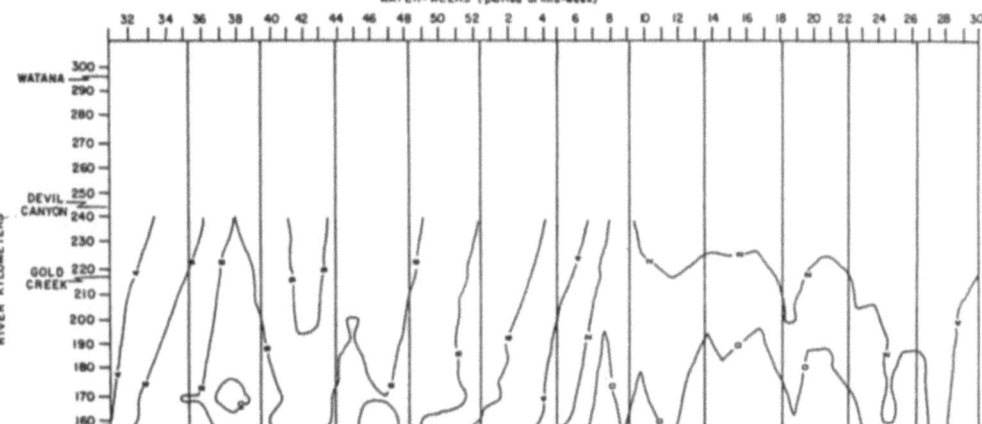

Figure 3. Isotherm plots of simulated instream temperature (°C) for natural and two-dam Devil Canyon) conditions, May 1981 - April 1982.

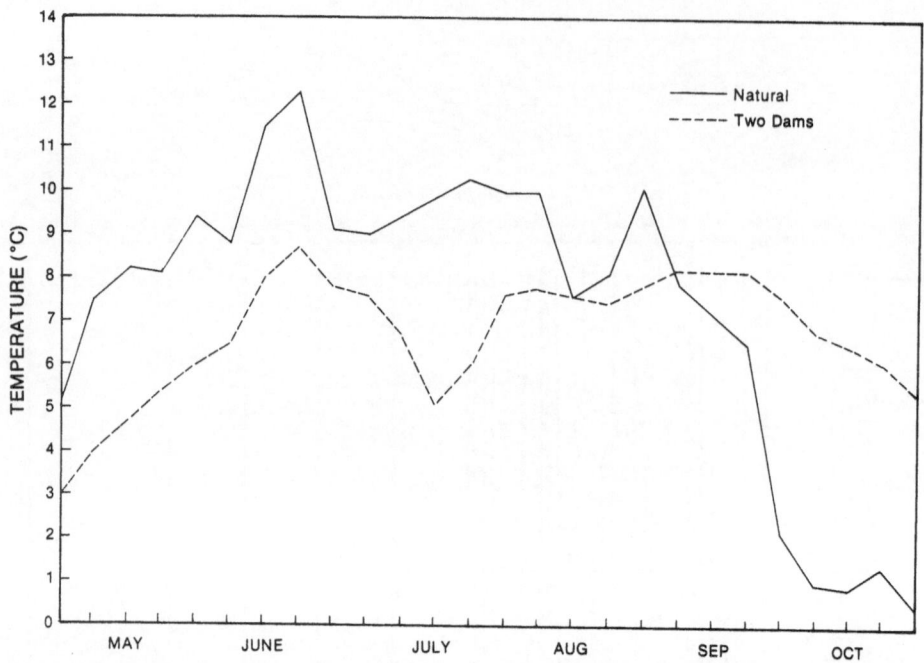

Figure 4. Natural and two-dam with project stream temperatures at RK 209.

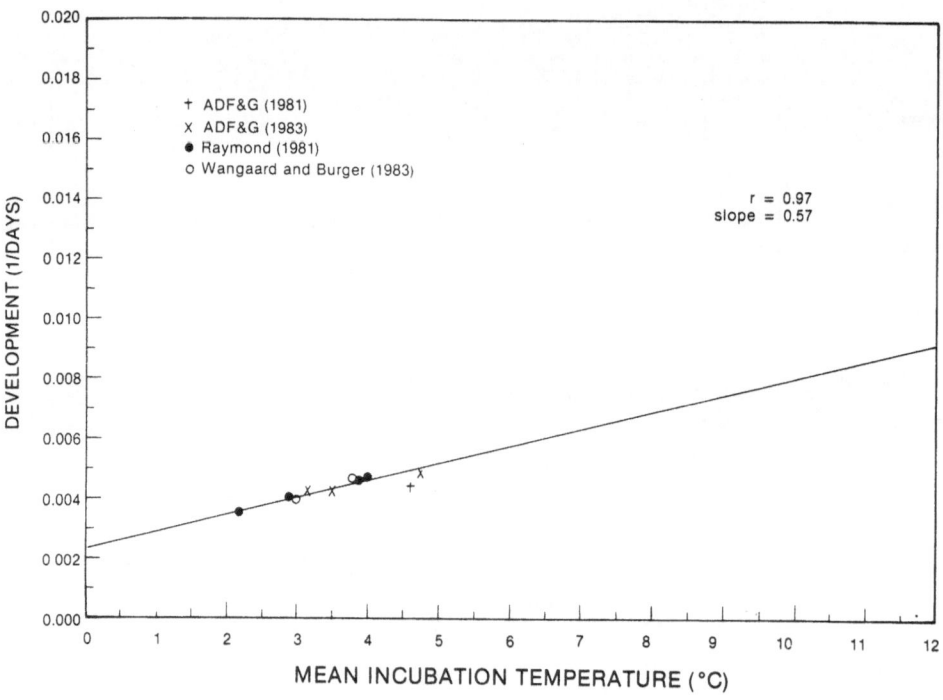

Figure 5. Relationship between mean water temperature and Alaskan chum salmon embryo development-to-emergence times.

Table 2. Observed temperature ranges for various life stages of Pacific Salmon from literature review and laboratory investigations.

SPECIES OF SALMON	LIFE STAGE	LITERATURE SOURCE	LOCATION	TEMPERATURE RANGE °C			
				MIGRATION	SPAWNING	INCUBATION	REARING
Oncorhynchus keta	Adult	Bell 1980	General	8.3-21.0	7.2-12.8		
		ADF&G 1981a	Kuskokwim Tributaries, AK	5.0-12.8			
		Mattson & Hobart 1962	Southeast, AK	4.4-19.4			
		McNeil & Bailey 1975	Southeast, AK		7.0-13.0		
		Wilson et al. 1981	Kodiak Island, AK		6.5-12.5		
		Neave 1966	British Columbia		4.0-16.0		
		Rukhlov 1969	Sakhalin, USSR		1.8-8.2		
		Merritt & Raymond 1983	Noatak R, AK		2.5		
		ADF&G 1984	Susitna R. AK	5.6-15.5	4.5-13.2		
	Juvenile	Trasky 1974	Salcha R., AK	5.0-7.0			
		Sano 1966	Bolshaia R, USSR	6.0-10.0			
		Bell 1980	General	6.7-13.5			
		McNeil & Bailey 1975	Southeast, AK				11.2-15.7
		Wilson et al. 1979	Kodiak Island, AK				4.4-15.7
		Raymond 1981	Delta R, AK	3.0-5.5			
		Merritt & Raymond 1983	Noatak R, AK	5.0-12.0			
		ADF&G 1984	Susitna R, AK	4.2-14.5			1.3-16.2
	Egg/Alevin	Bell 1980	General		4.4-13.3	0-15.0	
		McNeil 1969	Southeast, AK			0.2-9.0	
		Merritt & Raymond 1983	Noatak R, AK			4	
		Sano 1966	Japan			4.4	
		McNeil & Bailey 1975	Southeast, AK			0.5-4.5	
		Kogl 1965	Chena R, AK			0.4-6.7	
		Francisco 1977	Delta R, AK			2.0-4.5	
		ADF&G 1981	Clear, AK			0-7.4	
		Wangaard & Burger 1983	Susitna R, AK			0.5-8.0	
		ADF&G 1984	Susitna R, AK			2.0-4.3[5]	
O. kisutch	Adult	Bell 1980	General	7.2-15.6	4.4-9.5		
		McNeil & Bailey 1975	Southeast, AK		7.0-13.0		
		McMahon 1983	General	5-19.5-11[3]	2-17.5-13[3]		
		Wallis & Balland 1983	Anchor R, AK	2-15.7-14[4]			
		ADF&G 1984	Susitna R, AK	5.8-15.5			
	Juvenile	Cederholm & Scarlett 1982	Washington St.	6			
		Bustard & Narver 1975	Vancouver Is., BC	7			
		Bell 1980	General	7.0-16.5			
		McNeil & Bailey 1975	Southeast, AK				11.8-14.6
		McMahon 1983	General	4.16.6-12[3]			4.4-15.7
		Wallis & Ballard 1983	Anchor R, AK	2-15.7-14[4]			4-21.7-15[3]
		Whitmore et al. 1979	Caribou L, AK	11-15.5			
			Seldovia L, AK	3.0-5.7			
		ADF&G 1984	Susitna R, AK	4.2-14.5			
	Egg/Alevin	Bell 1980	General			4.4-13.3	
		McMahon 1983	General			4-14.4-10[3]	
		Dong 1981	Washington St.			1.3-12.4,4.4-6.5[3]	

Table 2. Observed temperature ranges for various life stages of Pacific Salmon from literature review and laboratory investigations. (continued)

SPECIES OF SALMON	LIFE STAGE	LITERATURE SOURCE	LOCATION	MIGRATION	SPAWNING	INCUBATION	REARING
O. gorbuscha	Adult	Bell 1980	General	7.2-15.6			
		McNeil & Bailey 1975	Southeast, AK		7.0-13		
		Sheridan 1962	Southeast, AK		7.2-18.4		
		McNeil et al. 1964	Southeast, AK		10.0-13.0		
		ADF&G 1984	Susitna R, AK	7.8-15.5	8.0-11.0		
	Juvenile	Bell 1980	General	5.0-7.0			
		McNeil & Bailey 1975	Southeast, AK	4.0-5.0			
		Wilson et al. 1979	Kodiak Island, AK				5.6-14.6
		Wickett 1958	British Columbia				4.4-15.7
		ADF&G 1984	Susitna R, AK	4.2-14.5			
	Egg/Alevin	Bell 1980	General			4.4-13.3	
		Bailey & Evans 1971	Southeast, AK			4.5	
		Combs & Burrows 1957	Laboratory			0.5-5.5	
		McNeil et al. 1964	Southeast, AK			1.0-8.0	
		Godin 1980	Laboratory			3.4-15.0	
O. nerka	Adult	Bell 1980	General	7.2-15.6	10.6-12.2		
		McNeil & Bailey 1975	Southeast, AK	8.3-14.3	7.0-13.0		
		Nelson 1983	Southeast, AK	5.8-15.5			
		ADF&G 1984	Susitna R, AK		4.9-10.5		
	Juvenile	McCart 1967	British Columbia	5.0-17.0			
		Raleigh 1971	Laboratory	4.5			
		Bell 1980	General				11.2-14.6
		McNeil & Bailey 1975	Southeast, AK	4.0-7.0			
		Fried & Laner 1981	Bristol Bay, AK	4.4-17.8			
		Bucher 1981	Briston Bay, AK	4.5-10.0			
		Hartman et al. 1967	Alaskawide				4.4-15.7
		Flagg 1983	Kasilof R, AK	6.7-14.4			
		ADF&G 1984	Susitna R, AK	4.2-14.0			
	Egg/Alevin	Bell 1980	General			4.4-13.3	
		Combs 1965	Laboratory			4.5-14.3,3.1.5[2]	
		ADF&G 1983	Susitna R, AK			2.9-7.4	
		Wangaard & Burger 1983	Laboratory			2.0-6.5	
		ADF&G 1984	Susitna R, AK			2.0-4.3[5]	
O. tshawytscha	Adult	Bell 1980	General	3.3-13.9	5.6-13.9		
		McNeil & Bailey 1975	Southeast, AK	2-14.5-10[4]	7.0-13.0		
		Wallis 6Balland 1983	Anchor R, AK	6.6-15.6	7.8-13.6		
		ADF&G 1984	Susitna R, AK				
	Juvenile	Raymond 1979	Columbia R, OR	7			
		Bell 1980	General				7.3-14.6
		McNeil & Bailey 1975	Southeast, AK	4.5			
		ADF&G 1982	Southcentral, AK	6-16.8-16[4]			
		Wallis & Balland 1083	Anchor R, AK				4.4-15.7
		ADF&G 1984	Susitna R, AK	4.2-14.5			

Table 2. Observed temperature ranges for various life stages of Pacific Salmon from literature review and laboratory investigations. (concluded)

SPECIES OF SALMON	LIFE STAGE	LITERATURE SOURCE	LOCATION	TEMPERATURE RANGE °C			
				MIGRATION	SPAWNING	INCUBATION	REARING
O. tshawytscha	Egg/Alevin	Bell 1980	General			5.0-14.4	
		Combs 1965	Laboratory			1.5^{2}	
		Alderdice & Velsen 1978	General			2.5-16.0	

1 Single temperature values are lower observed thresholds
2 After eggs had developed to the 128-cell or early blastula stage at 5.5°C
3 Optimum range
4 Peak migration range
5 Mean temperature

Table 3. Salmon temperature tolerance criteria for Susitna River drainage.

| SPECIES | LIFE PHASE | TEMPERATURE RANGE (°C) | |
		TOLERANCE	PREFERRED
Chum	Adult Migration	1.5-18.0	6.0-13.0
	Spawning	1.0-14.0	6.0-13.0
	Incubation[1]	0-12.0	2.0- 8.0
	Rearing	1.5-16.0	5.0-15.0
	Smolt Migration	3.0-13.0	5.0-12.0
Sockeye	Adult Migration	2.5-16.0	6.0-12.0
	Spawning	4.0-14.0	6.0-12.0
	Incubation[1]	0-14.0	4.5- 8.0
	Rearing	2.0-16.0	7.0-14.0
	Smolt Migration	4.0-18.0	5.0-12.0
Pink	Adult Migration	5.0-18.0	7.0-13.0
	Spawning	7.0-18.0	8.0-13.0
	Incubation[1]	0-13.0	4.0-10.0
	Smolt Migration	4.0-13.0	5.0-12.0
Chinook	Adult Migration	2.0-16.0	7.0-13.0
	Spawning	5.0-14.0	7.0-12.0
	Incubation[1]	0-16.0	4.0-12.0
	Rearing	2.0-16.0	7.0-14.0
	Smolt Migration	4.0-16.0	7.0-14.0
Coho	Adult Migration	2.0-18.0	6.0-11.0
	Spawning	2.0-18.0	6.0-11.0
	Incubation[1]	0-14.0	4.0-10.0
	Rearing	2.0-18.0	7.0-15.0
	Smolt Migration	2.0-16.0	6.0-12.0

[1] Embryo incubation or development rate increased as temperature rises.
Accumulated temperature units or days to emergence was determined for
each species for the incubation phase.

incubation time data. A regression analysis of these data illustrated a
linear relationship between mean incubation temperature and development
rate (the inverse of the time to emergence) for chum salmon (Figure 5). A
nomograph was then prepared from these data which could predict the
date of emergence based upon the date of chum salmon spawning and the
average temperature over the incubation period (Figure 6). A nomograph was
prepared only for chum salmon since this is the principal species spawning
in the mainstem where project-related temperature changes are predicted.
Other species spawn in tributaries or side sloughs expected to be
unaffected by the temperature change.

Effects of Altered Temperatures on Fish

Using the graphic techniques for illustrating relationships between
the natural and with-project temperature regimes and the salmon life stage
temperature tolerance criteria, we evaluated over 100 one- and two-dam
development scenarios, each under different combinations of
meteorologic/hydrologic conditions. An examination of departures of

with-project temperatures from the "tolerance window" was made. In most cases, each with-project temperature simulation fell within the temperature tolerance criteria for all life phases. For example, while with-project temperatures are different from natural, they are within the tolerance range for chum salmon (Figures 7 & 8). Therefore, we assumed that no obvious adverse impacts would result from predicted with-project temperatures for this species at this location under these meteorological and hydrological conditions.

In general, this first step in the assessment demonstrated that the Susitna Hydroelectric Project would have few adverse effects from temperature on the five salmon species. One potential impact under the

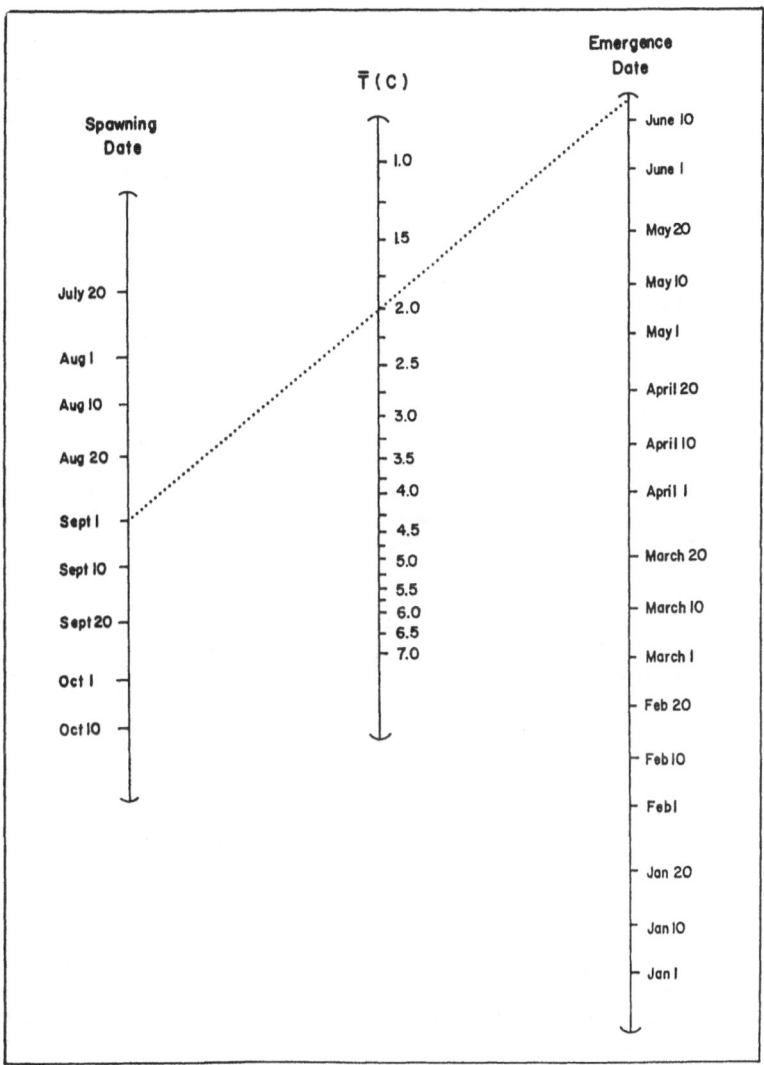

Figure 6. Nomograph for predicting Susitna River chum salmon fry emergence from spawning date and mean water temperature during the embryo incubation period. Line illustrates predicted fry emergence date from a September 1 spawning date and a mean incubation temperature of 2.0°C.

two-dam scenario involves adult pink and chinook salmon inmigration which may be delayed upstream of RK 209 in late June to mid-July as temperatures fall below the lower tolerance level for this life phase (Figures 9 & 10). The effects on pink salmon inmigration timing are greater than those on chinook because the potential thermal block would preclude access to more habitat, would occur nearer the time of peak pink salmon inmigration, and the period of exposure to temperatures below tolerance levels would be of longer duration. While adult chinook or pink salmon migration into this river reach could be delayed, we believe inmigration would ultimately occur 5 to 15 days later as temperatures rise. This may result in a shorter period between the time pink salmon occupy spawning grounds and the occurence of actual spawning.

Another situation was found where temperatures upstream of RK 209 in July also fall outside pink and chinook salmon spawning tolerance zones (Figures 9 & 10). Since this only occurs for about one week, we believe that this would temporarily delay this species' spawning migration but would pose no long-term impediment to the spawning act. Neither pink nor chinook salmon are presently known to use this habitat for spawning, and thus this is not a present concern. Mitigation studies are currently focusing on the potential increased suitability of mainstem habitats for chinook spawning after the project is operating due to improved hydraulic, turbidity, and winter ice conditions.

The second step in our analysis was a more in-depth examination of effects of temperature change on juvenile fish growth and on embryonic development. Even though the with-project temperature scenarios were found to be largely within the established thermal tolerance ranges for salmon, some reduction in juvenile salmon growth could occur due to cooler summer temperatures under with-project scenarios. Although unquantifiable, we believe effects on rearing chinook salmon could be the most severe as juveniles of this species are the most numerous in habitats directly under mainstem temperature influence. From spring to fall, juvenile chinook move from overwintering clearwater tributaries and side sloughs into turbid water side channels and mainstem habitats (Schmidt et al. 1984), presumably to forge on drift and benthic invertebrates and to utilize cover provided by the turbid conditions in these areas.

We made estimates of juvenile chinook salmon growth under natural and with-project temperature regimes using a growth table presented in Brett (1974). Our growth assessment indicates that, depending on climate and the temperature of reservoir-released waters, growth (measured by weight gain) of juveniles rearing in affected mainstem areas (above RK 209) could be substantially reduced (Figure 11). These estimates of growth reduction are based on the sum of increased growth during the warmer fall temperatures and decreased growth during cooler spring and summer temperatures They are also based in part on the assumption that affected juvenile fish would feed to satiation. Since we believe this may not occur in the wild, these estimates should be viewed as worst case scenarios.

Embryonic development time also is affected by changes in stream temperature, and was used as an estimator of project effect instead of tolerance criteria. With-project water temperatures are expected to be warmer during the salmon embryo incubation period of September to April. Simulated natural mainstem average water temperatures near RK 209 for the September to April period range from 0.8 to 1.2°C depending on meteorological conditions. Watana-only operational average water temperatures would be about 0.7 to 1.2°C warmer and Devil Canyon operational temperatures would be about 0.8 to 2.0°C warmer than natural (Table 4).

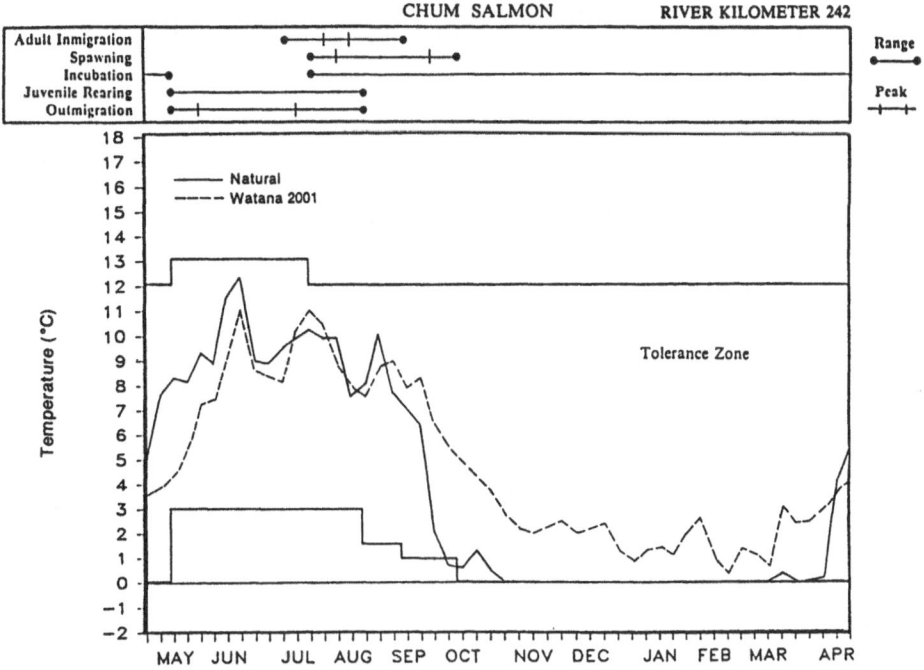

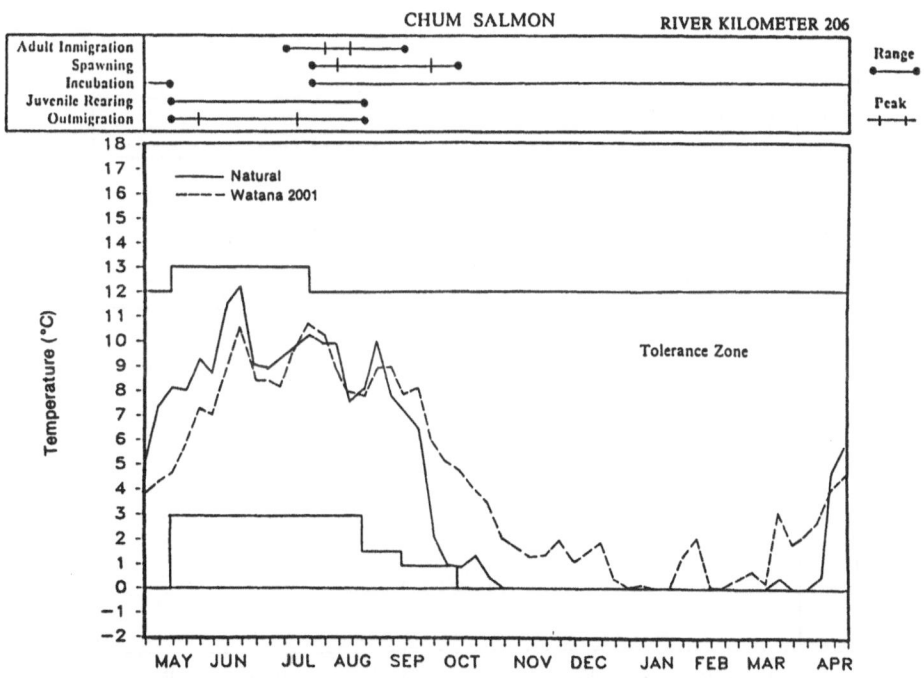

Figure 7. Natural and one-dam (Watana) with-project water temperature regimes in relation to thermal tolerance criteria for chum salmon at two locations on the Susitna River.

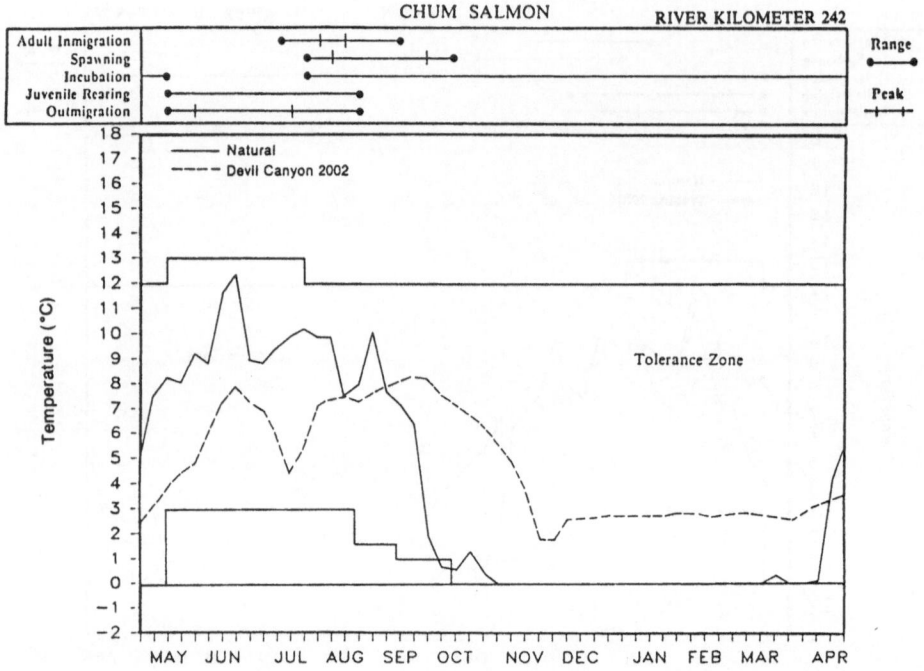

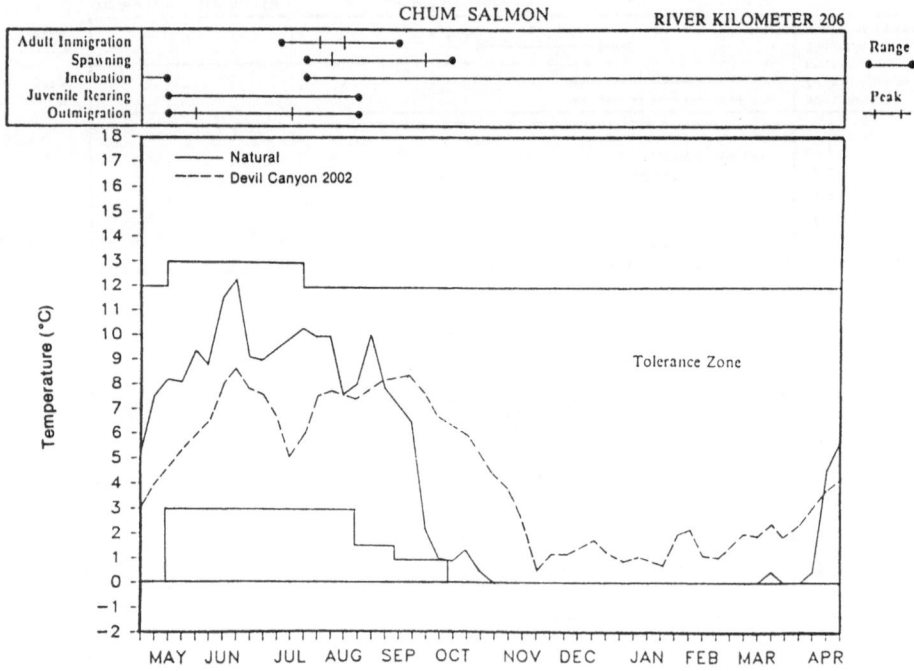

Figure 8. Natural and two-dam (Devil Canyon) with-project water temperature regimes in relation to thermal tolerance criteria for chum salmon at two locations on the Susitna River.

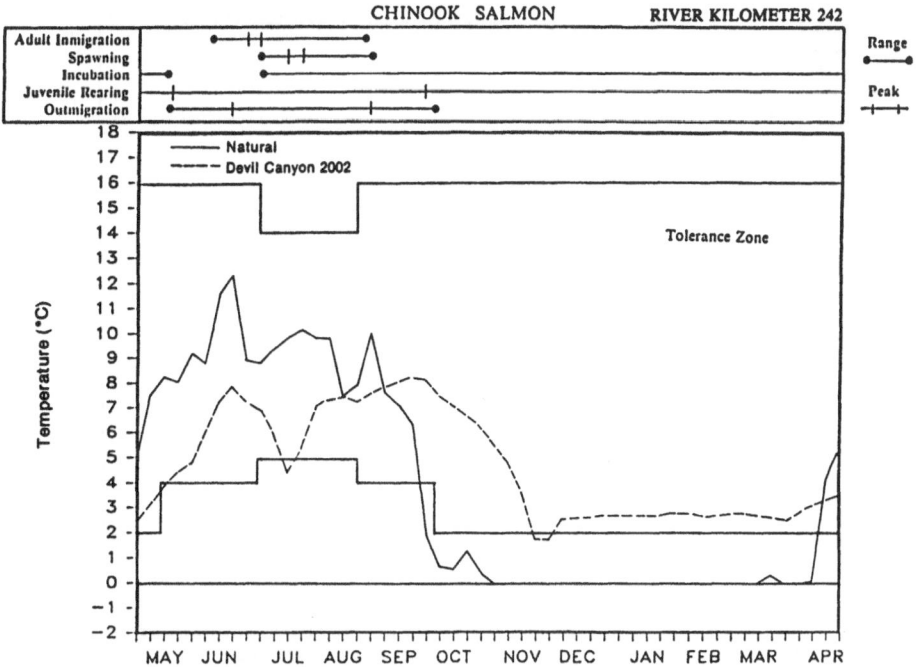

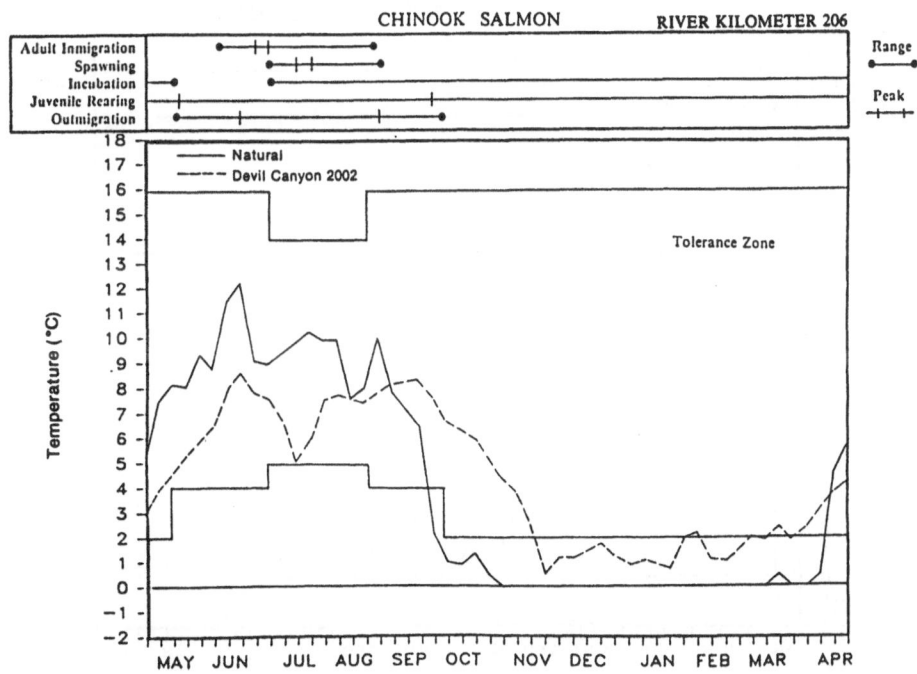

Figure 9. Natural and two-dam (Devil Canyon) with-project water temperature regimes in relation to thermal tolerance criteria for chinook salmon at two locations on the Susitna River.

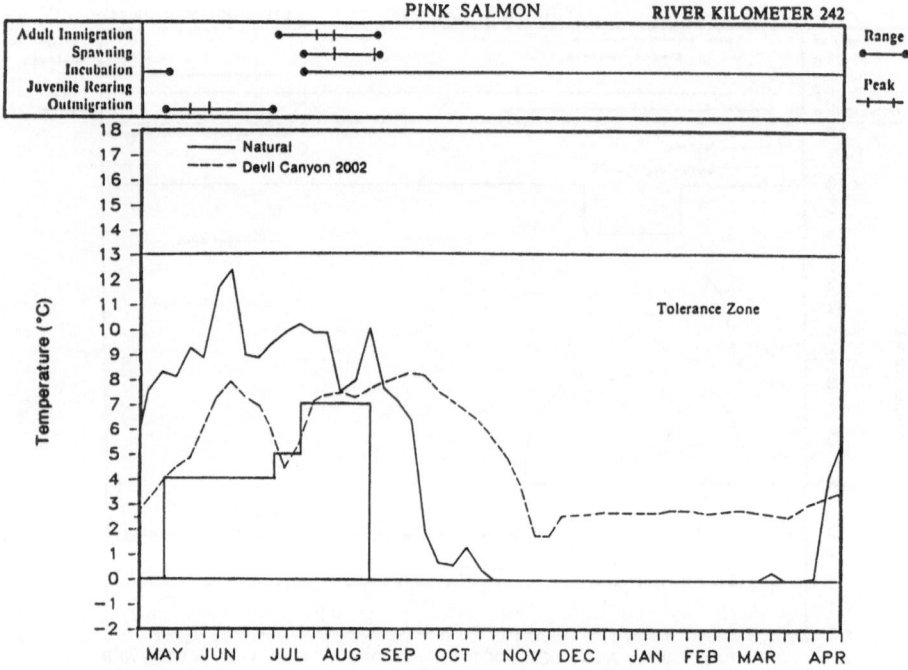

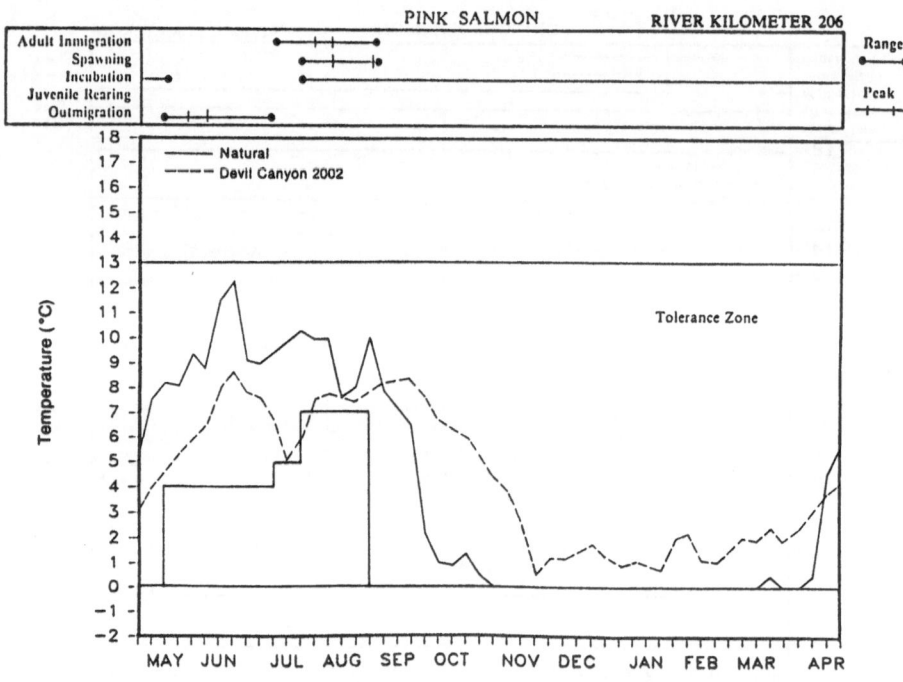

Figure 10. Natural and two-dam (Devil Canyon) with-project water temperature regimes in relation to thermal tolerance criteria for pink salmon at two locations on the Susitna River.

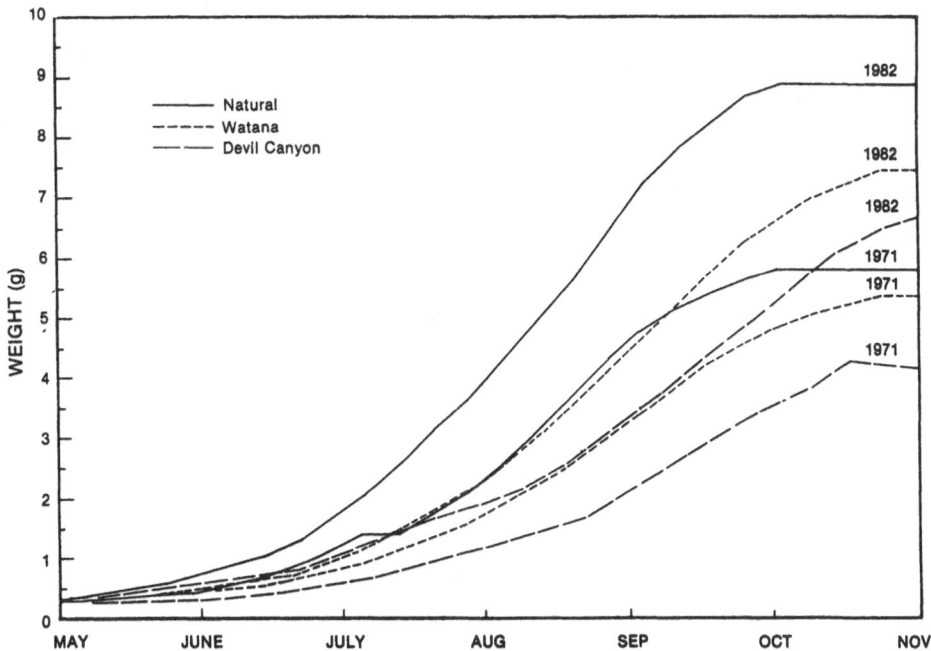

Figure 11. Estimates of juvenile salmon growth in the Susitna River near RK 209 under natural and with-project water temperature regimes comparing 1971 (cold) and 1982 (average) meteorological conditions.

Our assessment of these elevated winter incubation temperatures was based on the chum salmon nomograph previously described. Under natural conditions, only chum salmon have been found to spawn in mainstem habitats. In 1984 approximately 3,800 chum salmon used the mainstem for spawning; 14,600 spawned in side sloughs (Barrett et al. 1985) at a nearly constant 3 to 4°C where groundwater upwelling maintained elevated temperatures throughout the winter (ADF&G 1983). In the mainstem spawning areas, upwelling groundwater also maintains warm temperatures in the intragravel environment (ADF&G 1983). However, to illustrate effects of natural winter temperature regimes (c.1°C) on chum salmon incubation if warm groundwater is absent, our nomograph (Figure 6) shows chum fry emergence well into the summer from a spawning date of September 1, the time of peak spawning in Susitna River habitats. Under natural conditions, chum fry emerge in early May (ADF&G 1983). This illustrates that temperature may be a factor limiting successful production of chum salmon in mainstem habitats.

With either one or two dams in place, however, eggs deposited on September 1 at an average incubation temperature greater than 2.0 or 3.0°C should emerge in time to produce viable fry (Table 4 and Figure 6). Average mainstem temperatures under the Watana-only scenario are above 2.0°C in two of the four different meteorological scenarios and for three of the four Devil Canyon scenarios (Table 4). Mainstem temperatures near RK 209 in all but the coldest years average above 2.0°C for the incubation period and any eggs deposited under these warmer temperatures should produce viable fry. It appears, therefore, that better mainstem incubating habitat could exist under with-project scenarios due to the warmer temperatures.

Table 4. Natural and with-project Susitna River temperature ranges (°C) under four meteorological scenarios for the period September to April.

1971 - 1972 Meteorology (Cold, Wet)

RK	Natural		Watana 2001		Devil Canyon 2002	
	Range	Mean	Range	Mean	Range	Mean
242	0-6.8	0.7	0-8.4	1.7	0.7-8.4	2.3
209	0-6.9	0.8	0-8.3	1.5	0-8.4	1.6
161	0-7.1	0.8	0-8.5	1.3	0-8.5	1.4

1974 - 1975 Meteorology (Average, Dry)

RK	Natural		Watana 2001		Devil Canyon 2002	
	Range	Mean	Range	Mean	Range	Mean
242	0-8.5	0.9	0-9.8	2.2	1.2-9.4	3.0
209	0-8.6	1.0	0-9.6	1.8	0-9.4	1.9
161	0-9.1	1.1	0-10.0	1.6	0-9.9	1.9

1981 - 1982 Meteorology (Average, Wet)

RK	Natural		Watana 2001		Devil Canyon 2002	
	Range	Mean	Range	Mean	Range	Mean
242	0-7.7	1.1	0.4-9.0	3.0	1.8-8.3	4.0
209	0-7.9	1.1	0-9.0	2.5	0.7-8.2	3.2
161	0-8.4	1.3	0-9.4	2.1	0-8.6	2.4

1982 - 1983 Meteorology (Average, Average)

RK	Natural		Watana 2001		Devil Canyon 2002	
	Range	Mean	Range	Mean	Range	Mean
242	0-7.9	1.1	0-9.0	2.9	0.9-8.6	3.5
209	0-8.0	1.2	0-8.8	2.4	0-8.6	2.8
161	0-8.4	1.3	0-9.1	2.1	0-8.9	2.2

CONCLUSIONS

Our analysis of expected effects on salmon from altered water temperatures due to operation of the Susitna Hydroelectric Project is based on a comparison of available predictions from the SNTEMP model with fish thermal tolerance criteria. While the SNTEMP model served this analysis well, there are limitations in the available water temperature data and in the modeling system that affect the reliability of the absolute temperatures predicted. The temperature data to which the model was calibrated were available for only a few years and numerous discontinuities in these data exist. Additionally, water temperatures are taken at single points in the river, and are assumed representative of entire cross sections. The SNTEMP model itself, as used in this study for predicting with-project temperatures, relies on the results from a reservoir temperature model for upstream boundary conditions. Consequently, simulated temperatures include the possibility of a variety of combined errors.

While the ability of SNTEMP to predict absolute temperatures is uncertain, much greater reliance may be placed on the relative temperature differences resulting between different simulation scenarios. Thus, the ability to assess the temperature changes resulting from operation of the project remains good. We conclude that our analytical procedure, albeit largely nonquantitative, permitted a reasonable analysis of effects on salmon from temperature changes predicted to occur from operation of the Susitna Hydroelectric Project.

The available fish thermal tolerance information, while of sufficient scope for use in gauging effects on salmon generally, is biased to lower latitudes of North America, necessitating professional interpretation for use in Alaska. Also, salmon are poikilotherms, and thus their body functions are very influenced by environmental temperature. Yet salmon exhibit a degree of thermal plasticity, and are often able to maintain some degree of independence of environmental temperature through homeostatic mechanisms (Warren 1971). We believe the Susitna stocks are adapted to a temperature range of 0 to 18°C. Certainly, narrower tolerance ranges apply to each life phase, and ranges differ slightly among species. Due to the wide temperature range in which salmon can live and function, any project-induced change that remains within their tolerance range requires a subjective analysis.

Based on the SNTEMP model results, salmon thermal tolerance criteria, Susitna stock life history information, and professional judgement, we conclude that no direct mortality is anticipated to occur from with-project temperatures. Although unquantifiable, indirect mortality to some species may occur.

Foremost among these effects is our concern with rearing chinook salmon (in an 80 km mainstem reach downstream from the Devil Canyon dam). Regardless of operating scenario, we believe juvenile chinook salmon growth would be retarded; effects would be more acute under the two-dam configuration than with one. This may result in smaller than normal smolts and or a delay in out-migration, both of which are known to result in reduced survival (Groot 1982; Wedemeyer et al. 1982). The extent of this effect is unquantifiable without more specific information on Susitna salmon stock temperature versus growth relationships.

With-project water temperatures (for the two-dam scenario only) could also delay adult pink and chinook salmon inmigration (and hence, spawning) above RK 209. This could offset the normal timing of embryonic incubation, emergence, and outmigration of the progeny of these species.

Of lesser concern, with-project water temperatures (for the two-dam coldest climate scenarios only) could delay pink and chinook salmon outmigration from rearing habitats near RK 242.

ACKNOWLEDGEMENTS

Financial support for this work was provided by the Alaska Power Authority through its subcontractor Harza-Ebasco Susitna Joint Venture. Support also was provided for the authors' participation in this symposium by the Andrew W. Mellon Foundation and by the Arctic Environmental Information and Data Center, University of Alaska, Fairbanks.

REFERENCES

AEIDC (Alaska, Univ., Arctic Environmental Information and Data Center) 1982. Summary of environmental knowledge of the proposed Grant Lake hydroelectric project area. Alaska Power Authority, Anchorage, AK. Report for Ebasco Services. 212 pp.

Alaska Dept. of Fish & Game 1981a. Annual Salmon Management Report, 1980, Kuskokwim area. Commercial Fisheries Division, Anchorage, AK. 56 pp.

Alaska Dept. of Fish & Game 1981b. Freshwater habitat relationships. Unpublished report by Habitat Division for U.S. Fish and Wildlife Service, Anchorage, AK. 1 vol.

Alaska Dept. of Fish & Game 1983. Susitna hyrdo aquatic studies, phase 2 data report. Winter aquatic studies (October 1982 - May 1983). Final report for Alaska Power Authority. 137 pp.

Alaska Dept. of Fish & Game 1984. Observed water temperatures for salmon species life stages in the Susitna River drainage. Personal communication with Arctic Environmental Information and Data Center, University of Alaska, Anchorage, AK. April 18, 1984.

Alderdice, D. F. and Velsen, F. P. J. 1978. Relation between temperature and incubation time for eggs of chinook salmon (*Oncorhynchus tshawytscha*). J. Fish. Res. Board Can. 35(1): 69-75.

Bailey, J. E. and Evans, D. R. 1971. The low-temperature threshold for pink salmon eggs in relation to a proposed hydroelectric installation. Fish. Bull. 69(3): 587-593.

Barrett, B. M., Thompson, F. M. and Wick, S. N. 1985. Adult anadromous fish investigations (May - October 1984). Alaska Dept. of Fish & Game, Anchorage. Report for Alaska Power Authority. 1 vol.

Bell, M. C. 1980. Fisheries handbook of engineering requirements and biological criteria. Revised. Prepared for Fisheries Engineering Research Program, U.S. Army, Corps of Engineers, Portland, OR.

Brett, J. R. 1974. Tank experiments on the culture of pan-size sockeye (*Oncorhynchus nerka*) and pink (*O. gorbuscha*) using environmental control. Aquaculture. 4: 341-352.

Bucher, W. 1981. 1980 Wood River sockeye salmon smolt studies. In: 1980 Bristol Bay sockeye studies. (Ed by C.P. Meacham) pp. 28-34. Div. of Commercial Fisheries, Alaska Dept. of Fish & Game, Anchorage, AK.

Bustard, D. R., and Narver, D. W. 1975. Aspects of winter ecology of juvenile coho salmon (*Oncorhynchus kisutch*) and steelhead trout (*Salmo gairdneri*). J. Fish. Res. Board Can. 32(5): 667-680.

Cederholm, C. J. and Scarlett, W. J. 1982. Seasonal immigrations of juvenile salmonids into four small tributaries of the Clearwater River, Washington 1977-1981. In: Proceedings of the Salmon and Trout Migratory Behavior Symposium. (Ed by E. L. Brannon and E. O. Salo). pp. 98-100. School of Fisheries, Univ. of Washington, Seattle, WA.

Combs, B. D. 1965. Effects of temperature on the development of salmon eggs. Prog. Fish-Cult. 27: 134-137.

Combs, B. D. and Burrows, R. E. 1957. Threshold temperatures for the normal development of chinook salmon eggs. Prog. Fish-Cult. **19**(1): 3-6.

Dong, J. N. 1981. Thermal tolerance and rate of development of coho salmon embryos. M.S. Thesis. University of Washington, Seattle, WA. 51 pp.

Flagg, L. B. 1983. Sockeye salmon smolt studies, Kasilof River, Alaska 1981. FRED Div., Alaska Dept. of Fish & Game, Juneau, AK. Technical Data Report 11. 31 pp.

Francisco, K. 1977. Second interim report of the Commercial Fish-Technical Evaluation Study. Joint State/Federal Fish and Wildlife Advisory Team, Anchorage, AK. Special Report 9. 46 pp.

Fried, S. M. and Laner, J. J. 1981. 1980 Snake River sockeye salmon smolt studies. In: 1980 Bristol Bay sockeye studies. (Ed by C. P. Meacham). pp. 34-45. Div. of Commercial Fisheries, Alaska Dept. of Fish & Game, Anchorage, AK.

Godin, J. G. 1980. Temporal aspects of juvenile pink salmon (*Oncorhynchus gorbuscha*) emergence from a simulated gravel redd. Can. J. Zool. **58**(5): 735-744.

Groot, C. 1982. Modification on a theme - a perspective on migratory behavior of Pacific salmon. (Ed by E. L. Brannon and E. O. Salo). pp. 1 - 21. Proceedings of the salmon and trout migratory behavior symposium. University of Washington, Seattle, June 3-5.

Hartman, W. L., Heard, W. R. and Drucker, B. 1967. Migratory behavior of sockeye salmon fry and smolt. J. Fish. Res. Board Can. **24**(10): 2069-2099.

Kogl, D. R. 1965. Springs and groundwater as factors affecting survival of chum salmon spawn in a subarctic stream. M.S. Thesis. Univ. of Alaska, Fairbanks, AK. 59 pp.

Mattson, C. R. and Hobart, R. A. 1962. Chum salmon studies in southeastern Alaska, 1961. Bureau of Commercial Fisheries, U.S. Fish & Wildlife Service, Auke Bay, AK. Manuscript Report 62-5. 32 pp.

McCart, P. 1967. Behavior and ecology of sockeye salmon fry in the Babine River. J. Fish. Res. Board Can. **24**: 375-428.

McMahon, T. E. 1983. Habitat suitability index models: coho salmon. U.S. Fish & Wildlife Service. FWS/OBS-82/10.49. 29 pp.

McNeil, W. J. 1969. Survival of pink and chum salmon eggs and alevins. In: Symposium on Salmon and Trout in Streams. (Ed by T. G. Northcote). pp. 101-117. Univ. of British Columbia, Vancouver, B.C. H.R. MacMillan Lectures in Fisheries.

McNeil, W. J. & Bailey, J. E. 1975. Salmon rancher's manual. U.S. National Marine Fisheries Service, Auke Bay, AK. 95 pp.

McNeil, W. J., Wells, R. A. and Brickell, D. C. 1964. Disappearance of dead pink salmon eggs and larvae from Sashin Creek, Baranof Island, AK. U.S. Fish & Wildlife Service. Special Scientific Report - Fisheries 485. 13 pp.

Merritt, M. F. and Raymond, J. A. 1983. Early life history of chum salmon in the Noatak River and Kotzebue Sound. FRED Div., Alaska Dept. of Fish & Game, Juneau, AK. Techincal Bulletin 1. 56 pp.

Meyer, P. R., Kelley, M. D., Voos, K. A. and Wilson, W. J. 1984. Assessment of the effects of the proposed Susitna Hydroelectric Project on instream temperature and fishery resources in the Watana to Talkeetna reach. Vol. 1. Arctic Environmental Information and Data Center, University of Alaska, Anchorage. Report for Alaska Power Authority. 130 pp.

Neave, F. 1966. Salmon of the North Pacific Ocean - Part III. A review of the life history of North Pacific salmon. 6. Chum salmon in British Columbia. International North Pacific Fisheries Commission Bulletin 18. Vancouver, B.C.

Nelson, D. C. 1983. Russian River sockeye salmon. Sport Fish Div., Alaska Dept. of Fish & Game, Juneau, AK. Federal Aid in Fish Restoration. Vol. 24. Project AFS-44. Annual Report. 50 pp.

R&M Consultants, Inc. & Larry A. Peterson & Associates 1981. Review of
existing Susitna River basin water quality data. Report for Acres
American, Inc. 1 vol.

R&M Consultants, Inc., Harza-Ebasco Susitna Joint Venture, Arctic
Environmental Information and Data Center, University of Alaska,
LGL Alaska Research Associates, Inc. and Agriculture and Forestry
Experiment Station, University of Alaska 1985. Susitna River ice
processes: natural conditions and projected effects of hydroelectric
development. Unpublished report. Vol. 1. Alaska Power Authority,
Anchorage, AK. 305 pp.

Raleigh, R. F. 1971. Innate control of migration of salmon and trout fry
from natal gravels to rearing areas. Ecology 52: 291-297.

Raymond, H. L. 1979. Effects of dams and impoundments on migrations of
juvenile chinook salmon and steelhead from the Snake River, 1966 to
1975. Trans. Am. Fish Soc. 108(6): 505-529.

Raymond, J. A. 1981. Incubation of fall chum salmon (*Oncorhynchus keta*) at
Clear Air Force Station, AK. FRED Div., Alaska Dept. of Fish &
Game. Juneau, AK. 25 pp.

Rukhlov, F. N. 1969. The natural reproduction of the autumn chum salmon
(*Oncorhynchus keta*) on Sakahlin. Problems of Ichthyology 9(2):
217-223.

Sano, S. 1966. Salmon of the North Pacific Ocean - Part III. A review of
the life history of North Pacific salmon. Chum salmon in the Far
East. Pages 41-57 in International North Pacific Fisheries Commission
Bulletin 18.

Schmidt, D. C., Hale, S. S., Crawford, D. L. and Suchanek, P. M. 1984.
Resident and juvenile anadromous fish investigations (May - October
1983). Alaska Dept. of Fish & Game, Anchorage, AK. Susitna hydro
aquatic studies. Report 2 for the Alaska Power Authority. Document
1784. 1 vol.

Sheridan, W. L. 1962. Relation of stream temperatures to timing of pink
salmon escapements in southeast Alaska. In: Symposium on Pink
Salmon. (Ed by N. J. Wilimovsky). pp. 87-102. University of
British Columbia, Vancouver, B.C., 1960. H.R. MacMillan Lectures in
Fisheries.

Theurer, F., Voos, K. and Miller, W. 1983. Instream water temperature
model. Draft report. Instream Flow and Aquatic Systems Group,
U.S. Fish & Wildlife Service, Fort Collins, CO. Instream Flow
Information Paper No. 16. 263 pp.

Trasky, L. L. 1974. Yukon River anadromous fish investigations, July 1973
- June 1974. Div. of Commercial Fisheries, Alaska Dept. of Fish &
Game, Anchorage, AK.

Wallis, J. and Balland, D. T. 1983. Anchor River steelhead
investigations. Sport Fish Div., Alaska Dept. of Fish & Game,
Juneau, AK. Federal Aid in Fish Restoration. Vol. 24. Project
AFS-48. Annual Report. 44 pp.

Wangaard, D. B. and Burger, C. V. 1983. Effects of various water
temperature regimes on the egg and alevin incubation of Susitna River
chum and sockeye salmon. Final Report. National Fishery Research
Center, U.S. Fish & Wildlife Service, Anchorage, AK. 43 pp.

Warren, C. E. 1971. Biology and water pollution control. W.B. Saunders
Company, Philadelphia. 434 pp.

Wedemeyer, G. A., Saunders, R. L. and Clarke, W. C. 1982. Environmental
factors affecting smoltification and early marine survival of
anadromous salmonids. Mar. Fish. Rev. June: 1-14.

Whitmore, D. C., Dudiak, N. C. and Testor J. W. 1979. Coho enhancement on
the Kenai Peninsula. FRED Div., Alaska Dept. of Fish & Game, Juneau,
AK. Completion Report AFS-45-1. 54 pp.

Wickett, W. P. 1958. Review of certain environmental factors affecting the
production of pink and chum salmon. J. Fish. Res. Board Can. 15:
1103-1123.

HYDROPOWER DEVELOPMENT OF SALMON RIVERS: EFFECT OF CHANGES IN WATER

TEMPERATURE ON GROWTH OF BROWN TROUT (*SALMO TRUTTA*) PRESMOLTS

Arne Johan Jensen

Directorate for Nature Management
Tungasletta 2, N-7000 Trondheim Norway

INTRODUCTION

Several salmon rivers in Norway have been regulated for hydro-electric purposes. Large reservoirs are often established in the mountains, and power plants discharge into salmon bearing parts of a river resulting in lower water temperatures in summer and higher water temperatures in winter.

Hydrologists during the last few years have been able to predict changes in water temperature, while biologists have not succeeded to the same extent in estimating the corresponding responses of fish growth.

A model describing the growth rates of brown trout (*Salmo trutta*) at different temperatures when fed on maximum rations and on reduced rations has previously been constructed (Elliott 1975a, b). This model has been tested on parr of eleven Norwegian sea trout populations in the present paper, based on these results the model has been adjusted to fit actual growth in these rivers. Hydropower regulation in the River Orkla, in central Norway was completed in 1983. The effect of regulation on river temperature has been predicted by Berge et al. (1982). Based on these predicted temperature changes, the present growth model was used to predict the corresponding effect on growth of parr of the sea trout population.

METHODS

Increases in weights of sea trout presmolts were calculated from river temperatures using the equation given by Elliott (1975a).

$$W+ = [b_1 (a+b_2 T)t + Wo^{b_1}]^{1/b_1} \qquad (1)$$

where, $W+$ = final fresh weight (g)
b_1, a, b_2 = constants
t = time in days, at ____ T°C
Wo = initial fresh weight (g)

The values of a, b_1 and b_2 with their 95 % C.L. are given in Table 2 in Elliott (1975a). The growth rate, which was highest at 12.8°C, decreased below 12.8°C and increased above. In this equation b_1 is the weight exponent, while a and b_2 are the constants describing the linear relationship between specific growth rate and temperature for a trout of unit weight. Elliott's model was originally applied to trout in the size range 5-300 g (wet weight) and water temperatures in the range 3.8 - 19.5°C. The model has subsequently been found to be applicable to smaller trout (Elliott 1984).

The water temperature in most rivers was measured twice (morning and evening) on alternate days by the Norwegian Water Resources and Electricity Board, Hydrological Department. In some rivers continuous temperature measuring equipment was used. The mean river temperature for ten day intervals was used in the growth model.

Sea trout parr were usually collected by electrofishing in April or early May, just after ice has disappeared from the river, and at that time the river temperature was 0-3°C. The youngest year-class in the catch had hatched about 11 months earlier. They are called young-of-the-year or 0+ in this paper. Most of the fish from the River Orkla were downstream migrating smolts caught in traps. The fish were preserved in 70% alcohol, and their length and age determined in the laboratory. Age was usually determined both from scales and otoliths. Relations between length and weight were computed by using the condition factor k = 1.00.

Annual instantaneous growth rate may be calculated thus:

$$Gw = \ln W+ - \ln Wo \tag{2}$$

where, Gw = annual instantaneous growth rate ($g\ g^{-1}\ yr^{-1}$) and compared with the value of Gw derived from actual growth data of fish populations in those rivers where contemporaneous temperature and growth data exist. As Gw is related to fish size, the computed and actual growth data are compared for trout in their second year (1+).

In most rivers growth predicted from the model for growth on maximum rations did not exactly correspond to that observed. In these cases the corresponding model for growth on reduced rations was tried. In some rivers the growth rate also had to be increased or decreased to fit the model to the observed growth data. This was done by fitting the constants a and b_2 to the observed growth, supposing that the growth is zero at 3.5°C and 19.5°C, in accordance with Elliott (1975a).

RESULTS

Growth Rates for Maximum Rations Compared with Observed Growth

Table 1 shows the relationship between computed and observed annual instantaneous growth rates (Gw) using data from 11 Norwegian sea trout populations. The instantaneous growth rates of the trout in their second year of life (1+) are compared. In five of the rivers the actual mean growth rates were less than the computed mean rates at maximum ration, and in six rivers they were higher than the computed mean rates. In six cases the actual mean growth was also higher than the computed upper 95 % C.L., and in one case it was lower than the computed lower 95 % C.L.

The ratio of the observed growth rate and the computed rate at maximum rations varied from year to year in the same river. The lowest growth rate, compared to the computed, was found in the River Saltdalselva. In this river the ratio varied between 68 and 90 percent and the average was 80 percent. The highest observed growth rate when compared to the computed growth at maximum rations, was observed in the Rivers Beiarelva (179 percent in 1981) and Nidelva (172 percent in 1983).

TABLE 1. Comparison of observed and computed annual instantaneous growth
rates (Gw g g^{-1} yr^{-1}) with 95% C.L. on the computed values. Data
for sea trout in their second year.

River	Year	Observed annual Gw	Computed annual Gw	Computed lower 95% C.L.	Computed upper 95% C.L.	Ratio Observed/ computed (percent)
Altaelva	1980	1.28	1.50	1.05	1.95	85
"	1981	0.93	1.26	0.92	1.59	74
"	1982	1.25	1.54	0.98	1.91	81
"	1983	1.41	1.39	1.01	1.93	101
Beiarelva	1978	1.65	1.45	0.70	2.03	114
"	1979	1.44	1.24	0.56	1.79	116
"	1980	1.63	1.35	0.60	1.94	121
"	1981*	1.38	0.77	0.25	1.30	179
"	1982	1.22	0.78	0.23	1.37	156
"	1983	1.50	1.28	0.51	1.81	117
Lakselva	1977*	1.92	1.48	0.97	1.90	130
"	1978*	2.00	1.52	0.93	1.92	132
"	1980*	1.91	1.45	0.94	1.88	132
"	1981	1.69	1.63	1.01	2.11	104
Saltdalselva	1978	1.34	1,76	1.12	2.21	76
"	1979	1.38	1.71	0.93	2.25	81
"	1980	1.64	1.84	1.25	2.27	89
"	1981**	0.93	1.37	1.13	1.94	68
"	1982	1.17	1.30	1.08	1.92	90
"	1983	1.17	1.53	0.69	2.12	76
Vefsna	1979	1.77	1.77	1.14	2.25	100
"	1980	1.49	1.72	1.16	2.14	87
"	1981	1.58	1.62	0.83	2.21	98
"	1982	1.56	1.56	0.87	2.08	100
"	1983	1.16	1.54	0.79	2.09	75
Nidelva	1982	1.93	1.49	0.82	1.95	130
"	1983*	2.22	1.29	0.68	1.80	172
Orkla	1984	1.96	1.81	1.06	2.38	108
Strynselva	1975	1.56	1.82	1.20	2.32	86
"	1979	1.47	1.51	0.98	2.10	97
"	1982	1.62	1.94	1.22	2.49	84
"	1983	1.78	1.75	0.84	2.41	102
"	1984	1.28	2.04	1.05	2.73	63
Hjelledøla	1982	1.38	1.34	1.00	1.97	103
"	1983	1.57	1.07	0.38	1.70	147
"	1984	1.55	1.21	0.44	1.78	128
Loelva	1975	1.25	1.87	1.12	2.50	67
"	1979	1.64	1.53	0.96	2.01	107
Jostedøla	1976*	1.61	1.04	0.43	1.44	155

* Observed annual Gw higher than the upper 95% C.L. calculated Gw
** Observed annual Gw lower than the lower 95% C.L. calculated Gwd

The growth of sea trout presmolt in Norwegian rivers is hence often better than that described by Elliott´s model for trout fed on maximum rations (Elliott 1975a). The model is, therefore, not adequate for describing the growth of trout in these rivers without prior adjustments.

Predicting Growth of Young-of-the-year Trout

An estimate of the growth of young-of-the-year trout is required before any adjustments of the model can be made.

The growth model for trout fed on maximum rations was orginally applied to trout in the size range 5-300 g and water temperatures in the range 3.8-19.5 C (Elliott 1975a). Elliott later found that the model was also applicable to smaller trout kept in the laboratory. Therefore, Elliott (1984) used it to estimate the growth of young-of-the-year trout in Black Brows Beck, England from the start of the growth period, when the fish were still alevins.

The growth of young-of-the-year trout in the Norwegian rivers could not be computed in this way, however, since both initial weight and the time when the growth period started were unknown. Instead of this, a correlation was found between the computed growth of 1+ trout from Elliott´s growth model on maximum rations (adjusted for data on actual growth in Table 1) and the actual length of fry by the first winter (Figure 1; eg. 3):

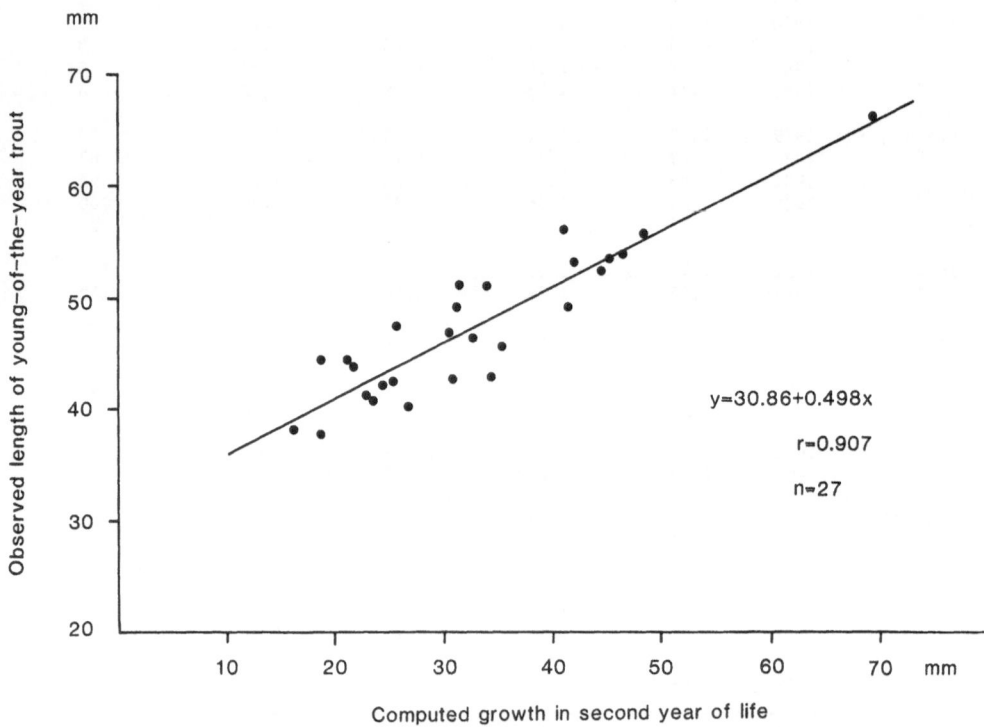

Figure 1. Correlation between the growth of 1+ trout computed from the growth model for trout fed on maximum rations (Elliott 1975a), and the observed length of young-of-the-year trout by the first winter in some Norwegian rivers.

$$y = 30.86 + 0.498 x \qquad\qquad (3)$$

where, y = actual length of young-of-the-year trout
       x = computed growth in second year of life

Actual Growth Adjustments for Elliott's Growth Model

Elliott (1975b) has shown that the optimum temperature for growth
decreased progressively from about 13°C at maximum rations to about 4°C at
a ration size just above the maintenance ration. If data on growth of
presmolt trout as well as water temperature data are available over several
years, an adjusted model may be constructed for a particular river. Growth
potential, actual optimum temperature, and the nutrient conditions in the
river must be considered. Three examples of growth model adjustments for
trout fed on maximum rations (Elliott 1975a) are given below (Figure 2.).

1. River Orkla. The River Orkla is situated at 63 N, 9 15′E, in central
Norway. The salmon carrying part of the river is 92 km long, and the
yearly catch of Atlantic salmon (*Salmo salar*) and sea trout is about 10
tons. Atlantic salmon is the predominating species. The trout smolt age
is approximately 3 years. The River Orkla was regulated for hydroelectric
purposes in 1983. Samples of downstream migrating smolts were made in five
years prior the regulation (Hesthagen and Garn s 1984). Additional samples
of presmolts are available for one year. Since the smolt age is about 3
years, the mean length of three-year smolts is compared with the growth
computed from Elliott's model. Growth is slightly better than that
calculated from Elliott's growth model (1975a) for trout fed on maximum
rations (Figure 3). A model with a somewhat higher growth rate, but with a
growth maximum at the same temperature (13°C), seems to be the most
suitable growth model for sea trout presmolts in the Orkla (Figure 2;
Figure 3).

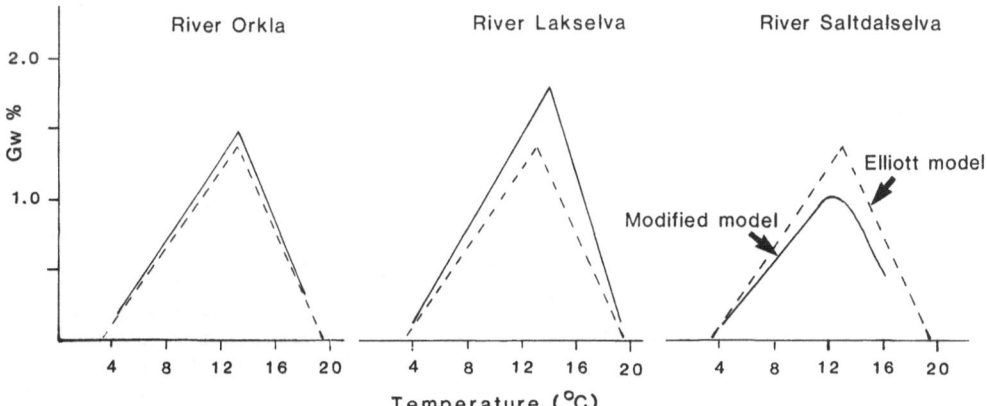

Figure 2. Three examples of growth model adjustments to the actual growth
observed in rivers. (Specific growth rate Gw % day^{-1} for a 10 g
trout).

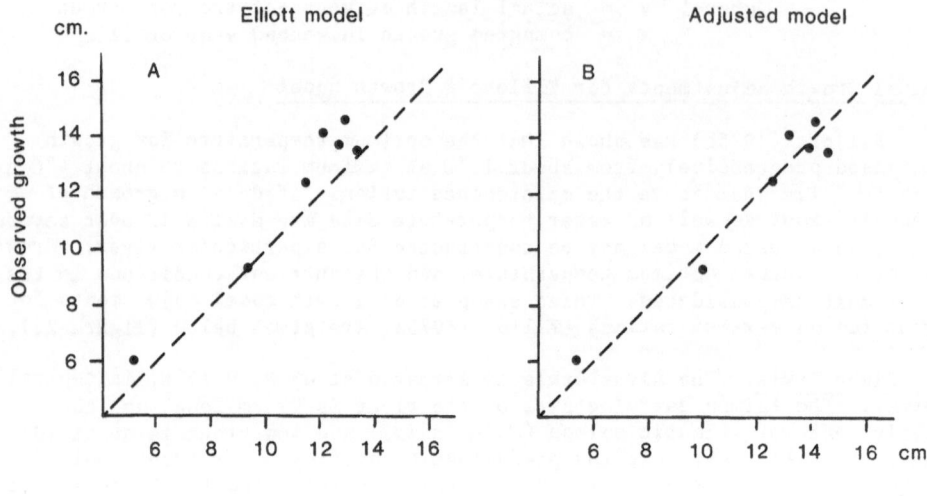

Figure 3. Correlation between computed and observed growth of 3 year old
trout smolts in the River Orkla. Data on 0+ and 1+ trout from
one year are also presented.
(A) Computed from the growth model for trout fed maximum rations
(Elliott 1975a),
(B) Computed from the adjusted growth model shown in Figure 2.

2. River Lakselva. The River Lakselva (67°5´N, 15 E), situated in Misvaer
in Nordland county, is a small salmon river. Salmon and sea trout can move
about 7 km upstream from the sea. The total catch is about 0.5 t yr^{-1}.
The river is rather warm for that latitude, with a mean temperature of
17.7°C in July 1980, and mean temperatures in July and August are about
14°C. The growing season is rather short, with only about 120 days above
5°C. This growth rate is significantly higher (p<0.05) than predicted by
the growth model for trout fed on maximum rations (Table 1; Figure 4).
The actual data was best fitted to a model which has a maximum growth at
14°C, and with a higher growth rate than predicted by Elliott´s model for
trout fed on maximum rations (Figure 2; Figure 4). A maximum growth rate
at 14°C is higher than that found by Elliott for trout fed on maximum
rations, and seems surprising, but the correlation between the model and
actual growth was considerably better with a growth maximum at 14°C, than
any model with a maximum growth at 13°C.

3. River Saltdalselva. The growth rate of trout in the River Saltdalselva
(67 N, 15° 20´E) was lower in all the years examined than that predicted by
the model for trout on maximum ration (Table 1; Figure 5). The
Saltdalselva is a cold river, with a mean smolt age for sea trout of about
4.5 years. The total catch of Altantic salmon sea trout and sea charr
(*Salvelinus alpinus*) is about 5-8 t yr^{-1}. The temperature for maximum
growth is lower than 13°C, and therefore, the nutrient conditions are
unsufficient. Growth at low temperatures was also lower than predicted by
the model. The actual data were best fitted to a model with a maximum
growth at 12°C, and with a somewhat lower growth rate than that found by
Elliott. When fitting the data to the modified model for Saltdalselva

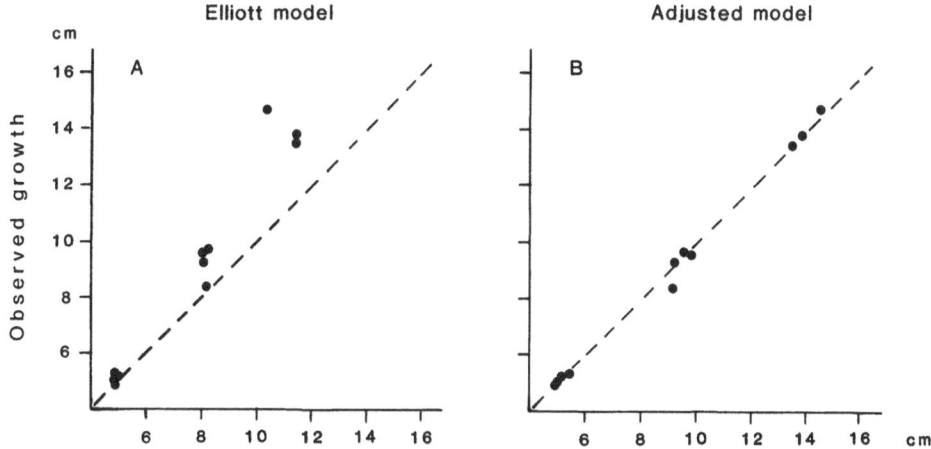

Figure 4. Correlation between computed and observed growth of 0+, 1+ and 2+
trout in the River Lakselva.
(A)   Computed from the growth model for trout fed maximum rations
      (Elliott   1975a),
(B)   Computed from the adjusted growth model shown in Figure 2.

(Figure 2), a growth curve corresponding to a 50 g trout fed a ration of
500 mg d^{-1} (Figure 3 in Elliott 1975b), was used, and the growth rate was
reduced, as shown in Figure 2.

Use of the Adjusted Model to Predict the Effect of Altered Water Temperature on the Growth of Trout Presmolts.

The proposed method of predicting the effect of altered water
temperatures on the growth of trout presmolts will be illustrated by an
example, for which the regulation of the River Orkla is chosen.

The Orkla hydro-electric power scheme was completed in 1983. It is
owned by Kraftverkene i Orkla, and consists of three reservoirs and five
power plants (Figure 6).

The most important areas for growth and spawning of salmon and trout
are found between Meldal and Brattset. Farther downstream most life has
been exterminated by pollution from copper and zinc mines.

The river temperature at the outlet of Grana before hydropower
regulation was usually 12-15°C in July and August. It decreased gradually
in September and October, and from November to April the water temperatures
were less than 1°C, usually about 0.1°C. In early May temperatures again
increased, but usually did not exceed 5°C before the end of the month.

The effect of regulation on river temperature was predicted by Berge
et al. (1982). They simulated the effect of the river temperature for each
of the five power plants between June 20 and September 30, in one
particular year, and the corresponding temperature farther downstream from

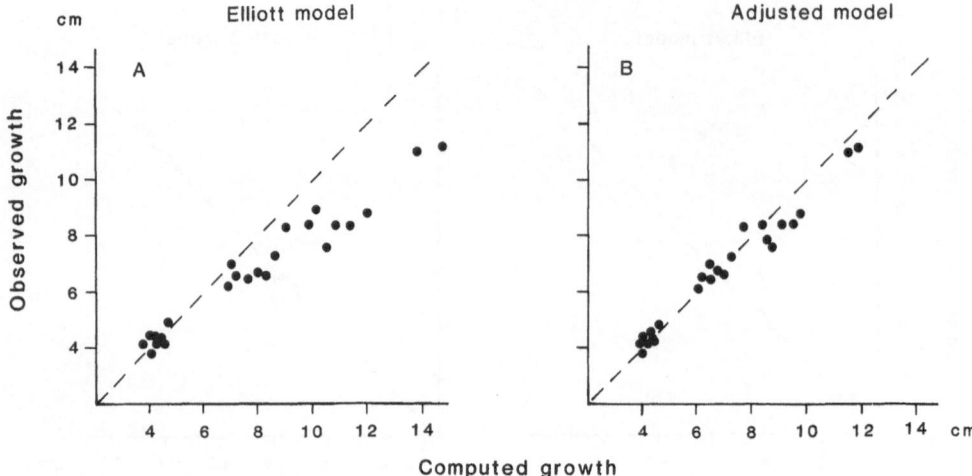

Figure 5. Correlation between computed and observed growth of 0+, 1+, 2+
and 3+ trout in the River Saltdalselva.
  (A) Computed from the growth model for trout fed maximum rations
       (Elliott 1975a),
  (B) Computed from the adjusted growth model shown in Figure 2.

hydrological models of the watershed. The year 1965 was chosen, as the air
temperatures at the meteorological station (Saeter i Kvikne) closest to the
river approximated the mean for the period 1931-1960. Precipitation during
this year was 85 percent of the mean.

The average decrease in summer river temperature was predicted at
about 2.5 °C downstream from both the Brattset power plant and the Grana
power plant. Mean river temperature for the stretch Brattset-Meldal for
the period June 20 to September 30 was calculated from predicted changes in
water temperature, and used in growth calculations. In October, May and
June (until June 20) mean river temperatures from before regulation were
used. River temperatures during the remainder of the year although higher
than previous to regulation, will be far below 3.5 °C, which is the minimum
temperature for growth of trout (Elliott 1975a, b). Berge et al. (1982)
suggested no changes in water temperature in October, while in the middle
of June they suggested a small decrease. Measurements from a few years
after regulation indicate a small increase in river temperatures in the
final days of May and the first days of June. These temperature changes
combined are expected to be mutually exclusive, therefore may be ignored.

According to the adjusted growth model for the River Orkla, the
predicted growth of 1+ trout was 43.8 mm in 1965. When using the mean
river temperature predicted for the strech Brattset-Meldal, if the
hydropower regulation had been completed before 1965, the corresponding
predicted growth of 1+ trout would have been 36.4 mm for that year (Table
2).

The mean length of trout after one, two and three years with the same
temperature conditions as 1965, is computed in Table 2 in the unregulated
as well as the regulated river. The mean length of trout after three years
like 1965 is calculated to be 140.9 mm, while the mean smolt length for sea

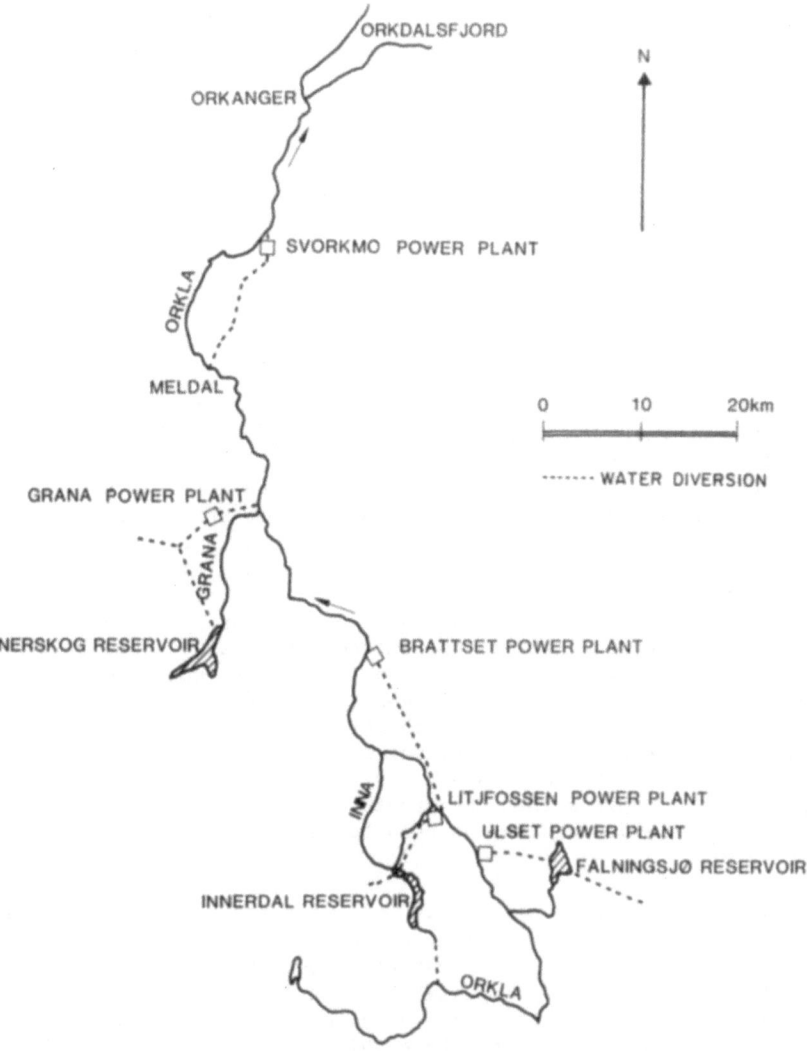

Figure 6. River Orkla with the hydroelectric power scheme.

Table 2. Computed length (mm) of trout after one, two and three years in
          the River Orkla, according to the adjusted growth model for this
          river. Water temperatures as predicted by Berge et al. (1982)
          for a "normal" year before and after hydro-electric prower
          regulation.

Age	Unregulated river	Regulated river
0+	52.7	49.0
1+	96.5	85.4
2+	140.9	121.2

trout caught before the regulation was 138.0 mm. The mean smolt age in these samples was 3.1 yr (n = 364). The mean length for 199 three-year old smolts was 137.9 mm.

In the regulated river the mean length of trout is predicted to be 121.2 mm after three years, which is 19.7 mm shorter than in the unregulated river. This corresponds to one half year's growth.

In the present calculations only the effect of water temperature is considered, while the nutrient conditions are considered as unchanged. Possible changes in nutrient conditions should be considered separately.

DISCUSSION

Elliott (1975a) concluded that growth rates calculated from the growth model for brown trout fed maximum rations represent the maximum growth rate at each temperature, and that it would be remarkable if these rates were ever attained by wild trout under natural conditions. The results from nine English rivers examined by Edwards et al. (1979) are in accordance with this conclusion. Comparisons of trout growth in these nine rivers revealed growth rates between 60 and 90 percent of the growth model. In six of the eleven Norwegian rivers studied the growth rates exceeded the growth rate calculated from Elliott's growth model. Allen (1985) also found a higher growth rate for the brown trout population in the Horokiwi Stream, New Zealand. In a man-made reservoir in Norway the growth rates both for brown trout and Arctic char exceeded the established limits of maximum growth for all salmonids (Jensen 1985). The source of variation between populations is most probably genetic. Refstie and Steine (1978) found considerable variation in growth rate between strains of Atlantic salmon during the freshwater phase, and such variation obviously also appears in trout populations. Genetic variation in growth rates between trout populations implies that the growth model for trout fed maximum rations (Elliott 1975a) can not be used for every trout population without prior adjustments.

Elliott (1975a) found his growth model for trout fed maximum rations applicable at both constant and fluctuating temperature. As a result daily mean temperatures were used in my computations. Hokanson et al. (1977), however, found in laboratory experiments with Rainbow trout (*Salmo gairdneri*) that when yield was plotted against mean temperature, the curve of response to fluctuating temperature shifted horizontally an average 1.5 °C toward colder temperatures than the curve of response to constant temperature treatments. This pattern of response to fluctuating treatments indicates that rainbow trout do not respond to mean temperature, but they acclimate to some value between the mean and maximum daily temperatures. The response of brown trout to fluctuating temperatures should be more thoroughly examined.

Supposing that the mean smolt length of brown trout in the River Orkla will be the same after hydropower regulation, half of the population must stay one more year in the river before smolt transformation in order to compensate for the lower growth. A considerable loss of smolts due to mortality will take place in this extra year. An annual mortality rate of 70 percent is normal in presmolt sea trout (Le Cren 1973, Egglishaw and Shackley 1977, Mortensen 1977). Smaller fish will have different space requirements, and hence, more fish may stay on the same area. Allen (1969) found a positive linear relationship between average area required by individual fish and fish length, when both were plotted on logaritmic scales. The space requirements of all age-classes of juvenile Atlantic salmon at their respective lengths in midsummer was estimated by Symons

(1979) for rivers producing smolts aged 1+ to 4+ from Allen's data. When using these data for the trout in the River Orkla, I computed that 10 percent higher densities could live in this river after regulation, supposing that the growth will be similar to that computed in Table 2. Hence, 1000 three year old smolts before regulation will change to 550 three year old smolts and to 165 four year old smolts (after regulation). The total smolt production would be about 71.5 percent of that before regulation, and the new smolt age could increase from 3.0 to 3.2 yr. Garnås and Hvidsten (1984) estimated the yearly production of trout smolts upstream from Meldal to be about 25,000 before regulation by finclipping wild smolts before downstream migration and trapping them during migration. It is still too early to chart the effect of regulation on the trout population, but the effect of the changed temperature regime is, according to the above computations, expected to result in a yearly decrease in smolt production of about 7000 individuals.

If a yearly mortality rate of 60 percent instead of 70 percent is chosen, the smolt production will still be reduced about 29 percent, as only 2 percent higher densities of trout (in this case) are expected to live in the river (Allen 1969, Symons 1979). Correspondingly, an annual mortality rate of 50 percent involves 25 percent loss of the smolt production. Hence, the value of the mortality rate chosen in the computations influences the results only to a slight degree. The new smolt age will, however, increase somewhat with decreasing mortality rate.

In this paper only the effect of the altered temperature is considered, while all other factors are considered to be unchanged. This is obviously not realistic. River flow is also usually altered, most often by greater discharge in winter and lesser in summer. Both a high, stable flow in winter and a lower peak flow in spring is expected to improve nutrient conditions for fish. Higher winter temperature results in earlier eggs hatching than prior to regulation, hence, the growth season for fry will be prolonged. Both of these factors are responsible for a lower growth reduction than predicted from river temperature alone. Density of fish, and the competition between trout and Atlantic salmon may also change after regulation because of altered flow. Hence, the present method is not expected to illustrate the precise magnitude of growth change, but handles one of the most important of several parameters which influence growth. Possible changes in nutrient conditions or other factors must be considered separately.

REFERENCES

Allen, K. R. 1969. Limitations on production in salmonid populations in streams. In: "Symposium on salmon and trout in streams". (Ed. by T. Northcote). pp. 3-18. Macmillan Lect. Br. Columb. Univ.
Allen, K. R. 1985. Comparison of growth rate of brown trout (*Salmo trutta*) in a New Zealand stream with experimental fish in Britian. J. Anim. Ecol. **54**: 487-495.
Berge, F. S., Stang, O. and Thendrup, A. 1982. Temperaturendringer i Orkla som følge av kraftutbygging - III. VHL, Trondheim, Norway. STF 60 A811091. (In Norwegian.)
Edwards, R. W., Densem, J. W., and Russell, P. A. 1979. An assessment of the importance of temperature as a factor controlling the growth rate of brown trout in streams. J. Anim. Ecol. **48**: 501-507.
Egglishaw, H. J. and Shackley, P. E. 1977. Growth, survival and production of juvenile salmon and trout in a Scottish stream, 1966-75. J. Fish Biol. **11**: 647-672.

Elliott, J. M. 1975a. The growth rate of brown trout (*Salmo trutta* L.) fed on maximum rations. J. Anim. Ecol. **44**: 805-821.

Elliott, J. M. 1975b. The growth rate of brown trout (*Salmo trutta* L.) fed on reduced rations. J. Anim. Ecol. **44**: 823-842.

Elliott, J. M. 1984. Growth, size, biomass and production of young migratory trout *Salmo trutta* in a Lake District stream, 1966-83. J. Anim. Ecol. **53**: 979-994.

Garnås, E. and Hvidsten, N. A. 1984. Utvandring og produksjon av smolt hos laks og aure i Orkla fra 1979 til 1983. Direktoratet for vilt og ferskvannsfisk. Regulerings undersokelsene. Trondheim. Rapport nr. 7-1984. (In Norwegian.)

Hesthagen, T. and Garn s, E. 1984. Smolt age and size of Atlantic salmon *Salmo salar* L. and sea trout *Salmo trutta* L. in a Norwegian river. Fauna norv. Ser. A **5**: 46-49.

Hokanson, K. E. F., Kleiner, C. F., and Thorslund T. W. 1977. Effects of constant temperatures and diel temperature fluctuations on specific growth and mortality rates and yield of juvenile Rainbow trout, *Salmo gairdneri*. J. Fish. Res. Board. Can. **34**: 639-648.

Jensen, J. W. 1985. The potential growth of salmonids. Aquaculture **48**: 221-231.

LeCren, E. D. 1973. The population dynamics of young trout (*Salmo trutta*) in relation to density and territorial behaviour. Rap. P.-V. Reun. Cons. int. Explor. Mer 164: 241-246.

Mortensen, E. 1977. Population, survival, growth and production of trout *Salmo trutta* in a small Danish stream. Oikos **28**: 9-15.

Refstie, T., and Steine, T. A. 1978. Selection experiments with salmon III. Genetic and environmental sources of variation in length and weight of Atlantic salmon in the freshwater phase. Aquaculture 14: 221-234.

Symons, P. E. K. 1979. Estimated escapement of Atlantic salmon (*Salmo salar*) for maximum smolt production in rivers of different productivity. J. Fish. Res. Board Can. **36**: 132-140.

PREDICTING THE EFFECTS OF A POSSIBLE TEMPERATURE INCREASE DUE
TO STREAM REGULATION ON THE EGGS OF WHITEFISH (*COREGONUS LAVARETUS*)
- A LABORATORY APPROACH

Svein Jakob Saltveit and Åge Brabrand

Laboratory of Freshwater Ecology and Inland Fisheries (LFI)
Zoological Museum, University of Oslo
Sarsgt. 1, 0562 Oslo 5, Norway

INTRODUCTION

In the fjord lake Randsfjorden, Norway, the whitefish (*Coregonus
lavaretus*) has four sympatric populations (Enge 1959). One of these spawns
in the lower parts of the main inlet River Dokka-Etna, while the others
spawn in the lake.

Recapture of tagged whitefish (Styrvold et al. 1981) and quantitative
echosounding (Lindem 1980 a, b) have shown that the river spawning
population is evenly distributed in the lake during the summer, but
migrates to the river in late summer and autumn. Spawning occurs during
October and the eggs are spread randomly over the river bed (Styrvold et
al. 1981). Egg development takes place during the winter, with hatching in
the second half of April, just before the main spring spate. The larvae
immediately drift into the lake (Figure 1).

In this region, with very low winter temperatures, most egg
development takes place in the autumn and spring. Regulation plans for the
inlet river and its catchment will lead to changes in discharge and
autumnal temperature increase below the power station, since water will be
taken from a deep release reservoir and subsequently released to the river
as minimum instream flow. Reduced water flow and increased temperature
during autumn and winter are expected to affect life history parameters of
the river spawning whitefish, including migration, spawning time, available
spawning area, egg development and hatching time.

Little information is available on the effects of river regulation on
whitefish. However, a relationship between temperature and egg incubation
period for *Coregonus wartmanni*, which is closely related to *C. lavaretus*,
indicated that a very small increase in temperature close to 0°C would lead
to much more rapid egg development (Braum 1964). We therefore carried out
experiments to determine the magnitude of the effect of the temperature
increase due to regulation on egg development. The intention was to
postulate possible effects of increase in temperature during the autumn,
and not to deduce any relationship between egg incubation and temperature.
The results of these laboratory experiments are presented here.

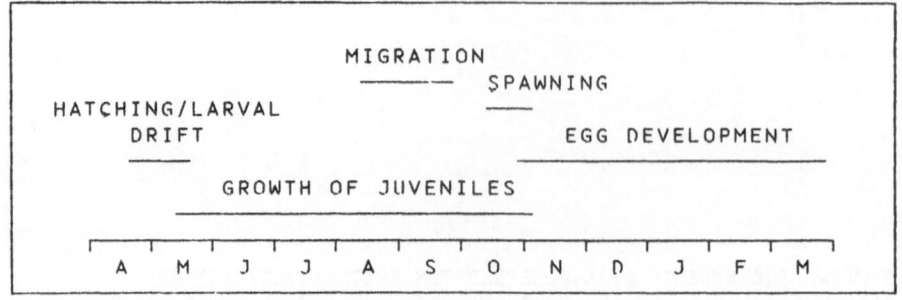

Figure 1. Life history of the river spawning population of whitefish
in Lake Randsfjorden.

STUDY AREA

Lake Randsfjorden, 132 m above sea level is the fourth largest lake in
Norway, with surface area of 136 km^2. The lake is situated in the central
part of south Norway (Figure 2). The two largest tributaries are the
rivers Dokka and Etna. The confluence is approximately 3 km from the lake
and they enter the lake from the north.

The two rivers drain a mountain area to the north-west of the lake.
Both are typical of eastern watercourses in Norway, with a high spring
discharge in May-June (highest weekly mean flow approximately 190 m^3 s^{-1})
and some smaller spates in summer and autumn. During the winter the
discharge is low (mean flow 2.5-6 m^3 s^{-1}).

The pelagic fish community in the lake is dominated by whitefish,
Coregonus lavaretus, and smelt, *Osmerus eperlanus*, which are the only true
planktivorous fish species. Brown trout, *Salmo trutta*, and charr,
Salvelinus alpinus, are palagic predators feeding mainly on smelt.

The littoral fish community is rich in species, with perch (*Perca
fluviatilis*) dominating.

Fish exploition is concentrated on whitefish and brown trout.
Offshore, fishing with pelagic gill nets is unrestricted and the calculated
whitefish yield is about 90 tonnes per year, of which 10-20 tonnes are
caught during spawning in the main inflow river.

MATERIAL AND METHODS

Egg Incubation

Eggs from mature whitefish caught in the river during spawning were
artificially fertilized and transported to the laboratory in glass jars kept
within iceboxes. Eggs and sperm from twenty individuals of both sexes were
mixed to reduce genetic variation.

Rearing of whitefish eggs seems to be more successful in running water
and an experimental setup to provide this was therefore used. This
included a thermostatically controlled reservoir (15 1) to control
temperature to the nearest 0.1°C. The water was pumped from the reservoir
into three glass bulbs from below. The fish eggs were placed in the glass
bulbs, and temperature recorded daily. After passing the glass bulbs, the

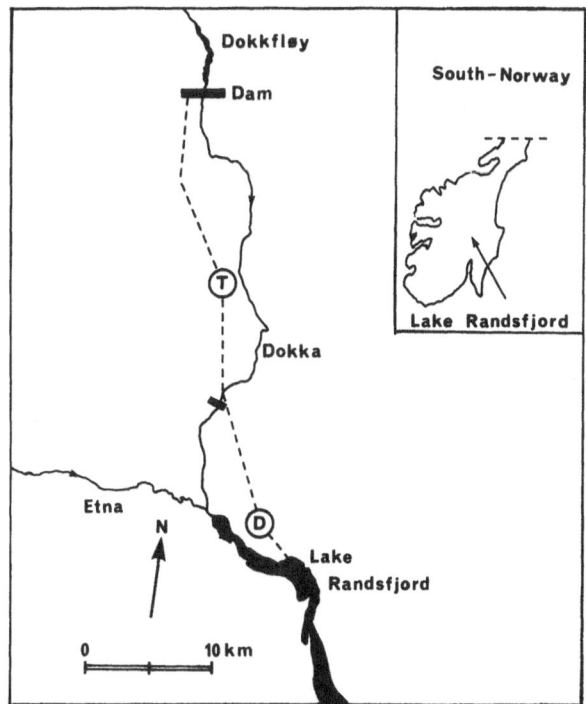

Figure 2.   Map of Lake Randsfjorden. The two main trubutaries included in
the regulation plans are the Rivers Etna and Dokka. The planned
power stations are marked T and D, T giving rise to a minimum
instream flow in the lower parts of the rivers.

water was returned to the reservoir.

The experiments were conducted in a cold room (c. 0.5°C). Dead eggs
were removed weekly and methylene blue oxalate was used to prevent fungus
development. The duration of egg incubation was taken as the period in days
from fertilization until 50% of the eggs that eventually hatched had
hatched.

SIMULATED TEMPERATURES

The general temperature regime during egg development is characterized
by falling temperatures during spawning (phase I), a stable winter period
with water temperature close to 0°C (phase II) and a temperature increase
in the second half of April (phase III). The shape of the cumulative
day-degree curves (Figures 5, 6) reflects these three phases.

The expected increase in river temperature after regultion depends
upon the water flow, air temperature and distance from the power station.
Figure 3 shows the simulated increase in river temperature with three
different water flows. In our experiments eggs of whitefish were exposed to
temperature regimes which occurred with a planned minimum instream water
flow of 3.0 $m^3 s^{-1}$ during September and October and 1.5 $m^3 s^{-1}$ from 1
November. The temperature difference attributed to regulation is +3.0°C
during maximum spawning (15-20 October). However, one week later, the post
regulation temperature is +1.5 to +2.0°C higher. By 1 November the

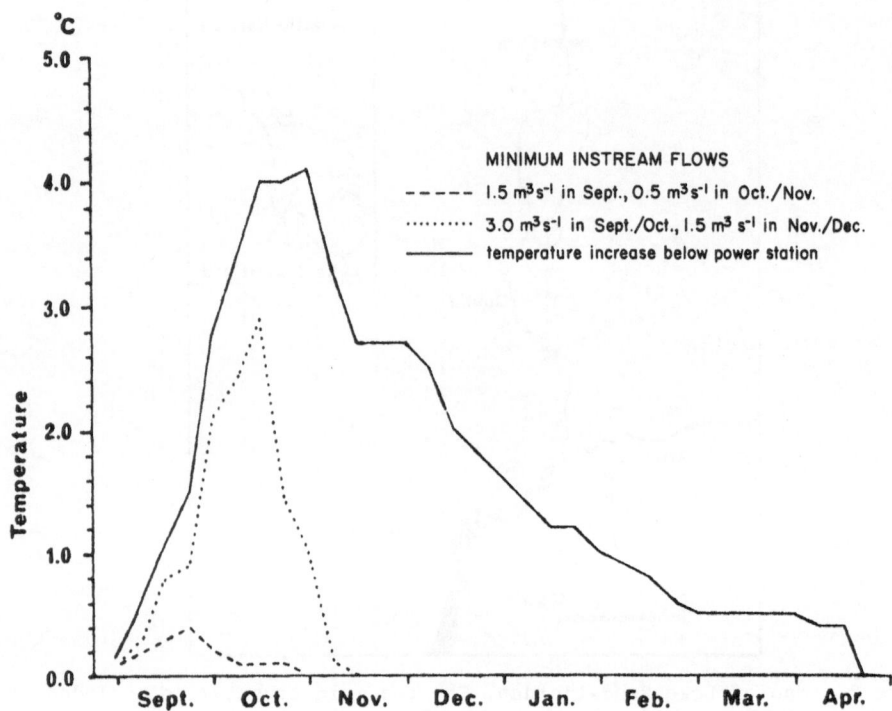

Figure 3. Temperature increase in the River Dokka just before its
confluence with the river Etna based on simulated temperature
regimes before and after regulation (SINTEF 1982).

Table 1. Temperature regimes for laboratory experiments carried out on
egg development of whitefish.

Experiment	Autumn (Phase I) Falling Temp.	Winter (Phase II) Constant Temp.	Spring (Phase III) Increasing Temp.
A	Natural	Constant $0.1°C$	Rise in April to $5.5°C$
B	Regulated	Constant $0.1°C$	Rise in April to $5.5°C$
C	Natural	Constant $0.5°C$	Stepwise increase to $5.0°C$
D	Regulated	Constant $0.5°C$	Stepwise increase to $5.0°C$

temperature in the unregulated river reaches +0.1°C, while after regulation this stable winter temperature is reached 14 days later. During winter the river will have normal river temperatures, and no further difference will occur.

The experiments were based on simulated river temperatures (estimated by SINTEF (1982)), starting with one simulated natural temperature regime and one regulated temperature regime during the autumn (see Table 1). The actual temperature regimes are given in Figure 4. In December each of the two starting experiments were divided into two groups, giving rise to a total of four different experiments during winter until hatching in spring.

During the winter, eggs were exposed to two sets of winter temperature regimes. In one, eggs were kept at 0.1°C throughout the winter until 9 April, when the natural increase in temperature was followed. In the other, eggs were kept at 0.5°C until 18 February, then increased to 1.0 until 15 March, when the natural increase in temperature was followed.

RESULTS

The number of day degrees to which the eggs were exposed is shown in Figure 5 and Figure 6. There was a relatively rapid increase in degree days just after spawning (Phase I) (Table 1), while during the winter the gain in degree days was low due to the extremely low temperatures (Phase II). It was not until March/April that a relatively large increase in the cumulative degree days occurred.

Eye-spots appeared in 28 days (at about 100 degree days) in eggs given post regulation temperatures, while eye-spots in eggs reared under the natural temperature regime appeared 3 weeks later, after having received only 75 degree days (Table 2, Figure 5 and Figure 6).

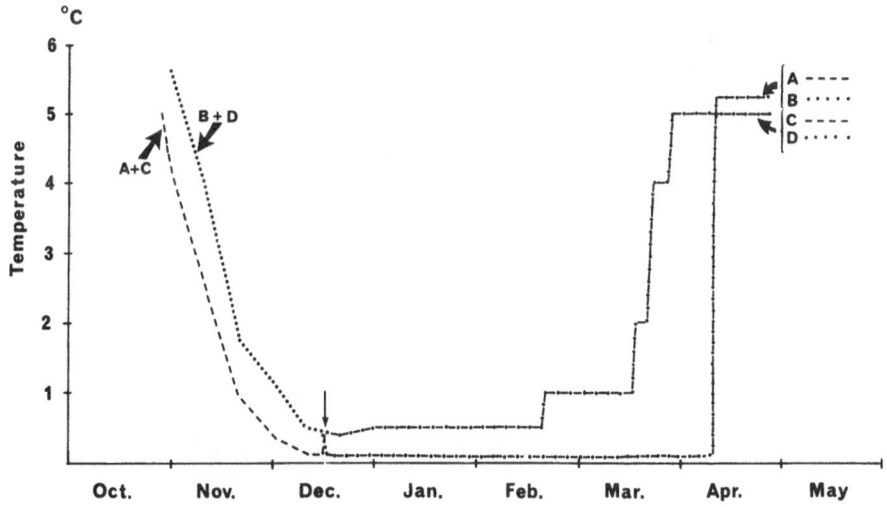

Figure 4.   Temperature regimes used for eggs of whitefish during laboratory experiments. In the middle of December eggs exposed to natural river temperature (A and C) and to regulated river temperature (B and D) were given two sets of winter temperature and different spring temperature regimes.

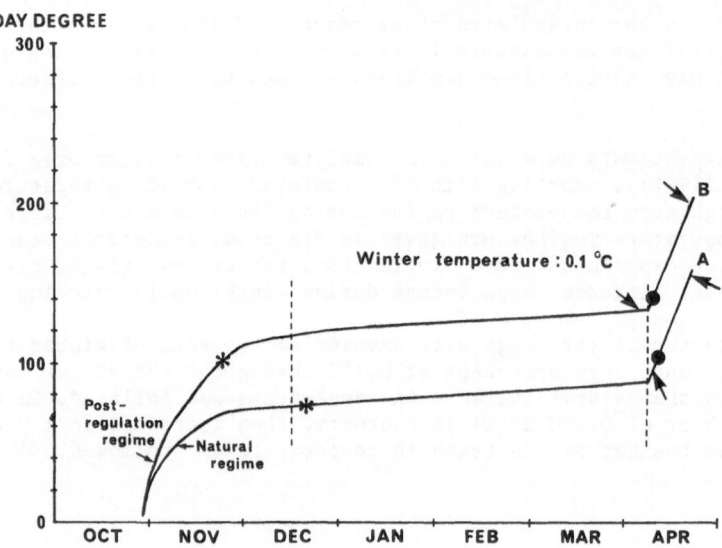

Figure 5. Cumulative degree days for eggs of whitefish under natural
temperature regime (A) and after regulation (B), both at a
constant winter temperature of 0.1°C. Spring temperature was
increased suddenly from 0.1°C to 5.5°C in April. Eyespots are
indicated with an asterix. Hatching periods are indicated
between the arrows and 50% hatching with filled circle.

Eggs kept at 0.1°C throughout the winter followed by an increase to
5.5°C at the beginning of April, began hatching by 9 April under natural
temperature conditions and by 5 April under regulated conditions. All eggs
kept at 0.5°C increasing to 1.0°C during the winter followed by a natural
temperature in March, started hatching at the same time, 18 March. The
hatching pattern and duration of hatching are given in Figure 7 and Figure
8. Despite small difference in the number of days required for hatching,
the mean incubation times were significantly different (ANOVA) both with
respect to winter temperatures (p<0.01) and under regulated and unregulated
conditions given a winter temperature of 0.1°C (p<0.01) (Table 3). In eggs
kept at higher winter temperatures, no significant difference (p<0.05) in
mean hatching time was found.

Table 2. Days and number of degree days necessary for development of
eye-spots and for 50% hatching of whitefish eggs reared at
natural and regulated autumn temperature conditions and two
different winter temperatures.

| Autumn Condition | Eye-Spots | | 50% Hatching Winter Temperatures | | | |
| | | | 0.1°C | | 0.5°C | |
	Days	Degree Days	Days	Degree Days	Days	Degree Days
Natural Temp.	54	75.4	162	97.9	155	223.5
Regulated Temp.	28	101.1	160	143.8	154	263.5

224

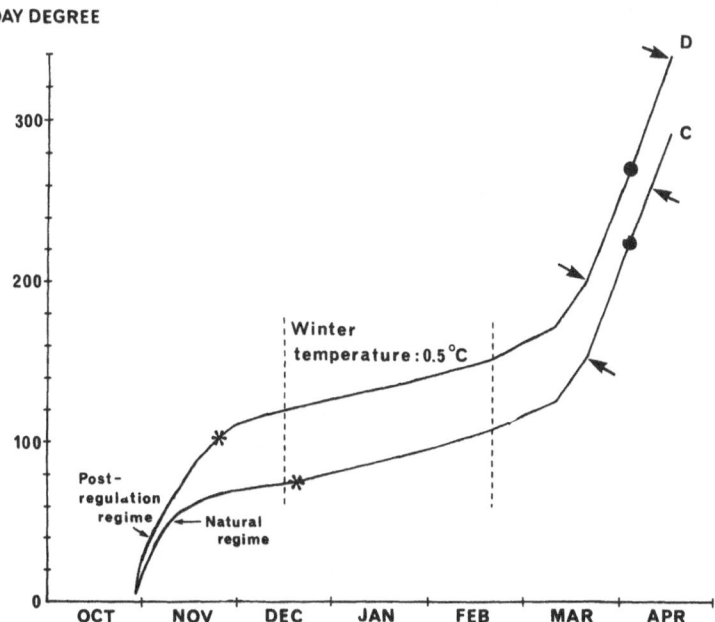

Figure 6. Cumulative degree days given to eggs of whitefish under the natural temperature regime (C) and after regulation (D), and with constant winter temperature of 0.5°C. In the middle of February temperature was increased stepwise up to a maximum of 5.0°C in late March. Eye-spots are indicated with an asterix. Hatching periods are indicated between the arrows and 50% hatching with filled circles.

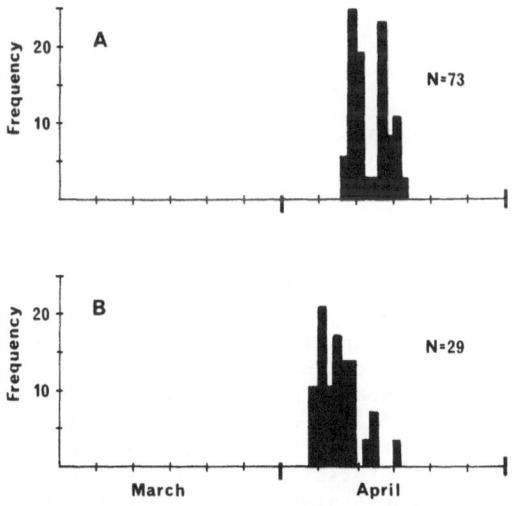

Figure 7. Hatching frequency of whitefish eggs exposed to natural temperature regimes (A) and after regulation (B) of the spawning river, and a constant winter temperature of 0.1°C, increasing suddenly until first part of April.

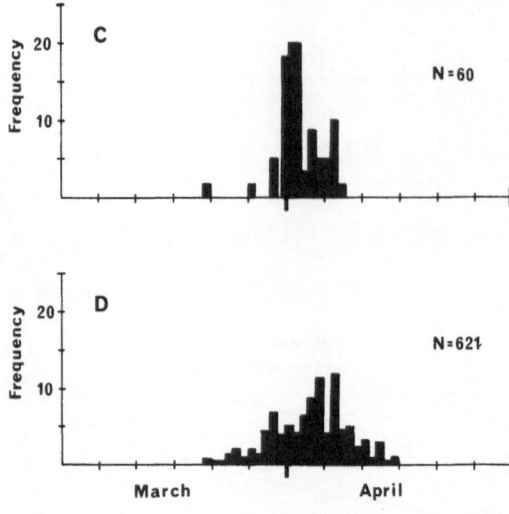

Figure 8. Hatching frequency of whitefish eggs exposed to natural
temperature regimes (C) and after regulation (D) of the
spawning river, and a constant winter temperature of 0.5°C,
increasing stepwise to 5.0°C until late March.

Table 3. Analysis of variance in mean incubation time in experiments of
hatching time in eggs of whitefish. T: temperature.

	Source	D.F.	F-Value	Significance Level
Autumn T	Winter T			
Regulated	0.1/0.5°C	1	36.1	p < 0.01
Natural	0.1/0.5°C	1	359.7	p < 0.01
Winter T	Autumn T			
0.1°C	Regulated/natural	1	33.2	p < 0.01
0.5°C	Regulated/natural	1	0.4	N.S.

DISCUSSION

The eggs exposed to the higher post regulation autumn temperature
conditions developed much faster towards the eye-spot stage and were nearly
one month ahead of eggs exposed to normal river temperatures. On this basis
earlier hatching in the spring was expected. This did not occur for eggs
reared at the winter temperature of 0.1°C, but the difference was not as
large as expected, although statistically significant. However, in eggs
reared at higher winter temperatures no significant differences in mean
hatching time were found.

By increasing the winter temperature from 0.1°C to 0.5°C and further
to 1.0°C during the late winter and increasing it further a little earlier
in spring, a slight (approximately 6 days for 50% hatching), but

significantly (p <0.01) earlier hatching was found both between natural and regulated temperature conditions. Although statistical differences were found in the mean hatching time, the differences are likely to be within the natural range due to variation in required day degrees between single eggs and the fact that spawning occurs over a 2-3 week period in the river. The predicted alteration in temperature after regulation during spawning and the first period of egg development in autumn and early winter is not therefore expected to produce any effect on the river spawning whitefish population, as no increase in temperature during winter is predicted.

In addition to changes in temperature, the planned regulation will lead to a major reduction in discharge throughout the year. This will affect both the spawning run and the area of stream bed covered by water. Decreased discharge during winter will increase egg mortality as spawning takes place at higher flows. Therefore, we concluded earlier that discharge was the main factor affecting the changes in whitefish population (Styrvold et al. 1981), and the proposal for instream flow was based on these data. The regulation plans have now been changed which will lead to even less difference in the temperature regime in the river.

The present studies agree with experiments that have shown that less degree days are required by fish eggs for embryonic development at lower water temperatures (Heggberget & Wallace 1984). Eggs of Atlantic salmon (*Salmo salar*) given different degree days, 280 and 240 respectively, hatched at the same time. This is in contrast to the values expected from models evaluated by Crisp (1981). For whitefish the differences in degree days were even larger, and the relationship between temperature and incubation time given by Braum (1964) probably does not fit well at very low temperatures.

Gray (1928) distinguished two major independent processes in embryonic development and hatching. After exposure to low temperatures a thermal cue seems necessary to induce hatching rather than cumulative degree days. Therefore river regulation giving rise to increased temperatures during winter and an earlier temperature increase in spring, will probably have a more pronounced effect than an autumnal increase. Preliminary results also indicate a mechanical effect of suspended material, producing earlier hatching compared to unexposed eggs. This seems to be most important when hatching occurs at low temperatures (below 5°C). Further studies are now being carried out on egg development at low temperatures and the factors which induce hatching.

ACKNOWLEDGEMENTS

We are grateful to N. Rønningen for his practical help during fishing, to J.-O. Styrvold for advice and assistance in the statistical analysis and to J. E Brittain for correcting the English. Financial support was given by "Oppland Energiverk".

REFERENCES

Braum, E. 1964. Experimentelle Untersuchungen zur ersten Nahrungsaufnahme und Biologie an Jungfischen von Blaufelchen (*Coregonus wartmanni* Bloch), Weissfelchen (*C. fera* Jurine) und Hechten (*Esox lucius* L.). Arch. Hydrobiol. **28**, Supple. 5: 183-266.
Crisp, D. T. 1981. A desk study of the relationship between temperature and hatching time for the eggs of five species of salmonid fishes. Freshwat. Biol. **11**: 361-368.

Enge, K. 1959. Om siken i Randsfjorden. Fauna **12**: 123-135.

Gray, J. 1928. The growth of fish. III. The effects of temperature on the development of eggs of *Salmo fario*. Br. Exp. Biol. **6**: 110-124.

Heggberget, T. G. and Wallace J. C. 1984. Incubation of the eggs of Atlantic Salmon, *Salmo salar* at low temperatures. Can. J. Fish. Aquat. Sci. **41**: 389-391.

Lindem, T. 1980a. The application of hydroacoustical methods in monitoring the spawning migration of whitefish (*Coregonus lavaretus*) in Lake Randsfjorden, Norway. Contr. Joint USA-USSR Meet. Hydroacoust. Methods Estimat. Mar. Fish Populat. Cambr., M, 25-29 June 1979.

Lindem, T. 1980b. Fiskeribiologiske undersøkelser i forbindelse med reguleringsplanene for vassdragene Etna og Dokka, Oppland. II. Registrering av fisk i Randsfjorden ved hjelp av hydroakustisk utstyr. Rapp. Lab. Ferskv. Økol. Innlandsfiske, **45**: 1-9.

SINTEF. 1982. Temperaturvurderinger og utslippsberegninger for reguleringsplanene i Etna-Dokka-vassdragene. NHI 2-83025: 1-72.

Styrvold, J. O., Brabrand, Å. and Saltveit, S. J. 1981. Fiskeribiologiske undersøkelser i forbindelse med reguleringsplanene for vassdragene Etna og Dokka, Oppland. III. Studier pa ørret og sik i Randsfjorden og elvene Etna og Dokka. Rapp. Lab. Ferskv. Økol. Innlandsfiske, Oslo. **46**: 1-103.

NATURAL SILTATION OF BROWN TROUT (*Salmo trutta* L.) SPAWNING

GRAVELS DURING LOW-FLOW CONDITIONS

P.A. Carling and C.P. McCahon[1]

Freshwater Biological Association, The Ferry House
Ambleside, Cumbria   LA22 0LP   U.K.

INTRODUCTION

Deposits of gravel in upland streams are a valuable ecological resource used by salmonid fish for spawning. Most of these gravels exhibit two distinct modes in the grain-size distribution. Generally the coarse mode is dominant and forms the framework of the deposit whilst a secondary mode, referred to as the matrix, consists of finer sediments (Carling & Reader 1982) which fill or partially fill the interstitial spaces between the framework particles. These fine sediments, if present in sufficient quantity, will cause a reduction in porosity and hydrostatic permeability, so that the volume of water held within the deposit and the intragravel velocity will be reduced. Consequently, oxygen supply-rates to fish eggs and the rate of removal of metabolic waste products will fall; possibly to lethal levels. Although some of these latter aspects have been investigated (Milner et al. 1981) the mechanisms by which bimodal gravel deposits form are poorly understood. Gravels characterized by an absence of matrix are termed open-work gravels (Cary 1951). Deposition of coarse and fine sediments may occur contemporaneously but this is not common (Fraser 1935); usually such deposits are characterised by a dilated framework whereby the matrix forms at least 20 to 30% of the deposit and would require a sudden reduction in sediment transport energy to develop (Dyer 1972). More usually in fluvial systems subject to steady changes in transportation energy, coarse particles are deposited first whilst fine material is held in suspension. Consequently the matrix in bimodal deposits usually fills the void space within the coarse deposit by a process of secondary infiltration of the stable bed (Smith 1974).

From an ecological point of view it has frequently been stressed that the percentage of fine material should not be too high (Fraser 1972), because fines have been shown to be detrimental to spawning success of salmonid fishes (Iwamato et al. 1978). Fine sediments also effect the distribution, specific composition and behaviour of invertebrates (Egglishaw 1964; Luedthe & Brusven 1976). Nevertheless there have been few investigations of the mechanisms and rates whereby matrix sediments deposit within gravel beds. Before the ecological impact of enhanced

---

[1] Present address : Department of Applied Biology, U.W.I.S.T., King Edward VII Avenue, Cathays Park, Cardiff.

deposition owing to man's activities can be assessed, baseline data concerning silting rates and mechanisms of deposition under natural conditions are needed. These data should be representative both of natural gravel mixes and also graded mixes used as artifical bed-substrata to rehabilitate salmonid spawning riffles which have been degraded.

Laboratory investigations have been made to provide a physical understanding of the mechanisms involved in matrix deposition (Einstein 1968; Beschta & Jackson 1979; Ikeda 1982; Carling 1984a). These models were developed under controlled conditions utilizing restricted size-ranges of sediments and uniform, steady-flow. Field data are few (Adams & Beschta 1980; Carling & Reader 1982; Frostick et al. 1984) but are required to scale laboratory studies.

In this paper, the infiltration rate of matrix sediments into a graded gravel is quantified. The depositional environment was subject to variable hydraulic conditions and suspended sediment concentrations in a natural stream. Specifically the temporal and spatial patterns of deposition across a riffle and pool section during summer base-flow and small freshet conditions were examined. The results are especially pertinent to the use of graded artificial gravel-beds but also are applicable to natural open-work gravel beds which previously have been reworked and winnowed by flood flows.

FIELD SITE

Great Eggleshope Beck is a tributary stream of the River Tees in Northern England. Catchment and stream sediment details have been described (Carling & Reader 1982). The streambed consists of sandstone cobbles of average grain size 83 mm with a matrix of sediment finer than 2 mm typically constituting 10% of the deposits by weight. The surface layer of gravel is usually flushed clear of fine sediments except during the base flow periods in summer when fine sediment may blanket the coarser bed material.

The stream meanders through a restricted flood plain and pool and riffle sequences are poorly developed. Bedslope is 0.0100 and the channel width varies between 4 and 11 m whilst depths at bankfull are typically less than 1 m. The reach is particularly favoured by spawning brown trout (*Salmo trutta* L.). The structure of redds cut in G. Eggleshope Beck has been described by Ottaway et al. (1981) whilst Carling (1983; 1984b) considered the dynamics of the suspended inorganic and organic load.

The stream drains rough heather moor developed on till, solifluxion deposits and Carboniferous sandstones, limestones and shales. Hydrographs of streamflow rise steeply and there is considerable bedload transport (Carling & Hurley in press) during the winter months. The low flow periods in the summer months may be punctuated by high flows as a result of thunderstorm activity.

METHODS

Double-walled permeable pots with a surface are of 266 cm^2 and a depth of 10 cm were used to collect settled fine sediments. The accuracy of this method under differing hydraulic conditions has been considered elsewhere Carling 1984a). Hard road-stone gravels, with the size-range truncated at the lower end at 4 mm, were used to fill the pots. These gravels gave a graded distribution of mean grain-size, 28 mm, a sorting coefficient of 4.7 mm, zero skewness and a porosity of 0.48. Although finer than the natural

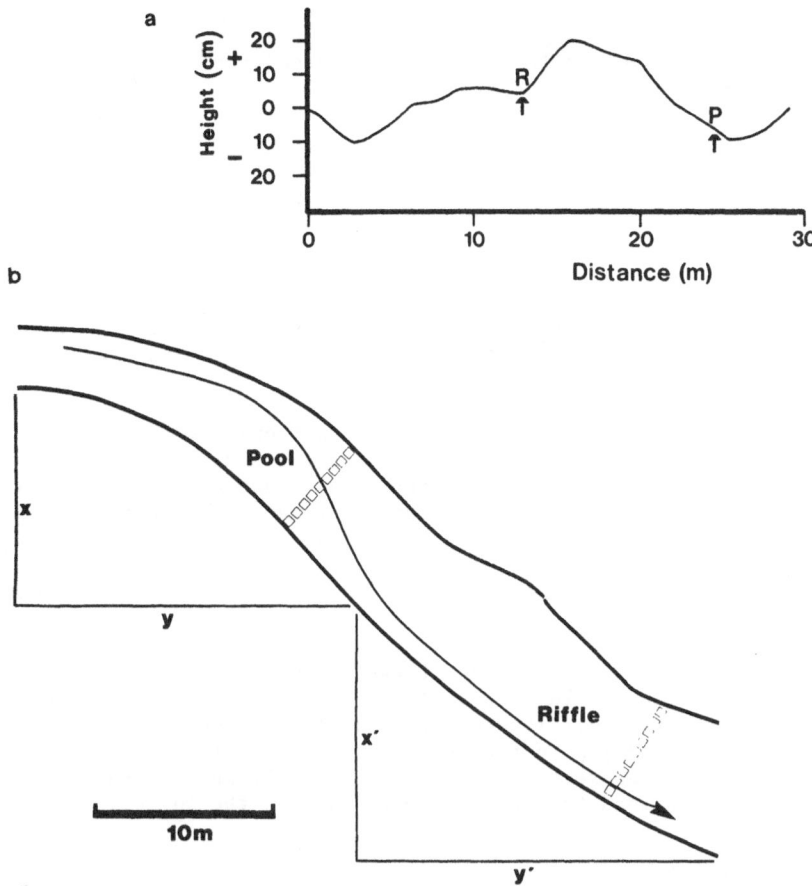

Figure 1.
  (a) Longitudinal profile of the variation in channel topography
       through the study reach in relation to an arbitrary datum. The
       letters P and R denote pool and riffle sampling sites. Flow was
       from right to left. Vertical exaggeration: x 20.
  (b) Plan view of the study reach. x,y; x´,y´ are the characteristic
       lengths used to calculate the radius of curvature (21 m) of the
       double bend (after Welch 1973). Arrow represents the main
       streamline and squares, the location of sampling sites.

bed material, tests demonstrated that the trapping efficiency was similar
to natural gravels. Some natural particles were friable when handled and
consequently the road-stone was preferred.

    Although porosity is a measure of the space within the framework
available for deposition, it is no measure of the likelihood of a
depositing grain being capable of entering the interstices of the
framework. A measure of the dimensions of the constrictions (pore throats)
between grains was required. These measures were obtained photographically
(details available upon request). In plan view the void density of the pot
gravel was 11.61% with a standard deviation of 1.46% for six replicate
determinations. The average throat length was 4.51 mm with a standard
deviation of 4.12 mm for 647 determinations of individual dimensions.

Grains smaller than c. 4 mm therefore are more likely to penetrate the interstitial space than larger grains. Once material has settled below the surface layer of gravel it is not resuspended and the pots accurately record the infiltration rate (Carling 1984a).

Pots were inserted into the streambed flush with the bed surface. In 1982 ten pots with a 0.5 m spacing were placed across the pool. On the riffle the spacing was 0.8 m (Figure 1). In 1983 the sampling interval was reduced; at each section fifteen pots were used.

At weekly intervals the inner pots were removed and the contents flushed into buckets. Clean gravel was put back and the pots replaced in the outer sleeves which were still inset in the stream-bed. In 1982 a total of 399 determinations of the deposition rate was made and in 1983 the total was 218.

Matrix-material, wet-sieved into the fractions 4 mm - 2 mm and less than 2 mm, was oven-dried at 80°C for 24 hours. Additional samples were retained to assess the basic geochemistry and organic content. Detailed particle-size distributions were measured for material finer than 63 μm using a Malvern 2200/3300 particle-sizer V3.1.

Data on discharge, water depth, stream-flow velocity and suspended sediment samples were also collected.

The sampling period consisted of twenty-two weeks in 1982 between 22 April and 23 September. In 1983 an eight week period between 5 August and 26 September was studied. It was not possible to sample during the months of October to March owing to frequent scour of the stream bed by floods. Although no high flows were sampled, several small summer freshets which induced limited bedload transport occurred during the investigation.

RESULTS

Hydraulic Aspects

Depths were shallow, typically between 3 and 15 cm over the riffle and depth-averaged velocities varied between 13 and 41 cm s^{-1}. In the pool, depths varied between 18 and 34 cm and velocities were below 1 to 2 cm s^{-1} except during a freshet at the end of September 1983 when depth-averaged velocities reached 27 cm s^{-1}.

The general spatial pattern of depth and velocity variations over the riffle in 1982 is shown in Figure 2 whilst the detailed cross-channel pattern of velocity variations throughout the study period in 1983 is shown as a response-surface in Figure 3a. In both figures the multi-filament nature of the velocity-field in the shallow stream is evident. During low flows in 1983 for example, two main filaments existed in the centre of the channel with a subsidiary filament closer to the left-bank (Figure 3a.) Areas of practically still water occurred close to the channel margins. As discharge increased, the flow became stronger close to the right-bank on the outside of the stream-bend. However during the highest flow examined (1.15 m^3 s^{-1} or about six times mean annual discharge) the main velocity filaments remained distinct but an additional thread developed in a small chute across the channel bar on the inside of the bend (Figure 3a).

Suspended sediment concentrations varied between less than 1 mg L^{-1} and 80 mg L^{-1} but typically fell within the range 1 to 3 mg L^{-1}. Variation with discharge throughout the year is described elsewhere (Carling 1984b) but no precise relationship can be obtained, data being scattered as

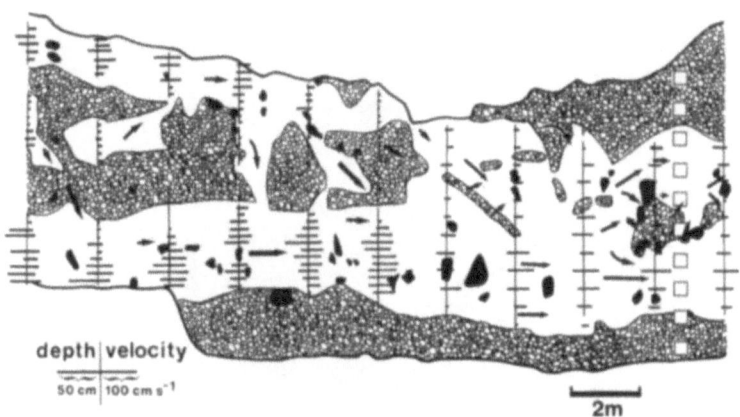

Figure 2. Detailed sketch of the generalized flow pattern over the riffle which occured on the 16 June 1982. Exposed gravel bars are shown as are large boulders (black areas). Approximate flow lines are shown as arrows. Upstream histogram bars represent water depths. Scale: 10 mm equivalent to 34 cm depth. Downstream histogram bars represent velocities. Scale: 10 mm equivalent to 84 cm s^{-1}. Reproduced from an unpublished survey conducted by Dr D.T. Crisp (F.B.A.).

suspended load is highly supply dependent. However Carling (1984b) found no evidence of seasonal variation in the sediment supply rate.

Deposition Rate

A two-way analysis of variance (Dixon & Massey 1969) was used to assess spatial and temporal trends in the deposition rate for material finer than 2 mm. Data for the pool and riffle were treated independently. On a week to week basis the analysis demonstrated that there was no significant statistical difference at the 10% level in the across-stream variation in the rate of accumulation in either pool or riffle. However, scatter in the data over short-time scales obscured an overall spatial pattern which became apparent when the weekly data were averaged for the total period of the investigation. Distinct temporal variation in accumulation rates was evident at both sites and was related to discharge fluctuations. In the week following an increase in discharge, deposition increased by one to two orders of magnitude. The time trends for the 1982 data (averaged across the section) are shown in Figure 4; the rate of deposition for < 2 mm material increasing from values typically of the order of 100 g m^{-2} week $^{-1}$ during the base flow conditions to 1 to 10 kg m^{-2} week $^{-1}$ during freshets. Similar peak deposition rates were noted for coarser material (2 to 4 mm) moving as bedload. However, during baseflow, bedload deposition rates were lower ~ 10 g m^{-2} week $^{-1}$. There is some evidence that for material finer than 2 mm the deposition rate was slightly greater in the pool than over the riffle area, both during base-flow and storm-flow. Nevertheless, the albeit short record for 1983 demonstrated a similar effect during storm flow but enhanced deposition over the riffle during base-flows.

The percentage of the available void-space filled by deposition each week was variable. Over twenty-two weeks in 1982 the average value for the pool section was 1.23% per week with a standard deviation of 3.73% per

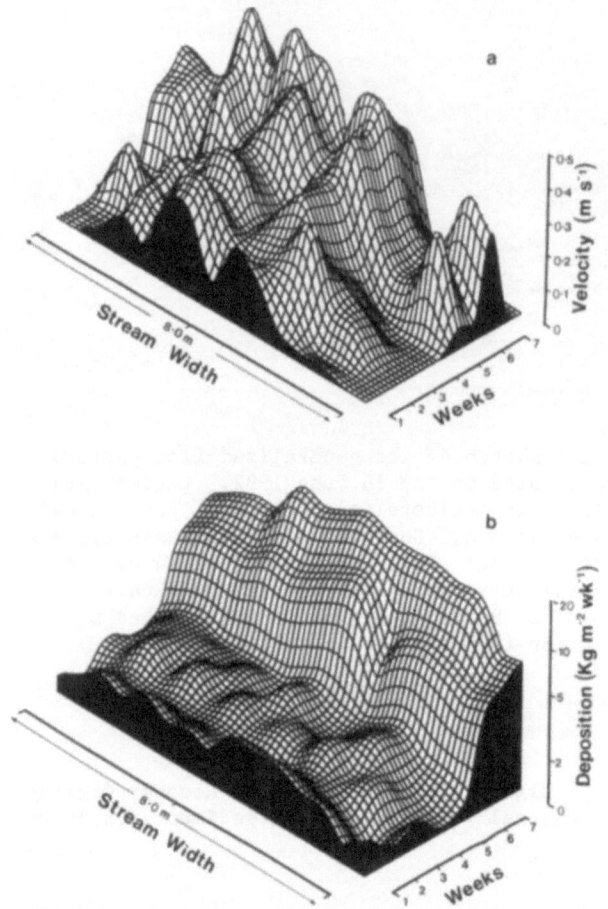

Figure 3.
    (a) Cross-channel variation in stream velocity as a function of time. Changes in time also correspond to a steady increase in discharge.
    (b) Cross-channel variation in deposition of material finer than 2 mm for the same period as (a). The vertical scale in (b) is logarithmic.

week; values ranging from a minimum of 0.06% to 17.62% during a freshet. For the riffle section the average value was 0.88% (δ = 2.24%) ranging between 0.04% and 10.58%.

Cross-channel Variation in Deposition

On a week-to-week basis no statistically significant difference in the rate of accumulation could be demonstrated, but cross-channel trends could be detected visually in weekly data sets. The trends became clear when data either were averaged for the period of the investigation (Figs. 5 and 6) or were plotted as a three dimensional response surface (Figure 3b)

In 1982 the deposition rate of material less than 2 mm across the stream width increased linearly from the outside of the bend (true

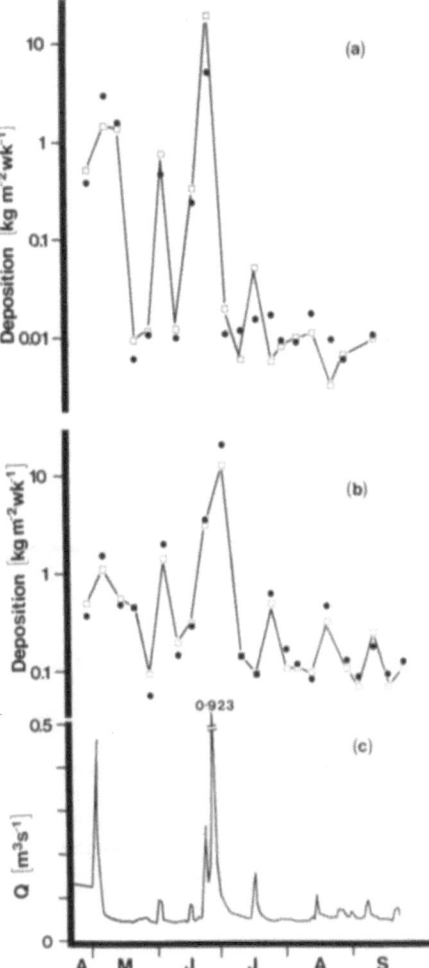

Figure 4. Time histories of average deposition across the section and discharge variations during 1982. Dots represent pool data and squares represent riffle data.
- (a) Deposition of material in the size range 4 mm to 2 mm i.e. bedload.
- (b) Deposition of material finer than 2 mm i.e. mainly suspended load.
- (c) Discharge fluctuations.

left-bank) to the inside (Figure 5). The rate of increase was similar across the pool and riffle; the rate increased, for a distance increment of 10% across the stream width, by 119 g in the pool and 104 g over the riffle. A linear model also described the depositional pattern of 2 - 4 mm material over the riffle but not in the pool. In the pool deposition was at a maximum in the centre of the section. Data were scattered owing to (i) the blocking effect of large boulders which diverted flow and (ii) the occurrence at very low flows of dry areas of gravel in the section. These effects are demonstrated in Figure 2.

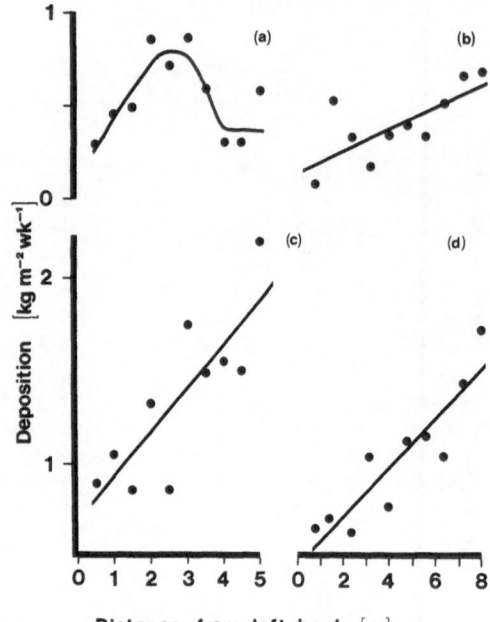

Figure 5. Average deposition rates in 1982 as a function of distance from the left bank.
(a)   Deposition of 4 mm to 2 mm material in the pool. Curve is equation 1 in text.
(b)   Deposition of 4 mm to 2 mm material over the riffle.
(c)   Deposition of material less than 2 mm in the pool.
(d)   Deposition of material less than 2 mm over the riffle. Curves in (b), (c) and (d) are linear least squares relationships between deposition rate and the unit width.

In addition to the blocking effect diverting flow-lines and locally accelerating velocities, the development of a central bar and two main current speed filaments can be seen in the upstream portion of the map. Although this bar has no discernable effect on the deposition rates in 1982 (with maximum velocity and deposition occurring towards the right-bank) by 1983 the bar had increased in height at the upstream end of the riffle and encroached into the pool. Consequently both velocity and deposition patterns were significantly altered in 1983.

Except during the freshets when the linear pattern of an increase in deposition across the pool was still evident, in 1983 the linear pattern for low flows had been disrupted by the development of the bar. Data for 1983 are considerably more scattered owing to transverse flows over the bar enhancing deposition at some locations in the section and reducing it at others (Figure 6). In each example, cross-channel patterns are shown by a smoothed curve obtained using a binomial filter of the form:

$$x_i = 0.25 x_{i-1} + 0.5 x_i + 0.25 x_{i+1} \qquad (1)$$

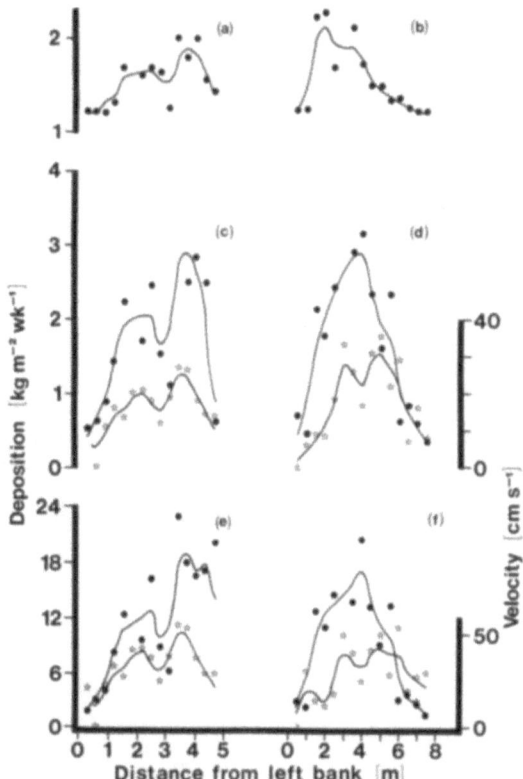

Figure 6. Average deposition rates and stream velocities in 1983
            as a function of distance from the left bank.
(a)   Deposition of 4 mm - 2 mm material in the pool.
(b)   Deposition of 4 mm - 2 mm material over the riffle.
(c)   Deposition of material finer than 2 mm across the pool, and
       associated velocities.
(d)   Deposition of material finer than 2 mm across the riffle.
(e) and (f) Patterns of deposition in pool and riffle during the
       freshet of the 19 September 1983. Solid symbols are
       deposition and starred symbols velocity. Smoothed curves
       are equation 1.

This method effectively removed local variability but preserved the major
cross-stream pattern and the apparent position of peaks in deposition. A
cross-channel pattern of two main stream-lines in the pool is evident both
in the velocity and depositional patterns. Over the riffle the main
stream-line had moved towards the centre of the channel although there was
some evidence in the velocity distribution of two to three weak residual
stream-lines, depending upon stream discharge.

Although distinct current filaments could be identified, the
depositional pattern did not correspond to the spatial or temporal velocity
pattern precisely. The fine nature of the suspended load, coupled with the
existence of strong lateral flows and turbulence resulted in a fairly even
distribution of deposition across the riffle, although at high discharge,

peak deposition was associated primarily with the two main current filaments (Figure 3b).

The velocity pattern dictated the pattern of cross-channel transport of bedload to a greater degree than it did the finer fraction which was transported in suspension. The 2 - 4 mm fraction, expressed as a percentage of the total load, showed enhanced deposition rates (and inferred transport rates) in the areas of the main velocity threads in the pool, whilst over the riffle there was only one main velocity thread with maximum deposition rates recorded approximately a third of the way across the section from the left-bank (Figure 7).

Organic Content

During 1982 165 samples (< 2 mm) were retained and the organic content assessed. The average organic content was 16.54% (δ = 6.85%; range 31.52% to 3.86%). There was no discernable difference in the organic content of sediments deposited in either pool or riffle. The percentage of organics deposited decreased during freshets and increased during low-flows. However, there was no evidence of cross-channel patterns in deposition.

Particle-size Analysis

Grain-size analyses of nine samples of suspended solids collected in 1983 demonstrated that a wide range of very fine particles were in

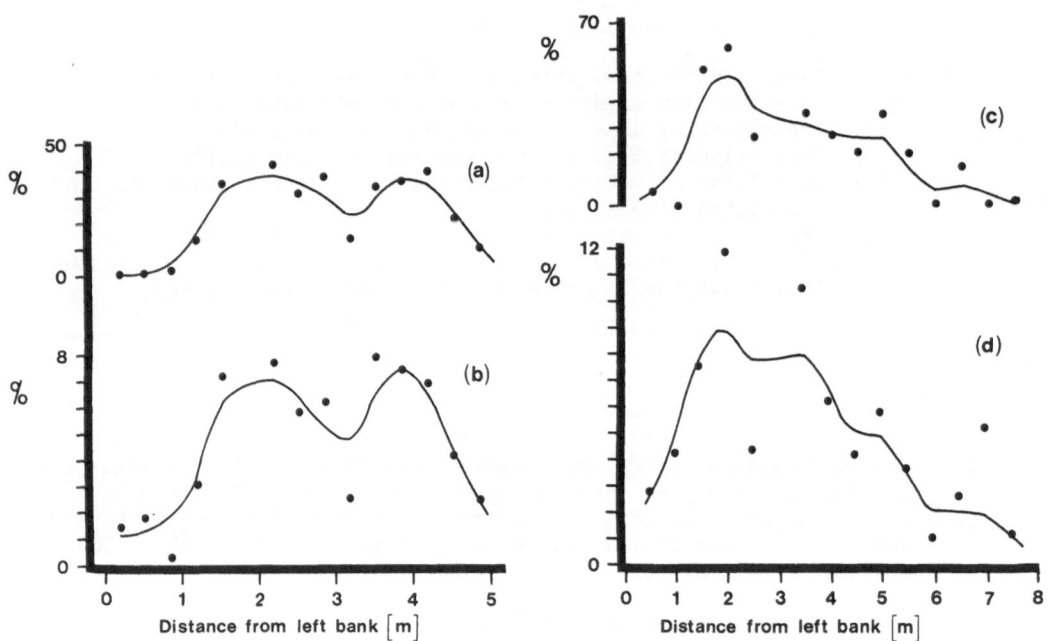

Figure 7. Spatial distribution of the deposition of bedload, 4 mm - 2 mm in size expressed as a percentage of the total deposition.
(a) Pattern across the pool for the freshet of the 19 September 1983.
(b) Average pattern for the pool for the whole of 1983 study period.
(c) Pattern across the riffle for the freshet of the 19 September 1983.
(d) Average pattern for the riffle for the whole of the 1983 study period.

238

suspension including abundant particles as fine as 1 μm. The size of material varied little during the period of investigation as flows were low; discharge at the time of sampling was between 0.28 m^3 s^{-1} and base-flow ~0.045 m^3 s^{-1}. Scatter envelopes of grain-size distribution curves had a median diameter in the range 10.75 to 15.75 μm although fine sands up to 85 μm were also found in suspension. These sands constituted between 4 and 9% of the suspended material but would constitute a significant proportion of the load during the peak discharge in 1983 of 1.15 m^3 s^{-1}.

Sediment deposited in the pots was much coarser. Fifty-seven samples were analysed, two to three per week in 1982, and the results of the size analysis averaged for each week. The average size of particles was 0.28 mm in the pool and 0.13 mm over the riffle with weekly values ranging between 0.13 mm and 0.62 mm. Very few particles coarser than 4 mm were recorded and none coarser than 8 mm. The fraction finer than 63 μm was analysed separately. With a median diameter in the range 8.6 to 10.4 μm the deposited silt was finer than the silts in suspension.

DISCUSSION

Deposition Rates

Frostick et al. (1984) compared deposition rates in artificial mixes of bed material but unfortunately, as in this investigation, the study provides little in the way of hydraulic interpretation. Collection of suitable field data to achieve this may prove problematic in streams lacking a steady discharge regime. Stream flow and suspended solids concentrations can vary over time scales of minutes to hours in flashy streams whilst temporal sampling frameworks to date have utilized intervals of days or weeks. It is difficult therefore to relate temporally-integrated results from sampling over a long time-base to detailed instantaneous hydraulic data purporting to represent unsteady and nonuniform flow. Consequently field derived depositional data can only be related to generalized bulk discharge or velocity data which by their intrinsic nature preclude precise mechanistic interpretation of depositional processes (Carling 1984a).

Because of the above observations no meaningful correlations can be obtained between determinants although useful patterns may be discerned. The suspended sediment data used in conjunction with the deposition rates indicated that for the fine silts and low density material in the natural stream the coefficient in the simple equation (Carling 1984a)

$$\Delta' = b \ C \qquad\qquad (2)$$

(where Δ' is the deposition rate (g min^{-1} per unit area), C is the suspended sediment concentration (mg L^{-1}) and b is a coefficient having the dimensions of a velocity) will be less than 0.60 found in laboratory experiments of quartz-density sands. However, on two occasions, when fine sand was in suspension in quantity, b was of the order of 0.60. Nevertheless small increases in suspended concentration (i.e. maximum recorded ~ 80 mg L^{-1}) led to rapid filling of the void space in the pot gravels. As there is no seasonal variation in suspended sediment concentration similar deposition rates might be anticipated in the autumn when trout spawn in the study reach. Ottaway et al. (1981) concluded that trout redds rapidly resilted after excavation owing to autumn flood-flows mobilizing sediments. It is now possible to estimate the rate using the data presented within this paper and that given by Ottaway et al. (1981). The average tail area of a redd in Eggleshope was 2379 cm^2 and egg pockets

were found to a depth of 14 cm (Ottaway et al. 1981). Assuming the egg
pocket is at the base of an elliptic excavation (such that the average
depth is 0.3 of the maximum) a total volume of some 10000 cm^3 might be
disturbed by redd excavation. For the gravels in the study reach,
porosities are about 0.16 but might locally be greater ~ 0.32 where
disturbed hydraulically or by the digging activity of fish. Consequently
the void volume within a redd might be somewhere between 1600 and 3200
cm^3. Given the bulk density of the matrix sediments is 0.45 g cm^{-3} this
volume of void space would fill in two to three days at the maximum rate of
deposition (10 kg m^{-2} week^{-1}). These calculations make no allowance for
the possibility of bed disturbance, elevated bedload transport rates and
lateral migration of matrix silts through the framework into the open
pore-space. The rate of resilting of redds during freshets may therefore
be less than calculated.

Cross-channel Variations in Deposition Rate

In a natural stream it might be expected that a correlation between
water velocity and deposition rate would occur, because the stream flow
would control the rate of sediment transport across the sediment traps.
Owing to the fine size and low density of the suspended particles, a close
correspondence was not always evident. In 1982, deposition over the riffle
had been greatest in the low velocity areas reflecting the marked
cross-channel velocity gradient (Figure 5). In 1983 with a more variable
cross-channel velocity pattern, deposition mimicked the velocity pattern,
though elevated deposition also occurred in sheltered areas. The
hydraulics therefore controlled the gross depositional pattern through the
supply rate. Even in high velocity areas deposition occurred through
turbulent transfer and the physical trapping of suspended particles in
openwork gravels (cf. Carling 1984a).

Bedload, being substantially coarser, reflected the cross-channel
velocity pattern more closely with greater deposition rates in strong
current areas; a similar result to that obtained by Frostick et al. (1984).

The question as to whether enhanced deposition occurs over riffles or
in pools during low flows is not resolved. Riffle areas may filter fines
from suspension in the shallow waters but particles do not have time to
settle to the bed during their passage through the reach. In contrast pool
waters are slow flowing and less turbulent and enhanced deposition might be
expected in the confined basins. The difference in pool and riffle
morphology in this investigation, a matter of only a few centimetres depth,
was sufficient nevertheless to produce variable patterns of deposition.

The Infiltration Process

Laboratory experiments indicated that resuspension of matrix did not
occur at depths greater than that approximately equal to the mean
grain-size of the bed material (Carling 1984a). In the present study pots
were not allowed to become completely filled so no measure of the depth to
which turbulent resuspension can occur was obtained. Frostick et
al. (1984) noted resuspension in their traps occurred to a depth of 80 mm
along the stream talweg. This measure has more physical meaning if related
to the mean grain size of their coarsest trap material i.e. 48 mm,
indicating that resuspension occurred to a depth typically between one and
two grain diameters. Despite the greater turbulence anticipated along the
main stream-line this result is in keeping with the observation of Carling
& Reader (1982; Figure 6) that spatially-averaged depths of hydraulic
winnowing in the natural stream were typically equal to between one and two
mean grain diameters of the framework gravels.

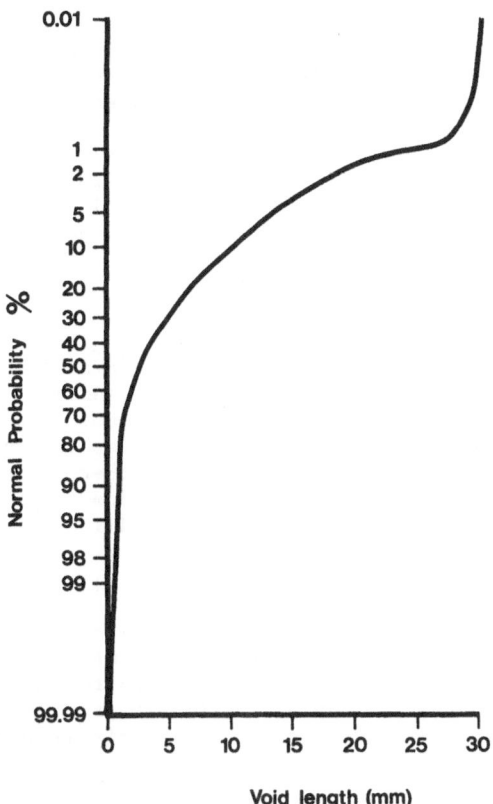

Figure 8. Probability plot of the likelihood of a given grain size
penetrating the interstices of the pot gravel (Schematic only).

The matrix material contained particles coarser than those found in
suspension indicating the importance of bedload transport of sand-size
particles at comparatively low discharges in the process of matrix
replenishment. Nevertheless, despite the threshold of motion for
pebble-sized material being exceeded during the highest discharges, the
paucity of matrix particles coarser than 4 mm reflects the inability of
these grains to pass down through the interstitial constrictions. The
ratio of the arithmetic average void throat dimension (4.51 mm) to the mean
grain size of the pot gravel (28 mm) is 0.161. Experimental studies by
Fraser (1935) on the packing of spheres indicated that the limiting maximum
diameter of a smaller grain for penetration of the interstices of the
second tightly-packed coarser population was 0.154 times the diameter
of the framework particles. The degree of penetration of various grain
sizes in the variable geometry pores of the natural gravels will clearly be
variable but the probability of penetration will remain size-dependent
(Culling 1963) as shown schematically in Figure 8.

Particle-size Analysis

The suspended sediments were slightly finer than those sampled in the
month of March during high discharges (Carling 1983) reflecting the lower
stream competence of moderate summer freshets. The overall grain-size of
the matrix silts (i.e. < 63 μm) nevertheless was finer than the suspended

silts. This indicates that although fine particles may have a very high
probability of being deposited deep within the interstitial space there is
still a selective process operating. Coarse silt particles being more
likely to settle near the surface either on framework particles or on
coarse bedload-derived matrix. At these elevated positions they are more
easily resuspended and flushed from the pots. The grain-size of the matrix
materials is very similar to that matrix sampled from natural riverbed
gravels by Carling & Reader (1982) implying a similar infiltration process
operates both within the natural openwork gravels and within the pot
gravels as was suggested by initial trials.

Organic Content

The organic content of the matrix deposits is similar to the organic
content of ephemeral surface deposits but is two and a half times the
average found in natural gravel matrix which has not been recently
deposited (Carling & Reader 1982). This reinforces the conclusion of
Carling & Reader that the organic fraction is either biodegraded or is
selectively resuspended. The present results allow the additional
observation to be made that the latter process will be size-selective.

Comparative Investigations

Measurement of infiltration rates in natural streams are few. In the
Turkey Brook, draining Eocene London clay, rates of deposition (Frostick et
al. 1984) were an order of magnitude greater than recorded in this
investigation during base-flow and up to three-times the maximum rate
during freshets. Particle-size range of the matrix however was similar to
the present experiment.

Welton (1980) measured deposition rates in the chalk stream, the
Tadnoll Brook in Dorset and recorded maximum deposition rates of sand of 35
kg m^{-2} week $^{-1}$ for a flood event in December in the main stream-line,
rising to 70 kg m^{-2} week^{-1} in marginal low velocity areas. In the summer
months rates were lower, typically between 2.6 to 6.5 kg m^{-2} week^{-1} during
baseflow rising to 24.7 kg m^{-2} week^{-1} in a summer storm. These values also
are substantially larger than recorded in this investigation of an upland
stream and reflect differences in sediment availability.

CONCLUSIONS

The unsteady and non-uniform flow characteristics of most natural
streams make it unlikely that field data of the desired quality can be
obtained to formulate hydraulic predictive models of matrix deposition in
shallow gravel-bedded streams. Field data are still required however to
give scale and pattern in order to design controlled laboratory
experiments.

Both suspended load and bedload contribute to the matrix material
infilling gravel frameworks. A selective depositional process, controlled
by the pore-size distribution of the framework gravels, dictates the actual
grain-size distribution of the matrix for a given suspended sediment
size-range.

The results of this investigation taken together with those of Carling
(1984a) and Frostick et al. (1984) suggest that artificial spawning gravels
should not be well graded. Rather, more poorly graded sediments with the
size range truncated at 2 mm would be preferable to prevent excessive
quantities of silt entering the void space. Despite the work of Frostick
et al. (1984) the exact size distribution of such a mix cannot be known

precisely as the optimum distribution will vary depending upon the size characteristics of the natural sediment load.

Despite considerable scatter in the data, spatial patterns in both matrix deposition from suspension and from bedload are evident. The pattern is more smoothed in the case of the finer suspended load compared with the bedload. Bed topography is very important in controlling the pattern of velocity and hence the sediment supply-rate and therefore the potential deposition. Nevertheless zones of low velocities may also demonstrate high deposition rates where sediment is entrapped in eddies.

Deposition rates are an order of magnitude less than rates recorded in lowland streams. Silting although temporally and spatially variable is nevertheless rapid. Trout redds are likely to be fully silted in a matter of a few days under flow regimes subject to small freshets with an adequate supply of fine sediments.

ACKNOWLEDGEMENTS

The project was funded by the Department of the Environment, the Natural Environment Research Council, the Northumbrian Water Authority, the Water Research Centre, the Ministry of Agriculture, Fisheries and Food, and the Welsh Office.

Dr D. T. Crisp and Ms S. Bidmead carried out the unpublished survey work from which Figure 2 is drawn with permission.

Mr A. Bale (I.M.E.R.) is thanked for the particle-size analyses of the fine particulates and Dr A.E. Irish (F.B.A.) produced Figure 3.

REFERENCES

Adams, J. N. & Beschta, R. L. 1980. Gravel bed composition in Oregon coastal streams. Can. J. Fish. Aquat. Sci. **37**: 1514-1521.
Beschta, R. L. & Jackson, W. L. 1979. The intrusion of fine sediments into a stable gravel bed. J. Fish. Res. Board Can. **36**: 204-210.
Carling, P. A. 1983. Particulate dynamics, dissolved and total load, in two small basins, northern Pennines, U.K.
   Hydrol. Sci. J. **28**: 355-375.
Carling, P. A. 1984a. Deposition of fine and coarse sand in an open-work gravel bed. Can. J. Fish. Aquat. Sci. **41**: 263-270.
Carling, P. A. 1984b. Comparison of suspended sediment rating curves obtained using two sampling methods. In: Channel Processes - Water, Sediment, Catchment Controls. (Ed by A.P. Schick). Catena Suppl. **5**: 43-49.
Carling, P. A. & Reader, N. A. 1982. Structure, composition and bulk properties of upland stream gravels. Earth Surf. Processes Landforms **7**: 349-365.
Carling, P. A. & Hurley, M. A. in press. A time-varying stochastic model of the frequency and magnitude of bedload transport events in two small trout streams. In: Problems of Sediment Transport in Gravel-Bed Rivers (Ed by C. R. Thorne, J. C. Bathurst & R. D. Hey). Wiley, Chichester.
Cary, A. S. 1951. Origin and significance of open-work gravel. Trans. Am. Soc. civ. Engrs. **116**: 1296-1308.
Culling, W. E. H. 1963. Soil creep and the development of hillside slopes. J. Geol. **71** 127-161.
Dixon, W. J. & Massey, F. J. 1969. Introduction to statistical analysis. McGraw-Hill, New York. 488 pp.

Dyer, K. R. 1972. Bed shear stresses and the sedimentation of sandy gravels. Mar. Geol. **13**: M31-M36.

Egglishaw, H. J. 1964. The distributional relationship between the bottom fauna and plant detritus in streams. J. Anim. Ecol. **33**: 463-476.

Einstein, H. A. 1968. Deposition of suspended particles in a gravel bed. J. Hydr. Div., Am. Soc. civ. Engr. **94**: 1197-1205.

Fraser, H. J. 1935. Experimental study of porosity and permeability of clastic sediments. J. Geol. Chicago **43**: 910-1010.

Fraser, J. C. 1972. Regulated stream discharge for fish and other aquatic resources - an annotated bibliography. F.A.O. Fish. Tech. Pap. **112**: 103 pp.

Frostick, L. E., Lucas, P. M. & Reid, I. 1984. The infiltration of fine matrices into coarse-gravel alluvial sediments and its implications for stratigraphical interpretation. J. Geol. Soc. **141**: 955-965.

Ikeda, H. 1982. An experimental study of the formation of open-work gravel layers under alluvial flow conditions. Trans. Japan Geomorph. Un. **3**: 57-65.

Iwamoto, R. N., Salo, E. O., Madej, M. A., McComas, R. L. & Rulifson, R.L. 1978. Sediment and water quality: a review of the literature including a suggested approach for water quality criteria with summary of workshop and conclusions and recommendations. U.S. Environ. Prot. Agency 910/9-78-048, 151 pp.

Luedthe, R. J. & Brusven, M. A. 1976. Effects of sand sedimentation on colonization of stream insects. J. Fish. Res. Board Can. **33**: 1181-1886.

Milner, N. J., Scullion, J., Carling, P. A. & Crisp, D. T. 1981. The effect of discharge on sediment dynamics and consequent effects on invertebrates and salmonids in upland rivers. Adv. Appl. Biol. **6**: 153-220.

Ottaway, E. M., Carling, P. A., Clarke, A., & Reader, N. A. 1981. Observations on the structure of brown trout, *Salmo trutta* Linnaeus, redds. J. Fish Biol. **19**: 593-607.

Smith, N. D. 1974. Sedimentology and bar formation in the upper Kicking Horse river, a braided outwash stream. J. Geol. **82**: 205-223.

Welch, D. M. 1973. Channel form and bank erosion, Red River, Manitoba, pp. 284-293. In: Fluvial Processes and Sedimentation. pp. 759. Proc. Hydrol. Symp. Univ. Alberta, Edmonton, Nat. Res. Council Canada.

Welton, J. S. 1980. Dynamics of sediment and organic detritus in a small chalk stream. Arch. Hydrobiol. **90**: 162-181.

SUSPENDED SOLIDS TRANSPORT WITHIN REGULATED RIVERS

EXPERIENCING PERIODIC RESERVOIR RELEASES

D. J. Gilvear

Department of Geography
University of Technology
Loughborough, Leicestershire, U.K.

INTRODUCTION

Few studies have examined variations in turbidity and suspended solids within impounded rivers during reservoir releases (Beschta et al. 1981; Eustis & Hillen 1954). Reservoir releases have, however, been shown to increase invertebrate drift (Brooker & Hemsworth 1978), entrain substratum bacteria (McDonald et al. 1982) and flush particulate matter from in-channel storages (Matter et al. 1983). Excess fine sediments and organic matter within stream channels can adversely affect the quality of salmonid spawning gravels and the habitats of insects and benthic organisms (Hall & Lantz 1979; Bjornn et al 1977). Reservoir releases have been used to remove channel-bed accumulations (Antonio 1969). Nevertheless, Wesche and Rechard (1980) state: "there is also a need for quantitative information about flushing flows. With the present `state of the art´ it is unknown if flushing flows are necessary to remove fines and maintain stream quality. If flushing flows are necessary to remove fines and maintain stream quality, methods for determining the proper, magnitude, time duration, as well as the time of the year, should become a valuable part of instream flow recommendations". Milhous (1982) provided a conceptual model of the processes operating, but could not acquire information on the optimum frequency for releases. If the time is too long, not only will excess accumulation occur within the channel but subsequent removal may lead to unacceptable suspended solids concentrations and turbidity values. High levels may limit photosynthesis, reduce macro-invertebrates and restrict fish populations (Alabaster 1972).

This paper uses a controlled reservoir release from Llyn Celyn to the regulated Afon Trywern, Wales, as a case study to illustrate turbidity and suspended solids changes both at-a-site and downstream. In addition, data collected during 24 other reservoir releases are used to examine the factors determining suspended solids loads. Eleven of these were also on the Afon Tryweryn, ten on the River Washburn, with one release on each of the rivers North Tyne, Garry and Sutton Brook (Figure 1).

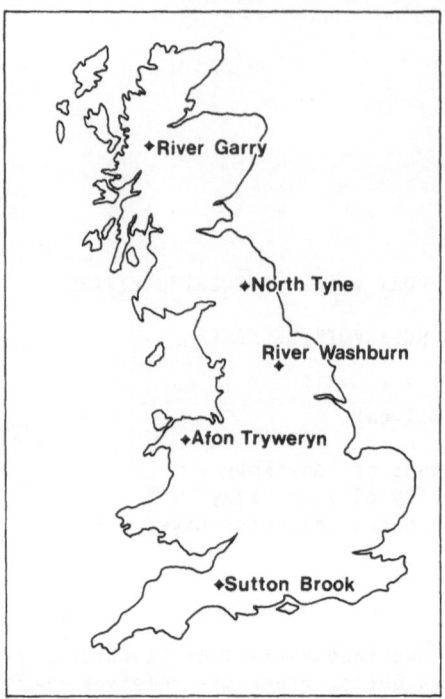

Figure 1. The location of the regulated rivers monitored in this study.

METHODOLOGY

During the reservoir releases, stage and turbidity were recorded continuously. Turbidity was monitored using Partech (Type 7000 3RP) monitors and recorders with either 0-1000 or 0-5000 twin-gap sensors set to the appropriate range. The monitors were calibrated with standard formazin solutions before each release. Water samples were also collected, by hand, at intervals of between one and five minutes during periods of elevated turbidity. These were selected relative to the time-of-rise of stage and previous knowledge on the rate of change of suspended solids concentrations (Foulger & Petts 1984). The water samples were collected in polyethylene containers of at least 350 ml capacity. Duplicate samples of different volumes showed no significant differences in suspended solids concentrations, despite larger samples having an increased probability of incorporating large organic particulate matter. The sampling was assumed to be representative of the suspended solids variations with changing discharge after a series of surveys of cross-channel variations had been undertaken. Three pre-release samples were taken to describe background suspended solids concentrations. In addition, pebbles previously collected from the channel substrata and selected to represent the particle-size range of bed material, were used to confirm the expected stability of substrata. None of the tracers were moved during the releases.

Samples were filtered by the methods of Eaton et al. (1969). No constant temperature room was available so control filter papers were used to correct for small weight changes arising from humidity effects. Weighing errors were estimated to be +2 mg l^{-1}. All samples were analyzed within 7 days of collection. Organic content was determined by ashing at 550°C for 2 hr (Barton 1977; Fisher & Lavoy 1972). Although this has been shown to be unreliable at low concentrations (Finlayson 1975), organic determination was only undertaken on the samples having high suspended solids concentrations or large water volumes.

Particle-size analysis in the range 4 to 80 μm was undertaken on the minerogenic fraction of the material in suspension, using a Coulter Counter (Model TAII with a 200 μm tube). These samples were dispersed ultrasonically in a 0.9% NaCl solution (Carling 1983). This material cannot be considered representative of the natural floccule-size distribution of particles in the river water but nevertheless, the technique has been used extensively to determine particle size distributions of the minergenic component.

RESULTS

The Afon Tryweryn Case Study

An 8 km reach of the Afon Tryweryn, Wales is regulated by Llyn Celyn reservoir (Figure 2). Compensation flows from Llyn Celyn equivalent to a natural spring discharge are required; and are 0.378 $m^3 s^{-1}$ during winter when tributary inflow is high and 0.737 $m^3 s^{-1}$ during summer. Controlled reservoir releases are made for water supply, hydro-electric power generation, flood and drought protection, and competitive canoeing events. The frequency and duration of releases varies from year to year depending on climatic conditions, hydro-electric power requirements and the needs of white water canoeists. The data relate to a reservoir release on the 23 April, 1983; no releases had been made prior to this date for 51 days. Rainfall within this period amounted to 182 mm.

Discharge during the release increased stepwise from 0.737 to 3.94 $m^3 s^{-1}$ to 5.78 $m^3 s^{-1}$ and then to 12.62 $m^3 s^{-1}$, each step requiring 15 minutes. The stilling pool immediately below the dam has the effect of reducing the stepped nature of the release and at site 1, 100 m downstream of the dam, the time of rise was reduced from 45 to 25 min, at site 2 maximum stage was attained in 15 min, at site 3 in 8 min, at site 4 in 9 min and at site 5 in 18 min. The difference in time to attain maximum stage for each site may be explained by different storage capacities of the channel reach above each site and by the changing roughness of the channel

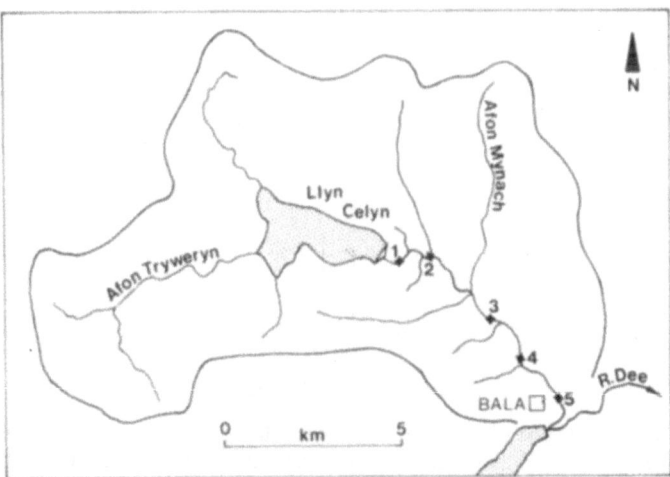

Figure 2. The River Tryweryn Catchment showing survey sites (1-5).

with increasing stage. At site 2 only 1.3 km downstream, the stepwise release pattern had been totally obscured. The change in wave steepness suggests that, at least to site 3, early release water fills the intricate network of storages created by the high channel roughness (Foulger & Petts 1984).

Suspended Solids Variations

Suspended solids concentrations increased with the start of stage rise at each site, peaked and then began to decline. In contrast to natural flood hydrographs, during controlled reservoir releases maximum stage is maintained for several hours and the decay of suspended solids concentrations relate directly to the depletion of in-channel sources. Peak concentrations increased downstream, particularly between sites 1 and 2 (100 to 190 mg l^{-1}), but differences between sites relate more to the duration of elevated suspended solids concentrations and to the variation of these concentrations with time (Figure 3).

Furthermore, the time difference between the attainment of peak stage and peak suspended solids concentrations varied with distance downstream. At site 1, peak concentrations preceded peak stage by 13 min but at sites 2 and 3 the peaks were synchronous and at site 4 peak concentration lagged peak stage by 16 min. The more rapid reduction in concentration at upstream sites relates, in part, to the load transported at each site. Downstream from the dam the total load increased rapidly to site 3 (640 kg at site 1, 2425 kg at site 2 and 8260 kg at site 3) and then more slowly to 9430 kg and 10,090 kg at sites 4 and 5 respectively. Total load was defined as that

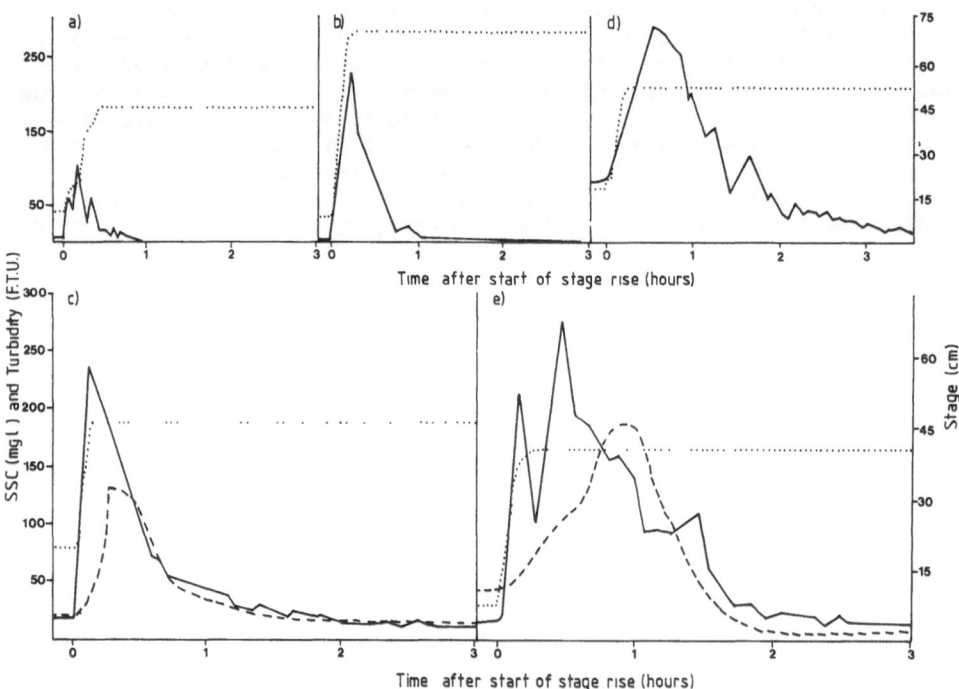

Figure 3. Suspended solids (———) and turbidity (-----) variations with stage (·······) for the five sites downstream of Llyn Celyn during the controlled reservoir release: a) site 1, b) site 2, c) site 3, d) site 4, and e) site 5.

material transported throughout the duration of high suspended solids
concentrations, an arbitrary cutoff being identified when concentrations
returned to a constant value (Gilvear & Petts 1985). However, the time of
stage rise and speed of `wave´ movement also affected the relationship
between suspended solids concentrations and discharge. The time of stage
rise increased downstream to site 3 but the water velocity was higher than
the speed of travel of the suspended solids and gave an increasing lag at
sites further downstream.

Turbidity Variations

Throughout the year Llyn Celyn is a low turbidity impoundment and
released water has a turbidity below 5 (FTU, Formazin Turbidity Units).
During the release, peak turbidity values of 130 FTU and 90 FTU at sites 3
and 5 respectively lagged peak stage by 10 and 31 minutes, with high values
maintained for longer at site 5 (Figure 3). Turbidity also lagged behind
peak suspended solids concentrations and the relationship showed marked
anti-clockwise hysteresis. This suggests temporal changes in the nature of
the material transported. For example, at site 3, turbidity values on the
rising limb of the hydrograph were half those on the falling limb, for a
given suspended solids concentration (Figure 4). This is associated, in
part, with the initiation of turbulent suspension on arrival of the
wavefront. Certainly, the flotation of coarse particulate organic debris
was the cause of the first peak at site 5 (Figure 3). An impulsive start of
motion can cause steep velocity gradients near the channel-bed (Petts et
al. 1985). On the rising limb of the hydrograph on average 0.1% of
minerogenic particles were coarser than 63 μm but during the period of
constant discharge less than 0.01% of particles were above this
value. However, this cannot account for the lag of turbidity behind

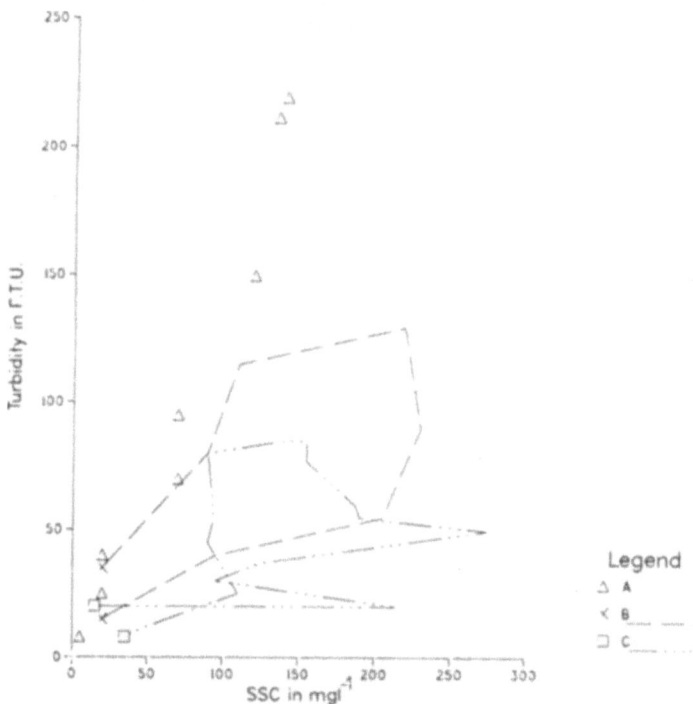

Figure 4. Suspended solids-turbidity relationships below Llyn Celyn:
A) tributary flood monitored at site 3, B) site 3 during the
release, and C) site 5 during the release.

peak stage which must relate to the routing downstream of discrete suspended solids sources of variable nature. The organic content of the seston, for example, generally decreased downstream from 50% at site 1 to 39%, 18%, 17% and 18% at sites 2 to 5 respectively. The variable proportions reflect the changing availability of inorganic supplies.

Sources for Suspended Material

During a storm event on the 22nd April, 1983 the Afon Mynach tributary had a peak suspended solids concentration of 250 mg 1^{-1} whilst concentrations were <3 mg 1^{-1} within the main river upstream from the confluence. Below the confluence, concentrations were reduced to 150 mg 1^{-1} at site 3 and to 40 mg 1^{-1} at site 5, some 3.5 km further downstream. Mixing of the clear reservoir water and the tributary-derived sediment-laden water accounted for a drop in concentration of only 50 mg 1^{-1}. The further fall of concentration might be attributed to the loss of material to the substratum by intrusion and infiltration and by deposition in lee areas created by boulders or along the channel margin. An estimated 1050 kg of material was deposited within one kilometre of the tributary junction. Analysis of downstream changes in the load per unit channel area during reservoir releases show an exponential decline. The interruption of this pattern on the Afon Tryweryn during the release can be related to the removal of this deposited material. Thus, the area between the two curves in Figure 5 represents this additional load of 1950 kg. At least 3 similar flood events occurred and, thus, an estimated load of more than 3000 kg may have been deposited. Some material may remain because disturbance of the substratum layer is needed to remove infiltrated fines (Beschta & Jackson 1979; Beschta et al. 1981). Higher turbidities were generated for a particular sediment concentration than during the reservoir releases (Figure 4), implying that the material from the Afon Mynach is relatively fine or of low specific gravity. SEM analysis showed the tributary sediment to be composed of fine minerogenic particles smaller than 16 μm (Figure 6a). Deposited tributary material may, therefore, be the source of the high turbidity late on the hydrograph at sites 3 and 5 (Figure 3).

Although tributary inputs appear to provide a major source of material, Figure 5 suggests that another source is important, at least

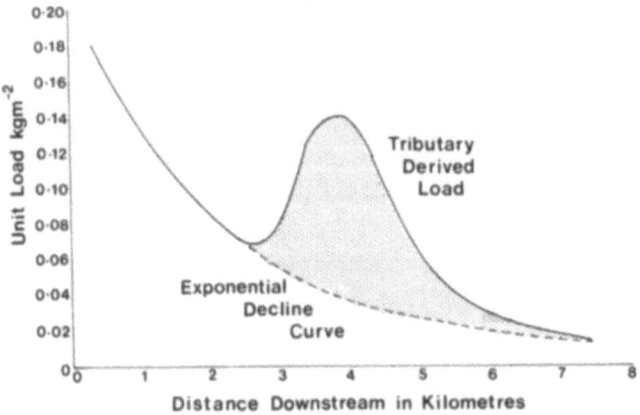

Figure 5.   Downstream changes in load per unit channel area transported during the reservoir release illustrating the contribution of tributary derived material.

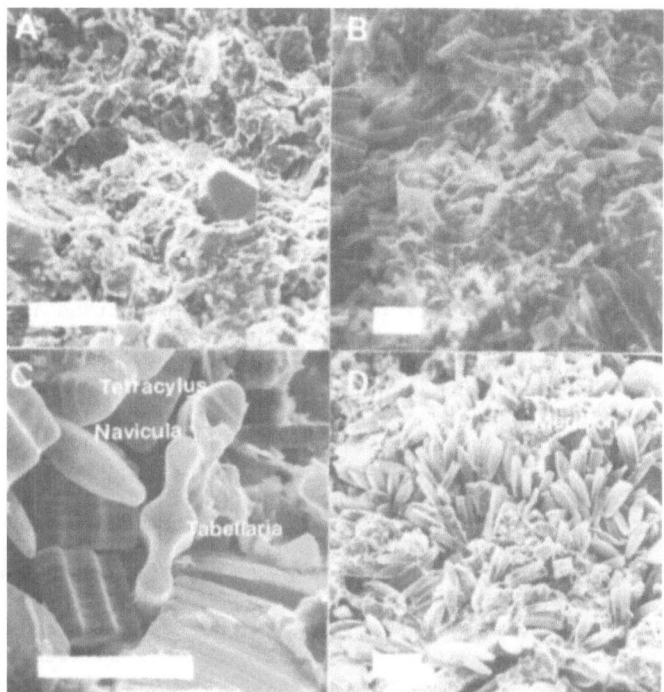

Figure 6. Scanning electron microscope photographs showing seston
composition (Scale bar equals 25 μm): a) tributary derived
suspended material, b) reservoir release site 1 seston,
c) reservoir release site 1 seston, and d) reservoir release
site 3 seston.

upstream. At site 1, above any tributary, the release transported a load of
over 600 kg. SEM analysis revealed that this seston was composed
predominantly of diatom frustules and fragments of filamentous algae
(Figure 6b & 6c), chiefly lotic forms similar to those within the dense
periphyton below the dam. Scouring of such material has previously been
reported (Brennan et al. 1978; Swanson & Bachmann 1976). Downstream the
periphyton continued to contribute but to a lesser degree, in terms of load
per unit channel area. Here other species of pennate diatoms predominated
(Figure 6d). These too were scoured from the periphyton.

Case Study Summary

The analysis of turbidity and suspended solids data during the release
from Llyn Celyn demonstrates the ability of such releases to remove
superficial coverings of tributary injected sediments. Disturbance of the
substratum is needed to remove infiltrated fines (Beschta & Jackson 1979).
Displacement of periphyton makes an important contribution to the suspended
load. Downstream the load increased with the downstream increase in channel
area (cf. Beschta et al. 1983) despite the progressive decrease in unit
load. The timing of the suspended solids pulse could be related to distance
downstream. The magnitude and duration of the high suspended solids
concentration and turbidity pulse may be limited by supply and by the
relative contributions of material from different sources along the
channel. In-channel supply is likely to be a function of tributary flood
frequency and of the time required for periphyton growth, relative to the
frequency of reservoir releases.

SUSPENDED SOLIDS LOADS AND ANTECEDENT CONDITIONS

Suspended solids loads increased during the reservoir releases studied. The material derived from the reservoir and stilling well was measured so that a simple input-output budget could be used to estimate the contribution from in-channel sources. The reservoir and stilling well contributions were significant in the load transported immediately below the reservoir but were exceeded at lower sites by material from within-channel sources. The exponential decline of load per unit area of channel below reservoirs has already been mentioned, though the pattern is disrupted by tributary inputs. To permit comparison of between-release and between-river loads, standardization of distance downstream is needed. For example, a change over 1 km downstream of the reservoir is, in relative terms, more significant in a first order stream than a fifth order stream. Monitoring sites between 200 and 235 channel widths from the reservoir were thus selected. Stage change immediately below the reservoirs generally varied between 60 and 70 cm. However, the stage change during the reservoir release from Sutton Bingham was only 20 cm.

Figure 7 demonstrates that the load per unit channel area at these sites correlated best with the number of days since the last release (R^2 = 0.86, significant at the 99.5% confidence level). Figure 8 demonstrates that peak suspended solids concentrations were generally correlated with the load per unit channel area (R^2 = 0.56, significant at the 99.5% confidence level). Scatter within the relationship is related to the time of rise of stage and the duration of elevated concentrations, the former, in part affecting the latter.

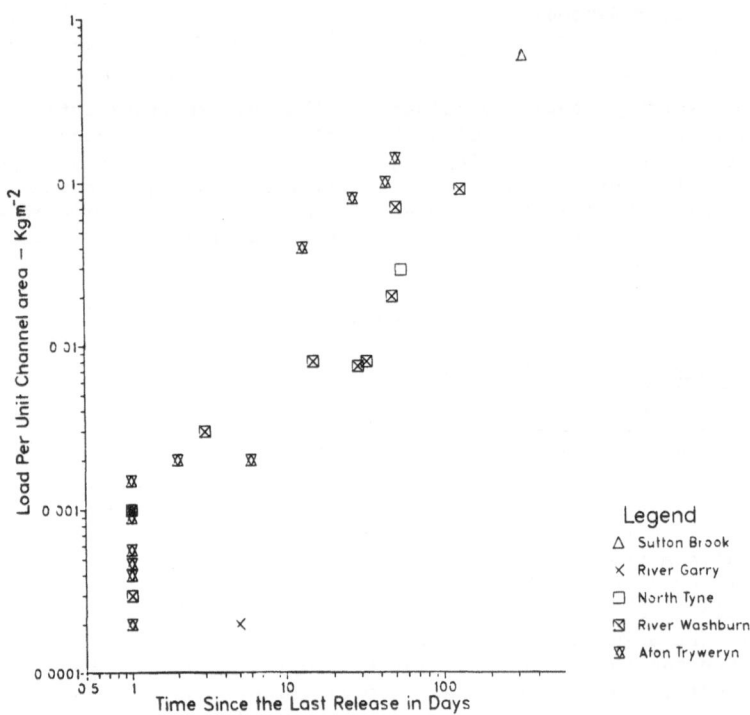

Figure 7. Reservoir release load per unit channel area versus time since the last release.

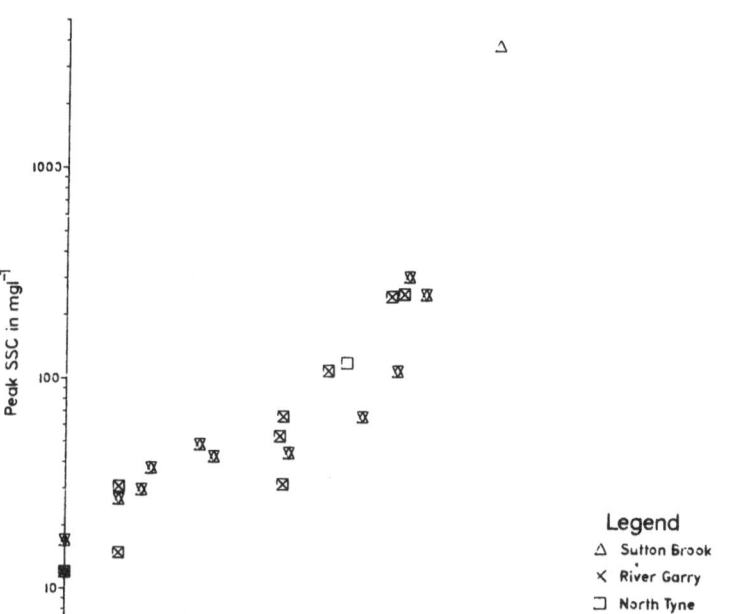

Figure 8.   Reservoir releases peak suspended solid concentration versus
load per unit channel area (Symbols as in Figure 7).

Long periods of stable flow and a stable substratum allow development
of an extensive periphyton which is characteristic of impounded rivers.
They also allow time for allochthonous organic matter to be an important
contributor. Leaf litter was important, especially during autumn. This
material accumulates in front of stones and in parts of the stream with low
water velocity (Fisher & Likens 1973). Thus, on the River Washburn, the
three releases with the highest organic content (29, 33 and 38%) each
occurred after at least 30 days of low flow, during releases in May, June
and September. The organic content in the first two examples reflects the
destruction of the periphyton and in the third example this is supplemented
by the contribution of leaf material. Minerogenic inputs were more
significant during the winter months (organic content ranged from 12 to
27%). On the Afon Tryweryn at a distance of 235 channel widths downstream,
minerogenic tributary inputs were appreciable and organic content ranged
from 6 to 27%. The minerogenic load was greatest after high rainfall totals
and a number of tributary flood events. In part, these were a function of
time since the previous release, but seasonal variation occurred, and
rainfall was greatest from December until April. In the winter months (8
releases studied during January, February and April) organic content was
less than 10%. Diatoms were also less numerous.

CONCLUSION

Removal of within-channel accumulations of material during reservoir
releases results in variations of both turbidity and suspended solids. With
greater loads, suspended solids concentrations and the duration of elevated
concentrations are increased. Turbidity is also increased, although this
relates not only to the concentrations of suspended solids but to the

nature of material in suspension. Thus, the relationship·between turbidity and suspended solids concentrations varies as discrete suspended solids sources are removed from within the impounded river. In addition, reservoir release routing alters the relationships between discharge, turbidity, and suspended solids.

The data demonstrate the potential of reservoir releases to remove fine superficial coverings of material that accumulate during low flows. Substratum disturbance is needed to remove material once it has infiltrated the channel bed. The loads transported and the suspended solids concentrations can be related to the time since the last release. Thus, if maximum levels of these parameters are defined, optimum time intervals can be quantified. Maintaining turbidity below critical levels is more problematic in that it varies with the nature of material being transported. The results presented here suggest that levels will be highest when fine minerogenic matter dominates the seston. This will occur wherever tributaries with high suspended sediment concentrations enter the channel. The interval between releases which is needed to maintain prescribed standards will then be dependent upon tributary flood frequency which will vary between climatic regions and seasonally. Thus, further research may be needed to fully utilize the conceptual model of Milhous (1982) and deem the statement of Wesche and Rechard (1980) incorrect.

ACKNOWLEDGEMENTS

I would like to acknowledge the cooperation of the Welsh Water Authority, and Dr. G. E. Petts for his numerous helpful comments. Assistance with diatom taxonomy was given by Dr. C. N. Roberts and Dr. R. Jones.

REFERENCES

Alabaster, J. S. 1972. Suspended solids and fisheries. Proc. R. Soc. B.L., **180**: 395-406.
Antonio, R. A. 1969. The Arkansas River Project civil engineering. American Society of Civil Engineering, **39**: 44-49.
Barton, B. A. 1977. The short-term effect of highway construction on the limnology of a small stream in Southern Ontario. Freshwat. Biol., **7**: 99-108.
Beschta, R. L. and Jackson, W. L. 1979. The intrusion of fine sediments into a stable gravel bed. J. Fish. Res. Board Can., **36**: 204-210.
Beschta, R. L., Jackson, W. L. and Knoop, K. D. 1981. Sediment transport during a controlled reservoir release. Wat. Res. Bull., **17**: 635-641.
Bjornn, T. C., Brusven, M. A., Molnau, M. P., Milligan, J. H., Klamt, R. A., Chaco, E. and Schaye, C. 1977. Transport of granitic sediments in streams and its effect on insects and fish. Bulletin No. 17, College of Forestry, Wildlife and Range Sciences, 43 pp.
Brennan, A., McLachland, A. J., and Wotton, R. S. 1978. Particulate material and midge larvae (Chironomidae: Diptera) in an upland river. Hydrobiologia, **59**: 67-73.
Brooker, M. P. and Hemsworth, R. J. 1978. The effect of the release of an artificial discharge of water on invertebrate drift in the R. Wye, Wales. Hydrobiologia, **59**: 155-163.
Carling, P. A. 1983. Particulate dynamics, dissolved and total load, in two small basins, northern Pennines, UK. Hydrol. S. J. **28**: 355-375.
Eaton, J. S., Likens, G. E. and Bormann, F. H. 1969. Use of membrane filters in gravimetric analysis of particulate matter in natural waters. Wat. Resources Res. **5**: 1151-1156.

Eustis, A. B. and Hillen, R. H. 1954. Stream sediment removal by controlled reservoir releases. Prog.-Fish Cult. **16**: 30-35.

Finlayson, B. L. 1975. Measurement of the organic content of suspended sediments at low concentrations. B.G.R.G. Tech. Bull., **17**: 21-26.

Fisher, S. G. and Lavoy, A. 1972. Differences in littoral fauna due to fluctuating levels below a hydro-electric dam. J. Fish. Res. Board Can., **29**: 1472-1476.

Fisher, S. G. and Likens, G. E. 1973. Energy flow in Bear Brook, New Hampshire: An integrative approach to stream ecosystem metabolism. Ecol. Monogr., **43**: 421-439.

Foulger, T. R. and Petts, G. E. 1984. Water quality implications of artificial flow fluctuations in regulated rivers. Science of the Total Environment, **37**: 177-185.

Gilvear, D. J. and Petts, G. E. 1985. Turbidity and suspended solids variations downstream of a regulating reservoir. Earth Surface Processes and Landforms, **10**: 363-373.

Hall, J. D. and Lantz, R. L. 1979. Effects of logging on the habitat of coho salmon and cutthroat trout in coastal streams. In: Proceedings of the Symposium on Salmon and Trout in Streams. (Ed. by G. Northcote). pp. 355-375. University of British Columbia, Vancouver.

Matter, J. W., Hudson, P. L. and Saul, G. E. 1983. Invertebrate drift and particulate organic matter transport below Lake Hartwell during a peak power generation cycle. In: Dynamics of Lotic Ecosystems. (Ed. by T. D. Fontane and S. M. Bartwell). pp. 357-370. Ann Arbor Science, Ann Arbor, Michigan.

McDonald, A., Kay, D. and Jenkins, A. 1982. Generation of fecal and total coliform surges by stream flow manipulation in the absence of normal hydrometeorological stimuli. Appl. and Environ. Microbiol., **44**: 292-300.

Milhous, R. T. 1982. Effect of sediment transport and flow regulation on the ecology of gravel-bed rivers. In: Gravel Bed Rivers. (Ed. by R. D. Hey, J. C. Bathurst and C. R. Thorne). John Wiley and Sons Ltd., Chichester, England.

Petts, G. E., Foulger, T. R., Gilvear, D. J., Pratts, J. D. and Thoms, M. 1985. Wave movement and water quality variations during a controlled release from Kielder reservoir, North Tyne River, U.K. J. Hydrol. **80**: 371-389.

Swanson, C. D. and Bachmann, R. W. 1976. A model of algal exports in some Iowa streams. Ecology, **57**: 1076-1080.

Wesche, A. T. and Rechard, P. A. 1980. A Summary of Instream Flow Methods for Fisheries and Related Research Needs. Eisenhower Consortium for Western Environmental Forestry Research, University of Wyoming, Laramie. 122 pp.

TIME-SCALES FOR ECOLOGICAL CHANGE IN REGULATED RIVERS

Geoffrey E. Petts

Department of Geography
University of Technology
Loughborough, Leicestershire, UK

INTRODUCTION

River impoundment causes dramatic changes in the physical, chemical
and biological characteristics of the river downstream; changes which are
not only confined to the lotic system but also extend to the riparian and
flood-plain systems, and, in some cases, to the near-shore zone. Dams
interrupt the pattern of downstream transfers involving water, sediments,
organic matter, nutrients etc., and river regulation has reduced the
diversity-enhancing effects of increasing physical heterogeneity from low-
to middle-order streams (Ward & Stanford 1983). However, one fundamental
problem in the development of a general model of system response to river
regulation is the failure of many studies to consider observed changes
within an appropriate time-scale (Petts 1984). This paper describes a
conceptual model which provides a temporal framework for the interpretation
and assessment of spatial patterns of change within regulated rivers.

A TIME-SCALE FOR READJUSTMENT

A wide range of physical, chemical and biological factors interact to
determine the character of a river and to maintain the river system in a
state of quasi-equilibrium. Different components of the system will respond
at different rates, not least because the controlling factors operate over
a range of frequencies. The hydrology and transfer of matter from upstream
sources will be altered immediately after the dam is closed. Flood-plain
vegetation (which is often dependent on regular inundation) can change
rapidly following flow regulation. For example, flood-plain dessication
quickly followed the impoundment of the Zambezi by Lake Kariba (Attwell
1970). However, 13 years after the closure of Glen Canyon Dam a new
ecological equilibrium had not been reached along the Colorado River within
Grand Canyon (Turner & Karpiscak 1980). The distribution of animals and
birds will also be influenced, directly and indirectly, by flow regulation
(Tinley 1975; Frith 1977), but increased human occupation and agricultural
or industrial activity, stimulated by flood regulation, commonly imparts an
even greater impact on the ecology of floodplains.

The time-lag required for a stable aquatic plant community to develop
within a regulated river may be dependent upon the frequency of tributary
sources for inoculation. Below Cow Green Reservoir, UK, on the River Tees a

steady state had not been reached 4 years after dam closure (Holmes & Whitton 1977). Young et al. (1976) and Williams and Winget (1979) considered that the successful reorganization of the benthic macroinvertebrate community, to fill the new set of niches created as a consequence of upstream dam construction, requires less than 5 years. However, the biotic responses are usually faster than the abiotic ones and Neuhold (1981) suggested that the ecosystem recovery process will keep pace with the physical recovery process.

The minimum time required for a system to adjust to the conditions imposed by an upstream impoundment is dependent upon those variables that require the longest time to achieve a new stable structure (Figure 1). The quality of impounded water, and of the discharges from the dam, can require a period of 5 to 10 years, on average, to stabilize (Grimard & Jones 1982) because of the trophic upsurge which takes place consequent upon the submergence of the soil-vegetation complex, at least in mid-latitude reservoirs. Physical readjustment of river channels below dams can require a time-scale measured in tens (or even hundreds) of years (Petts 1984). The effects of these changes on benthic communities have been illustrated respectively by Hilsenhoff (1971) and Petts and Greenwood (1985). However, the development of general models of system change have been constrained because as yet most studies have focussed on rivers which have been regulated for between 10 and 30 years. Completely adjusted systems have been observed only rarely and most studies relate to the period characterized by transient system states.

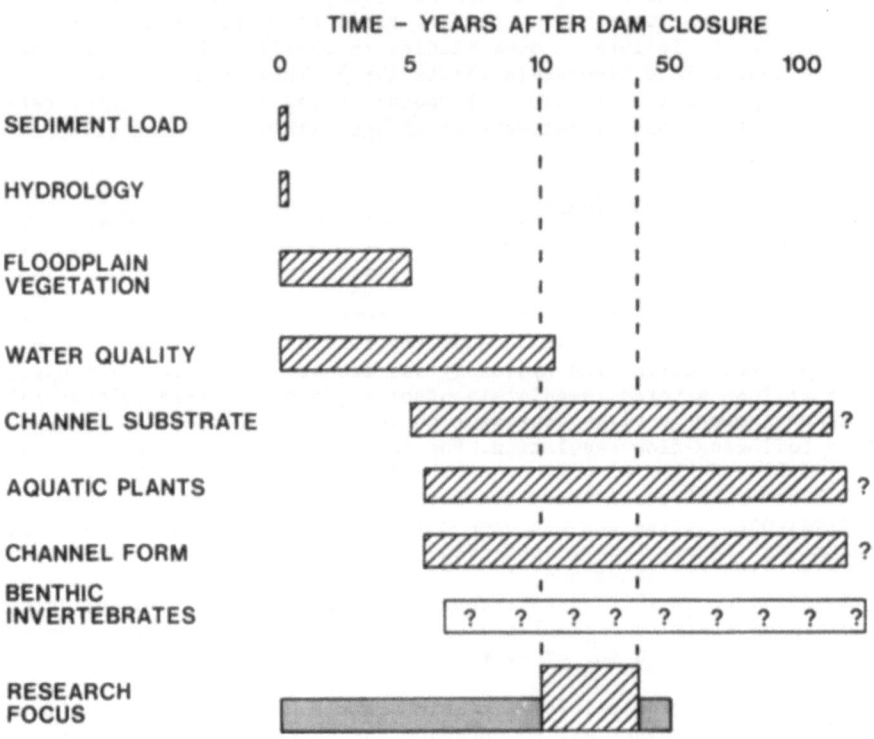

Figure 1. Time-scales for system readjustment (Based on Petts 1984).

THE TRANSIENT SYSTEM MODEL

During the period of physical readjustment, biotic communities may respond rapidly to the changing conditions. Consequently, at any time during this period, observed biological stability will relate to a temporary or transient state on the time-scale of system readjustment. The Transient System Model was developed in order to provide a conceptual framework for the interpretation of spatial patterns of change along regulated rivers. The Transient System Concept can be defined thus: at any time during the period of system readjustment, different parts of a regulated river will be at different stages of physical change and the biotic communities associated with these stages of change will be temporary states.

System Structure

The first assumption on which the model is based is that physical readjustment precedes biotic recovery and is a pre-requisite for system adjustment. The lotic ecosystem is viewed within a hierarchical framework (Figure 2) which has three levels of response. First-order changes take place consequent upon or shortly after dam closure and effect the transfer of energy and matter into and within the downstream river. Second-order variables respond to changes at the first level over a period of years; they relate to habitat structure and reflect local site conditions. Third-order responses will reflect all changes of the lower order variables as well as biotic interactions. System readjustment to a new quasi-equilibrium condition cannot proceed in advance of the third-order

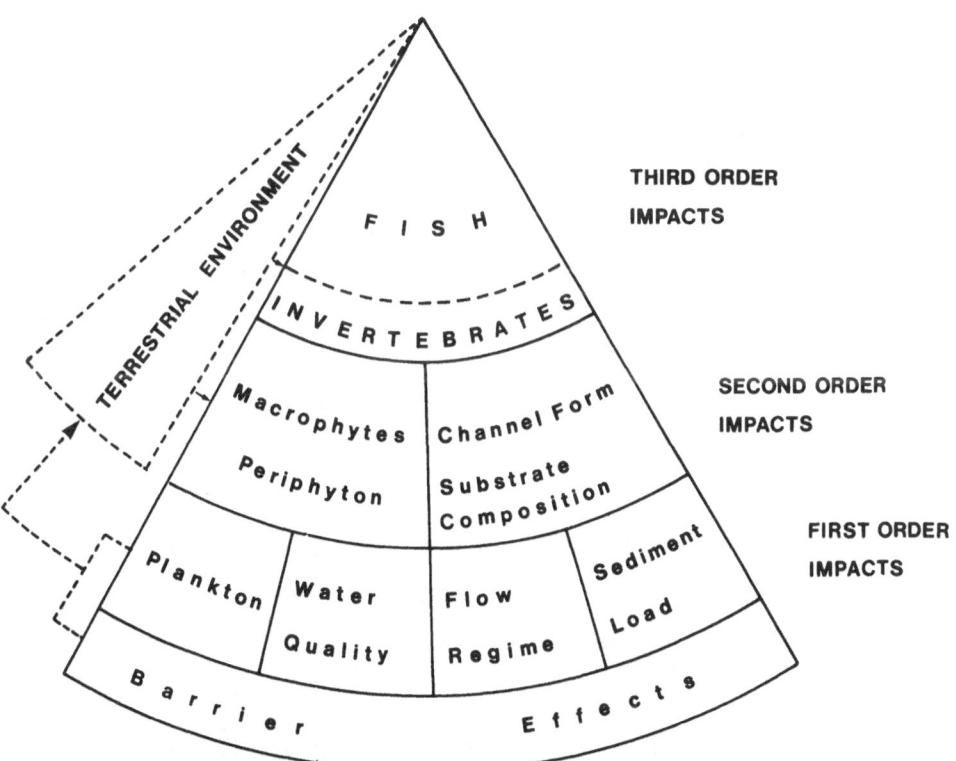

Figure 2.   A framework for the assessment of changes within regulated rivers (After Petts 1984).

changes and will require an appropriate time-scale. Several phases of biotic population adjustment can occur in response to the first- and second-order changes but these adjustments of the third-order variables over short time-scales will merely relate to TRANSIENT system states. Indeed, some parts or attributes of the system may remain as relic features for a long period of time whilst others can respond rapidly to the imposed regulated-flow conditions.

Location-for-Time Substitution Models

Because complete system readjustment requires a long period of time, and has rarely, if ever, been monitored, an approach is needed to interpret the observed, short-term, third-order variable responses in the context of long-term system readjustment. One approach to this problem is to use `location-for-time substitution` (Paine 1985) whereby field observations are placed in an assumed time sequence based on theory or their field relations. In this model an index of maturity is used as a substitute for absolute date control. Temporal readjustments are inferred from the assumed `development` across the spatial series in which they are placed. This approach has been used, both explicitly and implicitly, in a wide variety of geographical and biological studies but it introduces a second assumption into the Transient System Model, namely, that the observed sequence reflects exclusively the passage of time and that there is no other cause of the sequence. The recognition of both time-dependent and time-independent influences is, therefore, mandatory.

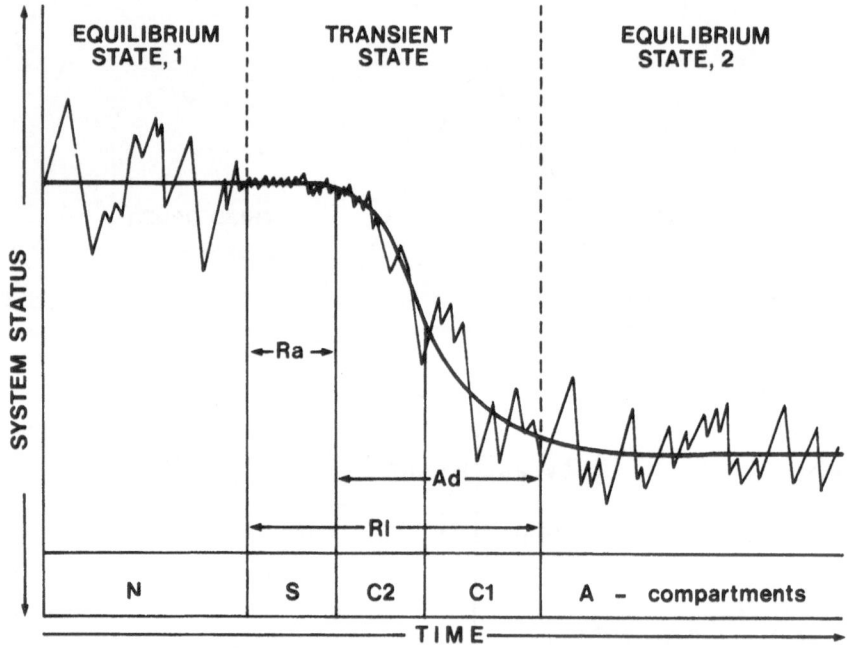

Figure 3. The Transient System Model. This conceptual model of system response to river regulation identifies transient system states during the relaxation period (Rl) between a natural equilibrium state (1) and the new equilibrium state adjusted to the regulated conditions (2). Two transient states are recognized: a stable state during the reaction period (Ra) and a phase of change during the adjustment period (Ad). The model is based on Allen (1974).

The Conceptual Model

Data for twenty regulated rivers located throughout the U.K. have been used to develop a general framework for the applicaton of `location-for-time substitution´ (Figure 3). The model is based on the recognition of channel COMPARTMENTS (reaches of channel which have a particular morphological and sedimentological character). Within each compartment a patchwork of micro-habitats can be identified, but the characteristic patchwork of each compartment is different from that of the other compartments. The system is seen to be in a transient state during the relaxation period, the time between dam closure and the attainment of a new quasi-equilibrium condition adjusted to the imposed character of the first-order variables. However, the transient phase may be divided into two periods, of stability (compartment S) and change (compartment C). For a time after dam closure (the reaction time) the second order variables describing channel structure will show no observable change although the third-order variables will change in response to the imposed conditions, especially, of flow and water quality. During the adjustment period, changes of channel structure will move the system to the new quasi-equilibrium state (compartment A).

For U.K. rivers, within which degradation is confined to a short reach, often only a few hundred metres immediately below the dam, due not least to the low sediment loads and to the coarse channel substrate, channel change is dominated by aggradation and accretion. Stable, changing and adjusted compartments have been recognized. Within the stable compartments, the regulated flows are incompetent to transport the bed materials and the substrate is stabilized allowing the development of a dense periphyton with patches of macrophytes. Slow sedimentation in slackwater areas may add to the enhanced micro-habitat diversity although the relatively constant velocity conditions are also influential. Changing compartments are characterized by an unstable (aggrading) substrate, with a relatively high proportion of fines and with a restricted periphyton cover. Gravel bars often form and pools are filled with fine minerogenic and organic deposits. Within the adjusted compartments, channel change has increased the annual range of stage (i.e., wetted area) and velocity due to changes of channel shape. This important characteristic of channel adjustment can produce higher velocities during the normal range of flows than prior to regulation (Table 1). The channel bed will develop a coarse, armoured, surface layer generally devoid of fines, above a compact substrate, and a moderate periphyton cover may establish.

Changing compartments are often associated with point sources of sediment from tributary streams. The spatial sequence of compartments downstream of a confluence, with compartments becoming progressively immature from adjusted, to changing, and then to stable, can be interpreted as a temporal sequence. Below a point source of sediment injection, the depositional front has been observed to move downstream at a rate which relates to the volume and grain-size distribution of the sediment injected and to the hydrological characteristics of the regulated river. Extreme rates for regulated sand-bed rivers exceed 20 km yr^{-1} but for gravel-bed rivers rates of less than 1 km yr^{-1} are common. An alternating sequence of compartments, often incomplete, may be identified along a regulated river associated with the frequency of important tributary streams.

MODEL APPLICATION

The model was first used to interpret system changes within the Afon Rheidol below Nant-y-Moch Reservoir, Wales (Figure 4), 21-years after dam closure (Petts & Greenwood 1985). Five compartments were identified (Table

Table 1. Flow characteristics of the River Okement below Meldon Reservoir, UK.

	Natural Channel	Adjusted Channel*
A At-a-station hydraulic geometry		
Width	$6.61 \ Q^{0.08}$	$4.36 \ Q^{0.20}$
Depth	$0.38 \ Q^{0.52}$	$0.46 \ Q^{0.47}$
Velocity	$0.40 \ Q^{0.40}$	$0.50 \ Q^{0.33}$
Bankfull Discharge	$12 \ m^3 \ s^{-1}$	$5 \ m^3 \ s^{-1}$

B Variation of flow width (m), mean depth (m) and mean velocity (m s^{-1})

	w	d	v	w	d	v
$Q = 0.5 \ m^3 \ s^{-1}$	6.25	0.27	0.30	3.80	0.33	0.40
$Q = 1 \ m^3 \ s^{-1}$	6.61	0.27	0.40	4.36	0.33	0.50
$Q = 5 \ m^3 \ s^{-1}$	7.51	0.88	0.76	6.02	0.98	0.85
$Q = 12 \ m^3 \ s^{-1}$	9.08	1.38	1.08	-	-	-

*Bankfull channel width reduced by 33% as a result of lateral bar accretion; mean depth reduced by 20%.

2). Four compartments (A, C1, C2 and S) form a clearly defined spatial sequence down-stream of the Peithnant tributary confluence; the first three of these represent stages of channel readjustment. Compartment A, the adjusted reach is near the tributary confluence, C2 is associated with fine substrate at the downstream limit of change, and C1 is an intermediate reach of coarse gravel aggradation. Each compartment is associated with an invertebrate fauna which differs in some way from the other compartments. In this example, the downstream limit of change is some 500 m below the confluence.

Two compartments have a greater invertebrate density than the natural reaches: the adjusted compartment, which is dominated numerically by Simuliidae, and the stable compartment which, in contrast, is dominated by Chironomidae, Plecoptera and Trichoptera. The Rheidol case-study, however, presents a relatively simple, and possibly unique, situation where local factors operate to create clearly delineated compartments. Therefore, the field methodology has been applied to another river, the River Daer, part of the upper Clyde system in Scotland.

The River Daer

The central Southern Uplands of the upper Clyde system (Figure 4) is bounded on the west and south by the Lowther Hills rising to 732 m and to the east by a line of hills rising to 748 m at Coulter Fell. The area is dominated by highly deformed Ordovician and Silurian sediments. The major valley systems were formed prior to the Quaternary and today contain

Figure 4.   Sketch map of mainland Britain to show the position of the
Rheidol and Daer Rivers.

thicknesses of glacial and fluvioglacial deposits. On the higher ground
peat attains a thickness of up to six metres. Mean annual rainfall is
approximately 1500 mm.

The Daer Reservoir commenced filling in 1956, impounding runoff from
the 47 km^2 catchment and inundating 4.33% of this area. Although
insufficient time has elapsed for the complete readjustment of channel
structure, stable and changing compartments have been identified (Table
3). The stable compartment is characterized by coarse, immobile, bed
sediments, with recent deposits of sand and fine-gravel in the slackwater
areas formed in the lee of boulders and coarse gravels. In addition there
is a relatively dense periphyton, and patchy developments of
macrophytes. These latter, which give rise to localized accumulations of
sand, are dominated by the ubiquitous moss *Fontinalis antipyretica*, and two
macrophytes associated with dystrophic waters, namely, *Myriophyllum
alterniflorum* and *Juncus bulbosus*. Such an intricate patchwork of micro-
habitats is characteristic of stable compartments within regulated rivers.
Two changing compartments can be defined within a 4 km reach below the
confluence of a major tributary, Portrail Water. Channel change has been
dominated by a reduction of width at riffle sites (C1), resulting from bar
accretion, and by bed aggradation, i.e., infilling, at pool sites (C2). In
both cases the grain-size of the substrate (and the diversity of
micro-habitats) is markedly reduced. Although the morphological and
sedimentological changes decrease progressively downstream of the
confluence, the data reveal a clear downstream limit to substrate
sedimentation where the proportion by weight of material within the
substrate finer than 2 mm falls below 10%.

Slight variations of the invertebrate fauna occur between the C1 and
C2 compartments, and between the changing compartments and the natural

Table 2. Compartments of the Afon Rheidol Below Nant-y-Moch Dam, UK.

Compartment	N	A	C1	C2	S	N
Number of Sample Sites	10	9	17	18	13	10
Length						
Channel Width (m)	14	8	19	19	14	14
Velocity (m s^{-1})	0.46	0.41	0.67	0.22	0.40	0.46
Substrate (mm)	20	16	12	4	14	20
Substrate, % sand	7.5	7.0	10	25	12.5	7.5
Periphyton, % cover	0	15	20	0	70	0
Invertebrate density (m^{-2})	336	463	205	303	485	336
Number of Taxa	10	13	14	22 ,	22	10
Diversity (H)	0.96	1.52	2.16	2.35	2.44	0.96

Dominant Taxa:
    N – Chironomidae, *Ecclisopteryx guttulata, Rhyacophila dorsalis,*
        Baetis rhodani, Diura bicaudata;
    A – Simuliidae, Chironomidae, *Leuctra hippopus,* E. *guttulata;*
    C1 – Simuliidae, Chironomidae, *PROTONEMURA MEYERI,* B. *rhodani,*
        E. *guttulata;*
    C2 – E. *guttulata,* Chironomidae, *L. hippopus, DICRANOTA* SP.,
        OLIGOCHAETA, TIPULINAE, *OREODYTES SANMARKI;*
    S – Chironomidae, *P. MEYERI, OXYETHIRA* SP., E. *guttulata,* R.
        dorsalis, Hydracarina, Simuliidae, *L. hippopus;*
The number of sample sites was proportional to the length of channel
    studied.
Velocity values are for compensation flow.
Substrate values refer to the mean grain-size.
Taxa NOT found in the `natural´ samples are shown in capitals.
Taxa are listed in order of numerical dominance.

river. The most significant difference, however, is between the stable
compartment and the other compartments, not least in terms of invertebrate
density. The higher invertebrate density of the stable compartment is a
characteristic of reaches having constant flows which are accommodated
within the existing channel form.

CONCLUSION

    A Transient System Model is described herein, using a hierarchical
system structure and location-for-time substitution, to provide a framework
for the consideration of spatial patterns of change along regulated rivers
within an appropriate time-scale. River channel compartments, defined by a
particular combination of micro-habitats, represent a temporal sequence of
system states. Within two British rivers, three compartments have been

Table 3. Compartments on the Regulated River Daer, UK.

Compartment	N	C1	C2	S	N
Number of Sample Sites	8	8	8	8	8
Flow Width (m)	17	12	17	17	17
Substrate (mm)	28	7.5	6	14	28
Periphyton, % cover	0	10	0	20	0
Invertebrate Density (m^{-2})	567	419	586	934	586
Number of Taxa	19	17	17	17	19
Diversity (H)	2.14	1.65	2.22	2.10	2.14

Dominant Taxa:
  N - *Heptagenia lateralis*, caseless caddis, Chironomidae, *Asellus*, *Chloroperla torrentium*;
  C1 - Chironomidae, Oligochaeta, cased caddis, *Heptagenia lateralis*, *Limmius volckmari*;
  C2 - *Heptagenia lateralis*, Chironomidae, *Perla bipunctata*, DICRANOTA SP., caseless caddis, *Baetis scambus*;
  S - *H. lateralis*, Chironomidae, CAENIS MOESTA, *L. volckmari*, caseless caddis, Oligochaeta.
Substrate values refer to mean grain size.
Taxa NOT found in the ˋnatural´ samples are shown in capitals.
Taxa are listed in order of numerical dominance.

identified: adjusted, changing and stable. The latter two compartments, representing transient system states, have been identified within both rivers using physical and biological criteria. Stable compartments typically have a higher invertebrate density than the other compartments. The changing compartments of both rivers have been divided into sites of gravel-bed sedimentation and aggrading sites dominated by fine minerogenic and organic sediments.

Some components of the river system will respond to flow regulation more readily than others, and some locations will readjust at a faster rate. In changing fluvial systems the continua describing downstream variations of individual parameters are characterized by discontinuities associated, not least, with tributary confluences. Consequently, the downstream pattern of change can relate to the structure of the drainage network. The conceptual model presented herein demonstrates that at any point in time, different parts of the regulated river can be in different phases of readjustment and that at a single location the biotic communities may experience a complex sequence of changes during the readjustment period. For the regulated coarse-gravel-bed rivers of Britain, a time-scale measured in tens of years is required for the assessment of third-order ecological change.

ACKNOWLEDGEMENTS

I am pleased to acknowledge the help offered in matters of invertebrate and plant taxonomy by Malcolm Greenwood and Max Wade, and the field and laboratory assistance of Martin Thoms and Vernon Poulter. The Daer study is part of a project funded by the Natural Environment Research Council.

REFERENCES

Allen, J. R. L. 1974. Reaction, relaxation and lag in natural sedimentary systems: general principles, examples and lessons. Earth-Science Reviews **10**: 263-342.

Attwell, R. I. G. 1970. Some effects of Lake Kariba on the ecology of a flood-plain of the mid-Zambezi Valley of Rhodesia. Biol. Conserv. **2**: 189-196.

Frith, H. J. 1977. Waterfowl in Australia. Reed, Sydney. 328 pp.

Grimard, Y. and Jones, H. G. 1982. Trophic upsurge in new reservoirs: a model for total phosphorus concentrations. Can. J. Fish. Aquat. Sci., **39**: 1473-83.

Hilsenhoff, W. L. 1971. Changes in the downstream insect and amphipod fauna caused by an impoundment with a hypolimnion drain. Ann. Entomol. Soc. Amer., **64**: 743-6.

Holmes, N. T. H. and Whitton, B. A. 1977. The macrophytic vegetation of the River Tees in 1975: observed and expected changes. Freshwat. Biol., **7**: 43-60.

Neuhold, J. H. 1981. Strategy of stream ecosystem recovery. In: Stress Effects on Natural Ecosystems. (Ed. by G. W. Barrett and R. Rosenberg) pp. 261-265. Wiley, Chichester.

Paine, A. D. M. 1985. Ergodic reasoning in geomorphology: time for a review of the term? Prog. Phys. Geog., **9**: 1-15.

Petts, G. E. 1984. Impounded Rivers. Wiley, Chichester.

Petts, G. E. and Greenwood, M. 1985. Channel changes and invertebrate faunas below Nant-y-Moch dam, River Rheidol, UK. Hydrobiol., **122**: 65-80.

Tinley, K. L. 1975. Marromeu wrecked by the big dam. Afric. Wildl., **29**: 22-5.

Turner, R. M. and Karpiscak, M. M. 1980. Recent vegetation changes along the Colorado River between Glen Canyon Dam and Lake Mead. US Geol. Surv. Prof. Pap., **1132**: 125.

Ward, J. V. and Stanford, J. A. 1983. The serial discontinuity concept of lotic ecosystems. In: Dynamics of Lotic Ecosystems. (Ed. by T. D. Fontaine and S. M. Barkeu) pp. 29-42. Ann Arbor, Michigan.

Williams, R. D. and Winget, R. H. 1979. Macroinvertebrate response to flow manipulation in the Strawberry River, Utah (USA). In: The Ecology of Regulated Streams. (Ed. by J. V. Ward and J. A. Stanford). pp. 365-377. Plenum, New York.

Young, W. C., Kent, D. H. & Whiteside, B. G. 1976. The influence of a deep storage reservoir on the species diversity of benthic macroinvertebrate communities of the Guadalupe River, Texas. Texas J. Sci., **27**: 213-24.

TOWARDS A RATIONAL ASSESSMENT OF RESIDUAL FLOWS BELOW RESERVOIRS

A. Gustard and G. A. Cole

Institute of Hydrology,
Wallingford, Oxfordshire, OX10 8BB, U.K.

INTRODUCTION

The growth in demand for public water supply in the U.K. has
traditionally been met by constructing new reservoirs and by developing
river or groundwater abstraction schemes. Direct supply reservoirs were
the most common surface water source until the post war period of
construction when an increasing number of pumped storage and regulating
reservoirs were introduced. Future increases in demand will initially be
met by improvements to the operating policies of existing reservoirs,
especially by linking sources with different refill characteristics. The
gross yield of reservoirs, is used to meet two demands; firstly the
provision of a public water supply, and secondly the maintenance of a
residual flow downstream of the reservoir for other water uses. In most
cases this will be the statutory compensation flow immediately below the
reservoir.

It is inevitable that part of the re-evaluation of some water
resources schemes will include a critical review of residual flows below
reservoirs. This paper describes the historical background to setting
these flows below U.K. reservoirs, the existing range of release patterns
and the impact of these patterns on the downstream flow regime. The
overall aim of the study is to enable water resource planners to arrive at
a more rational basis for setting downstream flow requirements.

HISTORICAL BACKGROUND

In order to construct a reservoir and abstract water an Act of
Parliament was required. Sheail (1984), using evidence from Parliamentary
Select Committees for a number of reservoir schemes completed in the UK in
the 19th and 20th century, has described how the concept of compensation
flow was as first perceived and illustrates the development of compensation
flow policy. The earliest reservoir schemes were designed to supply water
to maintain navigation on the expanding canal system and Acts of Parliament
were required to protect existing users of the river below each reservoir.
The same approach was used in granting powers for subsequent water supply
reservoirs and the influential position of industrial interests,
particularly mill owners, dominated the discussions in setting these early
compensation flows. For example a number of Acts of Parliament stipulated

zero compensation flows at weekends when mills were not working in order to maximize the residual flow during the working week. It was not until the close of the 19th century that consideration was given to other water uses such as fisheries and dilution of domestic and industrial effluent. The Elan valley scheme promoted in 1892 and the Haweswater scheme of 1919 were the earliest to consider seriously the problem of flow regulation on fisheries. In the latter case this provided for the periodic release of freshets in addition to a minimum discharge.

There have been several attempts to provide general guidelines for setting compensation flows. The first of these was proposed by Hawksley to the Royal Commission on Water Supply (1869). He suggested that the starting point of negotiation should be that two thirds of the gross yield should be used for supply purposes and the remainder for compensation flow. Hawksley's concept that downstream flows be related to reservoir yield rather than to the characteristics of the receiving stream was the basis of many schemes. It was not until 1930 that a technical subcommittee of the Ministry of Health recommended that both the natural character of the receiving stream and the use of the river should be considered. More recently the Water Resource Act of 1963 and Water Act of 1973 were introduced and a more coherent approach to assessing the environmental impact of new schemes has developed. However local precedents and `rules of thumb´ continue to be an important aspect of setting residual flows.

SURVEY OF RESIDUAL FLOWS

In the last 200 years more than 500 reservoirs have been constructed and compensation flows assessed for each one. A survey of these reservoirs was carried out to assess the pattern and magnitude of releases. A number of small reservoirs with catchment areas of less than 5 km^2 were excluded from the study. Figure 1 shows the location of the remaining 261 reservoirs for which information on compensation flows were obtained. 40 percent of these reservoirs had capacities of less than 2 000 Ml with only 15 percent having storage volumes greater than 20 000 Ml. Most of these reservoirs had catchment areas of less than 20 km^2 with only 30 being greater than 100 km^2.

Figure 2 illustrates the seven main patterns of residual flow which are combined into three groups. Some 70% of reservoirs release a constant discharge throughout the year although some of these also release artificial floods or freshets. The second group consists of a seasonally varying flow pattern that with one exception release a higher discharge in the summer than in winter. The third group have discharges maintained at a particular threshold at a point some distance below the dam. Thus if there is sufficient natural flow at this point releases may be stopped or at least reduced.

In order to compare compensation flows between different catchments the compensation discharge has been expressed as a percentage of the natural mean reservoir inflow. The average release is 18%, but, as shown by Figure 3, there is a wide variety in the level of compensation flow between different schemes. Those reservoirs having very high releases, in excess of 50% of the mean discharge, are compensation reservoirs whose function is solely to provide all the compensation water for a group of supply reservoirs with zero compensation flows.

Figure 4 shows the development of reservoir construction for four main types of reservoirs. An analysis of the historical trend of compensation flows over this period indicated that there have been lower compensation flows from 1920, with higher values in the industrialized areas and

Figure 1. Location of reservoirs.

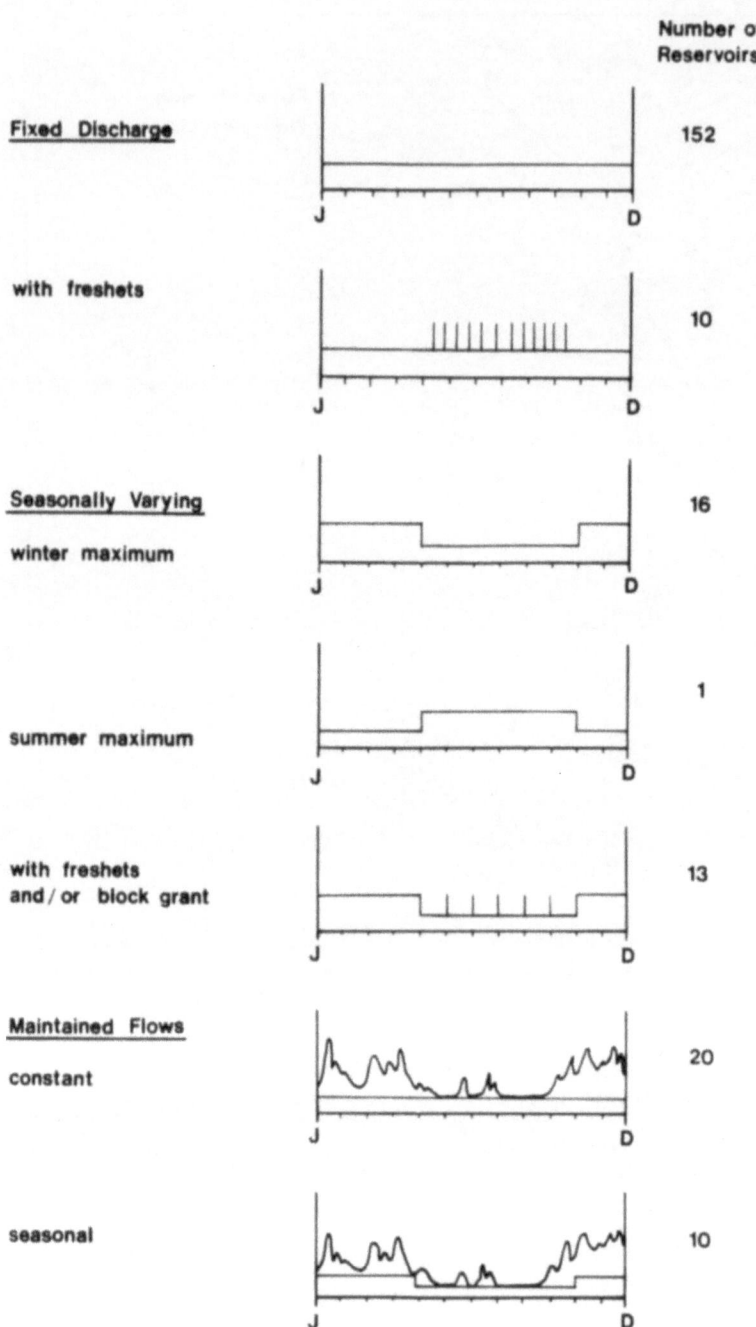

Figure 2. Pattern of residual flows.

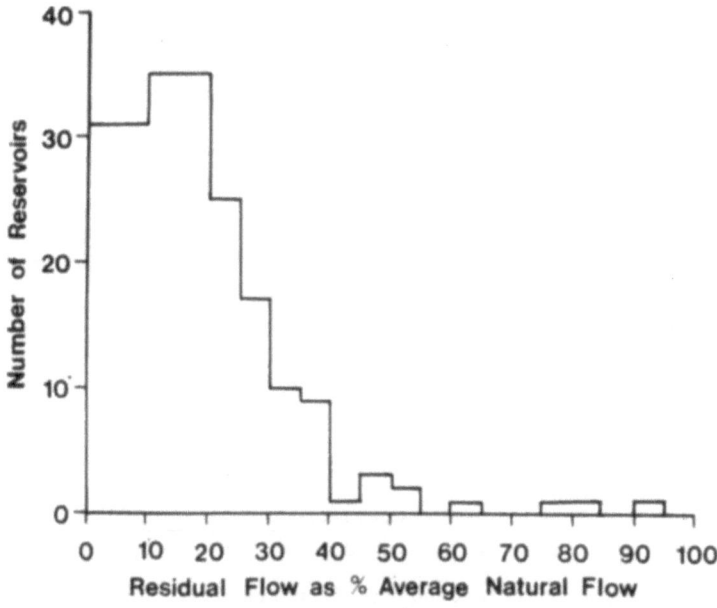

Figure 3. Residual flows for U.K. reservoirs.

generally lower ones elsewhere. These trends reflect the importance of the industrial interest in many of the 19th century water supply schemes.

RELATIONSHIP BETWEEN PRE AND POST RESERVOIR FLOW REGIMES

A number of disturbances to the natural flow regime following river regulation have been identified (Brooker 1981; Petts 1984). These include reductions in the mean annual discharge, in seasonal flow variability and in the frequency and magnitude of flood flows. In addition, the seasonal distribution of flows may change and, particularly for hydropower and regulating reservoirs, the incidence of rapid discharge fluctuations may

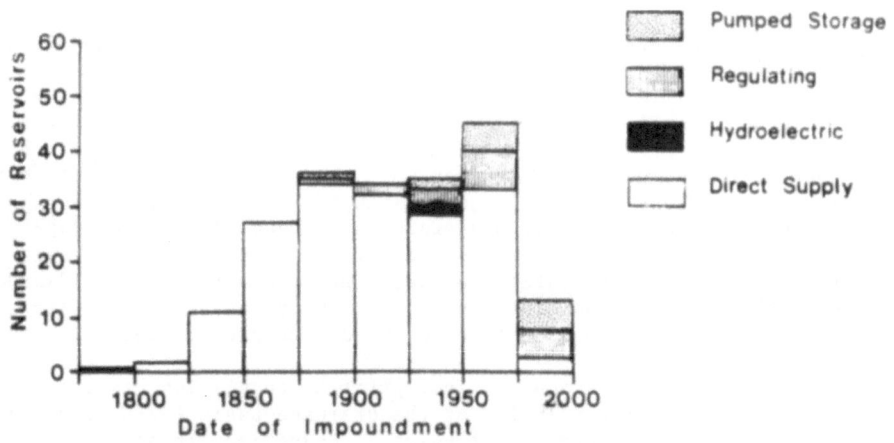

Figure 4. Date of impoundment of U.K. reservoirs.

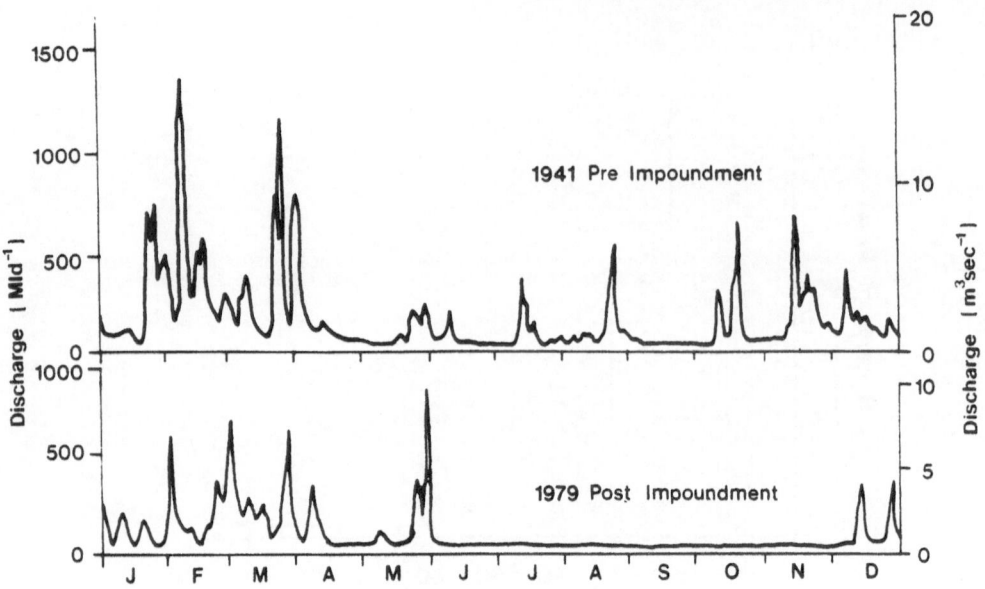

Figure 5. Pre and post impoundment flow regime for River Blithe.

increase. Figure 5 shows flow hydrographs for 1941 and 1979 for the River
Blithe and illustrates the change in the mean daily flow hydrograph
following impoundment which is typical of the majority of direct supply
reservoirs in the U.K. Changes in low flows are determined by the
statutory compensation discharge. The modified high flow regime results
from the relationship between reservoir inflows, abstraction, storage
capacity and spillway characteristics. Reservoirs with high yields in
relation to the mean annual runoff generally produce a greater reduction in
frequency of flood discharges than reservoirs with smaller yields because
the latter are full for much of the winter period and hence have less
capacity to attenuate inflow.

Hydrograph analysis can be used to summarize the changes in the flow
regime following impoundment (Gustard 1984) and this has been carried out
on 31 U.K. reservoirs shown on Figure 1. Mean discharge was on average
reduced to 60% of the natural flow with a maximum reduction to 15% of the
natural flow. Daily and monthly flow duration curves have been used to
summarize the changes in frequency of discharges standardized by the pre
impoundment natural mean discharge. As a typical example, Figure 6 shows a
change in the flow duration curve following impoundment of the R. Blithe.

The change in flow regime has been investigated by comparing the pre
and post impoundment flow for different exceedance frequencies. Figure 7
shows the 95 percentile discharge from the flow duration curves before and
after reservoir impoundment. Although the average 95 percentile discharge
over all schemes shows no significant change between pre (11.3% average
flow) and post impoundment (11.6% average flow) conditions, wide departures
from maintaining the natural low flows exist. This may provide a basis for
reassessing existing compensation awards. The greatest change in discharge
is between the 10 and 90 percentiles on the flow duration curve, with an
average reduction of the median flow to 45% of the pre impoundment flow.

The calendar day mean annual flood prior to regulation has been
estimated from the characteristics of the drainage basin (NERC 1975) and
compared to guaged post impoundment values. Reductions to 5% of the
natural flood discharge occur below some reservoirs with an average

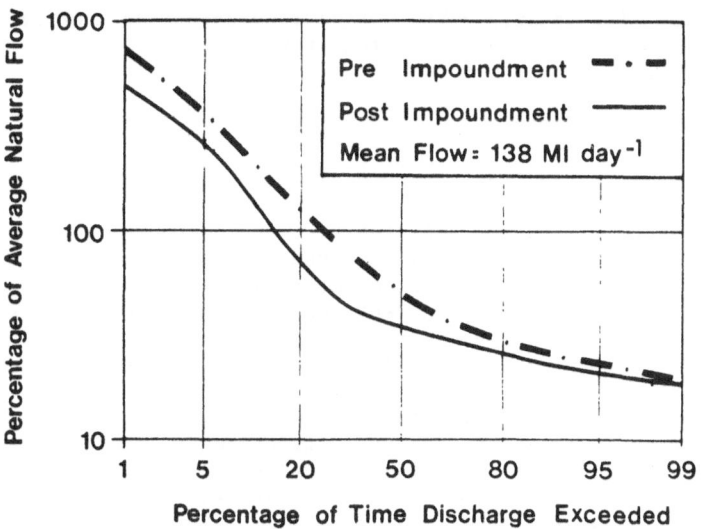

Figure 6. Pre and post impoundment flow duration curves for River Blithe.

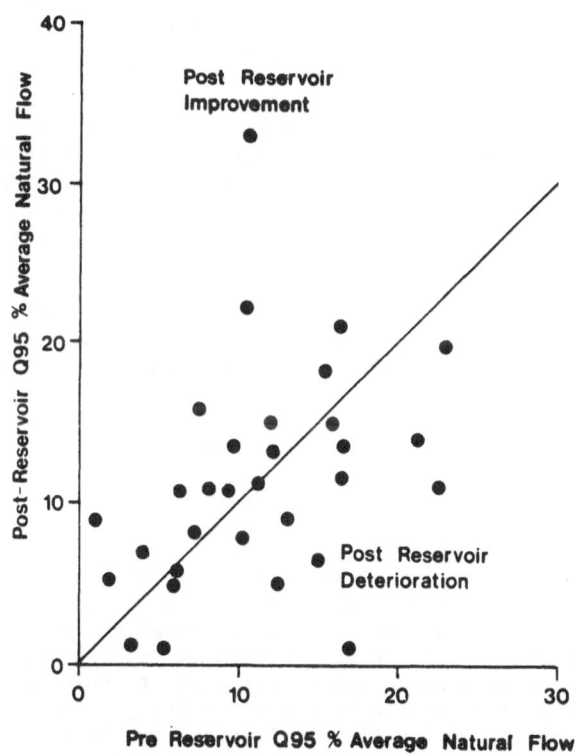

Figure 7. Change in 95 percentile discharge following reservoir impoundment.

reduction to 75% of the natural value. These changes exist close to the dam and will gradually reduce as one moves further downstream. Changes to the flow regime will be within the normal flow measurement error when the downstream point receives less than 10% of the natural flows from the reservoired catchment.

DISCUSSION AND CONCLUSIONS

Flow changes are an inevitable result of the primary function of a reservoir; to store water in periods of surplus and release to the consumer or river in times of deficiency. Many compensation flows were awarded at a time when there was little understanding of differences in catchment hydrology, and most occurred at a time when there was no knowledge of the impact of impoundments on downstream fauna and flora. This would suggest that a review of current releases below reservoirs is warranted. However any change in downstream releases will have an effect on, for example, the yield of a direct supply reservoir. An increase in a constant compensation flow will result in an equal reduction in reservoir yield or increase in the frequency of restrictions on supply. Maintaining the same annual release volume whilst changing the seasonal pattern to a summer maximum, will reduce the yield of a reservoir particularly if it has a small storage capacity.

Pressures to increase downstream flows would in the long term have to be met by additional water resource schemes whose environmental disbenefits may far outweigh the advantages of changing the flow pattern of an already regulated stream. However, a number of benefits may arise by making modest changes to existing release policies. For example, some reduction may be possible where the benefits of the compensation water are restricted to a short tributary of a much larger river. Techniques have been developed for estimating seasonal flow duration curves at unguaged sites (NERC 1985). These will enable more realistic seasonal compensation flows to be set which may improve the diversity of invertebrate communities and aid salmonid migration or spawning without serious reductions to reservoir yield. Greater flexibility in reservoir operating policy with compensation releases being a function of reservoir storage, may safeguard the water supply role of the reservoir in drought conditions, but enable more generous releases to be made at other times.

The survey of existing releases and studies of the influence of residual flows on invertebrate communities below U.K. reservoirs (Armitage 1985) will improve the basis for revising existing compensation awards. However, as a result of the large number of reservoirs impounding small catchment areas (over 300 less than 20 km^2) detailed environmental impact studies of the change in release policies cannot always be justified.

In conclusion, the historical development of setting residual flows has resulted in many reservoirs providing releases determined by industrial and political constraints which no longer apply. The average of these statutory minimum discharges is close to the natural low flow but there are wide departures from this trend. Any reassessment of residual flows could initially focus on these anomalous sites. Although the yield and storage capacities of existing reservoirs will provide a major constraint on the freedom to manipulate downstream flows, the water resource engineer will be looking for guidance from the freshwater biologist particularly for setting the seasonal variability of low flows. It will then be necessary to compare any change in reservoir yield with expected improvements or degradation of downstream ecology. Estimating the environmental impact of new regulation schemes is now of less importance in the U.K. than predicting the response to changes in the low flow regime of existing regulated rivers.

ACKNOWLEDGEMENTS

The authors would like to thank Regional Water Authorities, Purification Boards and Regional Councils for providing the river flow and reservoir data. The work was carried out as part of a research contract from the Department of Environment.

REFERENCES

Armitage, P. D. 1985. The classification of sites receiving residual flows from reservoirs using macro invertebrate data. This volume.

Brooker, M. P. 1981. The impact of impoundments on the downstream fisheries and general ecology of rivers. Adv. in App. Bio. **6**: 91-152.

Gustard, A. 1984. The characterization of flow regimes for assessing the impact of water resource management on river ecology. In: Regulated Rivers. (Ed. by A. Lillehammer and S. J. Saltveit). pp. 53-60. University Press.

NERC 1975. Flood Studies Report, London

NERC 1985. Seasonal Flow Duration Curves. Low Flow Studies Report. Institute of Hydrology, Wallingford.

Petts, G.E. 1984. Impounded Rivers - Perspective for Ecological Management. J. Wiley, Chichester, U.K.

Royal Commission on Water Supply 1869. Report, proceedings and minutes of evidence. 33.

Sheail, J. 1984. Constraints on Water Resource Development in England and Wales: the concept and management of compensation flows. J. Envir. Mgmt. **19**: 351-361.

EFFECTS OF SHORT-TERM REGULATION BY POWER PLANTS

ON EROSION AND WATER QUALITY OF A RIVER

Erkki Alasaarela and Markku Virtanen

Technical Research Centre of Finland/
Building Laboratory, P.O. Box 181
SF-90101 Oulu

INTRODUCTION

The demand for electricity varies with time of year, week and day. Electricity cannot be stored in Finland, so the output must be adjusted to meet the demand. Short-term regulation of regional hydropower plants is important. They generally account for 30-50% of the value of the annual production of power plants (Lehtonen 1985).

Short-term regulation brings about variations of discharge and water level in the river, which have been assumed to affect the other uses of the water course. The National Waters Board in Finland has conducted studies into these problems in the rivers of the Pohjanmaa region during 1982-1984. The major purpose of this work was to describe the transport of riverbed material during regulation.

RESEARCH AREA

The research sites are located in the drainage basins (F) of the Rivers Siikajoki (F = 4,395 km^2) and Kyrönjoki (F = 4,805 km^2) and in the Nurmonjoki River, which is a tributary of the River Lapuanjoki (F = 3,955 km^2) (Figure 1). The variations of water level at different values of discharge through the power plant are presented in Table 1.

The greatest variations occur in the Siikajoki River downstream of the Uljua power plant; as far as 90 km downstream of the power plant at the mouth of the river, the water height varies by one to five centimeters in summer and by 20-35 cm in winter as a consequence of short-term regulation. In the Rivers Siikajoki and Kyrönjoki there is generally one period of discharge through the power plant each day. Three discharge periods per day are used in the River Nurmonjoki in summer, to level out variations of water level. The power plants are used for two to four hours daily at the time of the low discharge in summer and for 10-15 h daily in winter. Discharge by-passing the power plant is small.

The riverbed material between 5 and 8 km downstream of the power plants mainly consists of fine material. Downstream of the Uljua power plant, there are rocky stretches and some sand banks in areas of slow flow, but most of the material consists of silt. Downstream of Pitkämö, in the

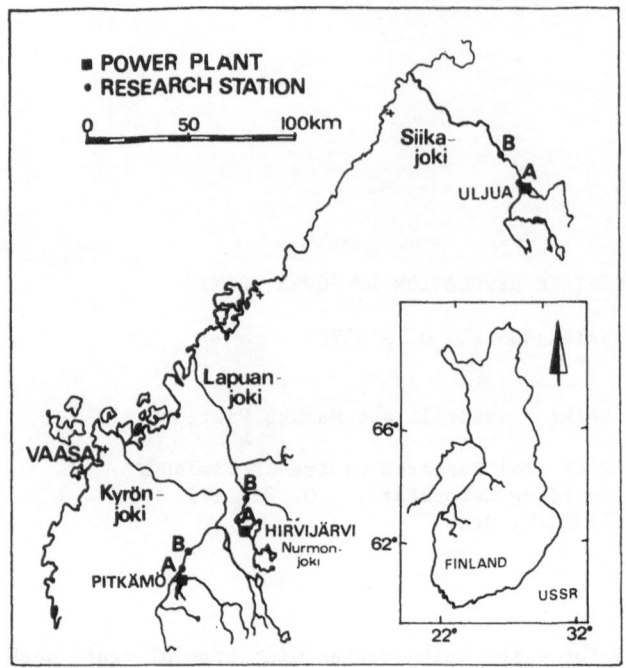

Figure 1. Power plants with short-term regulation and the most important research sites in the Siikajoki, Lapuanjoki and Kyrönjoki river basins.

Kyrönjoki River, the riverbed consists of 30-40% sand, 60-70% silt and 5% clay. Downstream of the Hirvijärvi power plant in the Nurmonjoki River, sand accounts for 10-20%, silt for 70-80% and clay for 5-10%.

RESEARCH METHODS

Quality of Water and Sedimentation

Turbidity was measured with an automatic recorder (HACH). The suspended solids content of river water was determined from water samples collected with an automatic sampler from a depth of 0.5 m in the middle of the river at one hour intervals. The amount of deposited material was measured in a series of vials (10 vials Φ12 mm; cf. Hargrave & Burns 1979) placed on the river bottom (3 series per cross section). The vials were left on the river bottom for a day and the samples were then analyzed by filtering. Most of the samples were analyzed in the laboratories of the Water Districts Offices of Oulu and Vaasa.

Numerical Model

For numerical simulation the river section is subdivided into segments; within each of them the flow and bed properties are selected to be as uniform as possible. The flow behaviour at each location is described in terms of the variation of depth H (m), cross-section A (m^2) and flow velocity U (m s^{-1}) as a function of time. These are estimated from the results of a separate flow model which predicts changes of water levels and flow from details of power plant operation (upstream discharge).

Table 1. The years when the power plants under study started operating (1), the maximum variation of discharge through the power plant during short-term regulation (2) at different distances from the power plant (3, see Figure 1) in the summer of 1982 and 1983 and in the winter of 1983. The table also shows the associated variations of water level (4) at the situations when the same daily amount of water was discharged through the power plant during a single (a), two (b) and three (c) discharge periods per day.

	Siikajoki			Nurmonjoki			Kyrönjoki		
	Uljua	A	B	Hirvijärvi	A	B	Pitkämö	A	B
1. Year of completion	1970	–	–	1973	–	–	1971	–	–
2. Maximum variation $m^3 s^{-1}$	0–32	–	–	0–18	–	–	0–25	–	–
3. Distance, km	0	4	20	0	4	18	0	4	10
4. Variation of water level, $m\ d^{-1}$ Summer									
a	1.0–1.4	0.55–0.75	0.15–0.20	0.8–1.3	0.20–0.55	0.12–0.27	0.6–0.9	0.22–0.44	0.22–0.44
b	–	–	0.10–0.15	–	0.20–0.35	0.04–0.05	–	–	–
c	–	–	0.04–0.09	–	0.10–0.15	0.02–0.03	–	–	0.12–0.42
Winter a	–	–	0.30–0.40	–	0.50–0.90		0.65–1.0	–	–

In the bed interaction model the change of concentration C (mg 1^{-1}) is computed from:

1) transport with the river flow
2) mixing with the velocity shear and turbulence
3) sediment deposition
4) erosion from the bottom induced by steady flow

and

5) additional erosion induced by the increase of flow velocity (together with the water level rise).

$$\frac{\partial(AC)}{\partial t} = -\frac{\partial(AUC)}{\partial x} + \frac{\partial}{\partial x}(AK\frac{\partial C}{\partial x}) - \frac{wC}{H} + \frac{b}{H}(U + pU^{+} - U_{cr})^{+} \qquad (1)$$

where K = dispersion coefficient (m^2 s^{-1})
                w = settling velocity (m d^{-1})
                b = erosion coefficient (g m^{-2} d^{-1} (m s^{-1})
                p = pulse factor (h)
                x = distance along the river (km)
                t = time (h)
                U^+ = change of flow velocity with time (ms^{-1} h^{-1})
                      where the superscript + indicates that only positive
                      values are taken into account.
                U_{cr} = threshold velocity needed for any erosion (m s^{-1})

The settling and cohesion of riverbed material depend on the particle size D (mm). Therefore, the model (1) is applied separately to different particle sizes $D_{i-1} < D \leqslant D_i$ and the final concentrations C are obtained as a sum of the fractional concentration C_i

$$C = \sum_{i=1}^{N} f_i \cdot C_i \qquad (2)$$

where N = the number of fractions considered (-) and
                f_i = proportion of the particle size in the bed (%).

In the present applications, four separate fractions of riverbed material are used, viz.

Fraction		D_i (mm)	w (m^{-1} day^{-1})	U_{cr} (m s^{-1})
1	clay	0.002	0.1	0.5
2	silt	0.02	2	0.3
3	fine sand	0.06	50	0.3
4	sand	0.2	1800	0.3

The settling velocity is estimated from the characteristic particle size $\bar{D}$ of each fraction as

$$W = 4.45 \text{ m } d^{-1} \times (\bar{D}/0.01 \text{ mm})^2 \qquad (3)$$

Parameters other than w and U_{cr} are taken to be independent of the particle size. The dispersion is assumed to increase with the flow velocity and with the water depth

$$K = \alpha \cdot U \cdot H \qquad (4)$$

with $\alpha = 20$. The erosion coefficient is assumed to equal b = 75 g m^{-2} d^{-1} (m s^{-1})$^{-1}$ and the pulse factor p = 5 h.

With these few coefficients the model is applied to several sections of two rivers with widely different discharge characteristics. The differences between the model and the observations arise from: (1) the stochastic nature of the erosion (2) the model simplicity. For example, the effects of a wide variety of phenomena are combined into a single "pulse factor", including the flow velocity, turbulence, waves, water level rise and weak cohesion of the dry shore areas. Furthermore, the average particle size distributions and the dispersion and erosion properties of each river section are used, while the vertical distributions of flow velocities and concentrations are not taken into account, and the effects of secondary current and transverse velocity variation in the river are disregarded. More detailed attention to these features, may improve the validity of the model.

RESULTS

Siikajoki

In the River Siikajoki, short-term regulation resulted in notable bank erosion over 7-10 km downstream of the Uljua power plant. At the Kerälänkylä village 20 km downstream of the power plant, the turbidity of river water varied with water level during the ice-free period (Figure 2). The turbidity was less conspicuous in winter and could not be related to fluctuations of water level.

Nurmonjoki

In the River Nurmonjoki, the plantless zone within the range of the variation of the water extended as far downstream as 4-6 km from the Hirvijärvi power plant. At some points erosion depth was 0.5-1.5 m.

The dependence of the turbidity of river water water level variation was striking (Figure 3). Turbidity was most evident when the water level was rising.

Variations in the suspended solids concentration of river water were also related to fluctuations of the water level. When, instead of using the conventional three discharge periods per day, the same daily amount of water was discharged through the power plant during a single period, the suspended solids concentration increased and the variation increased even

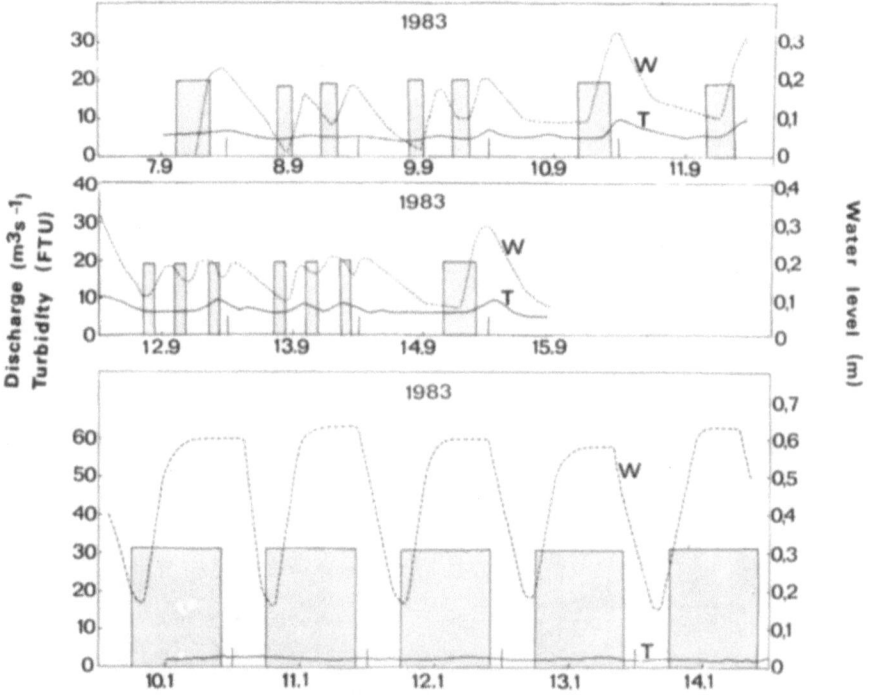

Figure 2. Discharge at the Uljua power plant (columns), the relative water level (W) and turbidity of river water (T) in the Siikajoki River at the Kerälänkylä village (B, 20 km from the power plant) during the ice-free period Sept. 8-15, 1983 and the ice-bound period Jan. 10-14, 1983.

at the mouth of the river (18 km from the power plant; Figure 4). The suspended solids concentration predicted by the mathematical model was reasonably consistent with the observed values.

Short-term regulation increased sedimentation down to 5 km from the power plant, where the variations of the water level and the maximum flow velocity were clearly greater than in the river segment further downstream (Figure 5). There was more sedimentation when the same daily amount of water was discharged through the power plant during one period than when the water was released during three separate periods.

Kyrönjoki

The River Kyrönjoki showed definite signs of bank erosion along a distance of 4-5 km downstream of the Pitkämö power plant. Turbidity of water at the time of increasing water height was obvious at a distance of 2.5 km from the power plant (Figure 6). At a distance of 4 km at Kurikka, the suspended solid concentration of the water varied in accordance with the water level. The results obtained with the model were quite well accordant with the concentrations revealed by the water analyses (Figure 7).

DISCUSSION

The changes of water quality caused by short-term regulation at the research sites were mainly attributable to inorganic matter eroded from the riverbed. This material is eroded from the bottom at the time of flow peaks

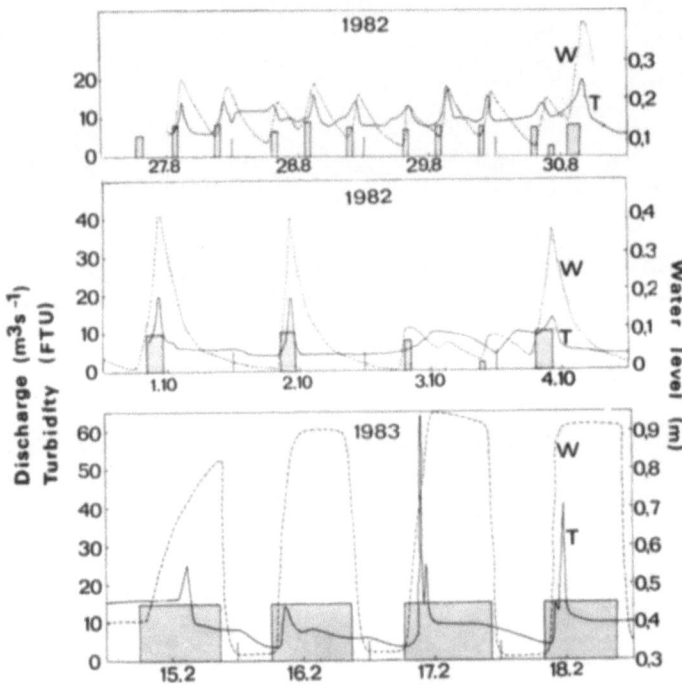

Figure 3. Discharge at the Hirvijärvi power plant (columns) and the relative water level (W) and turbidity of river water (T) at the Kirkkosilta bridge in the Nurmonjoki River (A, 4 km from the power plant) during the ice-free periods Aug. 27-30, 1983 and Oct. 1-4, 1983 and the ice-bound period Feb. 15-18, 1983.

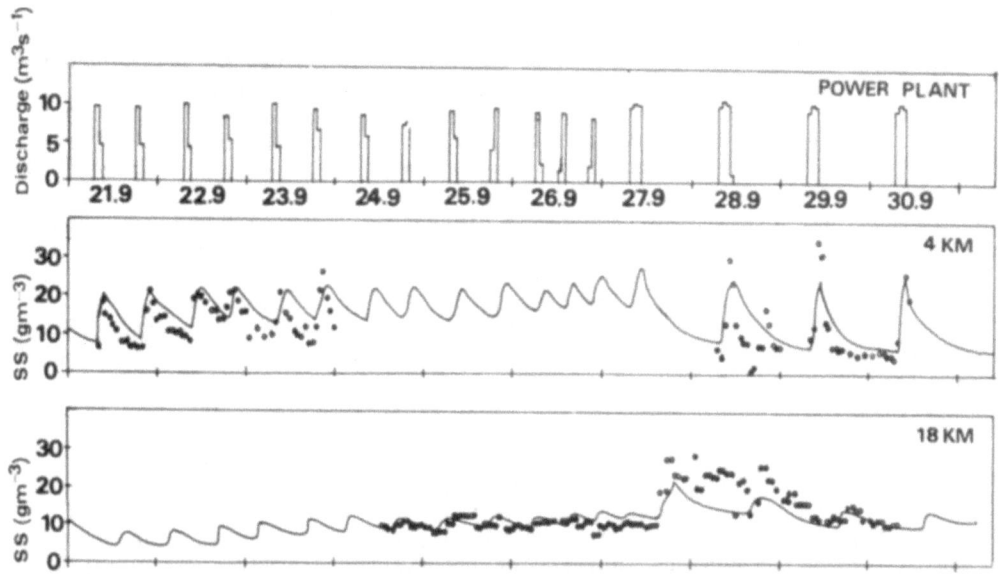

Figure 4. Discharge at the Hirvijärvi power plant and the variation of suspended solids concentration (SS) in the Nurmonjoki River at the Kirkkosilta bridge (A) and Kelloja (B, 18 km from the power plant) during Sept. 21-30, 1983. The suspended solid concentrations were obtained from water analyses (dots) from the mathematical model (line).

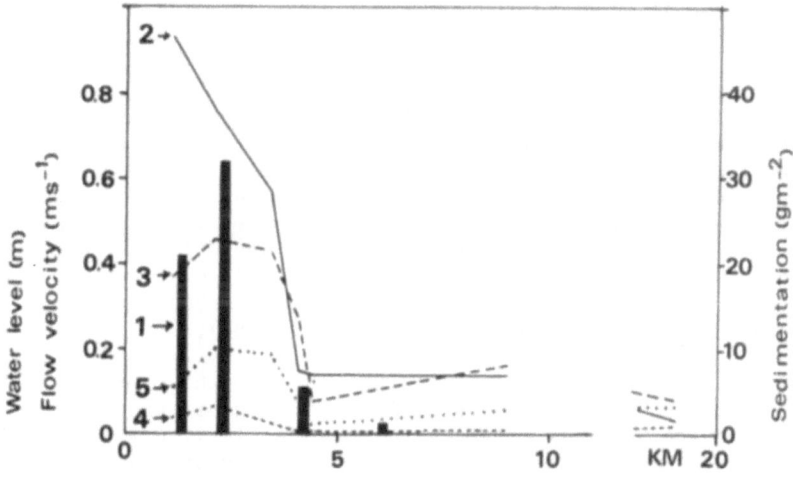

Figure 5. Amount of daily deposition of sediment (1, columns), relative change of water level (2), maximum flow velocity (3) and minimum flow velocity (4) in the Nurmonjoki River at different distances from the Hirvijärvi power plant on Sept. 21, 1984. The figure also shows the flow velocity in a situation where there is no short-term regulation in the river (5). The water velocities and levels have been calculated from the model for unsteady flow. The amount of water discharged through the power plant was 10 m^3 s^{-1}, and the discharge took place at 0800-1000 h and 1800-2000 h.

and transported until, when the flow slows down, it is redeposited.

Changes in the concentration of suspended solids can be related to the advance of the regulation wave. In variable flow the velocity of the regulation wave ($1-2$ m s^{-1}) is always greater than the flow velocity ($0-1$ m s^{-1}). In the Nurmonjoki River, for example, the regulation wave with the associated peak of the water solids concentration reached a distance of 20 km from the power plant in 4 h, while a tracer placed downstream of the power plant reached the corresponding distance in about 100 h (Alasaarela 1984).

The flow variation arising from short-term regulation reduces with distance downstream from the power plant. The variations of water level and velocity are related to this, but local factors within the cross-section of the river have an important influence. The greatest variations do not, necessarily, occur immediately below the power plant.

One special hydraulic feature of short-term regulation is the hysteresis of the water level – discharge curve. At a given water level, the discharge and the water velocity are greater when the water level is rising than when it is declining (Figure 8). Parallel to this, the turbidity of water and the increase of the suspended solids concentration are greatest when the water level is rising (cf. Forslin et al. 1980). The water velocity critical for the mobilization of the riverbed material is exceeded as the water level rises; and the effect is further intensified by the drying of the soil of the regulation zone at times of a low water level.

The mass balance calculations made for the Nurmonjoki River indicate that a major part of the riverbed material eroded into the water originates from the regulation zone (Forsius et al. 1983). The riverbed erosion typically consists of bank erosion. The most serious erosion occurs along the outer curve at bends of the river, where the main flow is directed (Chow 1959).

The eroding effect of short-term regulation on the riverbed reflects rapid variations of water velocity. In the Nurmonjoki River, for example, the highest summertime velocities over 5 km downstream of the power plant increased from $0.05-0.20$ to $0.10-0.50$ m s^{-1} (Figure 5). The critical velocity reported in the literature for the kind of finely granular riverbed material encountered at the research sites is $0.3-0.5$ m s^{-1}. During short-term regulation releases, the critical flow velocity is exceeded and the erosion increases. As conditions are so variable, the regulation zone is generally void of vegetation and bank erosion occasionally results in landslides (Alasaarela 1984).

Most of the eroded material is lifted up from the riverbed by the turbulent flow and transported as suspended particles, until, at a suitable point or when the velocity reduces, the material is deposited. At the research sites, most of the eroded material is deposited during the intervals between the discharges. Continuous exchange occurs between the bottom sediment and the suspended solids (wash load). Bed load (i.e. the material in continuous contact with the bottom) plays a minor role at the research sites, because suitable solids are scarce. The other changes in the quality of river water due to short-term regulation are associated with the substance concentrations of the solids. Owing to the unsteady flow, the waters entering the river downstream of the power plant (waste waters, etc.) are diluted non-uniformly and therefore bring about changes of water quality (Alasaarela 1984).

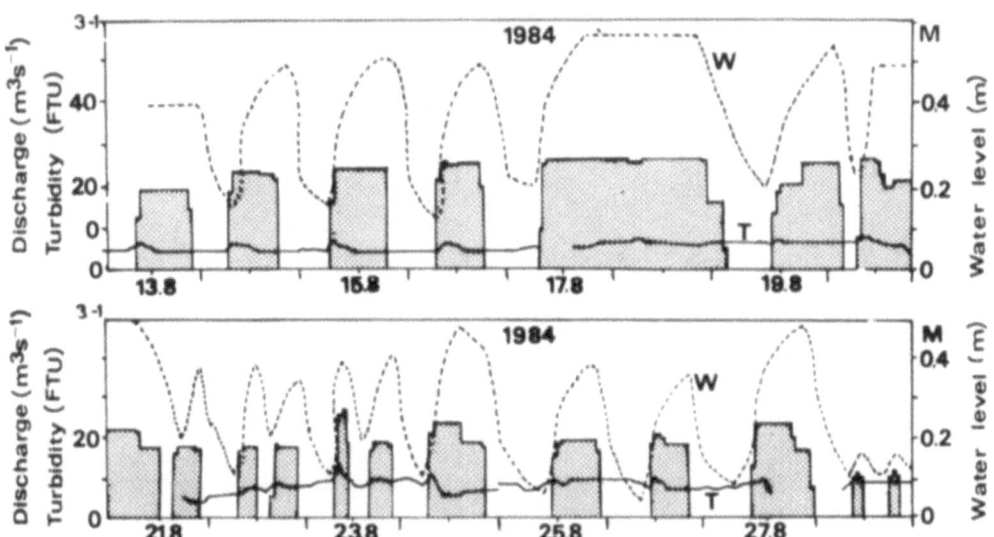

Figure 6. Discharge values and relative water levels of the Pitkämö power plant and the variations of turbidity in the water of the Kyrönjoki River 2.5 km downstream of the power plant during Aug. 13-28, 1984.

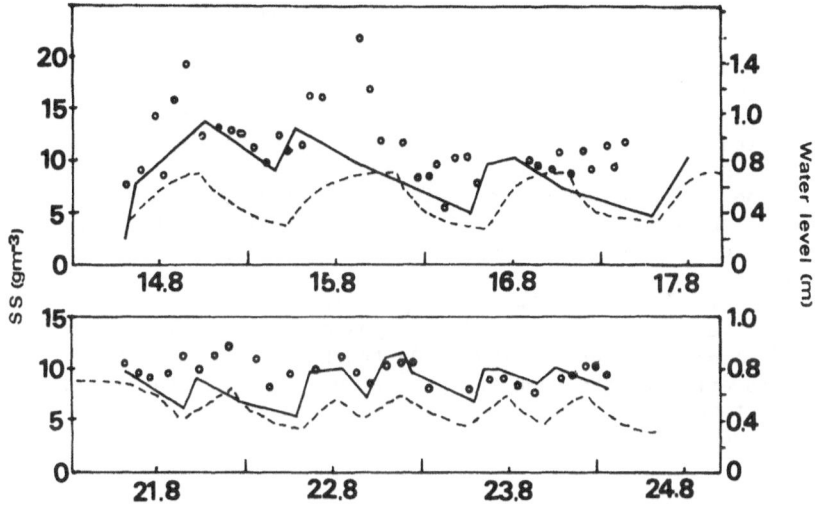

Figure 7. Observed (dots) and calculated (line) suspended solids concentrations of river water and the relative water level (dashed line) in the Kyrönjoki River at Kurikka (B, 4 km downstream of the power plant) during Aug. 14-24, 1984.

The thick ice cover in the Siikajoki River protected the zone susceptible to erosion in winter, and hardly any turbidity was observed. In the Nurmonjoki River the segment under study froze during the night, but the break-up of ice in the morning resulted in turbidity at the time when the water level rose, though notably less so than in summer. The erosion in all the rivers investigated was located at the level where the water height varies in summer and these observations suggest that the erosion of banks

285

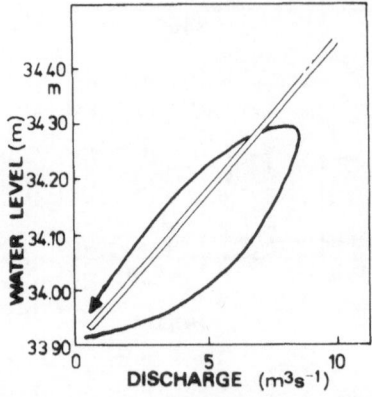

Figure 8. Relationship between water level and discharge during the discharge period of the Hirvijärvi power plant in the Nurmonjoki River at the Kirkkosilta bridge (A, 4 km downstream of the power plant, Forsius 1984).

would be most marked in summer.

The variations of water level and discharge caused by short-term regulation have been described quite well with mathematical models during the ice-free period. The results obtained from the models applied by Forsius (1984) agreed well with the measurements carried out in the Rivers Nurmonjoki and Kyrönjoki. Forsius et al. (1983) further applied a simple model to the Nurmonjoki River describing the transport of riverbed material, which in view of its simplicity was a reasonable representation of the motion of the suspended solids in the river. The model described in the present paper was supplemented by including the particle size distribution of the solid matter. Without changing the constants, the model was applied to both the Nurmonjoki River and the Kyrönjoki River, and the predictions of variation in suspended solid concentration and of deposition agreed quite well with observed values. The simplicity, generality and easy application of the model make it suitable for practical use. The model was applied to predict the effects of a proposed power plant on the downstream Kyrönjoki (Alasaarela et al. 1985). On the basis of the results, it is possible to estimate the need for protecting the river banks and the changes that may take place in the raw water qualities of the river water as a consequence of short-term regulation.

CONCLUSIONS

Downstream of power plants with short-term regulation, the flow velocities are higher than in natural streams and there are large peaks of discharge during each day. This results in bank erosion which leads to variation of the suspended solids concentration and of turbidity. This and the erosion of the riverbed are the main disadvantages of short-term regulation. These problems are most obvious in summer. Since the economic significance of short-term regulation is most marked in winter (Lehtonen 1985) the need for summertime regulation should be considered separately in each case.

The discharge of the diurnal quantity of water in several periods instead of one diminishes the erosion of the riverbed. As the erosion is most rapid during rising water level, the increase of suspended solids in the river water could be reduced by using lower rates of change of discharge. Moreover, the variations of water level can be reduced or shifted to suitable areas by means of bottom dams. Stone embankments and plantations can be used to prevent bank erosion.

The effects of new projects can be predicted and their disadvantages minimized by means of the unsteady flow model and the model of sediment interaction controlled by short-term regulation.

REFERENCES

Alasaarela, E. 1984. Lyhytaikaissäännöstelyn vaikutus joen ekologiaan. Vesihallituksen monistesarja **220**: 179-196.
Alasaarela, E., Virtanen, M. and Hyvönen, P. 1985. Pitkämön ja suunnitteilla olevan Kirkonkosken voimalaitoksen vaikutus uoma-aineksen kulkeutumiseen Kyrönjoessa. Technical Research Centre of Finland, Building Laboratory, Oulu. 13 pp.
Chow, V. T. 1959. Open-cannel hydraulics, New York. 680 pp.
Forsius, J. 1984. Computing unsteady flow and tracer movement in a river. Publications of the Water Research Institute. National Board of Waters. Finland, No. **60**: 3-21.
Forsius, J., Alasaarela, E. and Virtanen, M. 1983. Application of two transport models to a regulated river. Geophysica, **20**(1): 71-80.
Forslin, J., Nilsson, B., Nyman, L. and Willner, H. 1980. Nedströmseffecter av grumling från Halvfari kraftverk. UNGI Rapport **52**: 417-439.
Hargrave, B. T. and Burns, N. M. 1979. Assessment of sediment trap collection efficiency. Limnol. Oceanogr. **24**(6): 1124-1136.
Lehtonen, M. 1985. Alueellisilla vesivoimalaitoksilla harjoitetun lyhytaikaissäännöstelyn merkitys (Summary: Investigations on the importance of short-term regulation of local hydropower plants). Helsinki. 116 pp (mimeogr.).

STUDY DESIGN FOR FISHERIES AND HYDROLOGY ASSESSMENT

IN A GLACIAL WATERSHED IN BRITISH COLUMBIA

H. A. Smith,[1] S. P. Blachut[2] and B. Bengeyfield[3]

[1]Environmental Services, B.C. Hydro
1265 Howe St., Vancouver, B.C. V6Z 2C8
[2]Habitat Management Division, Department of Fisheries &
Oceans, 1090 West Pender St., Vancouver, B.C. V6E 2P1
[3]Global Fisheries Consultants Ltd., 13069 Marine Drive
Surrey, B.C. V4A 1E5

INTRODUCTION

Most instream flow assessment studies have been carried out on stable, single channel streams. Methodologies such as the incremental method or IFIM (Bovee 1982) are not designed for the unstable, shifting channels of a multi-channel system. This type of channel is characteristic of glacier-fed streams and although the distribution of such streams is geographically limited, they are numerically abundant. Major fisheries-hydrology studies which have addressed these problems include the Ohau River (Mosley 1982) and the Rakaia River (Glova & Duncan 1985) New Zealand, the Stikine River, British Columbia (Aquatic Environments Ltd. 1982; Jones & Associates 1984), the Maurice River, British Columbia (Envirocon 1984) and most notably, the Susitna River, Alaska (Alaska University 1983).

In Canada those rivers with glaciated catchments and salmonids primarily include major rivers which transect the Coast Mountains, of which the Homathko River is an example. These river basins are characterized by very rugged topography, steep gradients, high precipitation and heavy glaciation. The rivers are utilized by significant runs of wild stocks of anadromous and resident salmonids. This combination of important fisheries stocks utilizing complex mountainous glacial rivers in remote locations presents numerous problems beyond a typical instream flow assessment, including study design, logistics and lack of historical data.

As part of a feasibility study for hydroelectric development, hydrology and fisheries studies were designed and carried out by a study team of in-house staff and consultants to gain an understanding of interacting physical and biological processes.

A thorough understanding of the relationship between physical and biological processes in the natural environment is critical to developing the capability of predicting the influence of impacts on the fisheries resource. The first two years of study on the Homathko watershed were designed to acquire a solid data base and to establish some of the key physical-biological linkages governing the production of fish stocks in the watershed.

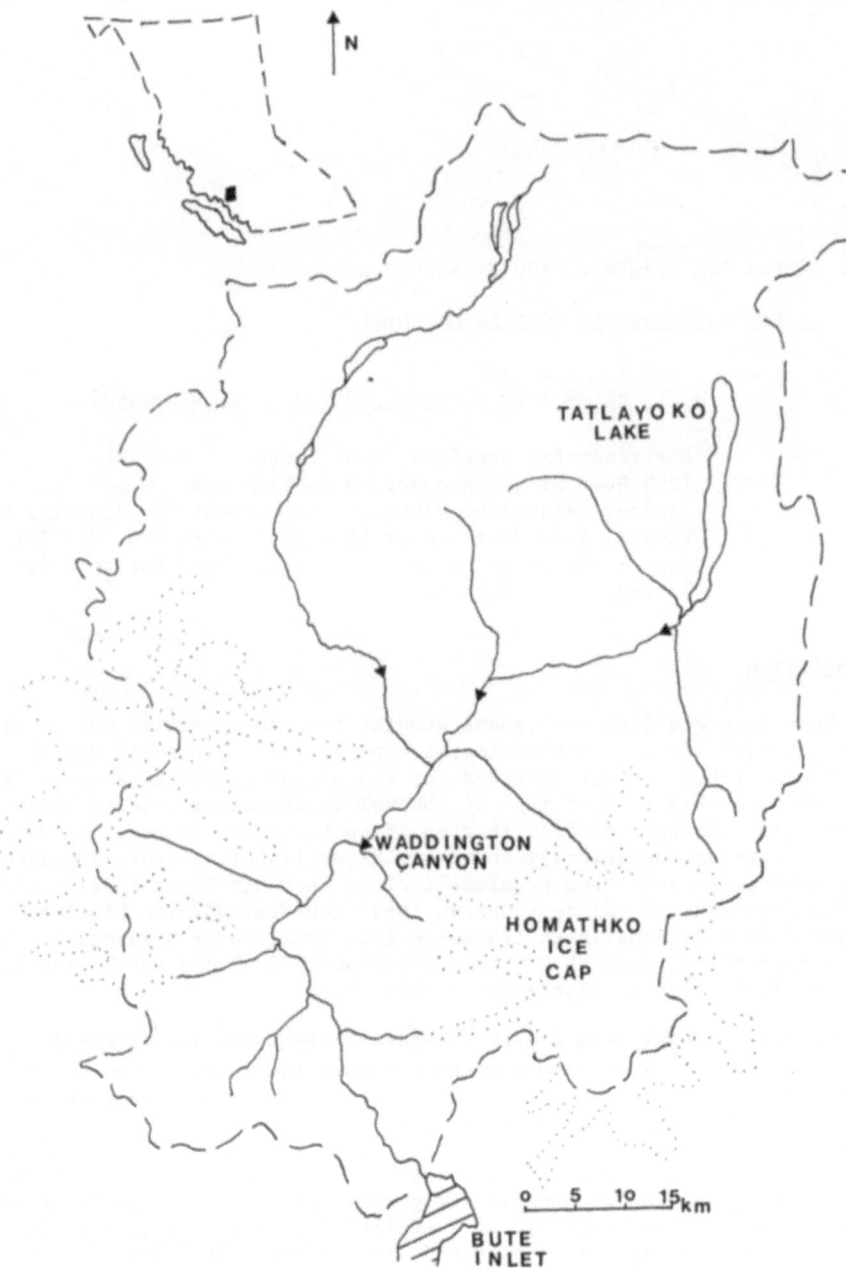

Figure 1. Sketch map of the Homathko River basin.

METHODOLOGY

The rationale behind the data collection program for the feasibility
study was based on acquisition and analysis of a physical data base and
collection of basic fisheries population and habitat utilization
estimates. The parameters identified for study were directly linked to
perceived potential impacts of hydroelectric development on a glacial river
system. The flow regime of the mainstem and major tributaries were

monitored using automatic stage recording equipment, in addition to spot
stage and discharge measurements. Sampling of suspended sediments and bed
material was carried out, as well as mapping of historical and present
channel morphology. Water temperatures were measured with automatic
recording thermographs at mainstem and secondary channel sites.

The lack of data on fisheries stocks necessitated the initiation of
inventory studies as a focus of biological studies. Populations of adult
chinook (*Oncorhynchus tshawytscha*), coho (*Oncorhynchus kisutch*) and chum
(*Oncorhynchus keta*) salmon were estimated by a combination of repetitive
aerial and ground enumerations. Coho salmon in tributaries to the Homathko
River were counted by repetitive stream walks. Angling, seining and
electrofishing were used to determine distribution of anadromous steelhead
(*Salmo gairdneri*), cutthroat trout (*Salmo clarki*) and Dolly Varden
(*Salvelinus malma*).

Redd surveys of chinook, coho and chum salmon were conducted
periodically throughout the spawning period. Parameters for determining
habitat preference criteria were measured in conjunction with inventory
surveys. A sample of early chinook redds was surveyed to local datum
points for later observation during low winter flows. The distribution of
redds throughout the watershed was mapped to facilitate comparisons
between years and water stage heights.

Estimates of migrant juvenile salmonid numbers were made using
floating trawl traps and inclined plane traps. A sample of captured fish
were marked and released upstream to determine trapping efficiency.

RESULTS: CASE STUDY - THE HOMATHKO RIVER

Hydrology

The Homathko River drains mostly mountainous, heavily glaciated
terrain of the central coast of British Columbia, with approximately 1250
km^2 (21%) of the basin covered by glaciers (Field 1975). The lower Homathko
River, a 50 km reach from the river mouth to the site of the proposed
downstream project at Waddington Canyon (see Figure 1), is approximately
44% glaciated. A large number of glacier-fed tributaries enter the Homathko
River below Waddington Canyon, contributing approximately 50% of the flow
to the Homathko at the mouth. This very high proportion of glaciers in the
catchment of the lower Homathko and the large percentage of the total
annual precipitation (70% of 4000 mm) which falls as snow result in a
glacial-nival (snowmelt) hydrologic regime. Such a flow regime is
characteristically highly variable, and highly seasonal, where winter
precipitation is usually stored as snow, to be released during the brief
summer months (Church & Gilbert 1975). The Homathko River at the mouth has
a long-term mean annual flow of 268 m^3 s^{-1} , based on the period of record
1957 to 1983 (Water Survey of Canada 1983).

The timing of the snowmelt flood can vary from late May to late June,
and it may represent the peak flow event of the year (Figure 2). The
summer period from June to September is characterized by glacial melt
contributions with occasional rainstorms. Ice dammed lakes are common in
the basin and a moraine dammed lake burst was recorded on the Homathko in
July 1983, generating a 5 m flood wave and causing considerable morphologic
change. Debris torrents and avalanches occur in steep tributary valleys in
areas of high precipitation. Fall rainstorms generated by cyclonic storms
commonly occur, resulting in very rapid runoff and the maximum floods on
record (e.g. September 1967 on Figure 2). The effect of rain on snow in
late fall often causes the highest floods.

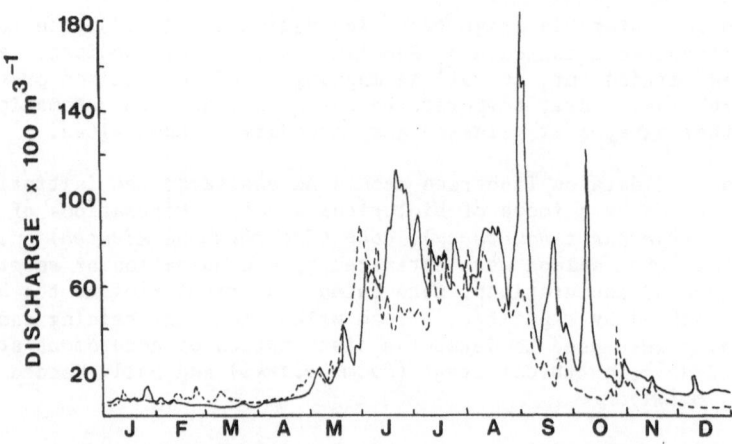

Figure 2. Discharge rates ($m^3 \ s^{-1}$) for years of high (1967) and low (1983
---) flow measured at the mouth of the Homathko River.

One of the characteristics of a glacial-fed river is the high degree
of flow variability, both short-term and long-term. In the short-term, this
characteristic can be demonstrated by the hourly and daily rate of change
of stage (Table 1). Fall rainstorms appear to generate more rapid changes
in stage than glacial melt. In the longer term, runoff in heavily glaciated
basins is subject to higher, possibly cyclic variation, as a function of
the mass balance (i.e., net ice volume change) regime of the glaciers. The
estimated glacier melt contribution to the Homathko, based on mass balance
data (Mokievsky-Zubok 1983) varied from 12 to 21% over a three year period.

The sediment and morphologic regime of the Homathko River is
characteristic of a glacier-fed river. The upper Homathko River is
primarily a confined and entrenched channel, flowing through bedrock
canyons with small active cobble-boulder bars. Below Waddington Canyon, the
lower Homathko is a wandering or anastomosing gravel-cobble bed river with
numerous bars and islands. As interpreted from a series of aerial
photographs spanning the last 40 years, the river is moderately to highly
laterally unstable. Numerous secondary channels exist both within the
active channel zone and on the vegetated floodplain. Tributaries to the
lower Homathko have built large alluvial fans into the main valley, often
deflecting the mainstem Homathko flow and altering the tributary
confluences.

Suspended sediment data are only available for one summer season (May
to September 1983) and the range of concentrations measured was 8 to 1593
mg l^{-1}. The diurnal fluctuation of sediment concentration common on
glacier-fed streams (Church & Gilbert 1975) was documented on the lower
Homathko. The glacial contribution of sediment complicates the use of a
simple concentration-discharge rating relationship to predict
concentrations (Walling & Webb 1981), as the concentration for a given
discharge was higher during the glacial melt period than during other
seasons.

Gravel bar sampling using tape grid and bulk sieve analysis
(Kellerhals & Bray 1971) displayed a gradual but not consistently
decreasing grain size with distance downstream from Waddington Canyon.
Near the river mouth, the B_{50} (B-axis dimension which 50% of the sample is
finer than) and B_{90} grain sizes were 38 mm and 96 mm respectively, and the
largest sample 3 km below the canyon had B_{50} and B_{90} dimensions of 290 mm

Table 1. Hourly, Daily and Mean Monthly Rates of Change in Stage (WL, m) and Discharge (Q, $m^3 s^{-1}$) at (a) WSC Station (Tragedy Canyon) and (b) WSC Station (Mouth). Based on preliminary hourly data available from Water Survey of Canada records using two years of data.

(a)		September 1 to 30, 1983	
		Q	WL
Maximum Change:			
in 1 h		21	0.058
in 6 h		62	0.174
in 12 h		94	0.307
in 24 h		100	0.514
Mean Monthly		232	2.765

(b)	Nov 1 - 22/82		Apr 1 - 30/83		June 4 - 30/83	
	Q	WL	Q	WL	Q	WL
Maximum Change:						
in 1 h	4.0	0.021	5.0	0.025	20.0	0.048
in 6 h	17.0	0.124	24.0	0.141	94.0	0.239
in 12 h	32.0	0.200	42.0	0.201	150.0	0.362
in 24 h	48.0	0.335	46.0	0.234	166.0	0.405
Mean Monthly	92.3	13.599	92.7	13.601	542.0	15.628

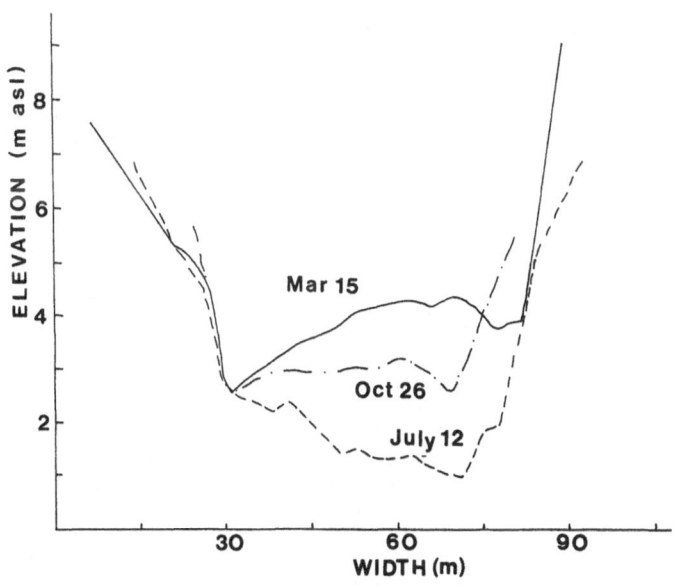

Figure 3. Changes in bed elevation at a cross section of the river during 1983.

and 760 mm respectively. The amount of bed elevation change recorded over the 1983 season indicates very high bed load transport rates (Figure 3). At a cross section approximately 25 km from the river mouth, the bed was scoured out to a depth of 3 m between March and July, with 2 m of redeposition by October.

Continuous recording thermographs were used both on the mainstem Homathko and several of the major tributaries for two years, and a highly variable temperature regime was documented. Mainstem winter temperatures varied between 0 and 3 C with a spatially and temporally discontinuous ice cover. In multi-channel reaches, secondary channels often demonstrated much warmer winter temperatures, attributed to groundwater inflow (Figure 4). Calculations of accumulated degree days required for incubation of salmonid eggs and emergence using mainstem water temperatures were complicated by the enhanced temperatures of secondary channels. Summer temperatures in the Homathko mainstem varied from 5 to 15°C, generally averaging less than 10°C. This low temperature and the observed diurnal cycle of up to 2°C are attributable to the heavy glacial melt period from June to August.

Fisheries Resource

Despite the harsh physical environment in the Homathko River, it supports substantial anadromous populations of chinook, chum and coho salmon, Dolly Varden and cutthroat trout, as well as smaller numbers of sockeye (*Oncorhynchus nerka*) and pink salmon (*O. gorbuscha*) and steelhead trout. The estimated average total run size over the last 10 years is 81000 fish. Canyons and steep gradients limit access of these stocks to estuarine, tributary and mainstem habitats in the lower 50 km of river valley. A velocity barrier to migrating fish exists on the mainstem Homathko at Waddington Canyon, when river discharge approaches 200 $m^3 s^{-1}$ (Bengeyfield et al. 1984). Few migrating fish pass the canyon, as flows of less than 200 $m^3 s^{-1}$ still present a difficult ascent. Also little habitat of suitable quality for spawning exists above the barrier due to continuous steep gradient and bedrock canyons.

Enumeration of spawning salmonids in glacial systems is seriously hampered by the large size and multi-channel complexity of the mainstem and glacial tributaries as well as the difficulty of seeing fish in turbid water. Population estimates for the principal salmon stocks in the Homathko system were derived from three independent methods, aerial spawning surveys by Fisheries and Oceans Canada personnel, mark-recapture analysis of out-migrating juveniles and spawner index counts using aerial and ground observations (Bengeyfield et al. 1984).

The first two methods indicate chinook and chum escapements in similar orders of magnitude, while coho escapements were estimated by Fisheries and Oceans Canada to be less by a factor of 5 (Table 2). Spawner enumeration in large glacial systems by aerial observation was found to underestimate actual population size due to high turbidity.

The lower Homathko River has 55 tributaries between the mouth and Waddington Canyon, which provide only another 55 km of total stream length accessible to fish. This low amount of tributary habitat is due to the narrow (1 to 3 km) valley floor and steeply rising valley sides. Homathko tributaries can be divided into three categories based on their respective water sources from May through September: glacial, mixed-glacial and non-glacial (Table 3). Glacial tributaries are cold and turbid during this period and support very few rearing fish; however, they provide good rearing conditions in early spring prior to the first glacial melt. Glacial streams also support spawning chum and coho salmon between October

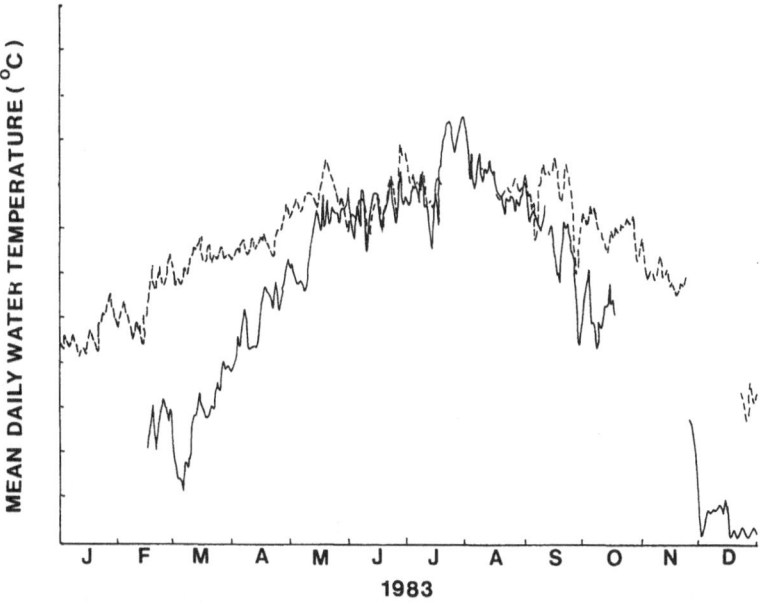

Figure 4. Water temperatures (°C) in the Homathko River (——) and in a nearby side channel (---) at a location 32 km from the mouth of the Homathko River.

Table 2. Estimates of numbers of juvenile and adult salmon in the Homathko River system.

Species	Juvenile Downstream Migrants	Marine Survival (%)*	Estimated Adult Production (numbers of fish)	Catch: Escapement Ratio	Calculated Adult Escapement	Average ** Recorded Escapements
Chinook	220,000	7.5	16,500	4:1	3,300	3,000
Coho	310,000	15.0	46,500	2:1	15,500	3,300
Chum	5,600,000	0.8	44,800	0.8:1	24,900	21,300

* Fisheries and Oceans unpublished data
**Fisheries and Oceans Canada escapement records, 1974-1983

Table 3. The distribution of juvenile fish and potential spawning area in
         three types of tributary

	Total Streams	Total Access. Length (km)	Potential Spawning Area (m^2)	No. of Streams Containing			
				Coho	Steel-head	Cut-throat	Dolly Varden
Glacial	8	32.2	500*	2	1	5	3
Mixed	4	1.4	3400	2	1	3	4
Non-glacial	43	21.5	2600	28	4	11	12
Total	55	55.1	6500	32	6	19	19

* No estimates were made for two large glacial tributaries the Heakamie
  and Jewakwa Rivers, due to high turbidities.

and January after glacial melt ceases. Mixed glacial and non-glacial
tributaries accounted for the majority of coho salmon spawners (75%) in the
watershed.

     Chum and chinook salmon rely almost exclusively on the mainstem river
for spawning and rearing. This amounted to 92% of all observed chum and 97%
of chinook spawners. Twenty-one per cent of coho salmon utilized the
mainstem for these purposes. Mainstem spawning habitats were classified
into the following channel types: primary channel, secondary channel,
side channel with continuous flows and side channel with intermittent
flows.

     During the months of peak salmon spawning, significantly different
hydrological regimes were evident in 1982 and 1983 as indicated by river
discharge measured upstream of Waddington Canyon (Table 4). September and
October 1982 were characterized by an extended warm period with
intermittent major rainstorms, whereas the same months in 1983 were drier,
cooler and had insufficient snowpack available to maintain the baseflow in
many tributaries. The November period of both years was similar in monthly
mean discharge, but water levels were lower through the first half of
November in 1983 than 1982.

     These hydrological conditions directly affected the spawning
distributions of the three principal salmon species (Table 4). In 1983,
intermittent side channels supported 76% of the early-run chinook which
spawned at relatively high flows, but only 40% of the main run chinook
which spawned during lower flows. The percentage of chum spawners in both
main channel categories increased from 10% at relatively high flows in 1982
to 36% at the lower flows of 1983. This shift probably reflects both
lack of access to side channels and more suitable velocities in the main
channel at lower flows. The spawning distribution of coho salmon increased
in all mainstem habitats in 1983 since access to tributaries was restricted
by low-flow conditions in both the mainstem and tributaries.

Chinook salmon redds established during high river flows in 1982 appeared susceptible to stranding during the winter low-flow period. In September 1983, 40 redds were surveyed at five sites along the Homathko main channel and in intermittent side channels. The degree of stranding was expected to be less severe in 1983 since redd elevations in the respective channel cross sections were generally lower due to the decreased discharge. At the time of survey, 31 redds were located in the wetted channel while nine were stranded at river flows of 68 to 121 m^3 s^{-1}. By January, ten additional redds were also stranded above the river level at flows of 20-30 m^3 s^{-1} with layers of ice and sand in the redd depressions. However, excavations indicated that only three of the 19 stranded redds were frozen or dewatered to the level of egg deposition. The remaining 16 stranded redds contained moist or wet gravels beneath a dry frozen surface layer. While the source of subsurface water was not determined, some sites apparently received groundwater percolating from valley sideslopes. Survival conditions were considered to be better in 1983 than in 1982 and the interannual variability of redd establishment and survival was well demonstrated.

Because of the direct relationship between incubating temperature and embryonic development, changes in development rates resulting from altered temperature regimes are a major area of potential concern to all species that spawn in the main river. Large differences were documented between mainstem temperatures and intragravel temperatures within the redds in secondary channels (Figure 4). Pools within a secondary channel were also found to contain considerable numbers of overwintering pre-smolt coho juveniles. Downstream migration by chum fry and coho and chinook smolts

Table 4. Number (n) of spawning salmon, 1982-1983 and the percentage distribution in various channel types.

	Tragedy Canyon Mean Flow (m^3 s^{-1})	Main Channel Primary (%)	Main Channel Secondary (%)	Side Channel Year-round (%)	Side Channel Intermittent (%)	Tributary (%)
Chinook						
1982 (n=704)	232	7	39	1	50	4
1983 early run (n=151)	252	9	15	0	76	0
main run (n=1016)	132	17	38	3	40	1
Chum						
1982 (n=4136)	83	1	9	2	75	14
1983 (n=1805)	51	12	24	6	57	2
Coho						
1982 (n=625)	102	0	2	1	10	87
1983 (n=611)	51	3	14	3	10	71

n = sample size

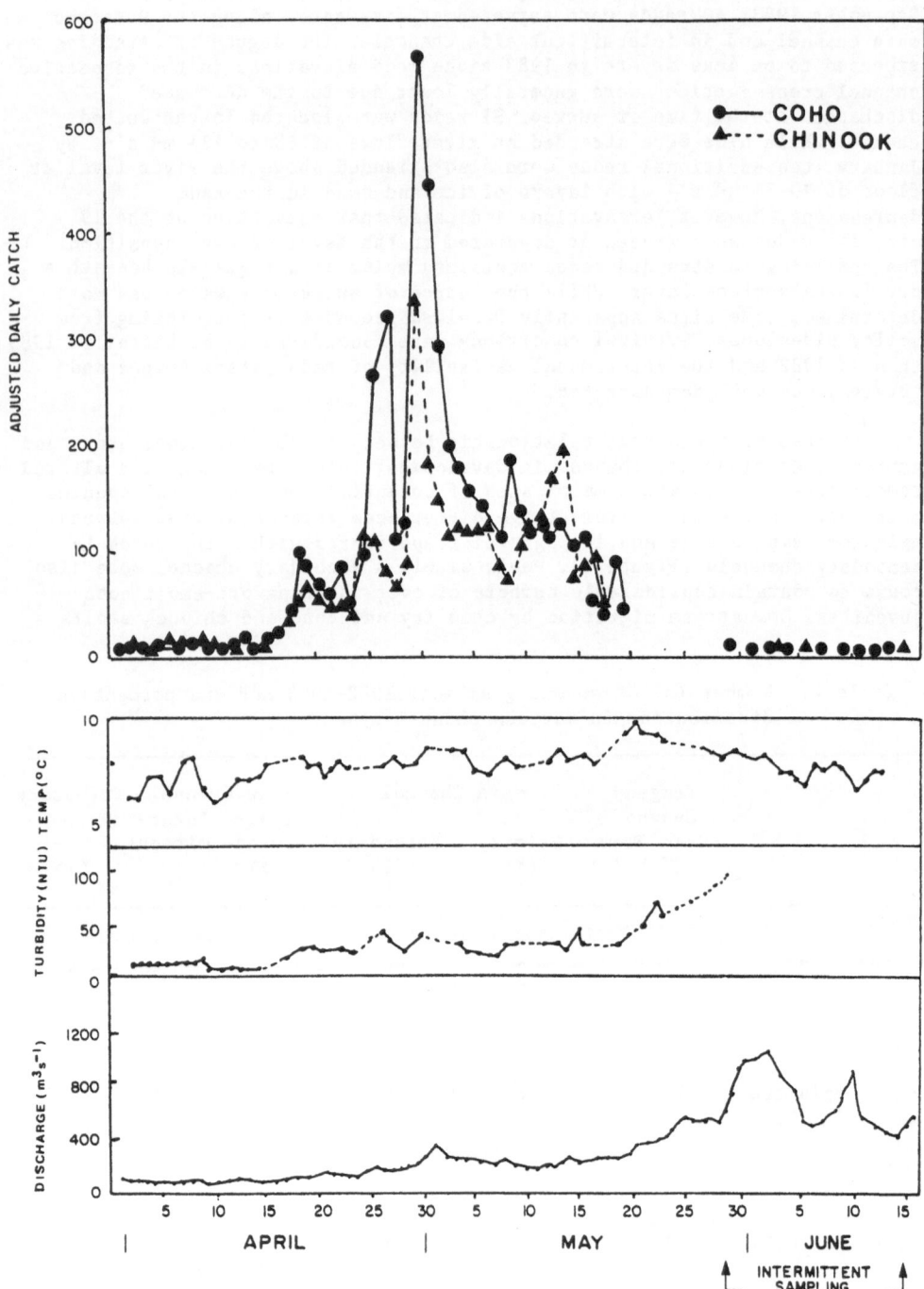

Figure 5. Relative abundance and time of migration of chinook and coho salmon smolts from the Homathko River in 1983. Also given are mean temperature (°C), mean turbidity (NTU) and river discharge (m³ s⁻¹).

occurred in April and May on a slowly rising discharge prior to the spring freshet. There appeared to be no strong correlation between outmigration timing and any other measured variables (Figure 5). Changing weather conditions and a light rainfall did, however, occur just prior to peak outmigration in late April. Photoperiod appears to have been the main factor in determining the initiation of outmigration.

DISCUSSION

On the Homathko project, the two years of study were valuable in establishing some relationships between physical and biological variables. The studies confirmed that glacial watersheds provide relatively harsh and unstable environments for the production of fish. This is exemplified by the fact that the bulk of production is in anadromous stocks which leave the watershed during much of their life cycle. In the natural environment the limiting factor in fish production appears to be the lack of a stable flow regime with associated morphological, temperature and sediment characteristics. Studies of the critical factors influencing the various life stages of fish are therefore primarily linked to discharge. Studies in the Homathko watershed have shown that anadromous stocks select spawning habitats according to ambient flow conditions. High river stage during spawning, however, does result in subsequent loss of stock due to dewatering of redds located on the main channel margins.

Secondary channels play a vital role in the spawning, incubation and rearing of Homathko River salmonid stocks. The relationship between flows in the main channel and in secondary channels, however, is not a simple relationship. Longitudinal profiles of channels and entrance sills would be one means of determining the influence of sills on flows at various stage heights.

The primary influence of a hydroelectric storage development is to create higher flows during winter and spring while reducing the peaks of freshets during summer. Under natural circumstances in the Homathko River, flows in secondary channels during low flow periods are apparently derived from seepage of water through gravel bars and sills and from groundwater sources resulting in warmer water temperatures. High winter discharges from upstream storage projects often flood secondary channels which normally contain only small volumes of flow. Such increases in cold water flow could result in increased incubation periods and decreased growth rate of fry. Preliminary modelling of water temperatures indicates that a storage reservoir would not result in significant changes in mainstem temperature, however, high flows into side channels would reduce temperatures there. The relationship between mainstem water temperature and temperatures in secondary channels and along the edge of the main channel is a topic requiring further study in glacial watersheds and would require modelling of the relationships between mainstem flow, channel morphology, groundwater flow and solar radiation.

The most frequently used salmon spawning habitat in the river is characterized by a relatively loose gravel-cobble substrate. High gradient and single channel portions of the river not used for spawning contain large cobbles and boulders which present an armoured surface layer. The most extensively used substrates in the main channel are contained in areas where gradient is low and aggradation occurs. During high flows, velocities are apparently not sufficient to move large volumes of substrate from these sections of the river. Peak flows resulting from upstream developments do not normally exceed natural flows, however, nor do they meet them as often. The influence of development on these habitats may therefore be the gradual inundation of substrates high on the channel side with sediments and gravels of smaller size. At the same time, however, sections of the river with higher gradient and continuous flow may armour due to the absence of freshet flows.

During the two years of study to date, approximately 83% of the major salmon stocks spawning in the river selected secondary channel habitat. This is due primarily to the fact that main channel habitats are of high velocity with significant bedload movement. Substrates in secondary channels are influenced by velocities and volumes of flow which are considerably lower. Many secondary channels are characterized by the presence of sand and fine gravel indicating that degradation does not generally occur under normal flows. In order to keep channels open and substrate loose and clean it is essential that some freshet flows pass through these channels. Average regulated flows during normal freshet periods, may be reduced by as much as 50%, in the upper portion of the river. Such a reduction in average freshet flows may be sufficient to reduce the viability of secondary channels to support spawning populations.

CONCLUSIONS

In conclusion, glacial river systems prevent a very dynamic physical environment and as a result the fish populations are adapted to such variability. Study methodologies require considerable assessment and modification to the specific river under investigation. "Off-the-shelf" models will not necessarily address all the significant parameters identified by hydrology-fisheries linkage studies. The importance of understanding the habitat-population linkages must not be underestimated, as the ultimate result of an impact assessment should be in terms of number of fish, area of habitat.

REFERENCES

Alaska University Arctic Environment and Data Center. 1983. Methodological approach to quantitative impact assessment for the proposed Susitna hydroelectric project. Report by Arctic Environmental Information and Data Center, University of Alaska for Harza/Ebasco Susitna Joint Venture and Alaska Power Authority.
Aquatic Environments Ltd. 1982. Fish populations associated with proposed hydroelectric dams on the Stikine and Iskut rivers. 2 vols. Report prepared for B. C. Hydro.
Bengeyfield, B., Lister, D. B., Fleming, O. and Burt, D. 1984. Preliminary Fisheries Inventory and Assessment in the Homathko River System. Report prepared for B.C. Hydro. Environmental Sciences Ltd. and D. B. Lister and Associates Ltd. 198 pp. and appendices.
Bovee, K. D. 1982. A guide to stream habitat analysis using the instream flow incremental methodology. US Fish and Wildlife Service. Instream Flow Information Paper 12. 248 pp.
Church, M. and Gilbert, R. 1975. Proglacial fluvial and lacustrine environments. In: Glaciofluvial and glaciolacustrine sedimentation. (Ed. by A. V. Jopling and B. C. McDonald). pp. 22-100. Society of Economic Paleontologists and Mineralogists. Special Publication No. 23.
Envirocon Ltd. 1984. Kemano Completion Hydroelectric Development Baseline Environmental Studies. Prepared for the Aluminum Co. of Canada. 22 vols.
Field, W. O. (Ed.). 1975. Mountain glaciers of the Northern Hemisphere. Cold Regions Research and Engineering Laboratory, U.S. Army Corps of Engineers, Hanover, N.H. 2 Volumes. 932 pp.

Glova, G. J. and Duncan, M. J. 1985. Potential effects of reduced flows on fish habitats in a large braided river. New Zealand. Trans. Am. Fish. Soc. **114** (2): 165-181.

Jones, D. P. and Associates and B. C. Hydro. 1984. Stikine-Iskut Development, Environmental Hydrology report. 2 vols. Reports prepared for B.C. Hydro.

Kellerhals, R. and Bray, D. I. 1971. Sampling procedures for coarse fluvial sediments. Proceedings of the American Society of Civil Engineers, Journal of Hydraulics Division. **97**: 1165-1179.

Mokievsky-Zubok, O. 1983. Glaciological studies in Homathko River basin in 1982. Internal report, National Hydrology Research Institute, Environment Canada.

Mosley, M. P. 1982. Analysis of the effect of changing discharge on channel morphology and instream uses in a braided river, Ohau River, New Zealand. Wat. Resour. Res. **18**: 800-812.

Walling, D. E. and Webb, B. W. 1981. The reliability of suspended sediment load data in erosion and sediment transport measurements. Proceedings of the Florence Symposium. International Association of Scientific Hydrology, Publication No. 133. 177-194.

Water Survey of Canada. 1983. Historical Streamflow Summary. British Columbia to 1982. Environment Canada. 940 pp.

SECTION IV
WATER QUALITY

PHOSPHORUS SPIRALLING IN RIVERS AND RIVER-RESERVOIR SYSTEMS:

IMPLICATIONS OF A MODEL

J. Denis Newbold

Stroud Water Research Center
Academy of Natural Sciences of Philadelphia
Box 512, R.D. #1
Avondale, PA 19311

INTRODUCTION

The term spiralling refers to the coupled processes of cycling and downstream transport of nutrients. The intuitive notion of the spiral is, most simply, that of a nutrient cycle that fails to close in place in a stream, but rather is stretched by transport along the longitudinal axis of the stream (Elwood et al. 1983). At a somewhat more complex level, we may envision the ecosystem as a set of compartments representing various organic and inorganic forms of the nutrient, but instead of being boxes, these compartments are strips or tubes lying along the length of the stream. Each tube slides downstream at some characteristic rate, some at the velocity of water, others with geologic slowness. Our smooth spiral is now replaced by an erratic pathway, consisting of lateral exchanges among compartments and varying rates of longitudinal transport.

The origins of the spiralling concept can be traced to Leopold (1941) who described the general downhill migration of nutrients in a watershed in terms of a "rolling motion", then to an excerpt from Leopold's paper used by Likens and Bormann (1974), and finally to Webster (1975; see also Webster & Patten 1979), who recognized its particular applicability to streams and introduced the term. The concept of spiralling has been applied to the study of phosphorus dynamics in a small woodland stream in Tennessee (Newbold et al. 1981; Newbold et al. 1983; Mullholland et al. 1985), to calcium and potassium dynamics in small North Carolina streams (Webster & Patten 1979), and to carbon dynamics in streams and small rivers (Wallace et al. 1977; Newbold et al. 1982a; Minshall et al. 1983).

There are two senses in which the spiralling concept is applicable to lotic ecosystems. First, it provides an alternative to traditional approaches for developing nutrient budgets and cycling indices for ecosystems. The spiralling approach views transport as an internal process of the ecosystem, rather than as inputs and outputs across ecosystem boundaries which, in the case of a stream, usually must be arbitrary (Elwood et al. 1983). Second, and more importantly in my view, it provides a conceptual starting point for asking how nutrient utilization (and metabolic activity in general) are regulated in lotic ecosystems. For example, Webster and Patten (1979) suggested that, "Internal regulation of the stream ecosystem can be mediated by characteristics of the stream which shorten or tighten spiralling, such as stream bed retention of particulate organic matter and insect filtering of fine particulate organic matter". This suggestion poses some fundamental questions: To what extent are lotic ecosystems regulated by nutrients supplied from upstream, as opposed to factors that have nothing to do with spiralling (e.g., light, temperature, substratum, etc.)? If nutrients are important, is their availability mediated from within the system, or is availability only a matter of nutrient supply from the watershed? Finally, if internal mediation is important what are the basic mechanisms by which this occurs? Note that these questions basically ask whether upstream biotic processes influence downstream processes, since any internal influence through nutrient availability will be transmitted via downstream transport.

Such internal mediation can occur only through effects on nutrients that are in limiting supply, or that are rendered limiting by internal ecosystem processes. Spiralling of a non-limiting nutrient must be regulated either by local (transport-independent) factors such as light, or available substratum, or by spiralling of a nutrient that is limiting. The question of spiralling is therefore, in a sense, also the question of nutrient limitation in lotic systems. Newbold et al. (1982b) developed a simple steady state model of spiralling to express the relationships between nutrient limitation and particulate transport in headwater streams. The model indicated both that the importance of ecosystem control over particulate transport increases with nutrient limitation, and also the converse: that low rates of particulate transport (high retention) diminish the likelihood of severe nutrient limitation. In other words, as the nutrient retention of a stream ecosystem increases, the regulatory role of spiralling diminishes.

Studies of phosphorus spiralling (Newbold et al. 1983; Mulholland et al. 1985) in Walker Branch, a second order Tennessee woodland stream, suggest that such headwater streams are highly retentive and that a strong regulatory role for spiralling is unlikely. With increasing stream order, however, particulate transport increases dramatically in relation to nutrient standing stocks (Minshall et al. 1983), which is to say that higher order streams are much less retentive than are low order streams. This decreasing retentiveness suggests an increasing role, along the river continuum (sensu Vannote 1981), for particulate transport in governing nutrient uptake, availability, and standing stocks, as well as an increasing degree of linkage between upstream and downstream systems.

In this paper, I address the questions of whether, and how, phosphorus spiralling might play an important role in mediating ecosystem processes in river systems, what this implies about our understanding of nutrient limitation in river systems, and how flow-regulation of a river system might affect the regulatory role of spiralling. To do this, I develop and analyze the behavior of a model applied to hypothetical rivers and reservoirs. The model is designed to describe the transition from a headwater stream, in which phosphorus dynamics is dominated by exchanges between the stream bottom and the water column, to a riverine environment in which phosphorus exchanges occur primarily in the water column. It is not a complex simulation model of a river ecosystem, but rather a very simple, highly schematic model applied to an idealized environment. This simplified approach helps to keep both the basic ideas and their limitations conceptually accessible, but requires a certain degree of forebearance from any reader familiar with the complexities of real rivers.

The model is in many respects an extension of that of Newbold et al. (1982b). Its basic components (shown in Figure 1) include a water compartment containing dissolved phosphate (PO_4), a benthic algae compartment (which can also be thought of as including benthic heterotrophic microbes), a suspended phytoplankton compartment, and compartments for suspended and benthic fine particulate organic matter (TFPOM and FPOM, respectively). The equations and parameter values for the model are presented in Table 1. The state variable for each transport compartment (PO_4, phytoplankton, and TFPOM) is a concentration of phosphorus expressed in mg P m^{-3}. Thus, for example, phytoplankton standing stock is expressed, not in terms of biomass, but in terms of the phosphorus in the phytoplankton. The benthic compartments are expressed as mg P m^{-2}.

Uptake of phosphorus from the water by benthic algae (Eq. T6, Table 1) is governed by saturation kinetics as used by Newbold et al. (1982b). At low P concentrations (C_W) and low standing stocks of benthic algae (B_A), uptake is proportional to both C_W, and B_A. As either C_W or B_A, or both increase, however, their respective influences on uptake approach a maximal rate, and at high values of both, the maximum uptake is given by aK_s/k_2. The k_2 parameter allows uptake rate to become self-limiting at high standing stocks. In effect, it represents the limiting effect of all factors other than phosphorus concentrations. It is necessary because, at steady state, benthic stocks deplete phosphorus concentrations only insofar as they contribute to particulate transport, and this depletion is unlikely to be sufficient to limit benthic stocks to reasonable levels (see discussion in Newbold et al. 1982b). Phosphorus uptake by phytoplankton is identical to uptake by benthic algae, except that no self-limitation parameter is used. Unlike attached benthic organisms, any phosphorus in planktonic organisms at steady state are balanced directly by a depletion in dissolved concentrations, and maximal standing stocks of phytoplankton are limited to the total concentration of phosphorus in the water column. The FPOM and TFPOM compartments play a peripheral role in this model, serving only as a repository for senescent algae. The uptake equations for FPOM and TFPOM are the same as for benthic algae and

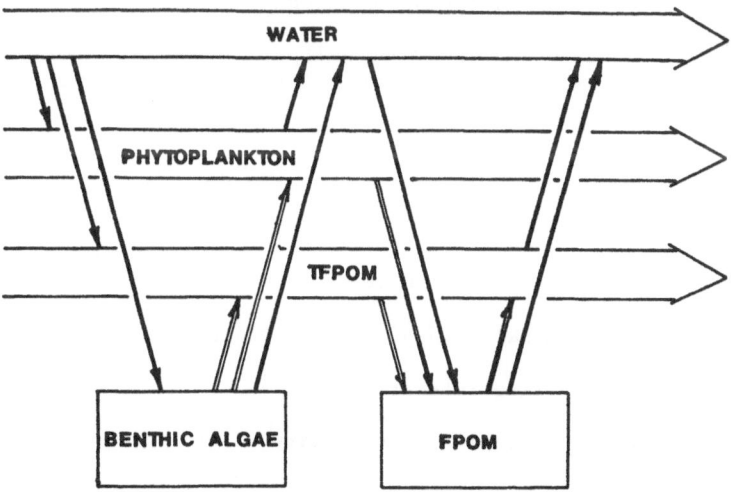

Figure 1. Compartments and transfers used in river model. Solid arrows represent biological transfers; open arrows are used for suspension and settling.

Table 1. Model equations, symbols and parameter values.

Equations for the model:

PO$_4$:

$$\frac{\partial C_W}{\partial t} = -\frac{1}{A}\frac{\partial}{\partial x}(QC_W) + \frac{G_W}{A}\frac{Q}{x} - (U_A + U_F)/d - U_P - U_T + r[(B_A + B_F)/d + C_P + C_T] \qquad T1$$

Phytoplankton:

$$\frac{\partial C_P}{\partial t} = -\frac{1}{A}\frac{\partial}{\partial x}(QC_P) + \frac{G_P}{A}\cdot\frac{Q}{x} - (r + q_p/d)\,C_P + u_{PA}B_A/d + U_P \qquad T2$$

TFPOM:

$$\frac{\partial C_T}{\partial t} = -\frac{1}{A}\frac{\partial}{\partial x}(QC_T) - (r + q_T/d)C_T + (u_{TF}B_F + u_{TA}B_A)/d + U_T \qquad T3$$

Benthic Algae:

$$\frac{\partial B_A}{\partial t} = -(r + u_{TA} + u_{PA})\,B_A + U_A \qquad T4$$

FPOM:

$$\frac{\partial B_F}{\partial t} = q_T\,C_T - (r + u_{TF})\,B_F + U_F \qquad T5$$

where phosphorus uptake fluxes, U are given by

$$U_X = \frac{a_X\,C_W\,Y}{1 + C_W/K_s + k_2\,Y)} \qquad T6$$

in which X = A, Y = B$_A$ for Benthic Algae

X = F, Y = B$_F$ for FPOM

X = P, Y = C$_P$, k$_2$ = 0 for Phytoplankton,

X = T, Y = C$_T$, k$_2$ = 0 for TFPOM

Symbols and parameter values:

C$_W$ — PO$_4$-P concentration in water mg m^{-3}

A — cross sectional area of river, at distance x, m^2

Q — volumetric flow rate at x, m^3 s^{-1}

G$_W$ — PO$_4$-P concentration in tributary water, mg m^{-3}

d — river depth, m

r — phosphorus regeneration rate, 2 x 10^{-6} s^{-1}

B$_A$ — Benthic Algae-P standing stock, mg m^{-2}

B_F – Fine Particulate Organic Matter (FPOM)-P standing stock, mg m^{-2}

G_P – Phytoplankton-P concentration in tributary water, mg m^{-3}

C_P – Phytoplankton-P concentration in water, mg m^{-3}

q_P – settling velocity of phytoplankton, 2×10^{-7} m s^{-1}

u_{PA} – suspension (sloughing) rate of benthic algae to phytoplankton 5×10^{-8} s^{-1}

q_T – settling velocity of TFPOM, 1.5×10^{-4} m s^{-1}

u_{TF} – suspension rate from FPOM to TFPOM, 1×10^{-6} s^{-1}

u_{TA} – suspension rate from Benthic Algae to TFPOM 2×10^{-6} s^{-1}

a_A – uptake coefficient for algae, 1.16×10^{-5} m^3 mg^{-1} s^{-1}

K_s – half saturation constant for growth, 1 mg P m^{-3}

a_F – uptake coefficient for FPOM 1.16×10^{-6} m^3 mg^{-1} s^{-1}

t – time, s

x – downstream distance, m

k_2 – self-limitation parameter for benthic algae, 0.05 m^2 g^{-1}

---

phytoplankton respectively, but the uptake parameter a_F was set to 10% of the value of a_A on the assumption that some phosphorus demand would be exerted by microbial decomposers of the senescent material. Nutrient regeneration is governed by the constant r, which is identical for all microbial compartments.

Suspension of benthic algae consists of two components, transfer of benthic algae to phytoplankton (u_{PA}), and transfer to TFPOM (u_{TA}). The former serves as an inoculum for supplying the phytoplankton compartment, while the latter represents a sloughing of senescent benthic algae. Constant settling velocities are used for deposition of phytoplankton and TFPOM. Phytoplankton is considered to be senescent on settling, and therefore enters the FPOM compartment rather than the benthic algae compartment.

Parameter values, represented in Table 1, were selected to represent reasonable or typical values from published reports and from studies of Walker Branch. The half saturation constant for growth (K_s) was taken to be 1 mg P m^{-3}, which is within a relatively wide range of reported values for planktonic algae (e.g. Rhee 1973; Tilman & Kilham 1976; Kilham 1978). Bothwell (1985) reported half saturation constants for uptake by periphyton in the range of 0.5 to 7.2 mg P m^{-3}, which suggests that the half saturation constants for growth (typically at least an order of magnitude lower than the uptake constant; Rhee 1973; Lehman et al. 1975) may be lower than 1 mg P m^{-3}. The phosphorus regeneration rate (r) of 0.17 day^{-1} was chosen as typical of phosphorus turnover rates measured in Walker Branch (Newbold et al. 1983; Mulholland et al. 1985).

Spiralling indices (see, e.g., Newbold et al. 1982b) were calculated at steady state, as follows: Uptake length, the expected distance traveled by a phosphorus atom in the water compartment before being taken up by an organism, is given by:

$$S_W = F_W/(U \ w) \tag{1}$$

in which F_W (mg s^{-1}) is the downstream flux of phosphorus in the water compartment, U (mg m^{-2} s^{-1}) is the uptake of phosphorus from the water to all other compartments, and w(m) is the river width.

Turnover length, the expected distance traveled downstream by a phosphorus atom between being taken up into any particulate compartment and being released back to the water, is given by:

$$S_B = F_B/(R \ w) \tag{2}$$

in which F_B (mg sec^{-1}) is the downstream flux of particulate phosphorus (phytoplankton and TFPOM), and R (mg m^{-2} s^{-1}) is the release of phosphorus from particulates back to the water.

Spiralling length, the total downstream travel distance associated with one phosphorus cycle, is the sum of the uptake and turnover lengths:

$$S = S_W + S_B \tag{3}$$

At steady state, assuming R=U, eqs. (1-3) yield $S = (F_W+F_B)/(U \ w)$. Therefore, letting $F_T=F_W+F_B$ where F_T is the total downstream phosphorus flux, gives:

$$S = F_T/(U \ w) \tag{4}$$

Thus spiralling length inversely measures total phosphorus uptake in relationship to total downstream phosphorus flux. From the above relationships it can also be seen that at steady state, the ratio of the uptake length to the turnover length is the same as the ratio of the downstream water flux to the downstream particulate flux:

$$\frac{S_W}{S_B} = \frac{F_W}{F_B} \tag{5}$$

These relationships are derived for a steady state stream, and while eqs. 1-3 are not particularly sensitive to departures from steady state, eqs. 4-5 depend heavily on the equivalence of phosphorus uptake (U) and release (R). It should also be noted that even at steady state, eqs. 4-5 are approximations because lateral phosphorus inputs along the length of the stream will produce a small imbalance in R and U (see Newbold et al. 1983).

The model was applied to two hypothetical rivers with the same flow but with a 2-fold difference in water velocity. The rivers will be referred to as "fast" and "slow," although even the fast river would be considered slow by most standards. To accommodate the slower velocity, the slow river is also somewhat wider and deeper than the fast river (Table 2). In effect, these are the same river, but in channels of differing gradient. However, the slow river might also be thought of as regulated by a series of small impoundments. As will become clear, the velocities were selected to illustrate the potentially dramatic effect of velocity on phosphorus spiralling. The modelled reaches begin near the headwaters of each stream and extend 200 km downstream. Differences in channel configuration of the

Table 2. Characteristics of fast and slow hypothetical rivers.

| | Upstream (0 km) | | Downstream (200 km) | |
	Fast	Slow	Fast	Slow
Flow ($m^3 s^{-1}$)	0.30	0.30	12.5	12.5
Velocity (m s^{-1})	0.40	0.20	0.48	0.24
Width (m)	4.4	5.0	28.5	32.5
Depth (m)	0.17	0.30	0.92	1.60
Travel Time (days)			5.2	10.4

two rivers and longitudinal variations within each river were established using relationships provided by Leopold et al. (1964), except that river flow (assumed proportional to watershed area) was computed as proportional to $L^{1.4}$, in which L is channel length, rather than as $L^{1.56}$ as suggested by Leopold et al. (1964). In effect, the modeled watersheds are narrower than a typical watershed. This was done to avoid modelling inputs from large tributaries, with their potentially large inputs of phytoplankton. It allowed a much simpler model in which channel growth occurred uniformly along the length of the stream, and all influent water carried the same load of dissolved and particulate material as the headwater stream.

The model was also applied to the fast river with a hypothetical impoundment located 100 km downstream. The dam was assumed to be 15.6 m high, and 200 m wide, making the reservoir about 22 km long. Depth along the length of the reservoir was determined by the original theoretical slope, while width was assumed to increase linearly from the beginning of the reservoir to the dam. Residence time for water in this reservoir would be about 21 days.

Both rivers were modeled as unpolluted systems with relatively low phosphorus inputs at the headwaters and along the river: 6 μg P L^{-1} total, consisting of 5.95 as dissolved, and 0.05 as phytoplankton inoculum. Both the steady state solutions and the temporal dynamics of the models were examined, although most of the subsequent presentation involves the steady state solutions. From somewhat arbitrary initial conditions of 7 mg P m^{-2} in benthic algae, 10 mg P m^{-2} in FPOM, 5.90 mg m^{-3} PO4-P, 0.05 mg P m^{-3} as phytoplankton, and 0.05 mg P m^{-3} as TFPOM, simulations required 50-60 days to approach steady state, with small adjustments thereafter. The steady state solutions presented below represent 90 days of simulation.

RESULTS AND DISCUSSION

Longitudinal Patterns in the Fast and Slow Rivers

Figure 2 shows the steady state solutions for the two rivers. FPOM and TFPOM are not shown, but are small relative to benthic algae and phytoplankton, respectively. Phytoplankton increases along the length of both rivers, but in the slow river this increase occurs much more rapidly (with distance) and results in higher downstream concentrations, than in

the fast river. The downstream increases in phytoplankton produce a
corresponding decline in the dissolved phosphorus (PO_4) concentrations in
both rivers, and in the slow river PO_4-P is depleted to <1 mg m^{-3} by
200 km. As a result of this depletion, benthic algae become phosphorus
limited, and fall to very low levels in the lower reaches of the slow
river. In the fast river, the depletion of PO_4 is less severe and does not
seriously limit the benthic algae. In neither river does the increase in
phytoplankton (on an areal basis) balance the decrease in benthic algae,
and as a result, there are downstream declines in total stock (the sum of
all compartments except PO_4, expressed on an areal basis). Total phosphorus
uptake on an areal basis also declines downstream. This is not shown, but
is directly proportional to total stock in Figure 2. (This proportionality
exists because, at steady state, phosphorus uptake from the water is
approximately equal to release back to the water, and phosphorus release is
in constant proportion, r, to total stock.) If we assume that metabolic
activity in the river is reflected in phosphorus uptake, then the model
also implies that the metabolism of the slow river ecosystem, on an areal
basis, declines substantially downstream, and, in the downstream reaches,
it is considerably lower than in the comparable reaches of the fast
river. If the slow river were to represent the effects of a series of small
regulating impoundments placed on the fast river, then the effect of this
regulation would be to reduce the productivity of the river. These results
are quite hypothetical, and I will address below the manner and extent to
which they might reflect patterns in real rivers. First, however, I
interpret the model results in the context of the spiralling concept.

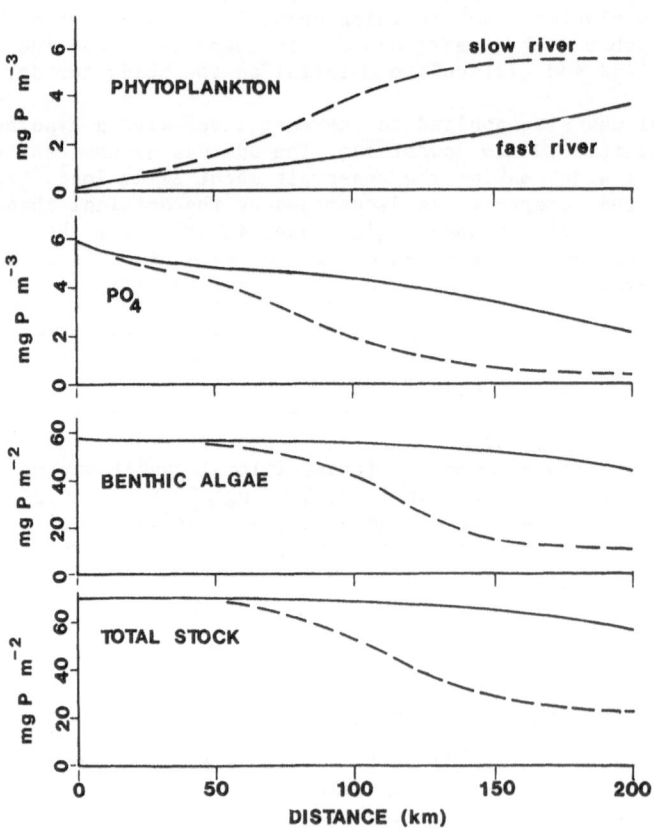

Figure 2.   Steady state model solutions for fast and slow rivers.

Spiralling length increases downstream in both rivers, but to a much greater degree in the slow river (Figure 3). As indicated by eq. (4) above, spiralling length (S) is the ratio of total downstream phosphorus flux (F_T) to phosphorus uptake per unit length of stream (U w). At steady state, $F_T = C_T Q$, in which Q is river flow, and C_T (6 mg P m^{-3}) is the total phosphorus concentration of influent water (referred to below as loading). Therefore, F_T increases identically along both rivers, and the much longer S in the downstream reaches of the slow river result from the much lower rate of phosphorus uptake (U), tempered by a somewhat greater width (w). Within each river, however, F_T increases longitudinally with Q, so that if U remains relatively constant, as it does in the fast river, S increases approximately as Q/w (or alternatively, as depth (d) x velocity (v), since Q=wdv). This suggests that in streams and rivers of comparable phosphorus loading and metabolic activity, we would expect stream size to be the major determinant of spiralling length. Support for this is indicated by Figure 4, which shows measurements of uptake length (from $^{32}PO_4$ releases) in several streams of varying size, along with a theoretical slope for v d vs Q (from Leopold et al. 1964). The comparison is made on the basis of uptake length rather than total spiralling length because turnover lengths are not available for several of the measurements. In those cases for which turnover length was estimated (Walker Branch), it was small in relation to uptake length. In the context of the river model presented here, all the points on Figure 4 represent headwater systems: the largest stream for which an uptake length is available (the Sturgeon River in Michigan) corresponds roughly to the model fast river at 12 km (shown as an open square).

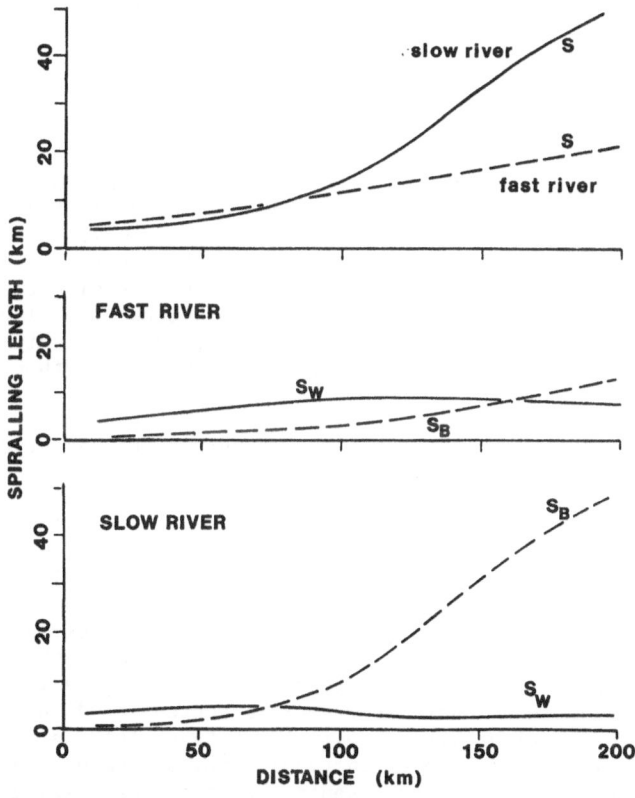

Figure 3. Spiralling lengths (S), uptake lengths (S_W), and turnover lengths (S_B) from the steady state model solutions for the fast and slow rivers.

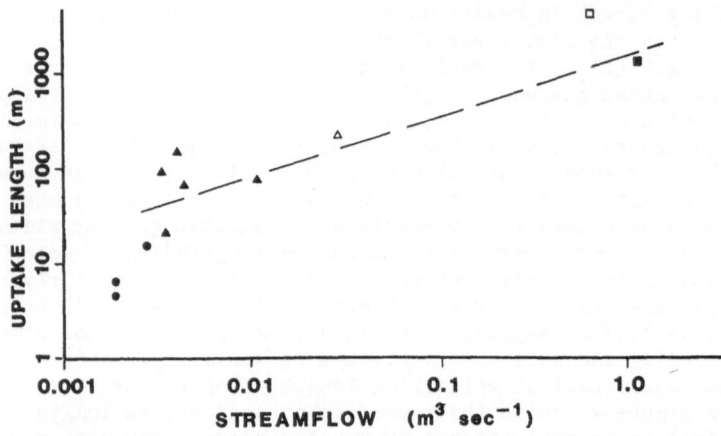

Figure 4. Uptake length for phosphorus in relation to stream flow. Solid
circles -- Streams at Coweeta Hydrologic Laboratory, Franklin,
North Carolina (J. W. Elwood and J. D. Newbold, unpublished
data); solid triangles -- First order reach of Walker Branch
Watershed, Oak Ridge, Tennessee (Newbold et al. 1983, Mulholland
et al. 1985); Open triangle -- Second-order reach of Walker
branch (J. W. Elwood and J. D. Newbold, unpublished data);
Solid square -- Sturgeon River, Cheboygan County, Michigan
(Ball and Hooper 1963); Open square -- Fast river model at 12
km. Broken line represents a slope of 0.6 for log(v d) vs log
(Q), representing average downstream relations for midwestern
rivers (from Leopold et al. 1964, p. 244).

The slow river undergoes a longitudinal transition from a nutrient
sufficient, retentive mode in the headwaters, to a nutrient depleted
transporting mode in downstream reaches, and this transition involves major
shifts in the nature and role of spiralling in the ecosystem. Near the
headwaters, the spiralling length of the slow river is slightly shorter
than that of the fast river (Figure 3a) simply because the slow river is
slightly wider (see eqs. 1 and 2). Below 100 km, however, spiralling
lengths in the two rivers diverge greatly, the longer length being in the
slow river. That is to say, that between successive cycles through the PO_4
compartment, phosphorus travels a longer distance in the river with the
slower velocity. This occurs because phosphorus in the downstream reaches
of the slow river, once taken up, is not retained on the bottom in the
benthic algae, but rather drifts downstream in the phytoplankton. The
downstream loss of retentiveness of the slow river is particularly evident
in the behavior of the turnover length, S_B, which is the dominant component
of the spiralling length below 75 km (Figure 3b). The nutrient
retentiveness of a lotic ecosystem can be described inversely by the
average velocity, v_B, at which the total particulate stock in the river
moves downstream (Newbold et al. 1982b); v_B is related to the turnover
length by the equation $v_B=rS_B$ (Ibid., see also Elwood et al. 1983). Thus,
the increase in turnover length represents a decreasing downstream
retentiveness in the river. The dramatic downstream increase in the slow
river S_B reflects effects both of increasing phytoplankton and of
decreasing benthic algae, and represents a kind of snowballing: the effect
of increasing particulate transport on the turnover length (50-150 km)
becomes amplified by the negative effect of particulate transport of
benthic algae (100-200 km).

The loss of retentiveness brings about a situation in which the slow river becomes phosphorus limited even while total downstream phosphorus flux, F_T remains the same as in the fast river, and total phosphorus uptake is lower than in the fast river. The phosphorus limitation occurs because most of the total downstream flux is in the form of particulates, i.e., as F_B, making both C_W and the ratio F_W/F_B quite small. By equation (5), the ratio of uptake to turnover length, S_W/S_B must also be small. As illustrated in Figure 3, the uptake length in the downstream reaches of the slow river is not only much shorter than the turnover length, but also shorter in absolute terms than the corresponding uptake lengths in the fast river and in the headwaters of the slow river. This shorter uptake length reflects a more rapid cycling rate through the PO_4 compartment as is commonly observed in lakes when PO_4 is depleted and nutrient limitation becomes severe (cf. Lean & Nalewajko 1979). The model calculated turnover time for PO_4 in the slow river at 200 km was about 3.5 hours, which is considerably longer than those observed in lakes under strong phosphorus limitation (Rigler 1973; Lean & Nalewajko 1979; Lean & Pick 1981; Prepas 1983). This discrepancy probably arises from the simple structure of the model, which leaves out a very rapid phosphorus turnover component in the algae (Rigler 1973; Lean 1973; Lean & Nalewajko 1976; Mulholland et al. 1983), and also ignores the production by algae of dissolved organic phosphorus (Lean 1973; Peters 1979); it might also reflect the use of too large a value for K_s.

There are two major points to be drawn from the preceding discussion: First, the downstream loss of retentiveness (and other changes) observed in the slow river represent a transition from an upstream system in which phosphorus spiralling is controlled by factors unrelated to nutrient supply or transport, to a downstream system in which spiralling is a major mediating influence on phosphorus utilization and, by inference, on productivity. This is because, most simply, PO_4 remains undepleted and unaffected by the biota in the headwaters, but becomes severely depleted in the downstream reaches, where its concentration both influences--and is influenced by--biotic processes. The second point is that this transition derives fundamentally from the biological structure of the ecosystem, through which the influence of abiotic factors is expressed. This point is perhaps best illustrated by the observation that the greater loss of retention occurs not in the fast river, as might be expected on purely physical grounds, but in the slow river, in which the conditions are provided for a biologically mediated increase in particulate transport. I will return to this somewhat paradoxical effect of velocity below.

I have so far suggested how the interaction of cycling and transport (spiralling) might be of central importance to the functioning of a river ecosystem, but I have used a highly simplified model of a hypothetical river to do this. In the following paragraphs I attempt to indicate the conditions under which spiralling may or may not be of importance in real rivers, and to address some of the major limitations of the model.

Effect of velocity. As has already been indicated by the comparison of the fast and slow rivers, the transition to a nutrient-limited, plankton dominated river system may only occur at low river velocity. A relationship between velocity and phytoplankton development in rivers has long been known (see Hynes 1970). Lack (1971), for example, observed marked declines in phytoplankton in the River Thames when velocity exceeded 1 km hr^{-1}, which is midway between the velocities of the fast and slow rivers modeled here. Although the slow river model produced the longer spiralling length, it is obvious that further slowing of the river must eventually produce a shorter spiralling length. Figure 5 shows the effect on spiralling length at 200 km of varying river velocity over a wide range. Channel shape was held constant. Depth was varied through the Manning equation (as though the

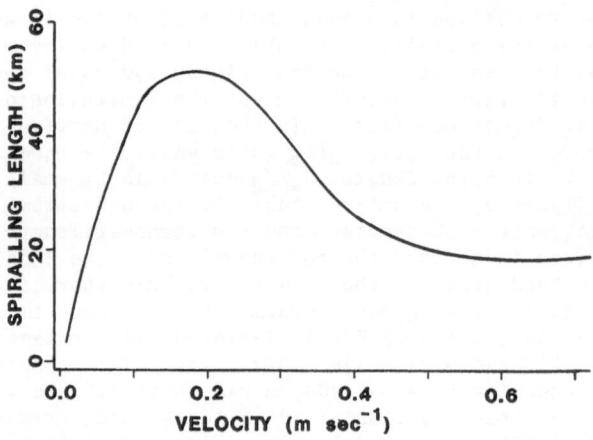

Figure 5. Effect of varying velocity on spiralling length of model river
at 200 km.

changes in velocity resulted from altering the gradient of the river), and
width varied so as to maintain a constant flow. Although it is impossible
to change the gradient of a river, the installation of a series of
impoundments can have a similar effect. Spiralling length is relatively
insensitive to velocity above 0.6 m s^{-1}, but becomes longer as velocity
is reduced and phytoplankton develop. Below about 0.18 m s^{-1}, however,
spiralling length declines with further reduction in velocity. The reason
for this is that maximal phytoplankton development is attained at about
0.18 m s^{-1}, at which point all relevant components of the ecosystem are
contained in the water column and move downstream at the water
velocity. Spiralling length therefore becomes directly proportional to
water velocity. At 100 km downstream, the model produces a similar,
but less pronounced relationship between spiralling length and velocity,
with spiralling length peaking at about 0.08 m s^{-1} (not shown). At high
nutrient loadings, the transition to phytoplankton domination does not
lengthen spiralling, for reasons discussed below, and under these
conditions, spiralling length increases monotonically with velocity.

Effect of phosphorus loading. The downstream declines in total stock
and phosphorus uptake produced by the slow river model depend on the fact
that the maximum phytoplankton standing stocks (limited by phosphorus) are
considerably lower than the maximum benthic standing stock (limited by the
parameter k_2). Model runs in which phosphorus loading of the slow river was
increased, showed that at a loading (C_T) of 50 mg P m^{-3}, downstream total
stocks (mostly phytoplankton) approximately equaled upstream stocks (mostly
benthic algae), while at 100 mg P m^{-3}, downstream stocks greatly exceeded
upstream stocks. In the latter case, the downstream phytoplankton
concentrations (nearly 100 mg P m^{-3}) were still well below maximal levels
observed in nutrient rich rivers (Megard 1981; Søballe & Bachmann 1984;
these comparisons assume that chl a:P is 1:1 or less). What this means is
that at phosphorus loadings in excess of about 50 mg P m^{-3}, the ability of
the plankton to utilize this load outweighs the adverse impact on benthic
algae. Upstream spiralling lengths become very much longer under high
nutrient loading, reflecting the inability of the benthos to respond to the
higher nutrient concentrations, while downstream spiralling lengths remain
about the same as under low nutrient load.

Under these higher loadings, the basic transition of the river to
higher particulate transport, longer turnover length, and phosphorus
limitation still occurs. Dècamps et al. (1984) report and model a similar

314

scavenging by phytoplankton of high phosphorus inputs in the River Lot (France), which is regulated by a series of small dams. Their model handles phytoplankton dynamics in considerably more detail than is used here. In particular, they demonstrate the value of incorporating cellular nutrient content into uptake and growth kinetics.

While phosphorus limits phytoplankton growth both in the Lot and in my hypothetical rivers, it is frequently the case that downstream river reaches are too highly enriched for nutrient limitation to occur. In this case, phytoplankton may grow to a level at which self shading becomes the limiting factor (Megard 1981). Alternatively nutrient loading may not be very high, but some other factor such as temperature, velocity, or inorganic turbidity limits phytoplankton growth (e.g. Cushing 1964; Swale 1964; Watts & Lamarra 1983; McCullough 1978; Crayton & Sommerfeld 1981). In either case, phosphorus would remain in excess, and the basic transitional scheme of the slow river model would not apply.

Phosphorus starvation of downstream benthic algae. The phenomenon is predicted by the model but to my knowledge has not been reported. Of course, downstream reductions in benthic algae may frequently result from other factors, such as light reduction due to phytoplankton growth, increasing depth and inorganic turbidity, or unsuitable substratum (e.g., Cushing 1964; Duffer & Dorris 1966). To the extent that these other factors are responsible for downstream declines, a downstream loss of retentiveness could occur that would not be ascribable to the spiralling model. There may, however, be cases in which nutrient depletion is an important factor that has been overlooked in favor of alternative explanations. Such depletion might also have important effects on nutrient utilization by benthic heterotrophs, or on abiotic phosphorus exchanges on the river bottom.

Source and development of phytoplankton. The manner in which river plankton develop is an old, but not fully answered question (e.g., see Margalef 1960; Ruttner 1963; Talling & Rzóska 1967; Hynes 1970; Swanson & Bachmann 1976; Søballe & Bachmann 1984). The issue has generally centered around the relative importance of in-channel growth, versus supply from the benthos, and from backwaters. The model uses both supply from the benthos and a small inoculum entering with all inflowing water. Phytoplankton development in the headwaters (0-40 km) is similar in the two rivers (Figure 2), and is controlled by accrual from benthic algae; in these reaches, the longitudinal rate of dilution by groundwater exceeds the rate of phytoplankton growth, effectively preventing in-channel growth from non-benthic inocula. Below 40 km, most of the phytoplankton increase results from in-channel growth. In model runs with no benthic algae, phytoplankton developed, but the growth curve was displaced considerably downstream. While phytoplankton in headwater areas is largely of benthic origin (e.g., Swanson & Bachmann 1976), lentic forms are frequently encountered downstream (e.g., Swale 1964). With neither inoculum nor benthic supply, the model cannot maintain a standing stock of phytoplankton, except at extremely low velocities (cf. Margalef 1960). The inoculum is therefore necessary to explain the presence of lentic forms, and while being quite imaginary, it serves as a substitute for the effects of small ponds or impoundments on tributary streams, or possibly backwater areas that mix only slowly with the main river.

Departures from steady-state. Given that real streams and rivers are, strictly speaking, never at steady state, the model results presented above are at best, an approximation. I have concentrated on the steady state solutions partly because spiralling, like cycling, is intuitively a concept involving the steady state. While it is possible to calculate non-steady state spiralling lengths, the fundamental relationship between downstream fluxes and spiralling lengths (eq. 5) is not valid unless the system is at steady state. Here I briefly examine some aspects of non-steady state behavior of the model to indicate how these alter the steady state interpretation.

Most of the model simulations were run using an initial standing stock of 7 mg P m^{-2} in benthic algae, which represents about 12% of the steady state upstream standing stock of benthic algae in the model rivers. Although this is an arbitrary figure, it might represent the effects of scour by a storm, or simply the conditions at the beginning of the growing season. From this initial condition, steady state is approached in 50-60 days, although minor adjustments continue beyond 100 days of simulation. (Initial conditions for PO_4 and phytoplankton had little influence on model dynamics). As shown in Figure 6, downstream PO_4-P concentrations (curve b) in the slow river equilibrate to near their final value in only 10 days, while upstream PO_4-P concentrations (curve a) and downstream phytoplankton (curve c) take much longer to equilibrate. (This is true for both rivers, although only the slow river is shown). The model indicates temporary, but strong depletion of PO_4 in the upstream waters, attributable to phosphorus accumulation in the growing stocks of benthic algae. This upstream depletion does not seriously affect growth of upstream benthic algae (not shown), because minimum PO_4-P > K_s. It does, however, appear to control the development of the downstream phytoplankton (curve c). Downstream benthic algae (curve d) grow rapidly in the first few days before PO_4 is fully depleted and then slowly decline to the equilibrium level.

The 50-60 days required for the river to equilibrate places the temporal dynamics of the river on a time scale that are roughly equivalent to seasonal changes, and, depending on season and geographic location, this

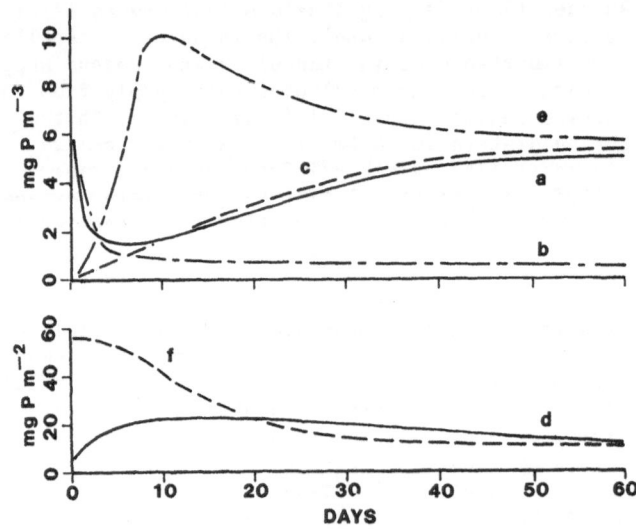

Figure 6. Temporal dynamics of PO_4 and phytoplankton in slow river model. Two cases are represented: (1) response to initial benthic algae stock of 7 mg P m^{-2} (curves a - d); and (2) response to initial benthic algae stock of 57 mg P m^{-2} (curves e - f). Individual curves are: a -- PO_4 at 200 km; c -- phytoplankton at 200 km; d -- benthic algae at 200 km; e -- phytoplankton at 200 km; f -- benthic algae at 200 km.

time scale may also be comparable to the frequency of significant flow fluctuations. In other words, this time scale does not rule out the achievement of steady state, but at the same time, it suggests that departures from steady state will be common.

Regular storms that scour the benthos (Fisher et al. 1982; Bott 1983) might keep the river in an accumulating mode with depressed upstream phosphorus concentrations. This suggests that headwater systems may be more phosphorus limited, and their influence on downstream dynamics greater than is indicated by the steady state model (see discussion of headwater phosphorus limitation below). There is considerable evidence that phosphorus accumulates on stream bottoms at normal flows, and then is flushed downstream in particulate form during storms (Meyer & Likens 1979; Leonard et al. 1979; Rigler 1979; Fisher et al. 1982). Meyer and Likens, however, attributed the accumulation to abiotic sorption rather than to biotic accumulation. There is similar evidence from rivers (Keup 1968; Johnson et al. 1976; Hill 1982), but these studies involve high levels of phosphorus inputs from waste discharges and abiotic, rather than biotic accumulation. Only in the studies of Rigler and of Meyer and Likens were phosphorus concentrations depleted sufficiently to be potentially limiting, and neither of these studies addressed the question of limitation. Mulholland et al. (1985) present indirect evidence that autumn phosphorus accumulation on fresh leaf detritus may produce a nutrient limiting phosphorus depletion. In general, however, the extent to which headwater streams may be nutrient limited via non-steady state nutrient depletion remains very much an open question.

Storms transport large quantities of fine sediments which are known to have a large capacity for inorganic phosphorus sorption (Taylor & Kunishi 1971; Meyer 1980; Hill 1982; Klotz 1985). In a steady state stream or river, sorbed phosphorus or the river bottom would, by definition, be in equilibrium with the ambient P concentration and have no effect. Under non-steady state conditions, however, the sorption/desorption processes may become quite important, either in removing phosphorus (cases cited above), or in providing a buffered supply of phosphorus (Mayer & Gloss 1980; Gloss et al. 1981; Klotz 1985).

The rapid development of phytoplankton in a river or shallow reservoir, with consequent depletion of dissolved P from the water column, could cause desorption of a large quantity of phosphorus from the bottom, and this would further enhance the bloom. A similar effect, but involving anoxic release of phosphorus from the bottom, has been observed in shallow lakes (Riley & Prepas 1984). Phosphorus stored in benthic organisms might similarly enhance a phytoplankton bloom under non-steady state conditions. To examine this question, the slow river was simulated with an initial benthic algae population equivalent to that in the fast river. As the phytoplankton developed, the benthic algae declined (Figure 6, curve f) releasing phosphorus to the water column. The phytoplankton density (Figure 6, curve e) rose rapidly to near double its equilibrium value (i.e, well in excess of C_T), and remained elevated for approximately 50 days. This situation is analogous to the trophic upsurge effect noted in new reservoirs (Ostrofsky & Duthie 1980; Grimard & Jones 1982).

Headwater nutrient limitation. While I have characterized headwater streams as being retentive, non-transporting systems in which nutrients are unlikely to be limiting, this is clearly an oversimplification. The modeled benthic algal populations in the headwaters equilibrate at a level near the maximum allowed by the self-limitation parameter k_2, and do not deplete PO_4 significantly. There is, in this case, a mild degree of phosphorus limitation, in that a large increase in PO_4 produces about a 10% increase in standing stocks. This mild response is consistent with responses observed by Elwood et al. (1981), and also with Bothwell's (1985) observation of no limitation of periphyton at 3-4 mg P m^{-3}. However, much larger responses to phosphorus enrichment have also been observed (Stockner & Shortreed 1978; Peterson et al. 1985), and studies by Wong and Clark (1976) and Horner and Welch (1981) both indicate that stream periphyton may remain phosphorus limited at concentrations exceeding 40 mg P m^{-3}. It seems likely that limitations at these high levels can be explained in terms of diffusion limitation to, and into, the periphytic mat. Wong and Clark (1976) and Horner and Welch (1981) both dealt with relatively dense periphyton growths, while Bothwell (1985) studied thin films. Further indication of an important role for diffusion is the strong influence of current reported by Horner and Welch (1981), and by Whitford and Schumacher (1961, 1964). If such diffusion limitation were incorporated into the model (cf. Fraleigh & Wiegert 1975; Rittmann & McCarty 1978), then the headwater streams would be phosphorus limited, but water column PO$_4$ would not be depleted (at steady state) unless the increased periphyton growth were accompanied by a large increase in sloughing. Without such depletion, phosphorus spiralling would remain in a non-regulatory mode in the headwaters, as postulated by the original model.

Alternatively, it is possible that headwater streams are not so retentive as modeled here, and it seems likely that some, at least, are regulated through transport-mediated phosphorus depletion. That is, the spiralling-regulated mode postulated for the downstream reaches of the slow river may apply to headwater streams as well. As discussed above, this situation could arise from a cyclic alternation between non-steady state accumulation and storm transport. At steady state, it can be simulated in the model by increasing the rate of algal sloughing, and simultaneously lowering k$_2$ (which allows higher algal standing stocks). It is clear that benthic algae can reach much higher densities than were modeled in the hypothetical rivers (e.g., Moore 1977; Fisher et al. 1982; Bott 1983; 60 mg P m^{-2} equates very roughly to 50 mg chl a m^{-2}, and 2.5 g C m^{-2}), and at these can generate high levels of particulate transport (Swanson & Bachmann 1976; Fisher et al. 1982; Peterson et al. 1985). What remains unclear, is whether these high levels of transport can actually deplete available phosphorus to limiting levels. Evidence for this is provided by Peterson et al. (1985) who observed a 10-fold response in benthic algae to a phosphorus enrichment of 10 mg P m^{-3} above an ambient level of <1-4 mg P m^{-3}, which in turn elevated chlorophyll in transport from 1-3 mg m^{-3} to 30 mg m^{-3}. This strongly suggests that nearly all the added P was accounted for in seston transport, although Peterson et al. did not address this question specifically.

Effect of a Reservoir on the River

I have suggested above that slowing a river through regulation by a series of small control dams might induce several downstream changes, including phytoplankton development, nutrient depletion, and depressed nutrient uptake. I now consider the effect of slowing not the whole river, but a portion of it. Figures 7 and 8 represent the steady state model results obtained from putting a reservoir on the faster river with the dam located at 100 km.

Spiralling in the reservoir. Within the reservoir, there is a transition to nutrient limitation and phytoplankton domination over a short distance. Total stock declines in the upper portion of the reservoir, as a result of the drop in benthic algae. In the lower part of the reservoir, however, it increases, because the increased depth supports a larger stock of phytoplankton. Spiralling length (Figure 8) in the reservoir is shorter than in the river. At the upstream end, spiralling shortens because the river widens, providing more habitat per unit length for benthic algae.

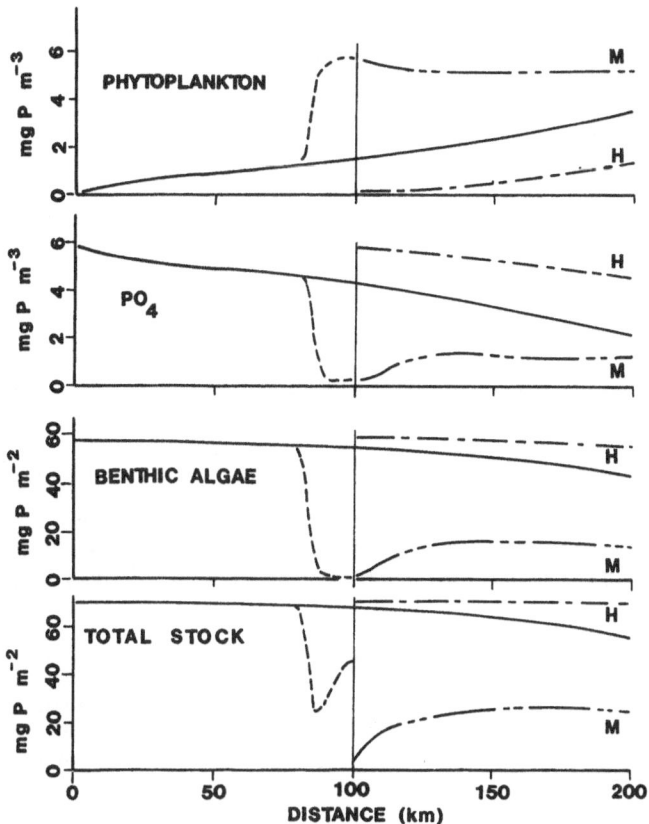

Figure 7. Steady state model solutions for effect of placing a reservoir on the fast river. Solid lines represent the original solution for the free flowing river. The reservoir starts at 78 km with the dam at 100 km. Two cases of downstream effects are represented as: M -- discharge from vertically mixed reservoir; H -- hypolimnetic discharge from a stratified reservoir.

Midway along the length of the reservoir, S increases slightly because the plankton develop and benthic algae decline. Near the dam, under fully lacustrine conditions, spiralling length becomes very short, reflecting the near-zero downstream velocity.

As this "reservoir" model is really no more than the river model with altered dimensions, it clearly leaves out many important characteristics of reservoirs, such as the effects of vertical stratification, density currents, sedimentation, and limiting effects of light penetration. Perhaps the greatest weakness of the model is that it does not account for the effects of reduced velocity on deposition of organic and inorganic materials. Sedimentation in the upstream portions of reservoirs can be important both in enhancing light penetration (Marzolf & Osborne 1972; Soltero & Wright 1975; McCullough 1978) and in trapping phosphorus (Soltero et al. 1975; Whalen et al. 1982; Paulson & Baker 1981). Chapra (1980) provides a simple model for the effects of sedimentation on phosphorus dynamics within a reservoir. Reduction of current can also increase the settling of phytoplankton that depend on riverine turbulence for suspension (Søballe & Bachmann 1984). The model does, however, provide a simple

319

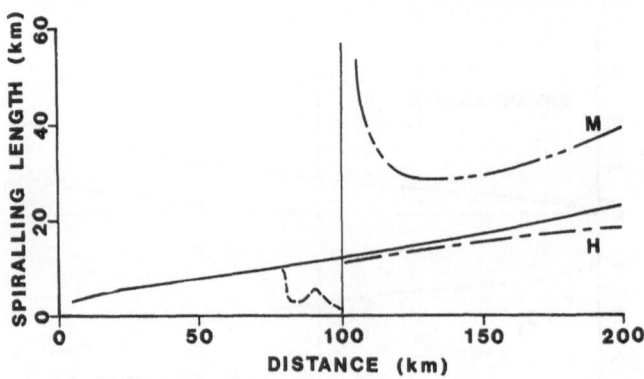

Figure 8. Effect of reservoir on spiralling length. See Figure 7 legend for explanation.

interpretation of spiralling within a lotic-to-lentic transition (cf. Kimmel et al. in press), in which the lentic portion is characterized by the high phytoplankton density and depleted phosphorus, commonly observed in phosphorus-limited lakes and reservoirs (e.g., Elser & Kimmel 1985).

Downstream effects of the reservoir. Figures 7 and 8 also show potential downstream effects of the reservoir. Two cases are shown in addition to the original unimpounded case. The first, labeled M for discharge from a vertically mixed reservoir, represents the downstream effects of discharges from the modeled reservoir. The second case, labeled H, for hypolimnetic discharge represents a case in which a reservoir were to discharge water with a high proportion of the phosphorus in solution, such as might occur in the hypolimnion of a stratified reservoir. No attempt was made to model the internal dynamics of the stratified reservoir. Rather, I simply assumed that the discharged water was of the same quality as the groundwater or headwaters to the river.

The river below the mixed reservoir (M) has low dissolved P, high phytoplankton, low benthic algae and a long spiralling length. In other words, the reservoir on the fast river converts the downstream reaches to a nutrient-limited, transporting mode similar to that of the downstream reaches of the slow river (Figure 2). In terms of Ward and Stanford's (1983) serial discontinuity concept, the effect is to displace the river downstream. It is interesting, however, that the extreme effects of the reservoir discharge attenuate somewhat over the first 25 km downstream: there is a small drop in phytoplankton population, a recovery of PO_4 and benthic algae, and a shortening of the spiralling length. These recovery effects are attributable in part to tributary inflows along the channel and in part to greater settling of phytoplankton in the shallower water (the latter phenomenon, an artifact of the model's constant settling rate, is of arguable validity; e.g., see Søballe & Bachman 1984).

The question of the fate of discharged plankton remains little understood (Kimmel et al. in press). Talling and Rzóska (1967) describe the role of a reservoir on the Blue Nile in providing a sufficient inoculum for subsequent growth in the downstream river, and similar cases have been observed by others (Hartman & Himes 1961; Damann 1951, cited by Ibid.; Greenberg 1964; Young et al. 1972). On the other hand, there are also many reports of downstream decreases in phytoplankton below a dam or lake outfall (Chandler 1937; Cushing 1964; Maciolek & Tunzi 1968). Such

320

decreases have been attributed to the fragility of lentic species in turbulent waters, and to entrapment by macrophytes, macrobenthos, and the epilithon. The model is oblivious to these mechanisms, although it does produce a slight downstream decrease as noted above. In general, decreasing downstream gradients are more widely reported from smaller rivers. A model run in which the upstream inputs to the rivers were plankton dominated (simulating a headwater lake or reservoir; results not shown) did produce substantial downstream declines in phytoplankton, simply as the result of longitudinal dilution by groundwater. This run also indicated that the effect of a small headwater reservoir on a river system would be negligible, even if it generates a considerable inoculum.

The prediction of depressed benthic standing stocks below the mixed release is subject to the same limitations as discussed above for the downstream reaches of the slow river. The situation below a reservoir discharge is further complicated by the issues of flow constancy, thermal alterations, inorganic turbidity, and effects on substrate (e.g., see Baxter 1977; Ward & Stanford 1979; Armitage 1984; Skulberg 1984), and therefore any effects arising from nutrient depletion may be difficult to document. Whether or not benthic primary productivity is depressed, there seems often to be a stimulatory effect on the benthic macroinvertebrates, particularly filter feeders (Armitage 1976; 1984; Henricsen & Müller 1979; Parker & Voshell 1983), which is presumably attributable to the high, nutrient rich, particulate load delivered from the reservoir. The effect of the filter-feeders would be to shorten the very long turnover length, regenerate nutrients, and ease the nutrient limitation. Parker and Voshell (1983) calculated that filter feeders below a dam in the North Anna River (Virginia) ingested sufficient seston to clear the river of seston in 4 km. This distance can be interpreted as an uptake length: it is the average distance a particle would travel downstream before being ingested by a filter-feeder. Average discharge at their North Anna River site is nearly identical to that of the model fast river below the dam, and a rough conversion of their consumption rate to phosphorus produces a very similar estimate of about 4.5 km for a seston uptake length in the fast river. Allowing that only 22% of ingestion was assimilated gives an estimate of 20 km for the distance traveled by a particle before assimilation. Given the calculated turnover length of about 50 km below the dam, it can be seen that this level of filter feeding activity would have a dramatic impact on downstream spiralling.

In terms of the serial discontinuity concept, the case of the hypolimnetic release ("H", Figures 7 and 8) corresponds to a negative or upstream displacement. This was, in effect, the assumption used to generate the results, so it is not a prediction of the model. Below the dam, PO_4 and benthic algae are somewhat higher, and phytoplankton lower, than in the unimpounded river. In relation to benthic stocks below the mixed release or at the same location in the unimpounded slow river (cf. Figure 2), benthic stocks below the hypolimnetic release are quite high. A downstream enriching effect of hypolimnetic releases on benthic productivity has often been observed (Neel 1963; Spence & Hynes 1971; Ward 1976; Marcus 1980; Stanford & Ward 1983). Downstream growth of phytoplankton is slower (in space) than in the headwaters because the benthic algae contribution is less significant in the deeper, faster water.

In a loose sense, these model results reflect Wright's (1967) suggestion that deep release reservoirs increase downstream nutrient supply, while surface release reservoirs tend to trap nutrients. Rather than trapping phosphorus, however, the model used here only alters the form downstream phosphorus, reducing its availability.

CONCLUDING REMARKS

Newbold et al. (1982b) characterized two extreme modes for spiralling in a lotic system, suggesting that real systems lie somewhere in between. At one extreme, the system is highly nutrient retentive, nutrients are neither depleted nor limiting, and therefore the spiralling of a nutrient plays no regulatory role. The headwater reaches of my hypothetical rivers approximate this mode. At the other extreme, the system is not retentive, nutrients are depleted to a limiting level, and therefore spiralling--or internal control over the form of downstream nutrient fluxes--plays a central role in mediating nutrient utilization. The downstream reaches of the slow river, the reservoir, and the fast river below the epilimnetic discharge approximate this mode. In the first mode, spiralling length is dominated by the uptake length, while in the second, it is dominated by the turnover length (cf. eq. 4). In this paper I have attempted to place the question of spiralling in the context of the river continuum, arguing that there may be a longitudinal transition between these modes, and that it is necessary to view this transition as itself a fundamentally important process of the ecosystem. The transition is essentially a longitudinal succession (sensu Fisher 1983) inasmuch as the upstream community serves as the necessary antecedent for the downstream ecosystem; it also represents the interplay between upstream influences and equilibration to the local environment that underlies the river continuum concept (Vannote et al. 1980; Vannote 1981). In addition to arguing that phosphorus spiralling can be important in river ecosystems, I hope I have also made clear that in many cases, it may not be important, for example when phosphorus loading is very high, or turbidity or some other nutrient is the primary limiting factor.

The model suggests that the transition to a nutrient-limited, transporting mode can be induced either through a single impoundment discharging plankton rich, nutrient poor water, or through a series of small regulatory impoundments whose primary effect is to slow the velocity of the river. This has two major practical implications: First, impoundment may reduce benthic activity, and perhaps total (water column plus benthic) productivity, as a result of nutrient depletion. Second, by inducing nutrient limitation, impoundments can produce a system in which response to nutrient loading is much greater than in the pristine benthic-dominated system. The latter phenomenon, at least in an extreme form, is well known: building a reservoir can create the potential for eutrophication which did not exist in the original river. The former (nutrient starvation of the benthos) has not to my knowledge been reported. As pointed out earlier, this phenomenon may easily be obscured or prevented by other factors such as turbidity, changing substratum, and high nutrient loading. Nonetheless, it could well occur without being observed or interpreted as such, and the phenomenon may well deserve further investigation. Although there may be few, if any, real-world instances in which phytoplankton development is responsible for actually decreasing total areal productivity, it does seem reasonable to suggest that some degree of depression of benthic activity may be widespread.

ACKNOWLEDGEMENTS

The support of the Francis Boyer Research Endowment during preparation of this manuscript is gratefully acknowledged. I thank B.L. Kimmel for helpful discussions in the early phases of this work, and L.A. Kaplan for comments that substantially improved the manuscript.

REFERENCES

Armitage, P. D. 1976. A quantitative study of the invertebrate fauna of the River Tees below Cow Green Reservoir. Freshwat. Biol. **6**: 229-240.

Armitage, P. D. 1984. Environmental changes induced by stream regulation and their effect on lotic macroinvertebrate communities. In: Regulated Rivers. (Ed. by A. Lillehammer and S. J. Saltveit). pp. 139-165. Oslo Univ. Press.

Ball, R. C. and Hooper, F. F. 1963. Translocation of phosphorus in a trout stream ecosystem. In: Radioecology. (Ed. by V. Schultz and A. W. Klement, Jr.).pp. 217-228. Reinhold Publ. Corp., New York.

Baxter, R. M. 1977. Environmental effects of dams and impoundments. Ann. Rev. Ecol. Syst. **8**: 255-283.

Bothwell, M. L. 1985. Phosphorus limitation of lotic periphyton growth rates: An intersite comparison using continuous-flow troughs (Thompson River System, British Columbia). Limnol. Oceonogr. **30**: 527-542.

Bott, T. L. 1983. Primary productivity in streams. In: Stream Ecology. Application and testing of general ecological theory. (Ed. by J. R. Barnes and G. W. Minshall). pp. 29-53. Plenum, N. Y.

Chandler, D. C. 1937. Fate of typical lake plankton in streams. Ecol. Monogr. **7**: 445-479.

Chapra, S. C. 1980. Application of phosphorus loading models to river-run lakes and other incompletely mixed systems. In: Restoration of Lakes and Inland Waters. pp. 329-334. USEPA 440/55-81-010.

Crayton, W. M. and Sommerfeld, M. R. 1981. Impacts of a desert impoundment on the phytoplankton community of the lower Colorado River. pp. 1608-1617. In: Symposium on surface water impoundments, June 2-5, 1980, Minneapolis, MN. (Ed. by H. G. Stefan). American Society of Civil Engineers, New York. 1682 pp.

Cushing, C. E. 1964. Plankton and water chemistry in the Montreal River lake-stream system, Saskatchewan. Ecology **45**: 306-313.

Dĕcamps, H. J., Capblancq, and Tourenq, J. N. 1984. Lot. In: Ecology of European Rivers. (Ed. by B. A. Whitton). pp. 207-235. Blackwell.

Duffer, W. R. and Dorris, T. C. 1966. Primary productivity in a southern Great Plains stream. Limnol. Oceanogr. **2**: 143-151.

Elser, J. J. and Kimmel, B. L. 1985. Nutrient availability for phytoplankton production in a multiple-impoundment series. Can. J. Fish. Aquat. Sci. **42**: 1359-1370.

Elwood, J. W., Newbold, J. D., Trimble, A. F. and Stark, R. W. 1981. The limiting role of phosphorus in a woodland stream ecosystem: Effects of P enrichment on leaf decomposition and primary producers. Ecology **62**: 146-158.

Elwood, J. W., Newbold, J. D., O'Neill, R. V. and Van Winkle, W. 1983. Resource spiralling: An operational paradigm for analyzing lotic ecosystems. In: The dynamics of lotic ecosystems. (Ed. by T. D. Fontaine, III., and S. M. Bartell) pp. 3-27. Ann Arbor Science.

Fisher, S. G. 1983. Succession in streams. In: Stream ecology. Application and testing of general ecological theory. (Ed. by J. R. Barnes and G. W. Minshall). pp. 7-27. Plenum, N.Y.

Fisher, S. G., Gray, L. J., Grimm, N. B. and Busch, D. E. 1982. Temporal succession in a desert stream ecosystem following flash flooding. Ecol. Monogr. **52**: 93-110.

Fraleigh, P. C. and Wiegert, R. G. 1975. A model explaining successional change in standing crop of thermal blue-green algae. Ecology **56**: 656-664.

Gloss, S. P., Reynolds, R. C., Jr., Mayer, L. M. and Kidd, D. E. 1981. Reservoir influences on salinity and nutrient fluxes in the arid Colorado River basin. In: Symposium on surface water impoundments, June 2-5, 1980, Minneapolis, MN. (Ed. by H. G. Stefan).

pp. 1618–1629. American Society of Civil Engineers, New York.

Greenberg, A. E. 1964. Plankton of the Sacramento River. Ecology **45**: 40–49.

Grimard, Y. and Jones, H. G. 1982. Trophic upsurge in new reservoirs: A model for total phosphorus concentrations. Can. J. Fish. Aquat. Sci. **39**: 1473–1483.

Hartman, R. T. and Himes, C. L. 1961. Phytoplankton from Pymatuning Reservoir in downstream areas of the Shenango River. Ecology **42**: 180–183.

Henricson, J. and Müller, K. 1979. Stream regulation in Sweden with some examples from Central Europe. In: The Ecology of Regulated Streams. (Ed. by J. V. Ward and J. A. Stanford). pp. 183–199. Plenum, New York.

Hill, A. R. 1982. Phosphorus and major cation mass balances for two rivers during low summer flows. Freshwat. Biol. **12**: 293–304.

Horner, R. R. and Welch, E. B. 1981. Stream periphyton development in relation to current velocity and nutrients. Can. J. Fish. Aquat. Sci. **38**: 449–457.

Hynes, H. B. N. 1970. The ecology of running waters. University of Toronto Press.

Johnson, A. H., Bouldin, D. R., Goyette, E. A. and Hedges, A. M. 1976. Phosphorus loss by stream transport from a rural watershed: Quantities, processes and sources. J. Environ. Qual. **5**: 148–157.

Keup, L. E. 1968. Phosphorus in flowing waters. Water Research **2**: 373–386.

Kilham, S. S. 1978. Nutrient kinetics of freshwater planktonic algae using batch and semi-continuous culture methods. Mitt. Int. Ver. Theor. Angew. Limnol. **21**, p. 147–157.

Kimmel, B. L., Lind, O. T. and Paulson, L. J. In press. Reservoir primary production. In: Perspectives on reservoir ecosystems. (Ed. by K. W. Thornton). John Wiley and Sons, New York.

Klotz, R. L. 1985. Factors controlling phosphorus limitation in stream sediments. Limnol. Oceanogr. **30**: 543–553.

Lack, T. J. 1971. Quantitative studies on the phytoplankton of the Rivers Thames and Kennet at Reading. Freshwat. Biol. **1**: 213–224.

Lean, D. R. S. 1973. Movements of phosphorus between its biologically important forms in lake water. J. Fish. Res. Board Can. **30**: 1525–1536.

Lean, D. R. S. and Nalewajko, C. 1976. Phosphate exchange and organic phosphorus excretion by freshwater algae. J. Fish. Res. Board Can. **33**: 1312–1323.

Lean, D. R. S. and Nalewajko, C. 1979. Phosphorus turnover time and phosphorus demand in large and small lakes. Arch. Hydrobiol. Beih. Ergebn. Limnol. **13**: 120–132.

Lean, D. R. S. and Pick, F. R. 1981. Photosynthetic response of lake plankton to nutrient enrichment: A test for nutrient limitation. Limnol. Oceanogr. **20**: 1001–1019.

Lehman, J. T., Botkin, D. B. and Likens, G. E. 1975. The assumptions and rationale of a computer model of phytoplankton dynamics. Limnol. Oceanogr. **20**: 343–364.

Leonard, R. L., Kaplan, L. A., Elder, J. F., Coats, R. N. and Goldman, C. R. 1979. Nutrient transport in surface runoff from a subalpine watershed, Lake Tahoe Basin, California. Ecol. Monogr. **49**: 281–310.

Leopold, A. 1941. Lakes in relation to terrestrial life patterns. In: A symposium on hydrobiology. pp. 17–22. Univ. Wisc. Press. Madison.

Leopold, L. B., Wolman, M. G. and Miller, J. P. 1964. Fluvial processes in geomorphology. W. H. Freeman, San Francisco.

Likens, G. E. and Bormann, F. H. 1974. Linkages between terrestrial and aquatic ecosystems. Bioscience **24**: 447–456.

Maciolek, J. A. and Tunzi, M. G. 1968. Microseston dynamics in a simple Sierra Nevada lake-stream system. Ecology **49**: 60–75.

Marcus, M. D. 1980. Periphytic community response to chronic nutrient enrichment by a reservoir discharge. Ecology **61**: 387–399.

Margalef, R. 1960. Ideas for a synthetic approach to the ecology of running waters. Int. revue ges. Hydrobiol. **45**: 133–153.

324

Marzolf, G. R. and Osborne, J. A. 1972. Primary production in a Great Plains reservoir. Verh. Int. Ver. Limnol. **18**: 126-133.

Mayer, L. M. and Gloss, S. P. 1980. Buffering of silica and phosphate in a turbid river. Limnol. Oceanogr. **25**: 12-22.

McCullough, J. D. 1978. A study of phytoplankton primary productivity and nutrient concentrations in Livingston Reservoir, Texas. Tex. J. Sci. **30**: 377-387.

Megard, R. O. 1981. Effects of planktonic algae on water quality in impoundments of the Mississippi River in Minnesota. In: Symposium on surface water impoundments, June 2-5, 1980, Minneapolis, MN. (Ed. by H. G. Stefan). pp. 1575-1584. American Society of Civil Engineers, New York.

Meyer, J. L. 1980. Dynamics of phosphorus and organic matter during leaf decomposition in a forest stream. Oikos **34**: 44-53.

Meyer, J. L. and Likens, G. E. 1979. Transport and transformation of phosphorus in a forest stream ecosystem. Ecology **60**: 1255-1269.

Minshall, G. W., Petersen, R. C., Cummins, K. W., Bott, T. L., Sedell, J. R., Cushing, C. E. and Vannote, R. L. 1983. Interbiome comparison of stream ecosystem dynamics. Ecol. Monogr. **53**: 1-25.

Moore, J. W. 1977. Some factors effecting algal densities in a eutrophic farmland stream. Oecologia **29**: 257-267.

Mulholland, P. J., Newbold, J. D., Elwood, J. W., Ferren, L. A. and Webster, J. R. 1985. Phosphorus spiralling in a woodland stream: seasonal variations. Ecology **66**: 1012-1023.

Mulholland, P. J., Newbold, J. D., Elwood, J. W. and Hom, C. L. 1983. The effect of grazing intensity on phosphorus spiralling in autotrophic streams. Oecologia **58**: 358-366.

Neel, J. K. 1963. Impact of reservoirs. In: Limnology in North America. (Ed. by D. G. Frey). pp. 575-593. Madison, Wisconsin.

Newbold, J. D., Elwood, J. W., O´Neill, R. V. and Sheldon, A. L. 1983. Phosphorus dynamics in a woodland stream ecosystem: A study of nutrient spiralling. Ecology **64**: 1249-1265.

Newbold, J. D., Elwood, J. W., O´Neill, R. V. and Van Winkle, W. 1981. Measuring nutrient spiralling in streams. Can. J. Fish. Aquat. Sci. **38**: 860-863.

Newbold, J. D., Mulholland, P. J., Elwood, J. W. and O´Neill, R. V. 1982a. Organic carbon spiralling in stream ecosystems. Oikos **38**: 266-272.

Newbold, J. D., O´Neill, R. V., Elwood, J. W. and Van Winkle, W. 1982b. Nutrient spiralling in streams: Implications for nutrient limitation and invertebrate activity. Amer. Natur. **120**: 628-652.

Ostrofsky, M. L. and Duthie, H. C. 1980. Trophic upsurge and the relationship between phytoplankton biomass and productivity in Smallwood Reservoir, Canada. Can. J. Bot. **58**: 1174-1180.

Parker, C. R. and Voshell, J. R., Jr. 1983. Production of filter-feeding Trichoptera in an impounded and a free-flowing river. Can. J. Zool. **61**: 70-87.

Paulson, L. J. and Baker, J. R. 1981. Nutrient interactions among reservoirs on the Colorado River. In: Symposium on surface water impoundments, June 2-5, 1980, Minneapolis, MN. (Ed. by H. G. Stefan). pp. 1647-1656. American Society of Civil Engineers, New York.

Peters, R. H. 1979. Concentrations and kinetics of phosphorus fractions along the trophic gradient of Lake Memphremagog. J. Fish. Res. Board Can. **36**: 970-979.

Peterson, B. J., Hobbie, J. E., Hershey, A. E., Lock, M. A., Ford, T. E., Vestal, J. R., McKinley, V. L., Hullar, M. A. J., Miller, M. C., Ventullo, R. M. and Volk, G. S. 1985. Transformation of a tundra river from heterotrophy to autotrophy by addition of phosphorus. Science **229**: 1383-1386.

Prepas, E. E. 1983. Orthophosphate turnover time in shallow productive lakes. Can. J. Fish. Aquat. Sci. **40**: 1412-1418.

Rhee, G.-Y. 1973. A continuous culture study of phosphate uptake, growth rate and polyphosphate in *Scenedesmus* sp. J. Phycol. **9**: 495-506.

Rigler, F. H. 1973. A dynamic view of the phosphorus cycle in lakes. In: Environmental phosphorus handbook. (Ed. by E. J. Griffith et al.). pp. 539-572. Wiley, New York.

Rigler, F. H. 1979. The export of phosphorus from Dartmoor catchments: A model to explain variations of phosphorus concentrations in streamwater. J. mar. biol. Assoc. U. K. **59**: 659-687.

Riley, E. T. and Prepas, E. E. 1984. Role of internal phosphorus loading in two shallow, productive lakes in Alberta, Canada. Can. J. Fish. Aquat. Sci. **41**: 845-855.

Rittmann, B. E. and McCarty, P. L. 1978. Variable-order model of bacterial-film kinetics. J. Environ. Engin. Div., ASCE **104**: 889-899.

Ruttner, F. 1963. Fundamentals of Limnology. Translated by D. G. Frey and F. E. J. Fry. Univ. of Toronto Press. 295 pp.

Skulberg, O. M. 1984. Effects of stream regulation on algal vegetation. In: Regulated rivers. (Ed. by A. Lillehammer and S. J. Saltveit). pp. 107-124. Oslo Univ. Press.

Søballe, D. M. and Bachmann, R. W. 1984. Influence of reservoir transit on riverine algal transport and abundance. Can. J. Fish. Aquat. Sci. **41**: 1803-1813.

Soltero, R. A., Gasperino, A. F. and Graham, W. G. 1975. Chemical and physical characteristics of a eutrophic reservoir and its tributaries: Long Lake, Washington--II. Water Resour. Res. **9**: 1059-1064.

Soltero, R. A. and Wright, J. C. 1975. Primary production studies on a new reservoir; Bighorn Lake - Yellowtail Dam, Montana, U. S. A. Freshwat. Biol. **5**: 407-421.

Spence, J. A. and Hynes, H. B. N. 1971. Differences in benthos upstream and downstream of an impoundment. J. Fish. Res. Board Can. **28**: 35-43.

Stanford, J. A. and Ward, J. V. 1983. The effects of mainstream dams on physicochemistry of the Gunnison River, Colorado. In: Aquatic resources management of the colorado River Ecosystem. (Ed. by V. D. Adams and V. A. Lamarra). pp. 43-56. Ann Arbor Science Publishers.

Stockner, J. G. and Shortreed, K. R. S. 1978. Enhancement of autotrophic production by nutrient addition in a coastal rainforest stream on Vancouver Island. J. Fish. Res. Board Can. **35**: 28-34.

Swale, E. M. F. 1964. A study of the phytoplankton of a calcareous river. J. Evol. **52**: 433-446.

Swanson, C. D. and Bachmann, R. W. 1976. A model of algal exports in some Iowa streams. Ecology **57**: 1076-1080.

Talling, J. F. and Rzoska, J. 1967. The development of plankton in relation to hydrological regime in the Blue Nile. J. Ecol. **55**: 637-662.

Taylor, A. W. and Kunishi, H. M. 1971. Phosphate equilibria on stream sediment and soil in a watershed draining an agricultural region. J. Agr. Food Chem. **19**: 827-831.

Tilman, D. and Kilham, S. S. 1976. Phosphate and silicate growth and uptake kinetics of the diatoms *Asterionella formosa* and *Cyclotella meneghiniana* in batch and semi-continuous culture. J. Phycol. **12**: 375-383.

Vannote, R. L., Minshall, G. W., Cummins, K. W., Sedell, J. R. and Cushing, C. E. 1980. The river continuum concept. Can. J. Fish. Aquat. Sci. **37**: 130-137.

Vannote, R. L. 1981. The River Continuum: A theoretical construct for analysis of river ecosystems. In: Proceedings of the National Symposium on Freshwater Inflow to Estuaries. Volume II. U. S. Department of the Interior. Fish and Wildlife Service. Biological Services Program. FWS/OBS-81/04 pp. 289-304.

Wallace, J. B., Webster, J. R. and Woodall, W. R. 1977. The role of filter feeders in flowing waters. Arch. Hydrobiol. **79**: 506-532.

Ward, J. V. 1976. Comparative limnology of differentially regulated sections of a Colorado mountain river. Arch. Hydrobiol. **78**: 319-342.

Ward, J. V. and Stanford, J. A. 1979. Ecological factors controlling stream zoobenthos with emphasis on thermal modification of regulated streams. In: The ecology of regulated streams. (Ed. by J. V. Ward and J. A. Stanford). pp. 35-55. Plenum Press, New York.

Ward, J. V. and Stanford, J. A. 1983. The serial discontinuity concept of lotic ecosystems. In: The dynamics of lotic ecosystems. (Ed. by T. D. Fontaine, III., and S. M. Bartell). pp. 29-41. Ann Arbor Science.

Watts, R. J. and Lamarra, V. A. 1983. The nature and availability of particulate phosphorus to algae in the Colorado River, Southeastern Utah. In: Aquatic resources management of the Colorado River Ecosystem. (Ed. by V. D. Adams and V. A. Lamarra). pp. 161-180. Ann Arbor Science Publishers.

Webster, J. R. 1975. Analysis of potassium and calcium dynamics in stream ecosystems on three southern Appalachian watersheds of contrasting vegetation. Ph.D. thesis, University of Georgia, Athens. 232 pp.

Webster, J. R. and Patten, B. C. 1979. Effects of watershed perturbation on stream potassium and calcium dynamics. Ecol. Monogr. **49**: 51-72.

Whalen, S. C., Leathe, S. A., Gregory, R. W. and Wright, J. C. 1982. Physicochemical limnology of the Tongue River Reservoir, Montana. Hydrobiologia **89**: 161-176.

Whitford, L. A. and Schumacher, G. J. 1961. Effect of current on mineral uptake and respiration by a fresh-water alga. Limnol. Oceanogr. **6**: 423-425.

Whitford, L. A. and Schumacher, G. J. 1964. Effect of a current on respiration and mineral uptake in *Spirogyra* and *Oedogonium*. Ecology **45**: 168-170.

Wong, S. L. and Clark, B. 1976. Field determination of the critical nutrient concentrations for *Cladophora* in streams. J. Fish. Res. Board Can. **33**: 85-92.

Wright, J. C. 1967. Effects of impoundments on productivity, water chemistry, and heat budgets of rivers. In: Reservoir fisheries resources. p. 188-199. Symp. Amer. Fish. Soc., Spec. Publ.

Young, W. C., Hannan, H. H. and Tatum, J. W. 1972. The physicochemical limnology of a stretch of the Guadalupe River, Texas, with five main-stream impoundments. Hydrobiologia **40**: 297-319.

TRADEOFFS BETWEEN STREAM REGULATION AND POINT SOURCE

TREATMENTS IN COST-EFFECTIVE WATER QUALITY MANAGEMENT

Gary E. Hauser and Richard J. Ruane

Water Systems Development Branch
Tennessee Valley Authority, Norris, Tennessee USA
Water Quality Branch
Tennessee Valley Authority, Chattanooga, Tennessee USA

INTRODUCTION

In some heavily developed areas, treatment of municipal and industrial wastewaters to desired effluent standards still does not allow stream water quality criteria to be met. If the water resource in these areas is developed with upstream reservoirs capable of streamflow regulation, a number of management alternatives may be available in addition to more stringent controls on point source waste discharges. This paper addresses the significant role that unsteady water quality models can play in evaluating tradeoffs between improvement strategies involving stream flow regulation and point sources effluent control.

This project supports a joint study by the U.S. Environmental Protection Agency (EPA) and the Tennessee Valley Authority (TVA) designed to identify the cost-effectiveness of different improvement strategies for obtaining various dissolved oxygen (DO) levels. EPA is currently studying an innovative concept referred to as the "water bubble" (Industrial Economics, Inc. 1984a). To achieve water quality goals at the least cost the bubble approach considers several potential pollution abatement options in a specific stream reach. The concept includes possibilities for trading of effluent loads between dischargers. Dischargers with high treatment costs would treat less, but would pay those with lower treatment costs to treat more. Thus, total BOD load to stream is unchanged but total cost is lower. Thus, with some compensation between dischargers (or other participants) for the new treatment levels, dischargers can comply with the same total load limit at a lower total cost. If a participant other than 2 discharger can improve DO by, say, pulsing flow or aerating, then that participant could be paid to pulse or aerate, improving DO at a total cost that is lower.

The study was also conducted in conjunction with TVA efforts to further enhance water quality downstream from TVA dams. The TVA Act directs TVA to aid in the proper use, conservation, and development of the Valley's resources to promote orderly and proper physical, economic, and social development of the region. The Act further directs TVA to regulate streamflow in the operation of its dams and reservoirs primarily for promoting navigation and controlling floods and, so far as may be consistent with such purposes, to generate hydroelectric power. TVA has a stated goal of demonstrating that resource conservation and environmental quality are not only compatible with but essential to economic development.

The cooperative efforts of these agencies resulted in initiation and completion of an exploratory study of water quality improvement strategies for a 30-kilometer reach of the Holston and South Fork Holston Rivers near Kingsport, Tennessee.

The 30 km study reach extends from South Fork Holston River Kilometer (SFHRKm) 13.1 to Holston River Kilometer (HRKm) 211.6 (Figure 1). The reach is regulated by the Fort Patrick Henry Dam at SFHRKm 13.1. Industrial and municipal waste discharges from Kingsport enter the Holston and South Fork Holston Rivers. The major dischargers are Tennessee Eastman Company (TEC), Mead Paper Company (Mead), Kingsport Publicly Owned Treatment Works (KPOTW), and the Holston Army Ammunition Plant (HAAP). At HRKm 228.8, the South Fork is joined by the smaller, unregulated North Fork Holston River that normally provides significant dilution of South Fork flow. A DO sag to levels below stream water quality criteria can develop under certain low flow conditions in the first few kilometers below the confluence of the North and South Forks. Some DO recovery is evident at the downstream end of the study reach.

Using DO as the primary management variable, the following improvement strategies were explored:

1) further restriction on discharge of oxygen-demanding wastes;
2) varying flow regimes with Fort Patrick Henry Dam;
3) turbine aeration at Fort Patrick Henry Dam;
4) instream oxygen injection; and
5) combinations of the above.

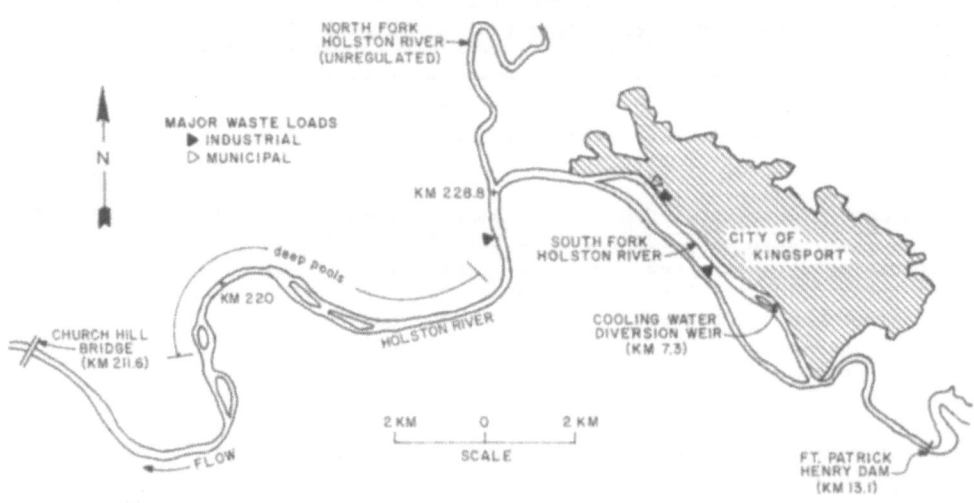

Figure 1. Location Map.

The study objective was to evaluate what mix of methods would achieve desired DO concentrations at key times and locations at the lowest total annual (incremental) cost.

MODELING APPROACH

Description of Mathematical Model

Response of the DO regime to improvement strategies was predicted by a mathematical model calibrated with field data and process rates from previous field surveys by Ruane (unpublished data) and Raschke (1978). The modeling system consisted of an unsteady flow model and an unsteady mass-transport water quality model (Hauser & Ruane 1985). The flow model provided flows, velocities, depths and surface widths at short time intervals for the water quality model. The water quality model predicted temperature, ultimate carbonaceous and nitrogenous biochemical oxygen demands (CBOD and NBOD, respectively), and DO. Temperatures were computed using a heat budget described by Ford et al. (1981). Solar radiation was an input to the model and was modified for shading by trees and banks according to Jobson (1980). Short time intervals were used to simulate diurnal DO variations. Modeled sources and sinks of DO included upstream and lateral inflow, natural reaeration, macrophyte photosynthesis and respiration, CBOD, NBOD, and residual oxygen demand of organic sediments (SOD).

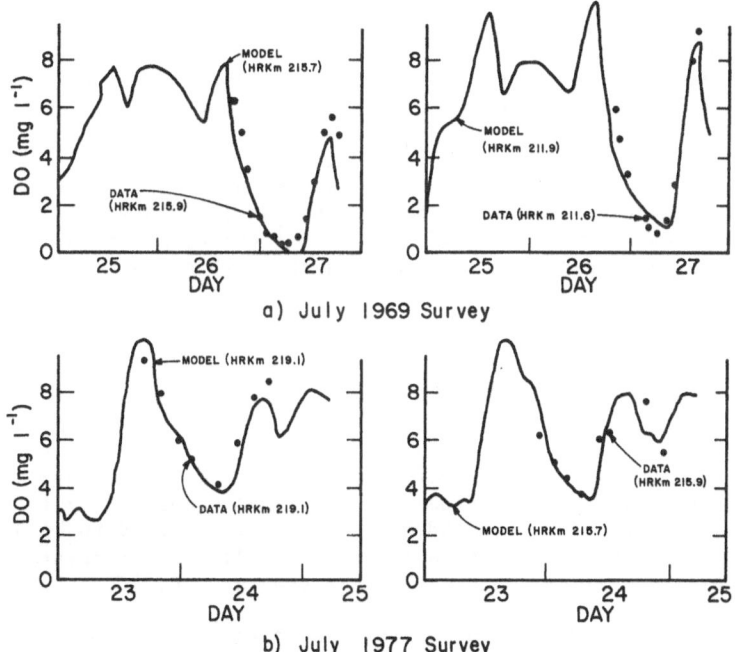

a) July 1969 Survey

b) July 1977 Survey

Figure 2. Calibration of model to field data.

Model Calibration

Figures 2a and 2b illustrate typical calibration results at stations near the DO sag using data measured during two-day low flow periods in 1969 and 1977. Model results are shown as continuous lines and field data are shown as discrete points. The extreme temporal variability in DO illustrated in these figures results primarily from transient flows and photosynthesis/ respiration by aquatic macrophytes. During late summer low flow conditions, minimum DO concentrations occur in early morning hours due to nighttime respiration without any compensating photosynthetic oxygen production. Provided flows remain steady, maximum DO concentrations occur concurrent with or just following peak light levels in mid-afternoon.

Format for Presentation of Model Results

For each improvement strategy, time series of water surface elevation, flow, temperature, DO, NBOD, and CBOD were computed by the model at 80 locations along the 30 km study reach. To reduce the volume of time-varying simulation output necessary to compare different improvement strategies, each simulation was summarized as:

   a) daily mean DO versus river mile; and
   b) six-hour minimum DO versus river mile.

For brevity, only type a) outputs are displayed in the following discussion.

Base Case Assumptions

The calibrated model was used to prepare a base simulation of current conditions with no improvement strategies. The base case represented a critical DO scenario that assumed simultaneous low flow on the North and South Forks of the Holston River, low DO and high temperature releases from Fort Patrick Henry Dam, moderately low solar radiation, and municipal and industrial waste loads at currently permitted levels. Each component of the base case is described below.

Base case flow on the North Fork was assumed to be the 3-day 20-year low flow of 1.5 m^3 s^{-1}. Base case flow on the South Fork was assumed to be minimum daily average flow provided by Fort Patrick Henry Dam. In recent years, a minimum daily average of 21.2 m^3 s^{-1} has been provided to satify cooling water requirements of a local industry. This was achieved by pulsing with one turbine (about 85 m^3 s^{-1}) for one hour every fourth hour. This pulsing schedule sustains minimum flows between all generating periods.

Water quality of flows from Fort Patrick Henry Dam and the North Fork Holston River were presumed constant over time. Fort Patrick Henry release temperature was assumed to be a summer high value of $17°C$ and release DO was assumed to be a summer low value of 3.0 mg l^{-1}. Temperature and DO of the unregulated North Fork inflow were set at $27°C$ and 7.0 mg l^{-1} respectively, while background CBOD and NBOD were assumed to be 3.0 mg l^{-1} amd 0.05 mg l^{-1}. Temperature of a 19 m^3 s^{-1} cooling water return downstream of the diversion weir at SFHRKm 7.3 was a constant $30.5°C$.

Permitted loads were used for the base case because they represent the upper limit of waste loads expected in the study area. Permits are normally established for an allowable monthly average load and a maximum daily load. The maximum day permitted level is the heaviest loading, but it is improbable that all dischargers would achieve maximum levels simultaneously. Therefore, we simulated one of the larger dischargers at

Table 1. Net Industrial and Municipal BOD (5-day) Discharges in the Study Area (kg day^{-1}).

		TEC	MEAD	KPOTW	HAAP	TOTAL
	SFHRKm	5.6	3.7	3.5		
	HRKm				227.9	
Permitted (current)						
Maximum day (May-Sept)		3850	2720*	2120	735	9425
Monthly average (May-Sept)		1815*	1590	1060*	865*	4830
Actual (current)						
Monthly average (May-Sept)		700	1320	270	100	2390

*Base case loadings (total = 5960 kg day^{-1})

maximum the daily permitted load and all others at a monthly average (Table 1).

Typical July meteorologic conditions were assumed, except solar radiation which was taken from a moderately dull day. Actual daily solar radiation exceeds this level 4 out of 5 days in the summer. Lower light yields less photosynthesis and lower afternoon DO maxima.

Base Case Simulation

Before simulating DO improvement strategies, it was important to understand the magnitude of each oxygen-demanding process contributing to the DO sag. Using the calibrated model the relative influence of each DO sink is shown in Figure 3.

Results of the base case simulation are shown as the lower line. Negative river kilometers indicate distance below the confluence of the North and South Forks. Moving downstream from Fort Patrick Henry Dam, daily average DO in the base case reaerated from 3.0 mg l^{-1} to around 4.5 mg l^{-1} near the diversion weir at SFHRKm 7.3. The small dip in DO at this location was due to a stagnating effect of the large industrial cooling water withdrawal from the diversion weir pool. The model predicted a significant drop in saturation DO just below the weir as cooling water returned at a temperature elevated 10 C.

Downstream, another pool extending from HRKm 227.1 to a shoals section at HRKm 221.9 contributed greatly to DO depletion because of the assumed high SOD and long residence time. The model predicted a 1.0 mg l^{-1} recovery in DO across the shoals due to natural reaeration, a value insensitive to the exact form of the reaeration formulation used. In the base case, a DO minimum of 2.5 mg l^{-1} (daily mean) occurred at HRKm 221.9 and a minimum of 1.0 mg l^{-1} (six-hour minimum) occurred at HRKm 217.6.

On subsequent model runs DO sinks were removed in the following order: waste loads, SOD, photosynthesis and respiration of macrophytes, and background CBOD and NBOD. The remaining deficit was due to insufficient cumulative reaeration to overcome the large upstream deficit. Under base case conditions, 30% of the deficit in daily mean DO at the predicted sag was due to waste loads, 25% was due to weed respiration, 20% because of

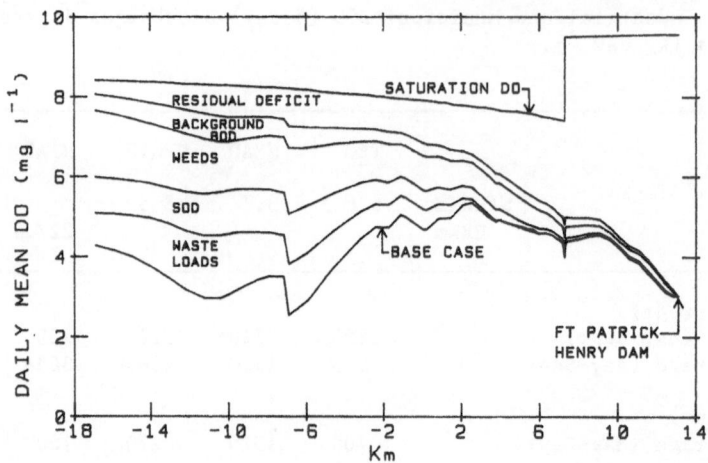

Figure 3. Sequential removal of modeled DO demands to show relative influences.

SOD, and 25% due to background BOD and the residual deficit. The weeds played a greater role in the 6 h minimum DO.

Measures of Effectiveness and Cost

The effectiveness of each improvement strategy was measured by comparing its simulated DO minimum to that of the base case. Incremental annual costs (1983 dollars) were determined for each improvement strategy by Industrial Economics, Inc. (1984b) using capital, operating, and maintenance costs submitted by each relevant participant. Options involving flow modifications or aeration at Fort Patrick Henry Dam required information on cost of power losses due to departure from optimal release patterns and decreased efficiency during turbine aeration. Costs associated with changes in upstream reservoir levels were not evaluated.

SIMULATION RESULTS

Effect of Altered Waste Loads on Dissolved Oxygen

In addition to providing cost estimates, each discharger provided an estimated percentage CBOD and NBOD decrease associated with each potential scheme for upgraded treatment. The sum of all the treatments reduced the total current CBOD load from 5960 kg day^{-1} (base case) to 2590 kg day^{-1}.

Figure 4a shows the improvement in DO from implementing improved treatment at all four industries simultaneously and from the best single treatment. None of the individual treatments improved mean or minimum DO at the sag by over 0.4 mg l^{-1}. Simultaneous treatments improved the mean and minimum DO by about 0.9 mg l^{-1}.

The upper limit of DO improvement achievable with point source treatments would result from the achievement of zero waste discharge at each industrial site. To simulate this, all but background CBOD and NBOD sources were eliminated from the model and the SOD downstream from the industrial sites was assumed not to change. The minimum DO at the sag increased to 3.8 mg l^{-1} (daily mean) and 2.7 mg l^{-1} (6 h minimum) as a result.

Effects of Altered Flow Patterns on Dissolved Oxygen

Flows from both the North and South Forks of the Holston River affect DO below Kingsport. Using Fort Patrick Henry Dam, flows on the South Fork can be increased to compensate for low flows on the unregulated North Fork at some cost for offpeak generation and impact on upstream reservoir levels.

The options considered for augmenting flow included increasing Fort Patrick Henry pulsing frequency to once every third hour, adding flow to the current fourth-hour pulses, and adding pulses between the fourth-hour pulses. The latter proved most effective because pulses could be timed to minimize offpeak generation and still provide flow at the sag during morning hours when DO was low. As shown in Figure 4b, adding a one-hour pulse just prior to the normal pulse at 9 p.m. increased the daily flow from 21.2 to 24.7 m^3 s^{-1} and increased the daily mean DO by 1 mg l^{-1} at the sag point. Two and three hour pulses (increasing flow 3.5 y m^3 s^{-1} for each hour) were less effective.

To meet daily average DO objectives of 3 to 5 mg l^{-1}, with a 6 h minimum of 2 mg l^{-1}, flow additions would be required some 10 to 30 days each year, depending upon variability in North Fork tributary flows. During non-drought years, the three flow pulsing options would result in combined drawdown of 0.15, 0.6, and 1.5 m, respectively, in upstream storage reservoirs.

Effects of Artificial Aeration on Dissolved Oxygen

Two types of artificial aeration were simulated: aeration of Fort Patrick Henry releases and instream aeration near the DO sag. Aeration of dam releases was simulated by increasing upstream boundary DO concentrations. Instream aeration involved adding oxygen as a mass loading (kg day^{-1}) between selected model nodes only during critical hours.

Figure 4c shows the effects of aerating releases from the base case level to 5 mg l^{-1} (technically feasible) and 8 mg l^{-1} (technically possible). In each case, the DO improvement from aeration at the dam was diminished by 90% at the DO sag point. This diminished improvement in the downstream direction is due to differences in the rate of natural reaeration. Because the reaeration rate is proportional to the DO deficit below saturation (about 9 mg l^{-1}), DO in the more deficient cases reaerates more quickly. The DO lines for all cases therefore converged in the lower river reaches.

Figure 4d shows the effect of adding oxygen at a rate of 7260 kg day^{-1} for a 12-hour period (net 3630 kg day^{-1}) at two locations above the DO sag. Trial simulations indicated that aerating between the hours of 6 p.m. and 6 a.m. provided the greatest DO improvement. Similar to aerating dam releases, DO improvement from instream aeration rapidly diminished downstream from the aeration source. A six-hour minimum DO of 2 mg l^{-1} was simulated with either one or two aerators (net 3630 kg day^{-1} each), and mean DO at the sag was improved by 1.5 mg l^{-1} with one aerator and 2.5 mg l^{-1} with two.

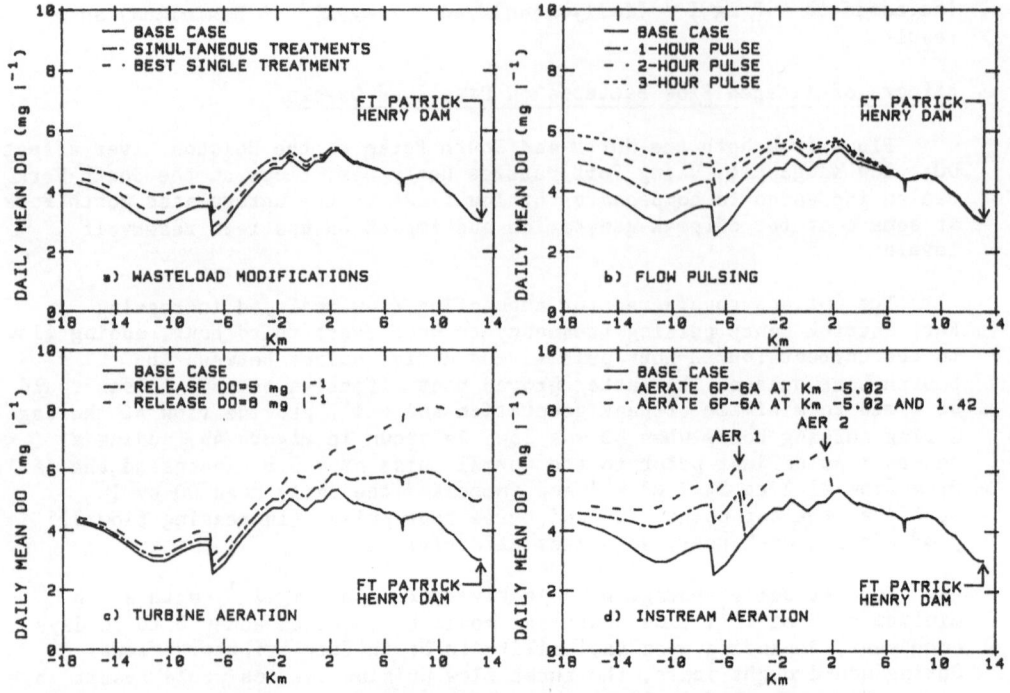

Figure 4. Simulated DO improvement strategies.

CONCLUSIONS

Insights on Water Quality in the Study Area

Significant improvements in municipal and industrial waste treatment in the Kingsport area have upgraded Holston River water quality since the early 1970s. Although EPA effluent based standards have been achieved or surpassed, stream water quality criteria have not been consistently attained, causing a water quality constraint on economic growth.

A DO sag was predicted to develop three to seven miles below the confluence of the North and South Forks of the Holston River under base case conditions that assumed simultaneous low flow on both, low DO and high temperature releases from the upstream dam, moderately low solar radiation, and municipal and industrial waste loads at permitted levels. Model simulations indicated that one-half of the DO deficit under base case conditions was from SOD and respiration of aquatic weeds, one-third was due to waste loads, and the remainder was from background BOD and residual deficit from insufficient reaeration of the dam releases. DO improvements were found to be required from 10 to 30 days a year.

Many creative proposals for improvement strategies were stimulated by insight gained using the model to improve understanding of the water quality in the study area. With the unsteady model, it was possible to identify critical windows in time at various locations downstream when important water quality conditions existed, such as early morning when flows were low and nighttime respiration had depressed DO, or during generation when DO was already high and continuing a treatment would have been wasteful. Although not costed out for this analysis, several other options such as discharging wastewaters proportionally with flow and reduction of heat loads were simulated during the course of the study.

Observations of Cost-Effectiveness of Improvement Strategies

Figure 5 is a cost-effectiveness diagram for all strategies showing costs versus results in DO improvement. A frontier of most cost-effective options exists along the points lowest and farthest to the right. The shaded area represents estimated inaccuracies in costing each option (plus or minus 50%) and in simulating DO (plus or minus 0.5 mg 1^{-1}). Even with the error envelope, two significant conclusions can be drawn.

First, the costs of upgraded waste treatment rise very steeply with DO improvement relative to that of the more innovative improvement strategies such as flow pulsing and aeration. Further restrictions on the four waste dischargers creates only modest water quality improvements because the dischargers have already installed equipment that removes over 90% of waste water BOD. Innovative techniques can achieve the same range of DO concentrations in the river at much lower costs.

Secondly, combinations of innovative methods can achieve DO concentrations that are not achievable using traditional methods. For example, even if the four industrial and municipal dischargers were advanced to zero discharge of BOD, the DO concentrations in some segments of the river would not reach that attainable with combined instream aeration and increased flow pulses.

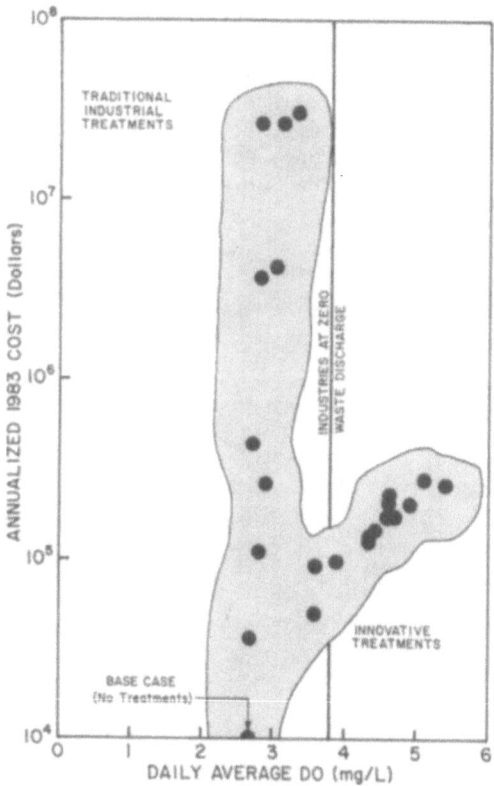

Figure 5. Cost-effectiveness of DO improvement strategies.

Next Steps

The realization of a complementary combination of methods for improving DO will require more detailed assessment of the operational impacts on each participant and considerable negotiation among participants, including regulatory authorities. Acceptance of any strategy will most likely be based upon cost-effectiveness, equity, enforceability, and ease of administration.

REFERENCES

Ford, D. E., Thornton, K. W., Lessem, A. S., Wlosinki, J. H. and Norton, J. L. 1981. "CE-QUAL-R1: A Numerical One-Dimensional Model of Reservoir Water Quality - User's Manual", Environmental Laboratory, USAE Waterways Experiment Station, Vicksburg, Mississippi.

Gulliver, J. S. and Stefan, H. G. 1984. Stream Productivity Analysis with DORM-II, Parameter Estimation and Sensitivity. Water Research, 18(12). New York, New York.

Hauser, G. E. and Ruane, R. J. 1985. Model Exploration of Holston River Water Quality Improvement Strategies. Report No. WR28-1-590-109, TVA Water Systems Development and Water Quality Branches, Norris, Tennessee.

Industrial Economics, Inc. 1984a. Case Studies on the Trading of Effluent Loads - Dillon Reservoir. Prepared for U.S. Environmental Protection Agency. IEC Cambridge, Massachusetts.

Industrial Economics, Inc. 1984b. Exploratory Study of Improving Dissolved Oxygen Concentrations in the Holston River Near Kingsport, Tennessee. Prepared for Office of Policy, Planning, and Evaluation, U.S. Environmental Protection Agency, Cambridge, Massachusetts.

Jobson, H. E. 1980. Temperature and Solute Transport Simulation in Streamflow Using a Lagrangian Reference Frame. USGS Water Resources Investigations 81-2, National Space Technical Laboratories, NSTL Station, Mississippi.

Raschke, R. L., Hall, S., Cavinder, T., Koenig, M., Frey, P., Howard, H., Shultz, D., Hicks, D., Murphy, P. and Barnwell, T. 1978. Holston River Study. EPA 904/9-78-019, EPA Surveillance and Analysis Division, Athens, Georgia.

WATER TEMPERATURE CONTROL AND AREAL OXYGEN CONSUMPTION RATES AT A NEW RESERVOIR, AND THE EFFECTS ON THE RELEASE WATERS

Richard A. Cassidy and Patrick E. Dunn

U.S. Army Corps of Engineers
Portland District, Portland, Oregon U.S.A.
North Pacific Division, Portland, Oregon U.S.A.

INTRODUCTION

The potential environmental effects of dams on the Rogue River have long been a specific concern to people of the Rogue Valley, Oregon (U.S.A.), and a general concern to the residents of the entire State of Oregon. Following serious flooding in the mid-1950's, efforts were begun to build large, multiple-purpose reservoirs (Cassidy & Johnson 1982). When the U.S. Army Corps of Engineers built two dams in the Rogue River Basin, an innovation in water release methods using selective withdrawal systems was utilized by the Corps' Portland District to reduce the harmful effects of the dams on the receiving streams. Selective withdrawal systems allow the removal of water from one of a number of different levels in a reservoir, or from a combination of levels, to utilize the best quality water in the reservoir for discharge downstream.

Conceptually, the intent of a selective withdrawal system is to provide the flexibility to choose better quality water for release compared to a traditional fixed, low-level withdrawal system. For instance, the temperature of water released from a reservoir, one important long-term environmental factor, can be more closely controlled to resemble natural temperatures. Control of the temperature of water being released can have a significant long-term effect on the biota of the regulated stream.

The use of this capability at Applegate Lake, the Portland District's second dam and reservoir project in the Rogue River Basin with a selective withdrawal structure, also unexpectedly reduced some short-term problems during the first year of impoundment. Degraded water quality conditions during the first years of reservoir impoundment have long been recognized (Saville 1925). Use of the selective withdrawal system at Applegate Lake during 1981 provided desirable 10°C to 15°C release water temperatures and oxic water (near to or above 5 mg l^{-1} throughout most of the critical summer season). By the fall season, the decision to change and withdraw water from near the reservoir bottom not only provided continued 10°C to 15°C release water, but also reduced the oxygen consumption rate in the lower level reservoir water.

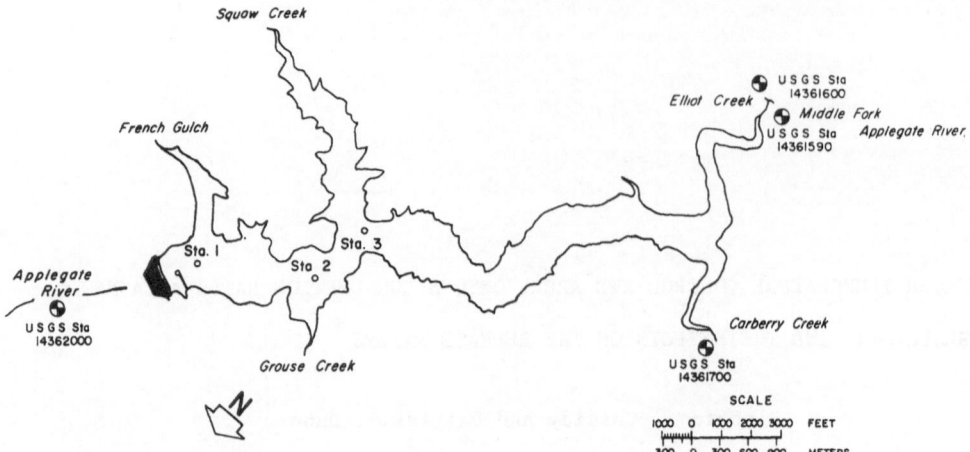

Figure 1. Water quality monitoring stations, Applegate Lake, Oregon, U.S.A.

PURPOSE

The purpose of this study was to monitor basic water quality changes, such as water temperature and dissolved oxygen concentration, in the new reservoir, and to evaluate the use of a dual wet well multiple level withdrawal structure to control the quality of the water being released to the regulated stream.

METHODS

Weekly water quality profile data were collected from three sampling stations in the reservoir (Figure 1), with station 1 serving as the main index station. Water temperature measurements were made with a Montedoro Whitney TC-5C telethermometer at 1 m intervals down to a depth of 20 m. Water temperature, dissolved oxygen, specific conductance, pH, and oxidation reduction potential were measured in situ with a Hydrolab 8,000 series water quality instrument at 1 m increments down to a depth of 25 m and at irregular intervals to the reservoir bottom. During the study period, the depth of the reservoir water varied from approximately 62.5 m to 39.6 m. Consequently, the measurement interval varied at the lower reservoir depths. In addition, grab samples were usually collected for dissolved oxygen analysis by the Winkler method at the 15 m and 30 m depths.

CATCHMENT DESCRIPTION

The Applegate River Basin is a subcatchment of the Rogue River, a coastal stream in southern Oregon. The confluence of the Applegate River with the Rogue River occurs approximately 8 km west of Grant's Pass, Oregon, at the upstream end of the 135 km portion of the Rogue River that is designated a national "wild and scenic" river (Figure 2).

Applegate Lake is located 74.5 km upstream on the Applegate River, near the California-Oregon border (Figure 2). It is a 100×10^6 m^3 multiple-purpose project authorized for flood control, irrigation, fish and

Figure 2. Applegate Lake and the Rogue River Basin, Oregon.

wildlife, recreation, and water quality enhancement in the Applegate
Valley.

The dam has a butterfly-shaped catchment that controls the rainfall
and snowmelt runoff from approximately 580 km^2 of the Applegate River
Basin. The headwaters of the Applegate Lake catchment are at the heavily
timbered crestline along the Siskiyou Range of California and Oregon,
usually above 1,525 m, National Geodetic Vertical Datum (NGVD). The
elevation at the dam site is approximately 536 m, NGVD. Streams have steep
gradients varying from 4.2 to 6.3 m km^{-1} and flow through narrow channels
cut deeply into intrusive rock.

The Applegate River Basin experiences mild, wet winters and warm, dry
summers. A predominately westerly flow of moist air from the Pacific Ocean
during the winter produces a large proportion of snow at the higher
elevations and rain at the low elevations (U.S. Army Engineers 1983). Rare
summer rainstorms of short duration and small areal coverage occur when
local convective activity causes thunderstorms. The normal annual
precipitation is approximately 114.3 cm, ranging from 76.2 cm at the
project to nearly 152.4 cm in the headwaters where significant orographic
lifts occur. Usually about 75 percent of the annual precipitation occurs
from November to March, and less than 2 percent occurs during July and
August. The reservoir did not fill during the first year because of
drought. The monthly mean flow of water entering the reservoir during the
first summer of impoundment was less than the 41-year monthly mean
pre-project flows (Table 1) for the summer period.

Fall chinook (*Oncorhynchus tshawytscha*), Coho Salmon (*O. kisutch*) and
steelhead trout (*Salmo gairdneri*), are the important anadromous fish that
utilize the Applegate River system. Resident rainbow (*S. gairdneri*) and
cutthroat (*S. clarki*) trout are present in the catchment. Before the
construction of Applegate Lake, approximately 15,000 fall chinook salmon
spawned annually in the mainstem Applegate River and approximately 5,000

Table 1. Applegate River monthly mean river flow.

Location (River Kilometer)	Month	1939-1980 Pre-Project River Flow/[1] (m^3 s^{-1})	1981 Post-Project River Flow/[1] (m^3 s^{-1})	1981 Project Inflow/[2] (m^3 s^{-1})
74.5 (dam site)	June			2.8
	July			1.7
	August			0.8
	September			0.8
73.5 (dam tail- water)	June	10.5	5.7	
	July	3.3	7.0	
	August	1.7	5.8	
	September	1.4	6.2	
12.2	June	13.8	5.4	
	July	2.8	5.4	
	August	0.7	4.0	
	September	1.1	4.8	

[1]U.S. Geological Survey Data
[2]U.S. Army Corps of Engineers Data

coho salmon spawned annually in the tributary streams. About 10,000 steelhead trout utilized both the mainstem and tributaries of the Applegate River for spawning before the dam was built (Fish and Wildlife Service 1961). Rainbow trout are stocked annually in the catchment.

In the recent past, the aquatic habitat of the drainage basin was affected by significant water diversions made for irrigation. Low flow and concomitant increases in water temperature were considered harmful to the fish. Consequently, the Applegate Dam was designed with a multiple level withdrawal structure to provide better quality water.

THE SELECTIVE WITHDRAWAL STRUCTURE

Applegate Dam was constructed with a dual wet well selective withdrawal structure capable of removing water from six different levels in the reservoir. The withdrawal structure consists of a 72.2 m high, freestanding, concrete tower and a 144.5 m long, 3.6 m wide service deck providing access to the intake tower (Figure 3).

The base of the tower contains two regulating outlets for control of high flows, a regulating gate chamber, and a trash rack structure. The tower consists of two vertical wet wells for water temperature control and a single dry well connecting the equipment room with the regulating gate chamber. The two water temperature control wet wells are served by five intake ports, two serving one wet well and three serving the other. The low level regulation outlet serves as the sixth level of withdrawal.

The bellmouth intakes for the regulating outlets are on the face of the base, behind the trash structure of vertical and horizontal trash bars. The outlets are controlled by two sets of 1.4 x 1.8 m slide gates

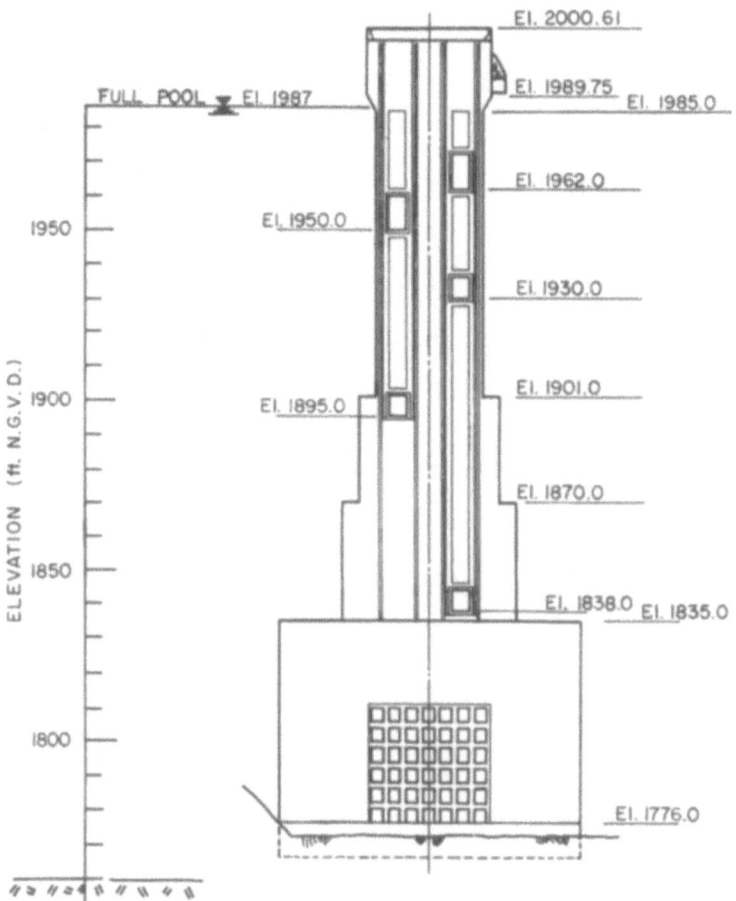

Figure 3. Schematic diagram of the Applegate Lake selective withdrawal
structure (specific elevations given in feet).

used for regulating closures and emergency closures. One steel bulkhead is
available for maintenance closure and is handled from the service deck by a
mobile crane. Water passes through the regulating outlet gates into twin
1.4 x 1.8 m rectangular conduits and into a 7.7 m to a 4.4 x 2.7 m oblong
concrete cut and cover conduit approximately 243 m long, connecting the
withdrawal structure to the downstream stilling basin through the
embankment. Flows of up to 161 m^3 s^{-1} will pass through the regulating
outlets. Downstream of the cut and cover conduit, an open channel flares
into a 9.1 x 85.3 m rectangular primary and secondary stilling basin.

Reservoir water used for temperature control can be withdrawn from any
one of up to five levels, or mixed from different levels. The top two
intakes are 1.5 x 3.0 m high and are rated at 14.1 m^3 s^{-1}. The other three
intakes are 1.5 x 1.8 m and are designed for 8.5 m^3 s^{-1}. They are operated
in either a fully open or closed position. Water passes through the ports
into one or both of the 2.1 x 1.4 m wet wells and then into a gate chamber
area regulated by 0.6 x 0.8 m high tandem sliding gates. The wet wells
merge into a common 24.4 m long exit channel in the centre of the
regulating outlet splitter pier.

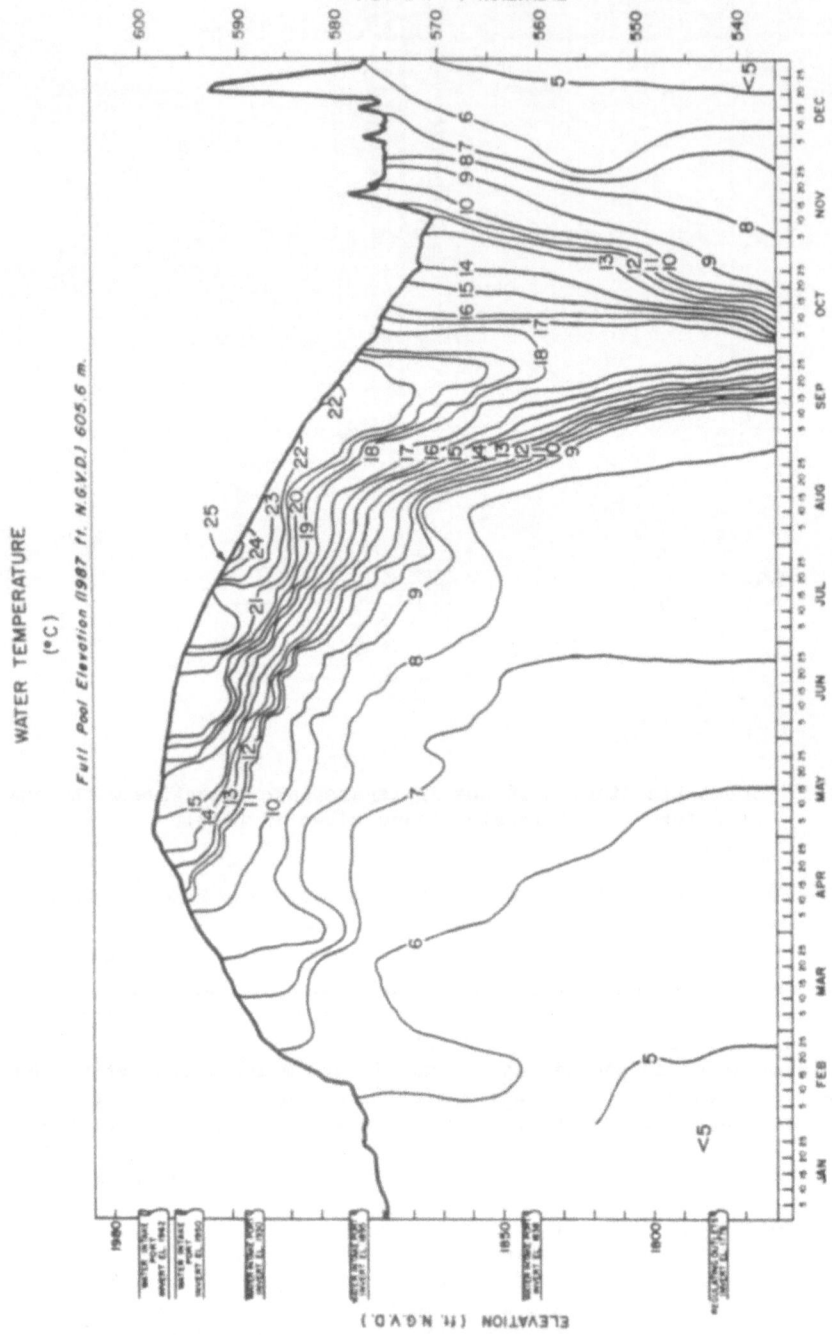

Figure 4. Water temperature isopleths and reservoir elevation for Applegate Lake, Oregon during 1981.

A 0.77 m diameter fish facility water supply pipeline takes water from the right wet well and delivers it to the fish collection facility adjacent to the stilling basin. The pipeline, with a capacity of 5.7 $m^3 s^{-1}$, is controlled by tandem 0.6 m diameter ball valves.

RESULTS AND DISCUSSION

Water Temperature

The establishment of maximum water temperature stratification at Applegate Lake (Figure 4) was similar to the stratification development of Lost Creek Lake, the other Corps reservoir in the Rogue River Basin (Cassidy et al. 1981). The Applegate impoundment is also a warm monomictic reservoir with water temperatures that are always above 4°C. Maximum surface temperatures occur during a period following the attainment of the maximum summer pool elevation (Figure 4). The 1981 summer heat content was 26,300 cal cm^{-1}. In contrast to Lost Creek Lake, definitive epilimnion, metalimnion, and hypolimnion did not persist throughout the stratification period at Applegate Lake because of the relatively smaller volume (1.0 x 10^8 m^3) of the impounded water compared to the 5.7 x 10^8 m^3 Lost Creek Lake. Water temperatures throughout most of the water column in the reservoir exceeded 15°C from mid-September through mid-October (Figure 4).

The Oregon Department of Fish and Wildlife requested experimental release discharge rates and water temperature combinations to evaluate downstream aquatic effects. The experiment was modified because of drought conditions during the 1981 water year. The average pre-project annual flow at a U.S. Geological Survey gauging station 0.6 km downstream of the dam was 16.4 $m^3 s^{-1}$ for a 41-year period while the 1981 average annual inflow to Applegate Lake was 13.7 $m^3 s^{-1}$. Since the reservoir was not expected to fill due to low runoff, release flow target temperatures were modified

Table 2. Applegate Lake experimental release schedule.

Date (1981)	Release Flow ($m^3 s^{-1}$)	Target Water Temperature (°C)
22 Jun – 28 Jun	5.9	10.0
29 Jun – 5 Jul	5.9	12.8
6 Jul – 12 Jul	7.1	12.8
13 Jul – 19 Jul	7.1	10.0
20 Jul – 31 Jul	7.1	10.0
1 Aug – 31 Aug	5.7	12.8
1 Sep – 30 Sep	5.7	10.0
1 Oct – 31 Oct	5.7	12.8

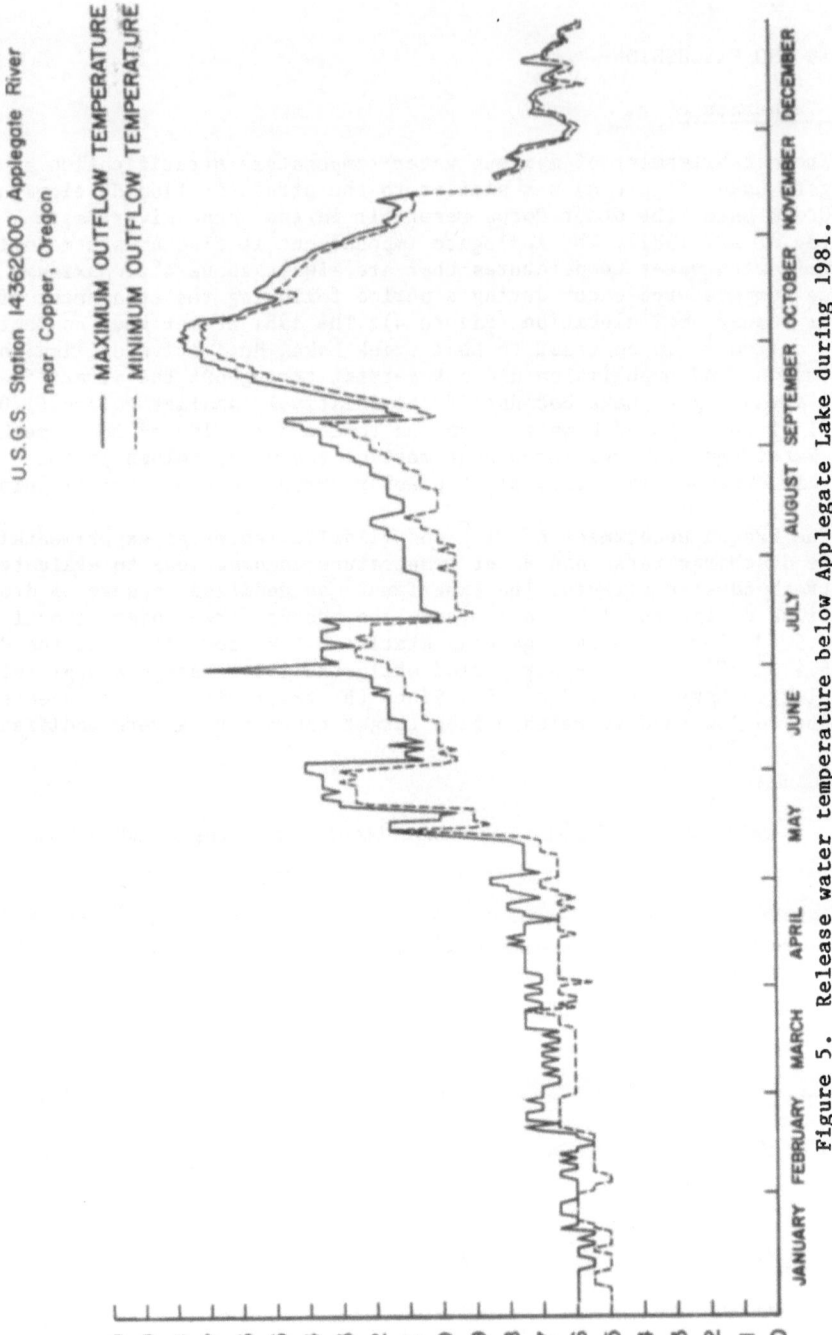

Figure 5. Release water temperature below Applegate Lake during 1981.

(Table 2), because cool waters were not available for the early fall
portion of the experiments (Figure 5).

Dissolved Oxygen

Walker (1979) suggested that the dissolved oxygen dynamics of natural
lakes may be significantly different than those of reservoirs. Certainly,
this is true in new reservoirs (Saville 1925; Miterev & Belova 1957;
Sylvester & Seabloom 1965; Maystrenko & Denisova 1972; Ball et al. 1975;
Duthie & Ostrosky 1975; Cassidy et al. 1981). The dissolved oxygen
isopleths for 1981 (Figure 6) at Applegate Lake show that the
concentrations during the first year of impoundment ranged from over 10 mg
1^{-1} during January and February to about 1 mg 1^{-1} by early
September. Before impoundment, much of the reservoir bottom was cleared of
organic debris and the reservoir bottom was used for construction roads,
stockpiles, heavy equipment areas, gravelwash settling ponds, and a
concrete batch plant. Vegetation along the steeper slopes of the reservoir
had been cut and burned at the site. Consequently, there was not a large
volume of organic material to deplete oxygen near the reservoir bottom.

The areal oxygen consumption rate at Applegate Reservoir cannot be
directly compared with the hypolimnetic areal oxygen consumption of natural
lakes (Hutchinson 1957; Lasenby 1975; Welch & Perkins 1979) because a true
hypolimnion did not remain throughout the first year of impoundment.

The dissolved oxygen concentration below 15 m indicates successful
inhibition of strong anoxic development in the reservoir (Figure 7). On 8
September, the 5.7 m^3 s^{-1} discharge of water from the reservoir was changed
from the withdrawal level at invert elevation 559.3 m NGVD to include
withdrawal from the regulating outlet at invert elevation 541.3 m
NGVD. Approximately 4.5 m^3 s^{-1} was released from the regulating outlet
while 2.0 m^3 s^{-1} was released from the elevation 559.3 m NGVD. The
reservoir bottom did not reach complete anoxia because the utilization of
the multiple level intake structure provided water flow there. Periodic
samples for hydrogen sulphide in the water being discharged (<7 g 1^{-1})
indicated that the gas did not occur at concentrations harmful to fish for
more than a few days. At the end of September, rainstorms introduced
sufficient water to the reservoir to increase the low dissolved oxygen
concentration to 5 mg 1^{-1}. The dissolved oxygen data below 15 m also show
that the areal consumption rate of the lower portion of the reservoir was
non-linear. A least squares second degree parabola was fitted to the data
between 12 May and 8 September 1981 (Figure 8). Early in the stratification
period, when conditions more closely approached those of a hypolimnion, the
consumption rate below the 15 m depth could be considered linear. The
consumption rate was 0.86 g O_2 m^{-2} day^{-1} from 12 May to 16 June 1981. By
late August and early September, the reservoir had rapidly exhausted its
supply of cool water below the 15 m depth, and the consumption rate from 6
August to 10 September was 1.69g O_2 m^{-2} day^{-1}. In general, however, the
oxygen consumption below the 15 m depth at Applegate Lake showed polynomial
decay characteristics (Figure 8).

Although reservoir dissolved oxygen concentrations approached 0 mg 1^{-1}
in the reservoir, tailwater dissolved oxygen levels measured approximately
1 km downstream of the project showed that the downstream dissolved oxygen
slump occurred for only a 3-week period (Figure 9). Because of reaeration
in the stream, lowest dissolved oxygen concentrations were near to, or
above, 5 mg 1^{-1}. The tailwater dissolved oxygen level remained near to, or
above, 10 mg 1^{-1} for most of the May through October period. The Portland
District considered these downstream results very encouraging for
discharges from a new reservoir.

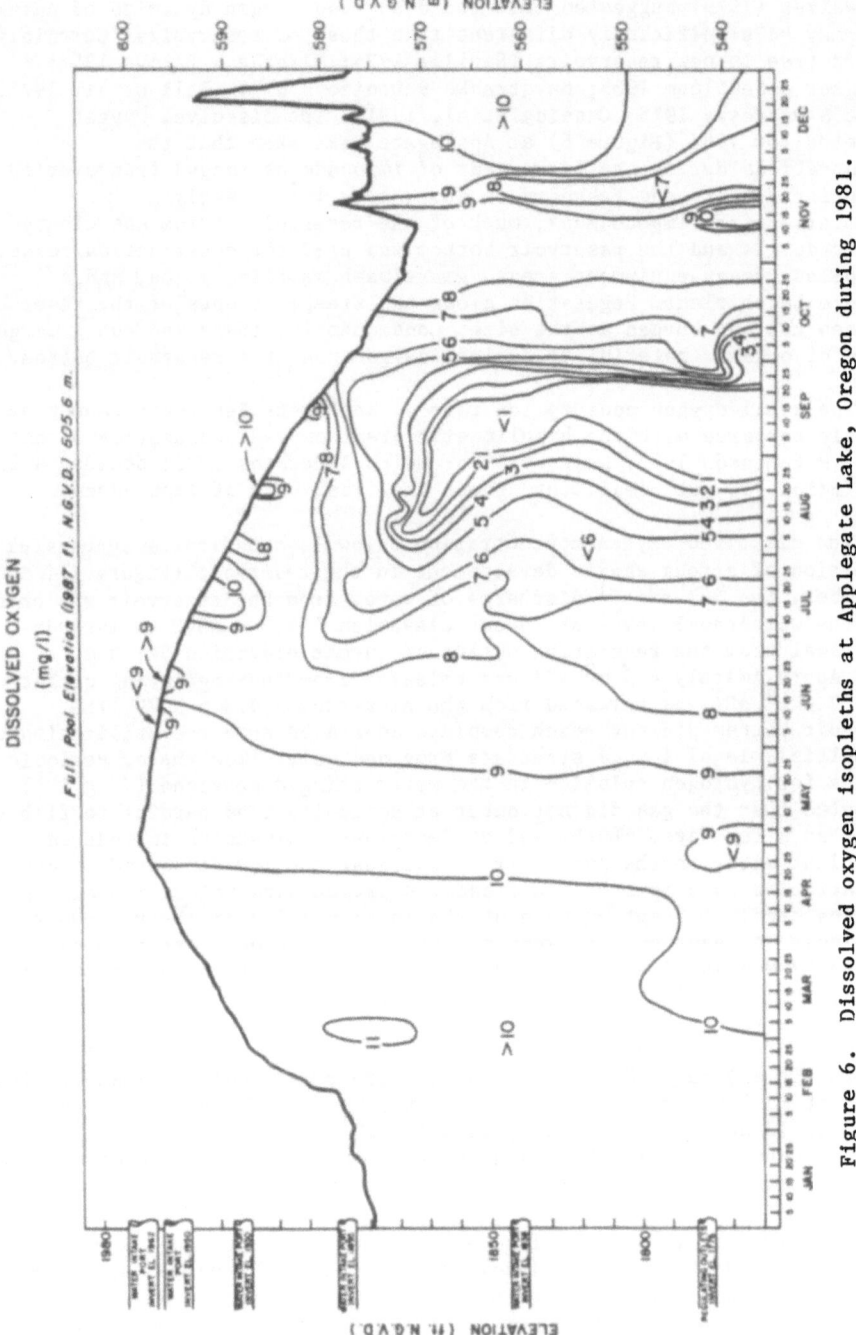

Figure 6. Dissolved oxygen isopleths at Applegate Lake, Oregon during 1981.

348

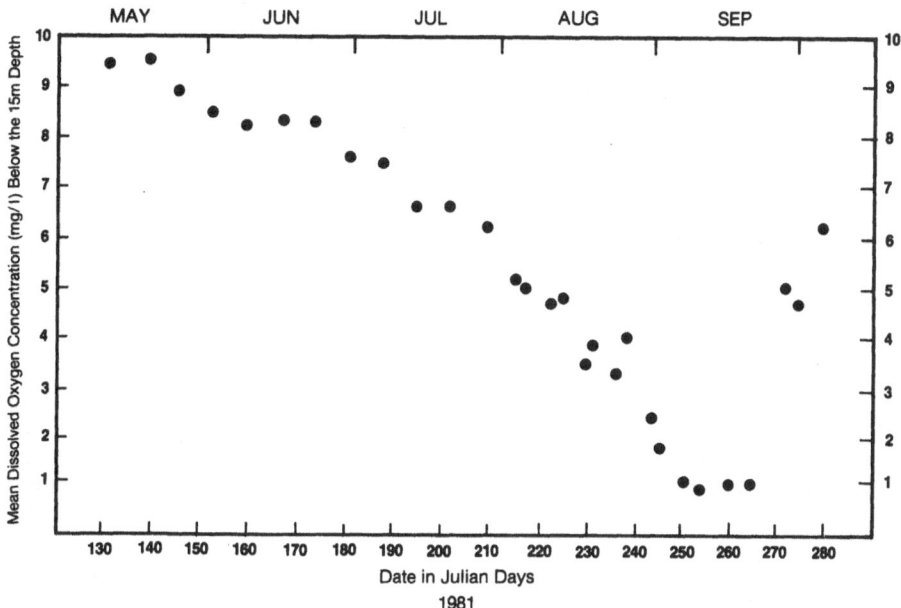

Figure 7. Volume-weighted mean dissolved oxygen levels below 15 m.

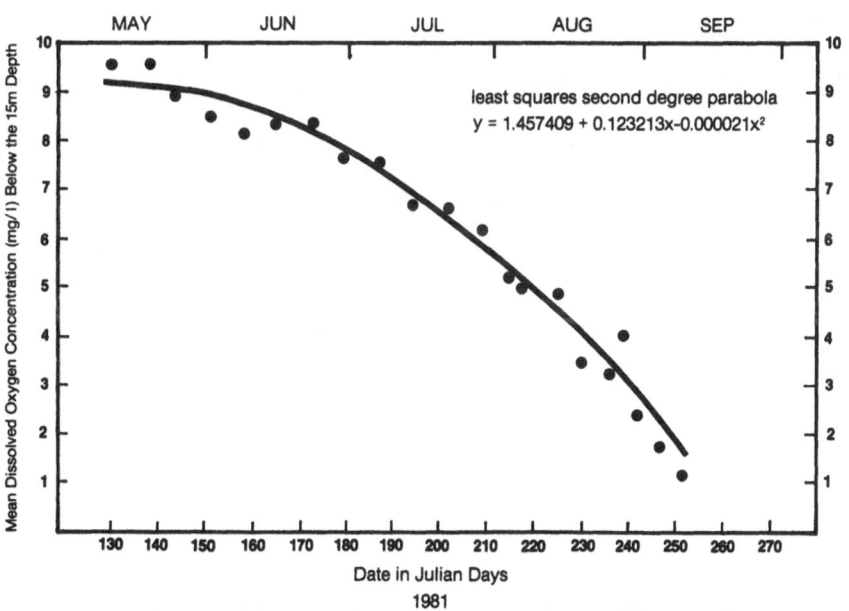

Figure 8. Second degree parabola, $r^2 = 0.94$, of dissolved oxygen consumption below 15 m.

349

The release of undesirable products of anoxic processes from reservoirs have been described by Keeley et al. (1978) and Gunnison and Brannon (1981). These include reduced forms of nitrogren, iron, and manganese, reduced organic compounds, and hydrogen sulphide than can produce toxicity, taste, odour, and staining problems. Use of the selective withdrawal structure at Applegate Lake during the first year of impoundment has helped to reduce or avoid water quality impacts that could have adversely affected the Applegate River fishery.

CONCLUSION

Use of the multiple level withdrawal structure at Applegate Lake has reduced the short-term harmful effects in the downstream river which are associated with a new reservoir. During the first year of impound, when significant water quality degradation can occur in a reservoir, water released through the selective withdrawal structure at Applegate Lake provided the receiving stream with summer water temperatures of 10°C to 15°C and reduced the development of a dissolved oxygen deficit. Although the dissolved oxygen concentrations in the reservoir during the first year of impoundment dropped to approximately 1 mg 1^{-1}, the reservoir did not become completely anoxic. This achievement prevented by-products of anoxia, such as hydrogen sulphide gas, from being released downstream to subject the aquatic organisms in the regulated river system to physiological stress.

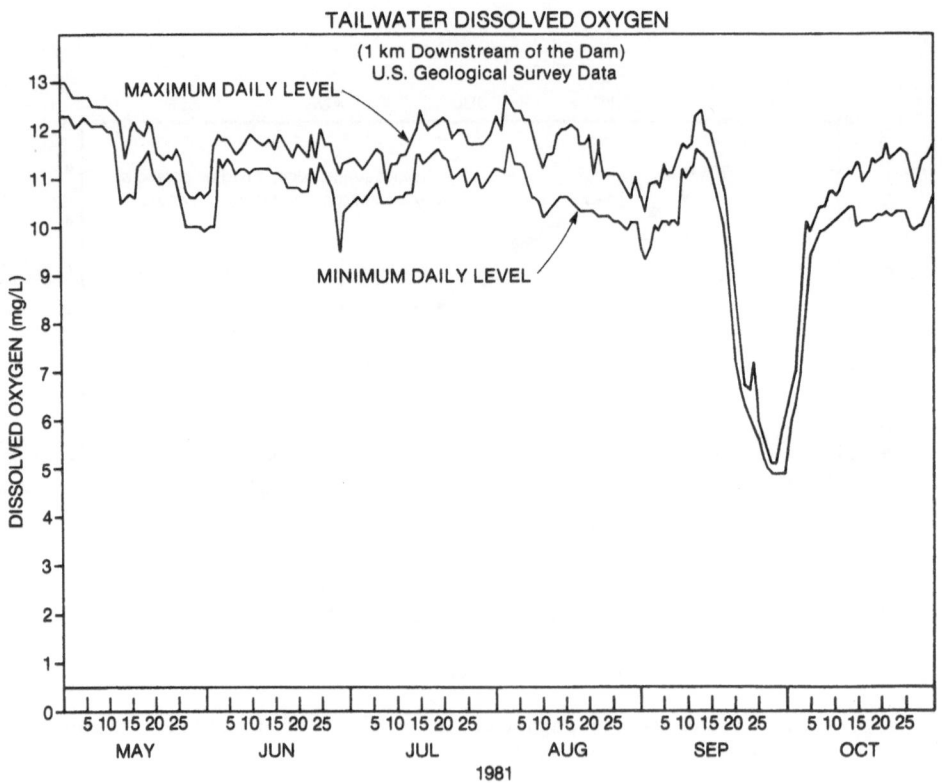

Figure 9. Dissolved oxygen levels in the Applegate River, 1 km downstream of Applegate Lake.

REFERENCES

Ball, J. C., Weldon, C. and Crocker, B. 1975. Effects of original vegetation in reservoir water quality. Water Resour. Inst. Texas A&M Univ. Tech. Rep. No. 64. p. 54.

Cassidy, R. A. and Johnson, E. B. 1982. A public participation decision to fill a Pacific Northwest reservoir. Can. Wat. Res. J. 7 (2): 90-111.

Cassidy, R. A., Larson, D. W. and Putney, M. T. 1981. Physiochemical limnology of a new reservoir. In: Proc. symposium on surface water impoundments. (Ed. by H. G. Stefan). 2: 1465-1473. A.S.C.E., NY.

Duthie, H. C. and Ostrosky, M. L. 1975. Environmental impact of the Churchill Falls (Laborador) hydroelectric project: a preliminary assessment. J. Fish. Res. Board Can. 32: 117-125

Fish and Wildlife Service. 1961. Water Development Plan, Rogue River Basin.

Gunnison, D. and Brannon, J. M. 1981. Characterization of anaerobic chemical processes in reservoirs: problem description and conceptual model formulation. Report E-81-6. U.S. Army Engineer Waterways Experiment Station, CE, Vicksburg, Miss.

Hutchinson, G. E. 1957. A Treatise on Limnology. 1. John Wiley and Sons, Inc. New York. p. 1015.

Keeley, J. W., Mahloch, J. L., Barks, John W., Gunnison, D. and Westhoff, J. D. 1978. Reservoir and Waterways. Indentification and assessment of environmental quality problems and research program development. Technical E-78-1. U.S. Army Engineer Waterways Experiment Station, C.E. Vicksburg, Miss.

Lasenby, D. C. 1975. Development of oxygen deficits in 14 southern Ontario lakes. Limnol. Oceanogr. 20: 993-999.

Maystrenko, Y. G. and Denisova, A. I. 1972. Method of forecasting the content of organic and biogenic substances in the water of existing and planned reservoirs. Hydrochemical Inst., Novocherkassk, Hydrochemical Materials (Gidgrokhimicheskiye materialy) 53: 86-114.

Miterev, G. A. and Belova, E. M. 1957. Influence of a submerged forest in water reservoir on quality of water. Sbornik Naveh. Rabot. Moskov. Farm. Inst. 1: 395-401.

Saville, C. M. 1925. Color and other phenomena of water from an unstripped reservoir in New England. J. New Eng. Water Works Assoc. 39: 145-170.

Sylvester, R. O. and Seabloom, R. W. 1965. Influence of site characteristics on quality of impounded water. J. Amer. Water Works Assoc. 57: 1528-1546.

U.S. Army Corps of Engineers, Portland District. 1983. Water control manual, Applegate Lake.

Walker, W. W. 1979. Use of hypolimnetic oxygen depletion rate as a trophic state index for lakes. Water Resources Res. 6(15): 1463-1470.

Welch, E. B. and Perkins, M. A. 1979. Oxygen deficit-phosphorus loading relation in lakes. Jour. WPCF 51 (12): 2823-2828.

INCREASING THE OXYGEN CONTENT OF THE KALAJOKI RIVER

Esko Lakso

Tec.Lich, National Board of Waters
Finland, Kokkola Water District
Torikatu 40 B, KOKKOLA

INTRODUCTION

In Finland, the need to raise the oxygen content of waters is most evident in late winter when the ice cover prevents the water from receiving oxygen from the air. In Central and Southern Finland, the watercourses become ice-bound in November-December, and the ice melts in April-May.

Although the water temperature in winter is only 0-4°C, the pollution loads put on the systems cause rapid consumption of the available oxygen. Despite waste water treatment, some waterways or parts of watercourses have such a high loading factor that the oxygen content drops too low for the use of these waters for fisheries or for some other purposes.

Low oxygen concentrations may also cause problems during early autumn, usually when large amounts of treated waste water are discharged into relatively small streams.

Temporary lowering of the oxygen content of the water has also been observed as the result of algal blooms. The probability of these occurring in Western Finland is not very great, due to the fairly cold climate as well as the characteristic brown colour of the waters there.

As the result of the thermal stratification which occurs in lakes and reservoirs, oxygen depletion problems generally occur in the bottom layers. The oxygen of the hypolimnion has been improved by pumping surface water rich in oxygen into the lower layers. In some instances, the hypolimnion has also been aerated. The oxygen content of river water has been improved by aerating at overflow dams and, in the Kalajoki River, at the power plants on an experimental basis.

The oxygen level has not been raised through the use of pure oxygen, except for a few very short trial runs as the use of oxygen is expensive.

THE KALAJOKI WATERCOURSE AND THE AERATION TRIALS CARRIED OUT IN THE SYSTEM

The Kalajoki River is a typical watercourse of Western Finland. Its drainage basin covers 4,300 km², and only c. 2% of the area is lakes

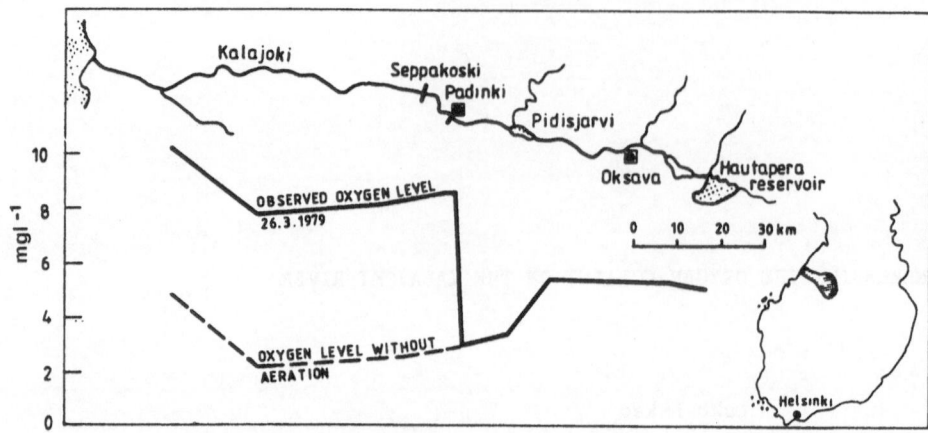

Figure 1. The Kalajoki Watercourse and the increases in oxygen content as
the result of the Padinki aeration dam in late winter 1979.

(Figure 1). Drainage ditches have been dug in most of the swamps and
forests located in the area. Therefore the runoff varies considerably along
the length of the river. At maximum, the flow in the river is c. 400
m^3 s^{-1}. In its natural state, the minimum flow is almost zero; and during
regulation the minimum flow is 3 m^3 s^{-1} (National Board of Waters 1980).

Flood control reservoirs have been built on the upper part of the
river basin and the river has been dredged and embankments built. The
system contains hydro-electric plants. The largest reservoir is Hautaperä,
built in 1975. Its regulated volume is c. 50×10^6 m^3 and its surface area
at maximum volume is 8 km^2.

The water released from the reservoir, particularly during the years
just after construction, had very low dissolved oxygen content in late
winter (Figure 2). The oxygen content decreased further as the water flowed
in the Kalajoki River. The consumption of the oxygen in the water in winter
was caused primarily by chemical and biological reactions in the bottom
sediment, as well as by rotting of the aquaflora in Lake Pidisjärvi. In
addition, in the 1970s, the waste water let into the river from the
population centres was insufficiently treated.

A reach of about 20-30 km on the lower part of the Kalajoki River is a
breeding ground for salmonid fishes, mostly whitefish (*Coregonus* sp.). An
additional c. 50 km stretch provides breeding grounds for river lampreys
(*Lampetra fluviatilis*). Studies indicate that a significant factor in the
reproductive failure of whitefish and lampreys in the lower reaches of the
river since the mid-1970s, has been the low oxygen content of the river
water. The recommended oxygen level in the lower reaches of the Kalajoki
River is over 7 mg l^{-1}, with the minimum being 5 mg l^{-1}.

Elsewhere in the river, as well as in the lakes and reservoirs, there
are spring-spawning fish and a limited number of crayfish (*Astacus
astacus*). Spring-spawning fish can accept a lower oxygen content than can
salmonid fishes.

The oxygen content of both the Kalajoki river and the Hautaperä
Reservoir has been raised by a number of different methods. The aeration of
the river system has been closely linked with experimental work, so that

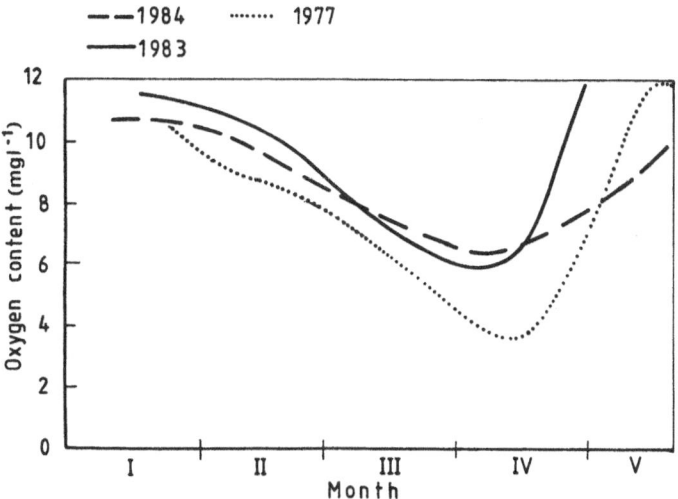

Figure 2. The oxygen content of water released from Hautaperä Reservoir.

the results obtained could be applied to other areas.

In the late winter of 1979, the Kalajoki River water was aerated at the Padinki overflow weir. In the winter of 1984-1985, an overflow weir was completed at Seppäkoski Rapids, from which the first data were obtained in the late winter of 1985 (March to April). During the years 1979 to 1982 the possibilities for using hydropower plants for aeration were examined at the Oksava and Padinki power plants. Beginning in 1983, the water quality of Hautaperä Reservoir has been improved by circulating the surface water through the hypolimnion (Lakso 1984).

A detailed presentation of the methods used to improve the oxygen content of the water in the Kalajoki Watercourse, as well as a description of related researches is given below.

Aeration at Overflow Weirs

The ability of overflow weirs to provide aeration has been studied at the Venetjoki Reservoir in the Perhonjoki Watercourse (Figure 3) and below the Vissavesi Reservoir, as well as in the Perhonjoki River. The surface area of the Venetjoki Reservoir is 18 km^2, and its regulated volume is 28 x 10^6 m^3. The corresponding figures for the Vissavesi Reservoir are 4 km^2 and 7 x 10^6 m^3. Both reservoirs are thermally stratified in the winter and in the late summer. During these times of stratification, the oxygen content of the hypolimnion is decreased considerably, and oxygen concentration may become negligible. In the Vissavesi and Venetjoki Reservoirs, aeration trials were carried out in 1983 to 1984 in both the late winter and early fall. Before aeration the oxygen content of the water varied between 0 to 80% of saturation. A total of 720 observations were made.

In the Venetjoki and Vissavesi Reservoirs, it was possible to manipulate the following factors which affected aeration:

```
Height of head            0 - 2.9 m
Depth of tail water       0.15 - 1.2 m
Flow                      0.04 - 1.2 m³s⁻¹
Form of weir crest (Figure 5).
```

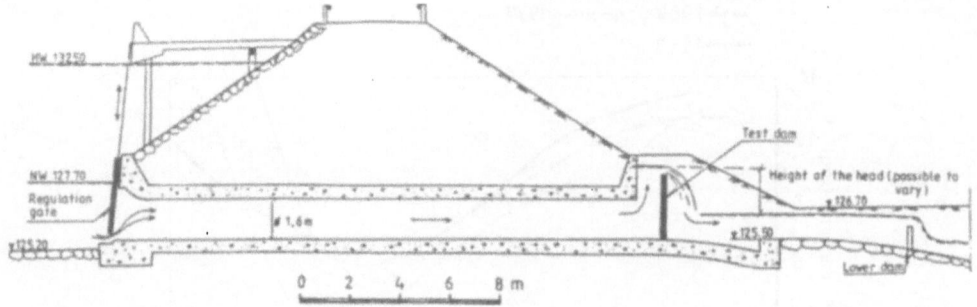

Figure 3.   The weir built in the mouth of the Venetjoki Reservoir outlet
            pipe.

When shallow tail water depths were used, the depth increased to some
extent with increasing flow. In the trials made in late summer, the
temperature of the water was 18.9°C on average. The corresponding value in
late winter was 1.2°C. At both Venetjoki and Vissavesi, the width of the
dam used in the trials was 1.7 m.

The observation data indicate that the rise in the oxygen content per
unit of fall was greatest when the head was between 0.6-1.1 m. When the
available head was >1.2 m, optimum results were obtained by dividing the
head into steps, in which the height of each step was between the above
values.

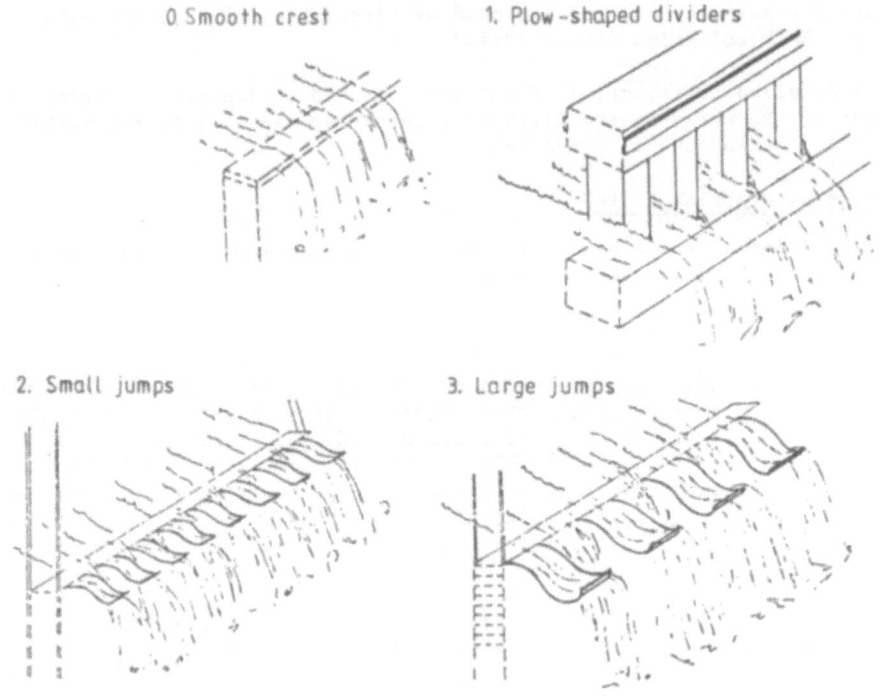

Figure 4.   A comparison of the results of aeration trials from deep tail
            water (0), shallow tail water (1) and shallow tail water and
            rocks (2), when the height of the head was between 1.4-2.2 m.

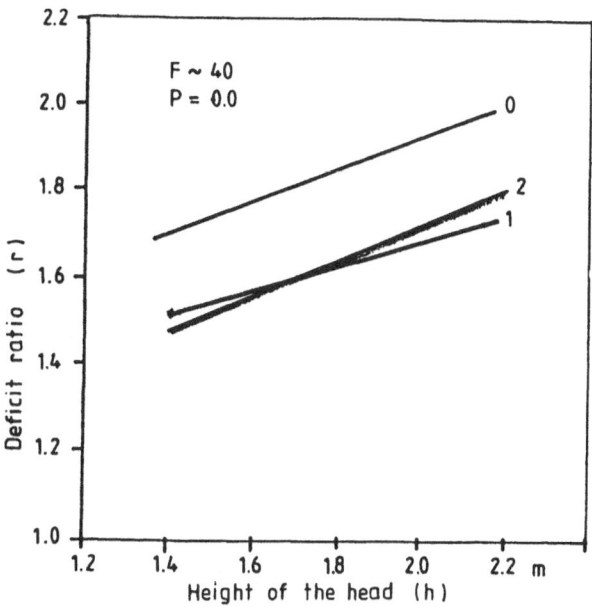

Figure 5. Water dividers (guide plates) used at the crest of the dams in both the Venetjoki and the Vissavesi aeration trials.

Deep tail water gives more efficient aeration than either shallow tail water or shallow water and stones (Figure 4).

Increased flow with constant height of head decreases the oxygen enriching capability of the weir. The minimum height at which aeration begins, increases as the flow increases. With a slow flow, this minimum height is 0.15 to 0.20 m, and with a large flow ($1 m^3 s^{-1}$ and over), the value is 0.4 to 0.5 m. Using dividers (Figure 5), at the crest of the dam, the head at which aeration begins can be reduced.

Water dividers located at the crest of the weir make it possible to improve aeration in comparison with a smooth crest (Kroom & Schram 1969). The plow-shaped dividers raise the oxygen content by nearly the same amount independent of flow. The drawback to this type of divider is that it freezes in the winter.

Aeration at overflow weirs is dependent on the quality of the water. When the results of the aeration trials carried out immediately below reservoirs were compared with trials carried out in the Kalajoki River in comparable conditions, the Kalajoki trials gave better results. In evaluting the aeration results from overflow weirs, the quality of the water being aerated should be taken into account (Gameson 1957; Pöpel 1976).

The temperature of the water had no significant effect on the aeration results at the weirs (Imhoff & Albrecht 1972). When the head was below one meter, the results obtained in summer were slightly better than those of the winter. The difference was so slight, however, that it can be disregarded in the aeration equation.

In the mathematical investigation of the results of the aeration trials of spill water from reservoirs, the following equations were used:

$$r = \frac{C_s - C_o}{C_s - C_e} \tag{1}$$

$$r = 0.886 + 0.602 \, h + P1 + P2 \tag{2}$$

where

r	is the deficit ratio
C_s	is the saturation value of the oxygen content at the temperature prevailing during the trial (mg 1^{-1})
C_o	is the oxygen content before aeration (mg 1^{-1})
C_e	is the oxygen content after aeration (mg 1^{-1})
h	is the height of the head (m)
$P1$	is the tail water depth and flow factor derived from

$$P1 = -0.107 \, q^{0.651} \, d^{-1.04} \, h^{0.602} \quad \text{in which}$$

q is the flow ($m^3 \, s^{-1} \, m^{-1}$)
d is the depth of tail water (m)

P2   is the shape factor of the dam crest and

P2 is 0 when the flow is unbroken
P2 is 0.06 - 0.08 when dividers are used (Figure 5).

The deficit ratio obtained from river water (Kalajoki River) gave a 0.1 better result than that from water below the trial weirs. The following limit values must be used in Equation 2:

Fall height 0.5 - 1.5 m
Depth of tail water below 1.2 m
Flow 0.04 - 1.2 $m^3 \, s^{-1} \, m^{-1}$
Dividers similar to those shown in Figure 5.

Aeration at Power Plants

At both the Oksava and Padinki Power Plants there are horizontal Kaplan turbines. At the power plants, trials were made introducing air in front of a screen, directly in front of the turbines and on the suction side. At Oksava use of the turbines own suction was found to be best. At Padinki the best results were obtained by feeding air into the front of the turbine. At Padinki the turbine's own suction did not pull sufficient air into the turbine. In addition, at this power station there were operational difficulties in the use of self-suction (Figure 6). When air was fed in through the screen, part of the air was lost upwards, before entering the turbine and this impaired aeration.

At the Oksava Power Plant, the height of the head is 10.5 m, and the measured flow 32 $m^3 \, s^{-1}$. The corresponding values for the Padinki Power Plant are 4 m and 30 $m^3 \, s^{-1}$. During the aeration trials, the flow at the power plants was between 15 and 22 $m^3 \, s^{-1}$. When the air was fed into the front of the turbine, the amount of air provided by the blower was 0.34 $m^3 \, s^{-1}$. At Oksava, using self-suction, the amount of air sucked into the turbine was at maximum 0.4 $m^3 \, s^{-1}$. For the transfer of air, steel pipes with a diameter of 150 mm were used.

The aeration results are shown in Table 1. The oxygen content of the water rose in the power plant aeration trials by 1.5 mg 1^{-1} on average. The oxygenation efficiency was calculated according to equation 3.

$$QE = -\frac{3.6 \; Q \; C_s \; \ln \left(\dfrac{C_s - C_e}{C_s - C_o}\right)}{N_G} \qquad (3)$$

where:

- QE is the oxygenation efficiency (kg kWh^{-1})
- Q is the flow (m^{3}s^{-1})
- C_s is the oxygen saturation value, when the water temperature is 0°C
- N_G is the total power of aeration (kW).
- C_e is the oxygen content after aeration (mg l^{-1})

Equation 3 gives the oxygenation efficiency, when the temperature of the water is 0°C and the original oxygen content of the water is 0 mg l^{-1}.

Using self-suction at Oksava, the loss in output at the power plant due to aeration was 130 kW. At the Padinki Power Plant, output loss was 65 kW. In addition, the blower at Padinki required 32 kW of power. Aeration is not needed in mid-winter, the time at which the demand for power output is greatest.

The results show that horizontal Kaplan turbines can be used with a relatively good efficiency ratio to raise oxygen content by c. 1 mg l^{-1}. If required, the oxygen content can be raised even more, to over 2 mg l^{-1}, but at this point the efficiency of the aeration drops. The possibility of using self-suction depends on the height of the head at the power plant and the position of the turbine relative to the tail water. At Padinki, sufficient suction was not generated. In addition, the automatic starter on the turbine did not function when the aeration vents on the suction side were open.

Vertical Kaplan turbines would appear to be better than horizontal Kaplan turbines for aeration purposes (Wagner 1958; Albrecht & Imhoff 1973). In vertical Kaplan turbines, the air can be evenly distributed throughout the body of the turbine, whereas in horizontal Kaplan turbines the air tended to collect in the upper part of the turbine and significantly decreased its efficiency. An attempt was made to prevent this by introducing as much of the air as possible into the bottom edge of the turbine.

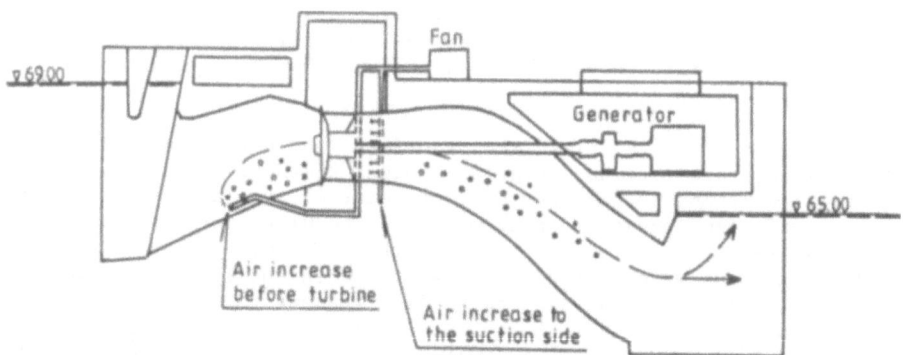

Figure 6. Air introduced into the Padinki Power Plant in front of the turbine or pressure side and after the turbine (suction side).

Table 1. The aeration results achieved at the Padinki and Oksava power plants.

Power plant and aeration system	Year	Flow $m^3 s^{-1}$	Amount of air $m^3 s^{-1}$	Av. oxygen content mg l^{-1} before	after	Oxygenation efficiency kg kWh^{-1}
Oksava						
feeding in front of the screen	1979	17.5	0.33	6.3	7.4	1.9
self-suction	1980	20	0.39	6.6	7.9	2.4
self-suction	1981	21	0.32	6.8	7.8	2.3
self-suction	1982	20	0.41	5.9	7.6	2.8
Padinki						
blower	1980	15	0.34	4.1	5.8	2.0
blower	1981	20	0.34	5.8	7.2	2.4
blower	1982	22	0.34	4.6	6.1	2.8

Water Circulation in Hautaperä Reservoir

In water circulation in lakes or reservoirs, oxygen-rich surface water was led into the hypolimnion through the current created by a propeller pump.

The energy required of internal pumping in a lake is fairly small. In Hautaperä Reservoir, which has a volume of 50×10^6 m^3, sufficient surface water to prevent oxygen depletion can be led into the hypolimnion using a pump that only requires 3 kW (Figure 7).

The propeller pump was located in the deepest part of the reservoir and marked with floats. The floats mark the pump's position in summer, in a mild winter the place is marked by a hole in the ice cover, but during long periods of extreme cold, the opening in the ice will close over. Particularly in winter, the area around the pump has been found to be an exceptionally good place to catch fish.

This method of oxygenation is appropriate when partial disturbance of the thermal layers will cause no great damage. Another, obvious requirement for using this method is that the surface water of the lake or reservoir must have sufficient oxygen. For example, in the Uljua Reservoir in the Siikajoki Watercourse, slightly to the north of the Kalajoki River, a similar method of water circulation was tried in the late winter of 1984. In that trial, oxygen disappeared from the entire volume of water and there was some mortality of fish and some dispersal of fish into streams.

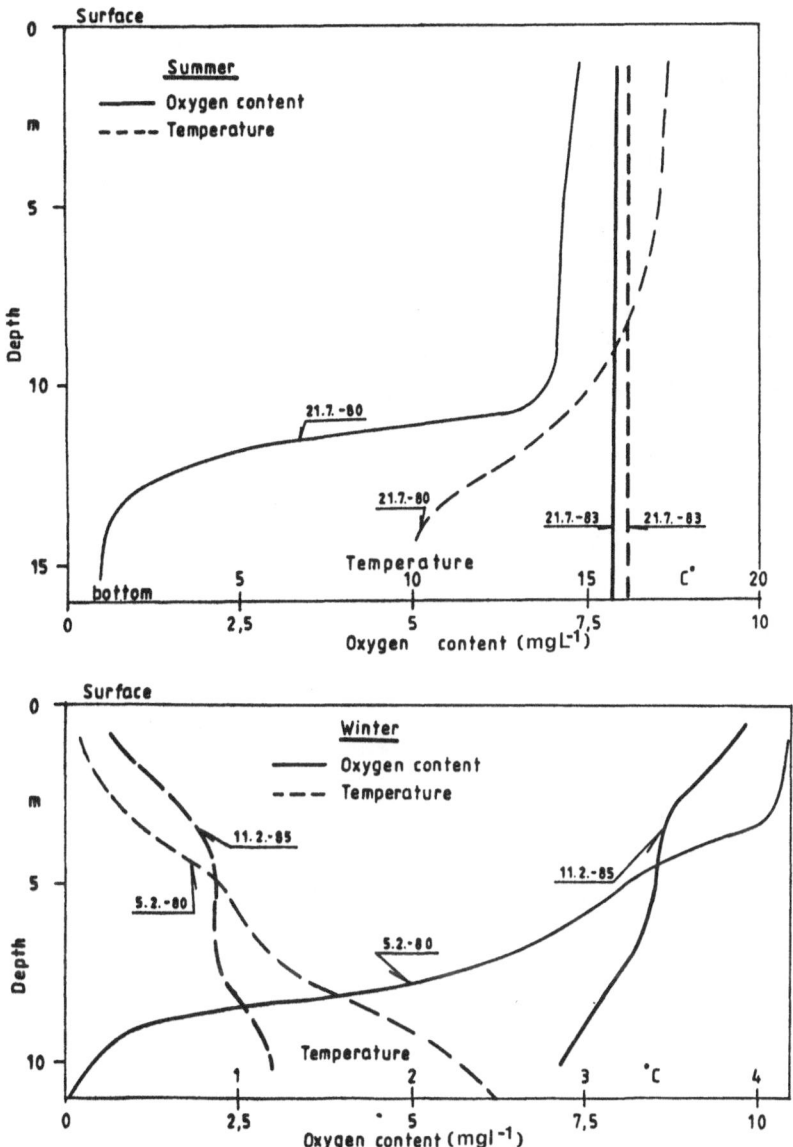

Figure 7. The effect of water circulation in Hautaperä Reservoir on the oxygen content of the water and on temperature in summer and winter. The water circulation has been used during the years 1983-1985.

REFERENCES

Albrecht, D. and Imhoff, K. R. 1973. Erfarungen mit der Künstlichen
    Ruhrbelüftung. GWF Gas Wasserfach Wasser/Abwasser. **114**(3):131-137.
Gameson, A. L. H. 1957. Weirs and the aeration of rivers. J.
    instn. Wat. Engrs. **11**: 477-490.
Imhoff, K. and Albrecht, D. 1972. Influence of temperature and turbulence
    on the oxygen transfer in water. Advances in Water Pollution Research,
    pp. 23-33. Proceedings of the 6th International Conference,
    Jerusalem.
van der Kroom, G. T. M. and Schram, A. H. 1969. Weir aeration - Part I:
    Single free fall. Water **2**(22): 528-537.
Lakso, E. 1984. Happitilanne ja sen parantaminen Kalajoen vesistossä (in
    Finnish). Helsinki National Board of Waters. p. 61-74.
    Vesihallituksen monistesarja 1984: 277. ISBN 951 - 46 - 8019 -7, ISSN
    0358 - 7169.
National Board of Waters. 1980. Hydrological yearbook 1976 - 1977.
    Publications of the Water Research Institute. Helsinki.
Pöpel, H. J. 1976. Aeration and gas transfer. Delft University of
    Technology. 169 p.
Wagner, H. 1958. Versuche mit der künstlichen Flusswasser Belüftung.
    Dtsch. Gewässerk. Mitt. **2**(4): 73-79.

SECTION V

SPECIAL TOPICS

POSSIBLE EFFECTS OF THE PROPOSED EASTERN ROUTE DIVERSION OF CHANGJIANG
(YANGTZE) RIVER WATER TO THE NORTHERN PROVINCES WITH EMPHASIS ON THE
HYDROBIOLOGICAL ENVIRONMENT OF THE MAIN WATER BODIES ALONG THE TRANSER
ROUTE

Huisheng Wu

Institute of Hydrobiology
Academia Sinica
Wuhan, Hebei Province, People's Republic of China

INTRODUCTION

In China there is an abundance of water in the south, the main sources
being the Changjiang (Yangtze) River and its tributaries, while in the
north, where the main supply is the Huanghe (Yellow) River, there is a
shortage of water. The average annual rainfall throughout south China is
more than 1000 mm while in the north it is only 500-600 mm (Figure 1). The
Changjiang River Basin and areas to the south receive 75% of the country's
total surface runoff, while the Huanghe Basin and areas to the northwest
receive only 8%. The Changjiang River has an annual flow of close to 100 x
10^{10} m^3 per year, while the Huanghe River which is almost the same length,
has an annual flow of less than 5 x 10^{10} m^3. This disparity in water
distribution indicates the need for eventual transfer of water from south
to north.

PROPOSED INTERBASIN TRANSFERS

Massive interbasin transfer schemes have been proposed and three
diversion routes from the Changjiang have been considered (Figure 2). One
alternative, known as the Western Route, would involve construction of a
high dam in the upper reaches of the Changjiang River and canals and
tunnels skirting the mountains to channel water to the northwest. A second
alternative, the Central Route, would divert water from the "Three Gorges"
area of the Changjiang to the Danjiang Reservoir on the Hanshui River, a
tributary of the Changjiang River. From there it would be diverted north,
crossing the Huanghe River near the city of Zheng-zhou in Henan Province.

In recent years a third alternative (Figure 3), the Eastern Route, has
been proposed which is the subject of this paper. This alternative would
involve the diversion of water from the lower reaches of the Changjiang
River near the city of Yangzhou in Jiangsu Province. The water would be
channelled through the Grand Canal and the Jiangdu flood prevention works
and would pass through a series of lakes including Gao-Shao Lake, Baoying
Lake, Baima Lake, Hongze Lake, Loama Lake, and Nansi Lake. It would pass
under the Huanghe River near Dongping Lake in Shantong Province and be
delivered to Tianjin Municipality in Hebei Province.

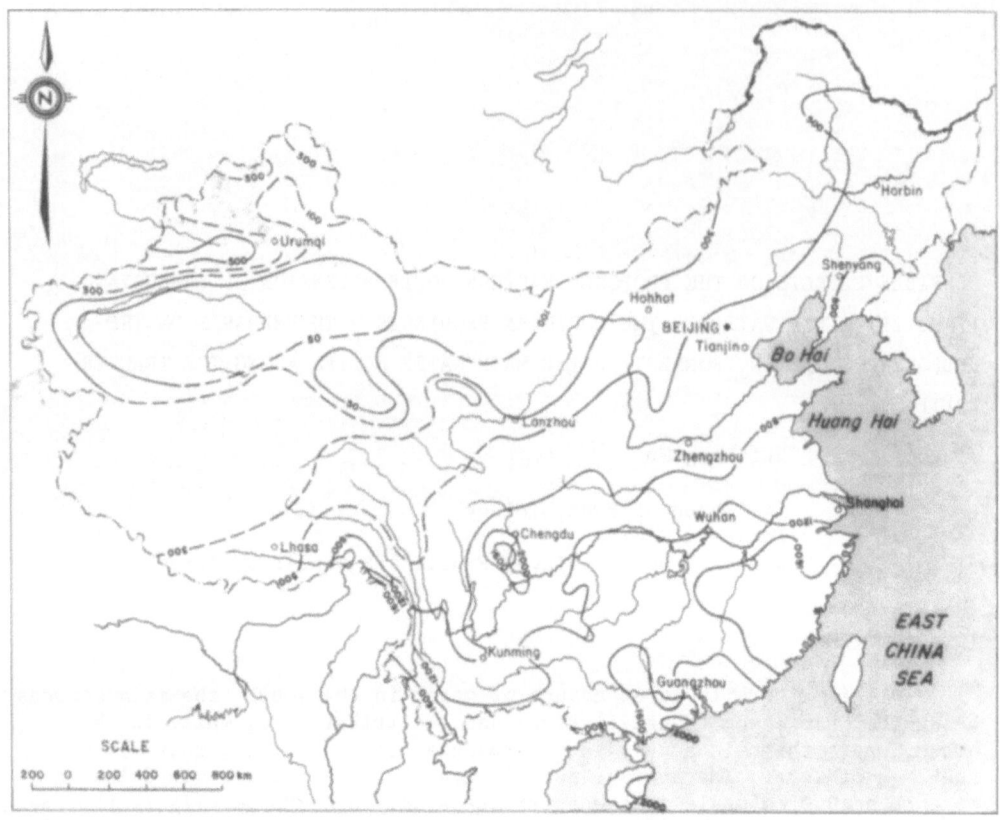

Figure 1. People's Republic of China, average annual precipitation (mm).

Water would be diverted at a rate of 1000 m^3 s^{-1} with a total annual diversion of 30 x 10^9 m^3. The total length of the watercourse would be 1150 km. Land elevation increases to the north, and in the vicinity of the Huanghe River is approximately 40 m above sea level. A series of 15 dams and 30 pumping stations, with an electrical power demand of about 1 million KW would be required to raise the water.

The water supplied by this project would effectively eliminate water shortages in North China, fulfilling demands for industry, irrigation, and navigation.

BIOLOGICAL IMPLICATIONS OF THE EASTERN ROUTE DIVERSION

The implementation of this project could have wide reaching biological effects on the contributing river system, the water bodies along the diversion route, and on the receiving water bodies.

Contributing Rivers

Following reduction of river flows due to diversion (Table 1), increased concentrations of salt water will alter hydrographic conditions in the Changjiang River estuary and the adjacent sea. Increased salinity and decreased nutrient levels in the water of the deltaic regions may have the following effects:

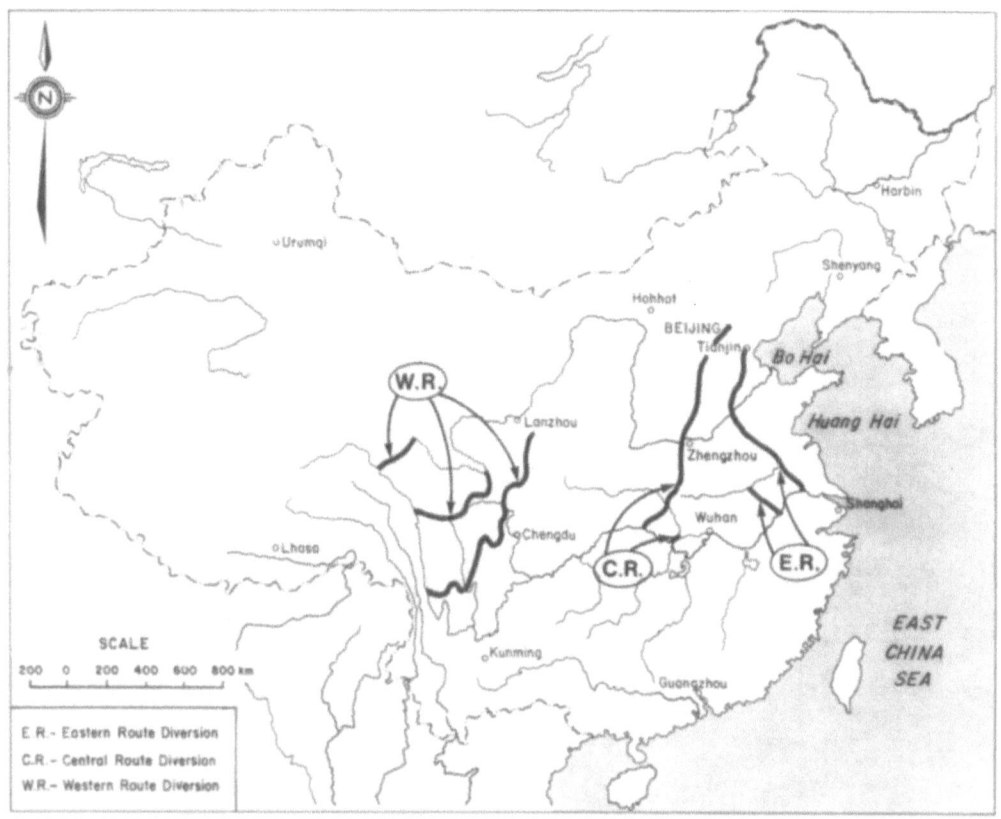

Figure 2. Alternative routes for the diversion of Changjiang (Yangtze) River water to northern China.

Table 1. Comparison of volume of water diverted with mean monthly flow in Changjiang River (based on 1958 to 1977 records) in dry and flood seasons.

Season	Month	Mean Monthly Flow (10^8 m^3)	Percent of Flow Diverted
Dry season	December	371.3	7.0
	January	363.4	10.0
	February	266.4	10.4
	March	379.4	7.0
Flood season	June	1,045.4	2.5
	July	1,312.6	2.0
	August	1,093.4	2.4
	September	958.1	2.7

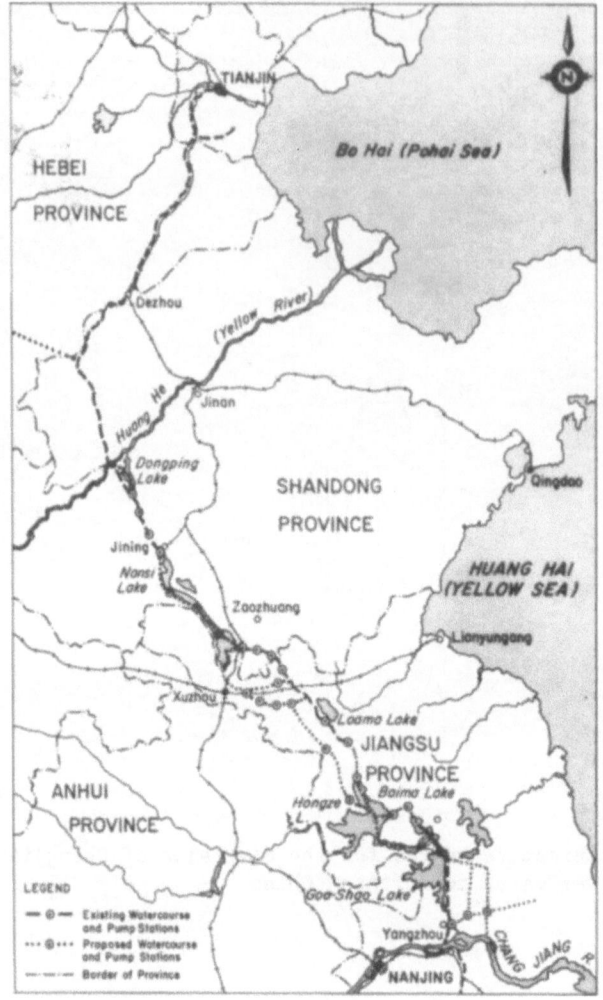

Figure 3. Proposed eastern route diversion indicating watercourse and related lakes.

1) There will be changes in conditions at the feeding grounds of fry and fingerlings of several fish species that may affect the number and vigour of recruits into the population. The species involved include *Hilsa reevsii* (Reeve's shad), *Coilia mystus, Coilia* spp. (anchovies), *Hemimisalanx* sp., *Protosalanx* sp. and *Acipenser sinensis* (Chinese sturgeon).

2) There may be an increase in abundance and diversity of halophilic species that provide forage in the estuary. Certain pelagic sea fishes, such as *Scomber japonicus* (common Japanese mackerel), *Scromberomerus* spp. and Carangidae (cavallas) may move further into the coastal region.

3) There would be changes in the hydrographic condition of the East Sea and the southern Yellow Sea, possibly affecting the hydrobiological environment of seven fishing grounds, including the Zhoushan, the most productive fishing ground in China. Size and composition of the catch

may be altered. The migratory routes of *Pseudosciaena crocea* (large yellow croaker), *Pseudosciaena polyactis* (little yellow croaker), and *Trichiurus humela* (hairtail) may also be affected.

4) Analysis of past fish catch and hydrographic data for the Changjiang River (Yu 1983) showed no significant correlation between catch of migratory fishes and flows in that year. It was found, however, that there was a relationship between flows and the relative abundance of Reeve's shad offspring. Similar changes could be expected following diversion.

The proposed diversion would begin at the Jiangdu Water Control Project. This project presently consists of the Jiangdu Pumping Station, eight sluice gates, four locks, and several canals. The system drains an area of 4000 km^2 and regulates the flow of the Changjiang, Huai, and Lihsia Rivers. The system has a pumping capacity of 460 m^3 s^{-1}, and lifts water 20 m. The effects of the present pumping station illustrate the impacts that the larger scale proposed diversion would have on fisheries.

1) When operating, the pumps suck up large number of eggs, fry, fingerlings and adult fish. Sudden water pressure changes in the turbine can cause the egg membranes to rupture.

2) Adult fish, including *Anguilla japonica* (common eel), *Ctenopharyngodon idellus* (grass carp), *Hypophthalmichthys molitrix* (silver carp), *Aristichthys nobilis* (bighead) and *Mylopharyngodon piceus* (black carp) have been found cut to pieces by impellers.

3) Certain numbers of fry and fingerlings pass through the pumps unharmed and may increase the fish stocks of rivers and lakes along the diversion route. Amphidromous fish such as Reeves' shad, however, may be prevented from migrating to the sea. Yu (1983) found that the Jiangdu pumping station sucked up large numbers of young Reeves' shad. As well, fishing boats from the Jiangdu Fishery brigade caught 200,000 to 300,000 shad per day at the project sluice gates.

4) The dam associated with the project presents a barrier to many migrating species. *Eriocheir sinensis* (the Chinese mitten crab) is delayed in its migration to the sea, while other species such as the common eel are unable to enter lakes where they are normally reared. As a rule, the grass carp, black carp, silver carp and bighead migrate from lakes to rivers during spring in order to spawn. The obstruction to lake-river passage is detrimental, therefore, to the populations of these species both in rivers and lakes.

5) During diversion, water passing through the pumping stations will become hypersatured with nitrogen. Sudden water pressure decreases can result in the release of dissolved nitrogen into the blood, termed "bubble disease", of fry and fingerlings passing through the turbines. Many cases of this disease have been reported in Japan and North America (Renfro 1963; Mathias & Barica 1985).

In a 1981 study, fry that had passed through the Gezhouba Dam on the Changjiang River were collected and examined by fish ecologists from the Institute of Hydrobiology, Academia Sineca. The results of this study are summarized in Table 2. Given the number of dams and pump stations involved in the proposed diversion mortality due to bubble disease in fry and fingerlings will probably be a major problem.

Table 2. Fry mortality due to bubble disease in the middle reaches of the Changjiang River near Yidu County, Hubei Province.

Date	Species	Total No. Examined	No. Living	No.Dead Due to Bubble Disease	Mortality Rate %
June 9, 1981	Grass carp	149	60	89	60
	Black carp	102	42	60	59
June 14, 1981	Grass carp	90	46	34	38
	Black carp	16	13	3	19

Diversion Route and Receiving Water Bodies

Water diverted from the Changjiang River will be rerouted through several shallow lakes. Wu and Jao (1958) examined the characteristics of many of the lakes along the middle and lower reaches of the Changjiang River. The climate throughout most of this area is sub-tropical and has one of the highest concentrations of lakes in the country. There are approximately 1,760 lakes covering an area of some 3,334,000 ha in the alluvial district of the Changjiang River. Those used for fish farming range in area from 300 to 30,000 ha, are generally from 2 to 6 m deep and rarely exceed 10 m in depth. These are temperature lakes of the third order (according to Forel, modified by Whipple 1927). They are old and smooth-bottomed with thick muddy sediments containing large amounts of decaying plant matter and dissolved organic compounds. The littoral zones have gently sloping bottoms with generally marshy bays.

The surfaces of these lakes may freeze over for short periods in winter, but during the rest of the year there is little temperature difference between surface and bottom water. Throughout the April to November fish farming period, the water temperature generally exceeds 20°C, but in April and November it may drop to as low as 15°C. Maximum temperatures, of approximately 38°C are reached in late August or early September.

Surface and bottom pH are similar, varying generally between 7.5 and 8.5. Nutrient concentrations tend to be high: phosphorus 0.002-0.0079 mg l^{-1}, ammonia (NH_3) 0.7-1.07 mg l^{-1}, nitrate (NO_3) 0.62-5.98 mg l^{-1}, silicate (SiO_2) 1.16-1.90 mg l^{-1}. Circulation of water throughout the lakes for most of the year aids in the distribution of the nutrients. This in combination with the thick muddy bottom sediments has resulted in most of these lakes being covered with macrophytes, predominantly submerged species, the most common of which are *Potamogeton maackianus*, *P. crispus*, *Hydrilla verticillata* and *Vallisneria spiralis*. Total macrophyte biomass may reach 20,000 kg ha^{-1} dry weight.

There is a low abundance of gastropods and bivalves with *Viviparus (Lodiopama) quadratus*, *Radix auricularia plicatula*, *Anodonta woodiana*, *Unio douglasiae*, *Corbicula fluminea* and *Hyriopsis cumingii* being the most common species. *Chironoms* and other insect larvae are common.

Most planktonic species present are cosmopolitan and calciphilous, and although their diversity is low, their abundance is high. Quantities of phytoplankton usually range from 200,000 to 300,000 L^{-1} and zooplankton from 2,000 to 5,000 L^{-1}. Plankton abundance tends to be low in the warm months of August and September. The most common phytoplankton species are: *Microcystis aerugenosa*, *Anabaena spiroides*, *Pandorina morum*, *Endorina elegans*, *Sphaerocystis schroeteri*, *Dictyosphaerium pulchellum*, *Ankistrodesmus falcatus*, *Synura uvella*, *Dinobryon setularia*, *Melosira granulata*, *Fragilaria* spp., *Tabellaria* spp., *Surirella splendida*, *Cryptomonas* spp., and *Ceratium hirundinella*. Common zooplankton species include: *Strombidium conicum*, *Strobilidium gyrans*, *S. velox*, *Halteria*, *grandinella*, *Askenasia volvox*, *Polyarthra platypera*, *Keratella cochlearis*, *Bosmina longirostris*, and *Encyclops serrulatus*.

The lakes are eutrophic and suitable for fish stocking. Particularly suitable species are *Ctenopharyngodon idellus*, *Mylopharyngodon piceus*, *Hypophthalmichthys molitrix*, and *Aristichthys nobilis*.

The diversion project will have diverse impacts on several of these lakes. Diversion water will be channelled through several very shallow (0.5 m deep) lakes, including Gou-Shao Lake and Baima Lake. As a deep watercourse will be dug through these lakes to accommodate diversion water, they will essentially be drained and will become swamps or totally dry. This will result in a considerable reduction in aquatic fauna productivity including some 800 tonnes annually of shrimp *Macrobrachium nipponense* and *M. asperulum*, and 2,630 tonnes annually of crab *Eriocheir sinensis*.

On the other hand, the water level of several other lakes will be increased. For example, levels of Hongze Lake and Nansi Lake will rise by 0.5 to 1.5 m. This will result in a significant decrease in aquatic macrophytes such as *Phragmites communis* (reed) and *Zizania caduciflora*. Reed and *Z. caduciflora* stands are estimated to presently cover some 27,000 ha in Hongze Lake and 20,000 ha in Nansi Lake. These are important commercial species in China. Reeds are used in the production of paper and mats. *Z. caduciflora* shoots are consumed as a vegetable, fed to cattle and are browsed by herbivorus fish species, such as grass carp, *Megalobrama terminalis* (bream), and common carp.

Sediment carried into the lakes from the Changjiang River will interfere with photosynthesis by aquatic plants. The abundance of many aquatic plant species such as *Hydrilla verticillata* and *Vallisneris spiralis* (eel grass) will be reduced. Similarly, the abundance of important commercial species including *Nelumbo nucifera* (lotus), *Euryale fexose*, and *Trapa natans var. bispinosa* (water chestnut) in Hongze and other lakes will be reduced. The annual production of lotus seed, *E. fexose* seed, and water chestnuts from Hongze Lake is presently 500 to 750 tonnes. These crops have both food and medicinal value and are important export commodities. In addition, the reduction in aquatic plants may adversely affect the commerical fishery in the lakes.

The influx of diversion water into the lakes will dilute existing nutrient concentrations and reduce water temperatures thereby resulting in a decrease in planktonic and mollusc biomass. Assuming a diversion flow rate of 800 m^3 s^{-1}, the flushing rate of the 253,000 ha Hongze Lake would be reduced to 55 days. As a result:

1) The food supply of herbivorous fish, such as grass carp, bream and filter feeders such as the silver carp and bighead, will be greatly reduced.

2) Spawning conditions for fish that lay adhesive eggs on aquatic macrophytes will be adversely affected. Thus, the survival rate of

eggs and, consequently, fish stocks of *Cyprinus carpio* (common carp), *Carassius auratus* (crucian carp) *Megalobrama terminalis* and *Parabramis pekinensis* (bream) will decline.

3) Conditions for small pelagic fish such as *Coilia brachygnathus*, *C. extenses*, *Protosalanx hyalocranius* and *Neosalanx tankankeii tuihuensis* will improve. These fishes may become dominant in affected lakes and replace economically important fishes such as the grass carp, black carp, silver carp, bighead, bream, common carp and crusian carp.

4) The reduction of mollusc populations in several lakes would have serious consequences. Snails and mussels are important high protein foods for Chinese and Japanese people. Hongze Lake contains 755.25 kg of mollusc ha^{-1}. Annual production is 170,000 tonnes. Loama Lake contains 3,000 kg ha^{-1} with an annual production of 80,000 tonnes. The 1973 and 1974 export of frozen snail meat from Jiangsu Province to Japan amounted to 265 tonnes.

In addition to supplying food for human consumption, *Viviperus chinensis*, *Parafossarulus striatules*, *Anodonta woodiana*, *Hyriopsis comingii*, and *Corbicula fluminea* are important food sources for black carp, common carp, and bream.

It is estimated that the total loss to the fresh water fishery in the affected lakes, not including the mollusc, will amount to about 20,000 tonnes per year. Measures could be taken to mitigate these impacts.

The creation of some 20 to 30 small lakes ranging in size from 1,340 to 3,440 ha could, under conditions of proper management, compensate for the productivity lost due to the drainage of the large shallow lakes. In order to achieve the necessary productivity, several measures that have been implemented at East Lake near Wuhan (Liu 1984) should be adopted in the management of the newly created lakes. These include stocking with appropriate fish species, rearing large numbers of sizable fingerlings for stocking, improvement of fish screens, control of predatory fish, and improvement of fish harvesting methods. The adoption of these measures at East Lake resulted in a yield of 801 tonnes in 1978 compared to 180 tonnes in 1971.

Concern has been expressed by the Jiangsu Provincial Institute of Schistosomiasis and other institutions that the diversion would result in the introduction of the snail, *Oncomelania hupensis*, the intermediate host of schistosomiasis, to north China. Prior to the founding of the People´s Republic of China, schistosomiasis was one of the most serious parasitic diseases in China, affecting over 7 million people, and was widely distributed throughout the lower reaches of the Changjiang River. Although the incidence of the disease has been reduced by 70% in the last 30 years, there is concern that the range of the disease will spread beyond its present northern limit of 33°15´ north latitude in Jiangsu Province. An increase in the area affected by schistosomiasis and a spread of other disease organisms was reported following construction of the Aswan Dam in Egypt.

Several studies have been undertaken to examine the risk of introducing schistosomiasis to northern China through the diversion project (Xiao et al. 1982). In 1978 and 1979 floating debris, including aquatic plants, reeds and wood chips, was collected from the Changjiang River and inspected. In 18,255.5 kg of debris, 5,095 individuals of *Oncomelania* were found, of which only 18 were alive. There are reports that *Oncomelania* cannot tolerate low temperatures. One hundred percent mortality was recorded in one study after 65 days of exposure to air temperatures of $-1^\circ C$ to $-2^\circ C$. In another study, snails were collected in Jiangsu Province and introduced to the city of Dezhou in Shandong Province. After 100 days, only 3.5% of those introduced had survived. The gonads of the surviving individuals had atrophied and propagation was prevented.

Furthermore, no living *Oncomelania* have ever been reported in northern Jiangsu Province although the Grand Canal, which was constructed in ancient China, provided a transport route for the spread of snails that has existed to the present day. Consequently the likelihood of schistosomiasis being introduced to north China through the diversion project appears slight.

The transport of some fish pathogens, *Reovirus* sp. of grass carp, *Myxococcus piscicola* and *Myxococcus* sp. (including the bacteria carrier and virus carrier) from the Changjiang basin to the northern provinces of Shandong and Hebei, however, may be a concern. Bacterial gill rot of freshwater fishes is caused by *Myxococcus piscicola* and "White head and mouth" of fry and fingerlings is caused by *Myxococcus* sp. These are serious fish diseases in China and in waters between $25^\circ C$ and $32^\circ C$, cause serious reductions in the survival of fresh water fishes. These diseases are epidemic to the Changjiang River Basin. Until now, *Reovirus* and *Myxococcus* have not been found in north China, however water temperatures in Shandong and Hebei Provinces from July to September usually reach $28^\circ C$ to $32^\circ C$ and the possibility of the spread of these diseases is a serious concern.

The proposed Chinese interbasin transfer is a colossal project, and it is very important that its benefits and disadvantages be assessed.

ACKNOWLEDGEMENTS

I am indebted to Prof. Ronald W. Davies and Prof. Jiankang K. Liu for helpful discussions and also to Prof. Rongwei Xiao and Dr. Zuefang Yu, who sent me reprints of their papers. Ron Middleton contributed helpful comments, organizational and English editing for which I am grateful.

REFERENCES

Liu, J. K. 1984. Lakes of the Middle and Lower Basins of the Chang-Jiang (China) Lakes and Reservoir. (Ed. by F. B. Taub). Chapter 14. Elsevier Science Publishers, B. V. Amsterdam.

Mathias, J. A. and Barica, J. 1985. Gas supersaturation as a case of early spring mortality of stocked trout. Can. J. Fish. Aquat. Sci. **42**: 268-279.

Renfro, W. C. 1963. Gas-bubble mortality of fishes in Galveston Bay, Texas. Trans. Am. Fish. Soc. **92**: 320-322.

Whipple, G. C. 1927. The Microscopy of Drinking Water. John Wiley & Sons, Inc. New York.

Wu, H. W. and Jao, C. C. 1958. Fish Farming in the Shallow Lakes in China. Verh. Int. Ver. Limnol., **13**: 765-769.

Yu, X. F. 1983. Possible effects of the partial diversion of Changjiang River water (the Eastern Route) to the northern provinces on the fish stock of the principal water-bodies along the course. In: Long Distance Transfer of Water, Diversion of the Yangtze River Water Transfer to the Northern Provinces in China and Experience of International Transfer of Water. (Ed. by Zuo Daikang, Biswas, A. K. et al.) pp. 223-230. Science Publishing House Beijiang (In Chinese).

Xiao, R. W., Sun, Q. Q. and Chen, Y. T. 1982. Study on the northward migration of *Oncomelania*, after the diversion water from the Yangtze River to the Northern Provinces. Geographical Research 1(4): 73-79. (In Chinese).

CHEMICAL AND BIOLOGICAL CHANGES IN THE

TER RIVER INDUCED BY A SERIES OF RESERVOIRS

M. A. Puig[1];   J. Armengol;   G. Gonzalez;   J. Peñuelas;
S. Sabater;   and F. Sabater

Departament d'Ecologia, Universitat de Barcelona, Spain
[1] Departamento de Ecologia, Fas. Biologia. Univ. de Murcia
Spain

INTRODUCTION

The climatological and topographical characteristics of the Iberian peninsula have made the construction of reservoirs a necessity. Thus, more than 700 reservoirs have been constructed since the beginning of the century, one hundred of which have been the object of intensive study (Margalef et al. 1976). Nevertheless, the effect of these reservoirs on rivers is practically unknown. Garcia de Jalon (1984) reached a similar conclusion in a survey of existing literature on this topic. In this study we attempt to quantify the biological and chemical effects of three reservoirs located in the middle section of the River Ter. An earlier study suggested that these reservoirs had affected the macroinvertebrate communities residing below the reservoirs (Prat 1981). Because a flood occurred during our study we were also able to monitor the regulatory efficiency of these reservoirs.

STUDY AREA

The River Ter originates in the eastern Pyrenees at an altitude of 2500m (a.s.l.) and flows into the Mediterranean Sea. It is 206 km in length with a drainage basin of 3010 km^2. The dominant geologic strata are calcareous, however the headwaters and portions of the Prelitoral mountains are siliceous.

The hydrological characteristics of the Ter are Mediterranean although the headwaters exhibit a Pyrenean influence (Sabater & Armengol 1985). Water levels are lowest in summer and increase in autumn due to an increase in rainfall. Rainstorms also cause occasional flooding. During the past 10 years flows average 29 m^3 s^{-1}. Rainfall in the drainage ranges from 1000 mm in the upper streches to 600 mm at the mouth.

The reservoir system examined in our study, the Sau system, stabilizes flows in the middle and lower portions of the river. Hydrographic and morphometric characteristics of these reservoirs are presented in Table 1.

Table 1. Hydrographic and morphometric characteristics of the Sau system reservoirs.

	SAU	SUSQUEDA	EL PASTERAL
Altitude (m a.s.l.)	410	300	185
Distance to the mouth (km)	91.8	76.8	68.6
Storage capacity (Hm^3)	177	233	2
Area (Ha)	760	463	35
Maximum depth (m)	73	110	33
Mean depth (m)	23.3	50.3	5.7
Residence time (day)	117	146	1.3

MATERIALS AND METHODS

Fifty-two stations were established throughout the length of the drainage. Of these 30 were sampled monthly (October 1982 - October 1983) whereas the remainder were sampled partly. During each sampling period we obtained detailed measurements on: 1. physico-chemical composition of surface and hyporreic water (Sabater & Armengol 1985); 2. periphyton; 3. macrophytes; 4. macroinvertebrates (Puig et al. 1985); and 5. hyporreic fauna.

In this study we present monthly data from three stations above the reservoirs and two stations below the reservoirs (Figure 1).

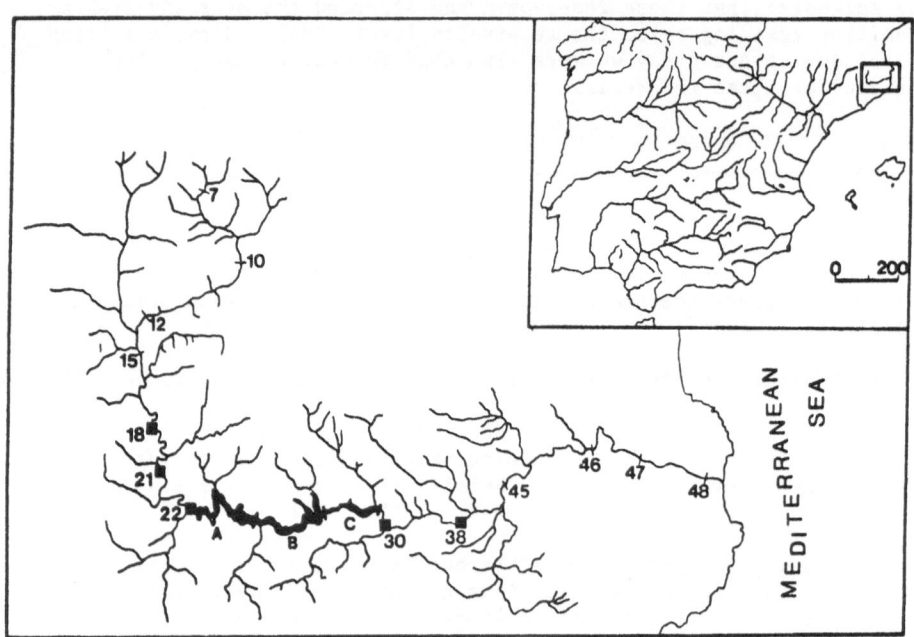

Figure 1. Location of the sampling sites on the Ter River basin. A) Sau Reservoir; B) Susqueda Reservoir and C) El Pasteral.

RESULTS

River Flow and Temperature

The reservoirs have a strong regulatory effect on the lower section of the Ter (Figure 2). They reduce water fluctuations substantially and only floods exceeding discharges >40 m^3 s^{-1} reach the lower section of the Ter. In addition 6 m^3 s^{-1} are diverted from the downstream reservoir, El Pasteral, for domestic consumption in cities located outside this hydrographic region. This represents a reduction of 30% in average annual flows.

Due to the regulations of flow we expected to observe related changes in water temperatures. Annual temperature profiles for station 22 (located just above the reservoirs) and 30 (located just below the reservoirs) are presented in Figure 3. Water temperatures in the river are affected by water diversions because water is diverted from different depths during the year. In contrast to maximum and minimum temperatures observed in station 22 (Figure 3) water temperatures in station 30 were >8°C in winter and <24°C in summer. These temperatures were identical to hypolimnetic temperatures during winter and thermocline temperatures in summer. These similarities exist because downstream releases are made from the hypolimnion in winter and thermocline in summer. Hence the effect of the reservoirs is to reduce the temperature range present in the river although seasonal patterns remain the same (Figure 3).

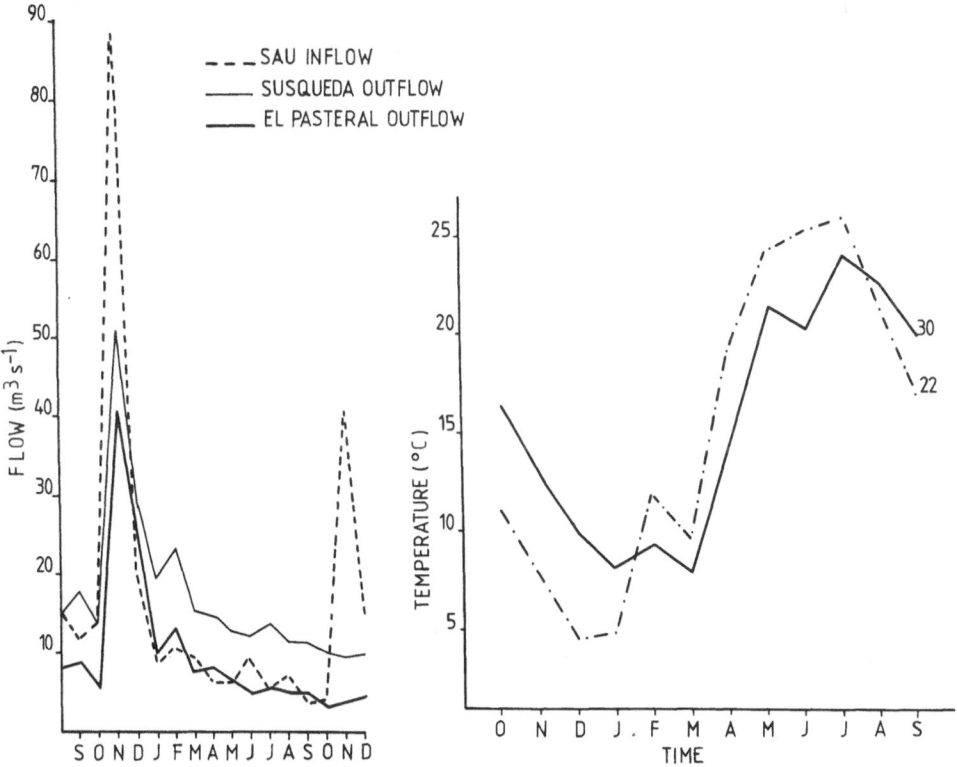

Figure 2. Annual variations of the Sau infow, the Susqueda outflow and the El Pasteral outflow during out study.

Figure 3. Annual temperature changes above the reservoirs (station 22) and below the reservoirs (station 30).

Changes in Major Ion Concentrations

Dissolved salts (Cl⁻, HCO₃, SO₄) increase in concentrations from the headwaters to the mouth of the Ter (Sabater & Armengol 1985). The Gurri River enters the Ter just upstream of station 22. It appears that this contributes considerable amounts of dissolved salts which also is expressed in an increase in conductivity (Table 2). These changes were not substantial in comparison to these observed in other compounds (e.g. nutrients). In general, sulfates and chlorides decreased in concentration whereas bicarbonates increased slightly.

Table 2. Changes in major ion concentrations and conductivity (the mean annual concentrations are indicated only.)

STATION	CONDUCTIVITY ($S.cm^{-1}$)	Cl^- ($mg.l^{-1}$)	HCO_3 ($mg.l^{-1}$)	$SO_4^=$ ($mg.l^{-1}$)
21	284.60	9.92	140.30	46.10
22	556.98	75.65	186.70	59.04
30	459.25	26.80	196.42	52.32

Nutrients

Increases in nutrients in the Ter can be attributed to anthropogenic sources. In general, however, increase in nutrients from anthropogenic sources are localized probably due to their rapid mobilization by bacteria and plants.

We observed substantial differences in nitrogen and phosphorus concentrations between stations (Figure 4 & 5). In station 22 (just

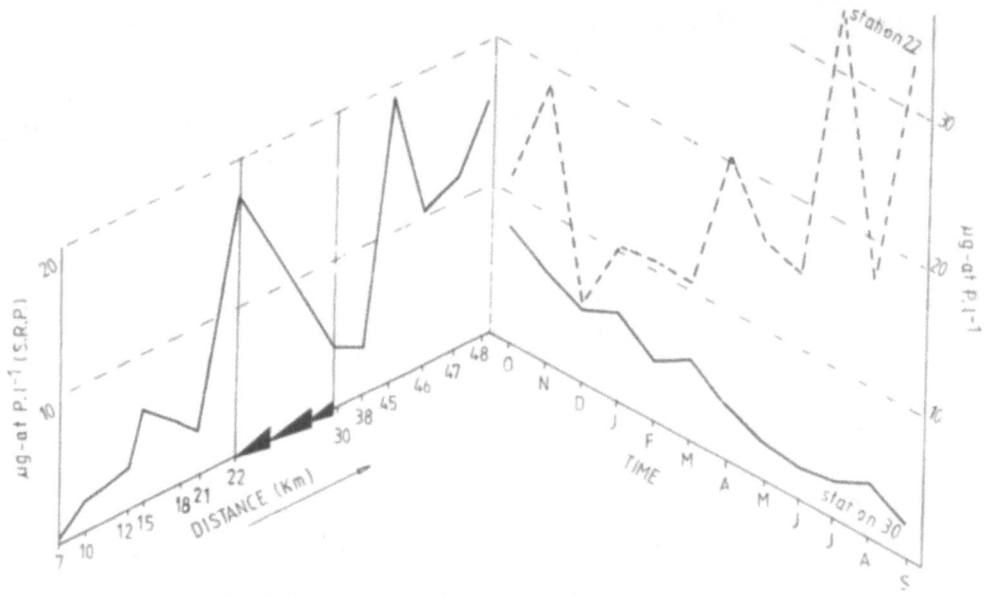

Figure 4. Annual fluctuations and spatial changes of the phosphate concentrations. The annual fluctuations are given for two stations only (station 22, above the reservoirs and station 30, below the reservoirs).

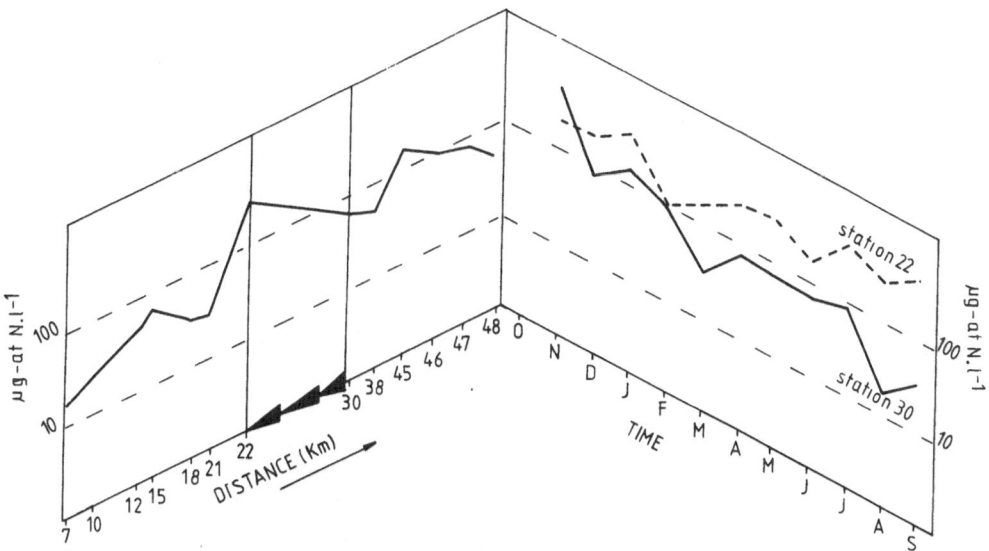

Figure 5. Annual fluctuations and spatial changes of the total nitrogen
concentrations. The annual fluctuations are given for two
stations only (station 22, above the reservoirs and station 30,
below the reservoirs).

upstream from the reservoirs) the river was subjected to intense pollution.
The nutrient inputs raised nitrogen and phosphorus concentrations to levels
higher than those observed elsewhere in the drainage. Eighty-one percent
of the dissolved P and N entering the reservoirs is retained, accumulating
phosphorus in the sediment (Table 3) and substantial amounts of nitrogen
also are released to the atmosphere via denitrification. Downstream from
the reservoirs concentrations of dissolved phosphorus and nitrogen
decrease. Annual fluctuations of these two compounds also are lower below
the reservoirs than above them (i.e. at station 22). There was one
exception, however in the flood of 1982 the concentration of total nitrogen
was higher below the reservoir than at station 22 (Figure 5).

Besides the decrease in total dissolved nitrogen content occurring
within the reservoirs, oxidative processes transform the majority of
dissolved nitrogen leaving the reservoirs to nitrates. In station 22
practically all of the detectable nitrogen was in the form of ammonia
(ratio of nitrates to ammonia was 10:1 (Figure 6).

Table 3. Input, output, and percent retained of P and N
for Sau system reservoirs.

	NITROGEN TOTAL	PHOSPHORUS (S.R.P.)
INPUT (up.at.1^{-1})	271.02	17.68
(ANNUAL MEAN) (Tm.year^{-1})	1918.10	277.00
OUTPUT (ug.at.1^{-1})	63.43	4.20
(ANNUAL MEAN) (Tm.year^{-1})	356.20	52.23
% RETAINED BY RESERVOIRS	81.50	81.00

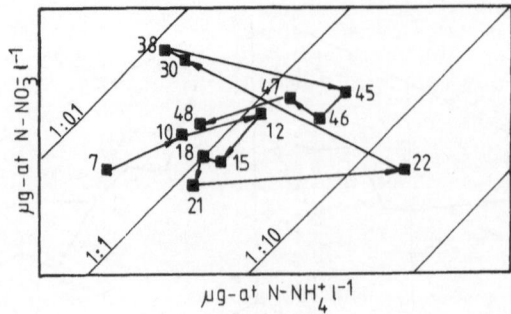

Figure 6. Ratio of nitrates to ammonia along the River Ter.

In conclusion, nutrient concentrations present in station 30 were similar to those observed in station 21, located upstream of the confluence of the River Gurri. The reservoirs act as a nutrient sink that stabilizes nutrient concentrations in the lower portion of the river.

Periphyton

The response of periphyton to various perturbations, and especially the diatom community, manifests the effects of the environmental changes (Patrick 1973). In the stations above the reservoirs (18, 21) the diatom community corresponds to a community typical of the middle reaches of the river, which is composed primarily of species of *Navicula* and *Nitzschia* (Figure 7). The contamination present in station 22, however, reduces the diversity of the community and two species (*Nitzschia palea* and *N. gandersheimensis*) become dominant. Below the reservoirs we encountered a completely different diatom community from that present above the reservoirs. This community was dominated by species of the genus *Fragilaria* (Table 4).

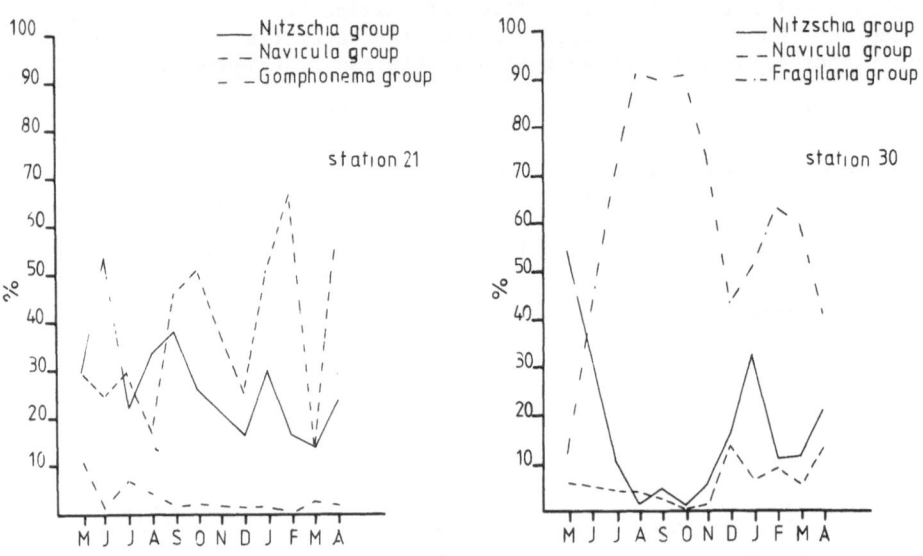

Figure 7. Changes in relative abundance for the dominant species of diatoms in station 21 (above the reservoirs) and station 30 (below the reservoirs).

Table 4. Biological communities observed above and below of the reservoirs in the River Ter.

COMMUNITIES	ABOVE RESERVOIRS 18 & 21	POLLUTED REACH 22	BELOW RESERVOIRS 30 & 38
DIATOM COMMUNITY	*Diatoma elongatum* *Navicula gregaria* *N. tripunctata* *Nitzschia frustulum* *N. romana*	*Nitzschia palea* *N. gandersheimensis*	*Fragilaria construens* *F. construens venter* *F. pinnata* *F. brevistriata*
MACROPHYTES COMMUNITY	*Leptodictyum riparium* *Rhynchostegium riparioides* *Eurhynchium speciosum* *Dialytrichia mucronatus* *Fissidens crasipes*		*Cinclidotus fontinaloides* *Fontinalis antipyretica* *Fontinalis hypnoides* *Fissidens crassipes* *Leptodictyum riparium* *Rhynchosteguim riparioides* *Potamogeton nodosus* *Potamogeton crispus* *Myriophyllum verticillatum* *Calitriche* sp.
MACROINVERTEBRATES COMMUNITY	*Baetis vardarensis* *Ecdyonurus lateralis* *Ephemerella ignita* *Rhyacophila evoluta* *Rhyacophila dorsalis* *Simulium ornatum*	*Chironomus* sp. *Tubificidae*	*Baetis fuscatus* *Ephemerella mesoleuca* *Caenis luctuosa* *Hydropsyche exocellata* *Hydropsyche siltalai* *Cheumatopsyche lepida* *Psychomyia pusilla* *Simulium erythrocephalum*

Chlorophyll $\underline{a}$ values for stations 18 and 21 approximated 130 mg m^{-2} whereas values were much higher, 314 mg m^{-2} in the lower station (30). The cause of this increase could have been due to the higher concentrations of nitrates liberated by the reservoirs (McConnell & Sigler 1959).

The effects of physical barriers (i.e. dams) on the dispersion of benthic diatoms is well known (Holmes & Whitton 1981). This effect is easily observed in the River Ter (Table 4). The autumn flood had a strong impact on the diatom communities of the upper stations which were completely destroyed. Below the reservoirs, the flood caused a restructuring of the community resulting in the loss of *Fragilaria* as the dominant genus and a diversity increase. In both cases, however, the diatom communities returned to their initial composition one month after the flood.

Macrophytes

Responses of the macrophytes to environmental changes attributable to the reservoirs were similar to those observed in the periphyton. The stations above and below reservoirs could be separated by the absence in the upper stations of higher plants (Table 4). Substrate diversity was greater below the reservoirs and resulted in a wide diversity of macrophytes at the stations. Areas with sand and silt substrates were dominated by *Potamogeton nodosus, P. crispus* and *Myriophyllum verticilatum* whereas habitats with stony substrates were dominated by bryophytes (*Cinclidotus fontinaloides, Fissidens crassipes, Fontinalis hypnoides, F. antipyretica, Leptodictyum riparium* and *Rhynchostegium riparioides*). Some differences were present, however, among the upper stations. For example, station 22, in contrast to stations 18 and 21 did not contain any macrophytes which could tolerate the intense pollution present there.

Macroinvertebrates

Ward & Stanford (1983 a; b) has noted that the maximum diversities observed in a river should be in areas experiencing intermediate disturbance. The impacts produced by reservoirs, however, cause a decrease in downstream diversity, which is a result of decreased environmental variability.

In the River Ter we did not observe these patterns. Macroinvertebrate diversity was similar in uncontaminated stations above and below the reservoirs. However the impact of the flood was apparent. We observed a decrease in diversity immediately after the flood which was probably caused by the massive downstream transport of macroinvertebrates from the upper stations. This impact was less pronounced below the reservoirs due to the attenuation of flood effects. Diversity values below the reservoirs returned to levels comparable to initial conditions within seven months; a shorter recovery period than that observed above the reservoirs (Figure 8).

As with periphyton and macrophyte communities, distinct macroinvertebrate communities were observed above and below the reservoirs (Table 4). Ephemeroptera (*Baetis vardarensis, Ecdyonurus lateralis, Ephemerella ignita*) were dominant above the reservoirs, whereas below the reservoirs the dominant species were generally Trichoptera (*Hydropsyche exocellata, H. siltalai, Cheumatopsyche lepida* and *Psychomyia pusilla*).

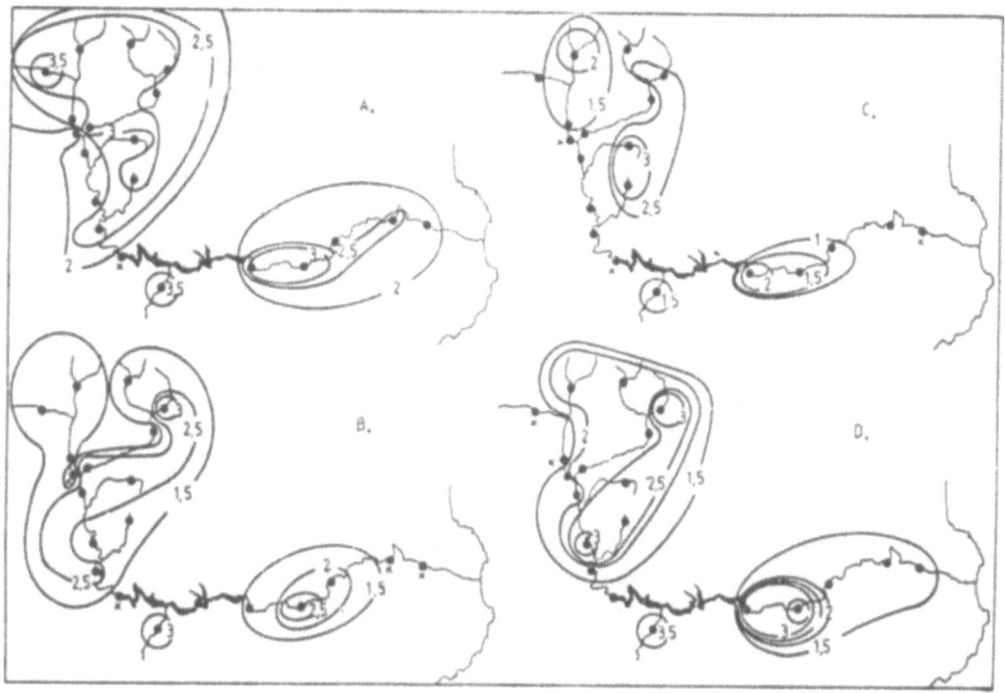

Figure 8. Longitudinal and temporal variation of specific macroinvertebrate diversity: A) October, 1982; B) November, 1982, after the flood; C) December, 1982; and D) June, 1983.

CONCLUSIONS

The interaction between the effects produced by the pollution above the reservoirs (station 22) and the ecological discontinuities caused by the reservoirs themselves resulted in effects which balanced one another. This compensation was very clear in the case of nutrients, where more than 81% of the phosphorus and nitrogen entering the reservoirs was retained. This resulted in the occurence of similar nutrient concentrations in both stations below the reservoirs and in unpolluted stations above the reservoirs.

Distinct biological communities were observed above and below the reservoirs. This situation was facilitated by the presence of a dispersion barrier (i.e. the dams). In the River Ter we did not observe a decrease in macroinvertebrate diversity below the reservoirs. We believe that this was a result of the lack of thermal stress as well as the presence of high substrate diversity below the reservoirs.

ACKNOWLEDGEMENTS

Financial support for this project was provided by the "Comisión Asesora de Investigación Científica y Técnica" (nº 478/81).

The manuscript was translated by Dr. Gary Grossman.

REFERENCES

Garcia de Jalón, D. 1984. Stream regulation in Spain. In: Regulated rivers. (Ed by A. Lillehammer and S. J. Salveit) pp. 481-484. Universitetsforlaget AS, Oslo.

Holmes, N. H. T. & Whitton, B. A. 1981. Phytobenthos of the river Tess and its tributaries. Freshwat. Biol., 11: 139-163.

Margalef, R., Planas, D., Armengol, J., Vidal, A., Prat, N., Guiset, A., Toja, J. and Estrada, M. 1976. Limnología de los embalses españoles. Dirección Gral de Obras Publicas, M.O.P. (eds), Publ. nº 123: 385 pp.

McConell, W. J. and Sigler, W. F. 1959. Chlorophyll and productivity in mountain rivers. Limnol. Oceanogr., 4: 335-351.

Patrick, R. 1973. Use of algae, especially diatoms, in the assessment of water quality. In: Biological methods for the assessment of water quality. (Ed by J. Cairns Jr and K. L. Dickson) pp. 23-47. American Soc. for Testing and Materials, Philadelphia.

Prat, N. 1981. The influence of reservoirs discharge on benthic fauna in the river Ter, N.E. Spain. Proc. of the 3rd. Int. Symp. on Trichoptera. (Ed by G. P. Moretti). Series Entomologica, 20: 293-301.

Puig, M. A., Gonzalez, G. and Recasens, L. 1985. Modelos de distribución de Plecópteros, Efemerópteros, Tricópteros y Simúlidos en el rio Ter. 3rd. Congreso de Limnología, León-Spain.

Sabater, F. and Armengol, J. 1985. Caracterización química del rio Ter. 3rd. Congreso de Limnología, León-Spain.

Ward, J. V. and Stanford, J. A. 1983a. The serial discontinuity concept of plotic ecosystems. In: Dynamics of Lotic Ecosystems. (Ed by T. D. Fontaine and S. M. Bartell) pp. 330-346. Ann Arbor Science Publishers, Ann Arbor.

Ward, J. V. and Stanford, J. A. 1983b. The intermediate-disturbance hypothesis: an explanation for biotic diversity patterns in lotic ecosystems. In: Dynamics of Lotic Ecosystems. (Ed by T. D. Fontaine and S. M. Bartell) pp. 347-356. Ann Arbor Science Publishers, Ann Arbor.

RESPONSES OF EPILITHIC ALGAE TO REGULATION OF

ROCKY MOUNTAIN STREAMS

R. G. Dufford[1], H. J. Zimmermann[2],
L. D. Cline[2], and J. V. Ward[2]

[1]Department of Botany and
[2]Department of Zoology
Colorado State University
Fort Collins, Colorado, U.S.A.

INTRODUCTION

Throughout Colorado many streams have already been impounded, and new reservoirs are currently in the planning stage. Previous research has shown that main stream reservoirs do change the composition and quality of the aquatic flora (Holmes & Whitton 1981; Lowe 1979; Marcus 1980; Lawson & Rushforth 1975; Ward 1982) which in turn affects the communities of aquatic fauna in regulated streams. However, detailed investigation of the influence of reservoirs on downstream algal communities in Colorado has only been accomplished for a site in the South Platte drainage system (Ward 1974). The present study of epilithic algae is part of a long-term effort to determine the effects of regulation on Colorado mountain streams. Zoobenthos data have been reported elsewhere (Zimmermann & Ward 1984).

STUDY AREAS

Nineteen regulated stream segments in six major drainage basins of the Rocky Mountains of Colorado, U.S.A., were studied (Figure 1, Table 1). Sample sites range in elevation between 1951 and 3153 m above sea level and are located on 3rd to 6th order streams. Most of these streams are regulated by large deep-release storage reservoirs. However, Brainard and Manitou are shallow surface-release storage reservoirs. Both Williams Fork and Estes Reservoirs have a limited hydroelectric function, and Estes Reservoir which is relatively shallow (13 m), releases water from the surface and bottom. Detailed descriptive data for these regulated stream segments have been published in other sources (Ward 1984; Ward & Short 1978; Cline & Ward 1984; Zimmerann & Ward 1984). Sampling was done in riffle areas below impoundments and at unimpacted riffle sites above the impoundments except in the case of Granby Reservoir, where an unregulated tributary was selected as a reference site. Although several sites within a study area were frequently sampled, results presented in this paper are based on comparisons between a single downstream site and a reference site above each reservoir.

Table 1. Study areas indicated by solid symbols in Figure 1.

Drainage Basin	Map Code	Stream	Reservoir
South Platte	J	Joe Wright Creek	Joe Wright
	G	S. St. Vrain Creek	Brainard
	H	N. St. Vrain Creek	Buttonrock
	I	Big Thompson River	Lake Estes
	A	South Platte River	Antero
	B	South Platte River	Eleven Mile
	F	Trout Creek	Manitou Lake
Arkansas	Z	Lake Fork of Arkansas	Turquoise
Rio Grande	V	Rio Grande River	Rio Grande
	W	Conejos River	Platoro
	X	Trincherra Creek	Mountain Home
San Juan	U	Los Pinos Creek	Vallecito
	T	Florida Creek	Lemon
Colorado	L	Colorado River	Granby
	M	Williams Fork	Williams Fork
	N	Blue River	Dillon
	O	Frying Pan River	Ruedi
	R	Rifle Creek	Rifle Gap
Yampa	S	Yampa River	Lake Catamount

METHODS

Epilithon, associated epiphytes and detritus were collected by the timed scraping technique (Ward 1974), or by scraping a known area (0.005 m^2) from the upper surfaces of rocks. Each sample was preserved in 5% formalin in the field. Percent compositions of the moss, algal microphytes, algal macrophytes and detritus components were determined concurrently using a square petri dish with a bottom grid. A Sedgwick-Rafter counting cell was used to estimate the percent composition of the algal microphytes and detritus. Relative contributions of the algal divisions were obtained by counting the number of organisms present in six horizontal strips. This method was also used to determine the percent composition of species within each algal division, except in the Bacillariophyta for which the Sedgwick-Rafter counting cell was used to determine the percents of living and dead diatoms identified to the lowest possible taxonomic level. Algal diversity was calculated by using the Shannon-Weaver diversity index (Shannon & Weaver 1949).

The diatoms were cleared of organic matter with 30% hydrogen peroxide. They were then placed on heated coverslips and the hydrogen peroxide allowed to evaporate. The hot coverslips with adhering valves were mounted on slides with Hyrax mounting medium. Five hundred valves per sample were counted to determine species and percent composition within the diatom groupings established during the Sedgwick-Rafter counts.

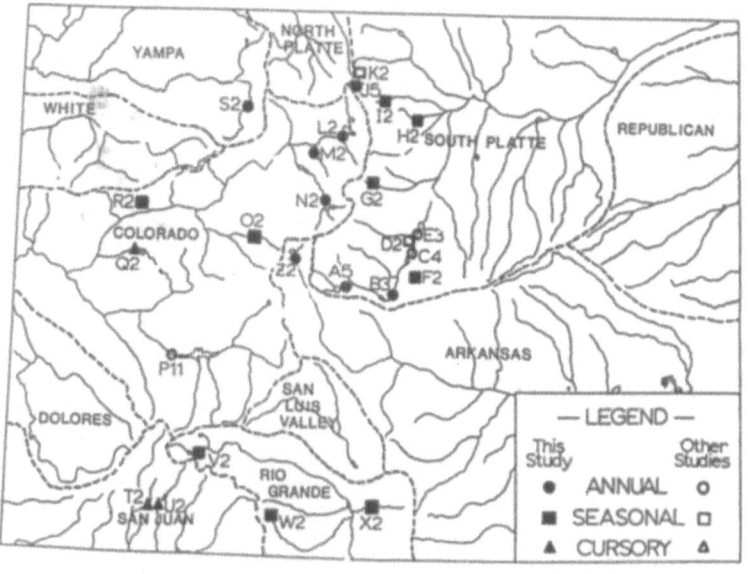

Figure 1.   Regulated stream research sites in Colorado, U.S.A. Dashed lines
            delimit major drainage basins. Letters designate study areas;
            accompanying numerals indicate the number of sites sampled
            within the study area. Solid symbols indicate sites which
            provided data for this study; open symbols represent sites which
            have been described in other studies. (Ward 1974; 1976; Ward &
            Short 1978; Stanford & Ward 1984).

     A Lietz phase contrast microscope and a Lietz orthoplan microscope,
equipped with a Normarski phase interference attachment, were utilized for
algal identification. For standard taxonomic references used for algal
identifications see Ward and Dufford (1979).

     Samples were dried at 60°C until constant weight was attained, after
which they were weighed and then fired in a muffle furnace at 550°C for 24
hours to obtain ash-free dry weight.

RESULTS AND DISCUSSION

     Five algal divisions comprised the epilithic algae: Bacillariophyta,
Chlorophyta, Chrysophyta, Rhodophyta and Cyanophyta. A total of 133
diatoms, 2 golden-brown algae, 12 blue-green algae, 17 green algae and 3
red algae were identified during the study. The aquatic moss, *Fontinalis*,

was also present at several of the stations.

Lotic flora was significantly affected below the deep-release
reservoirs (Figure 2). In the unregulated stream segments, environmental
conditions favored the growth of blue-green algae, diatoms and the
chrysophyte, *Hydrurus foetidus*. The latter was previously reported to
prefer cool, swift-flowing streams (Blum 1956; Prescott 1970; Parker et
al. 1973). At the stations below the reservoirs, periphyton was comprised
predominantly of green algae. Dense populations of *H. foetidus* may also
occur below deep-release reservoirs at low elevations, resulting in a
considerable lowering of the normal altitudinal distribution of this
species (Ward 1974).

Holmes and Whitton (1981) also noted increased importance of
blue-green algae in the upper reaches of streams in England. These
investigators believed that the ability to fix atmospheric nitrogen may
have been a factor which influenced the distribution of the cyanophytes.
Since headwater streams are generally low in nitrates, an organism
possessing this ability would definitely have a competitive advantage. In
the present study, the predominant blue-green at the reference sites was
Nostoc parmeloides (Table 2). This alga has a specialized cell or
heterocyst whose primary function is nitrogen fixation (Steward 1973).

Generally, diatoms considered eutrophic indicators either increased in
importance in the diatom communities below the reservoirs, or were found
only at those locations. *Diatoma vulgare*, which Lowe (1974) considered a
eutrophic taxon, appeared only in association with *Cladophora glomerata*
below the reservoirs where at times it comprised 98% of the diatom
composition. However, other diatoms such as *Asterionella formosa*,
Cyclotella bodanica and *Stephanodiscus niagarae* which also only occurred at
the regulated stations, were actually euplanktonic forms derived from the
reservoirs (Table 2). Diatom communities above the impoundments consisted
mainly of taxa which are typically found in oligotrophic, non-acidic waters
(Cholnoky 1968; Lowe 1974; Patrick 1977). A similar phenomenon occurred
above and below Deer Creek Reservoir, Utah (Lawson & Rushforth 1975).
However, Ross and Rushforth (1980) failed to detect any significant changes
in the periphyton of Huntington Canyon, Utah following upstream
impoundment. Holmes and Whitton (1981) recorded fewer species of diatoms at

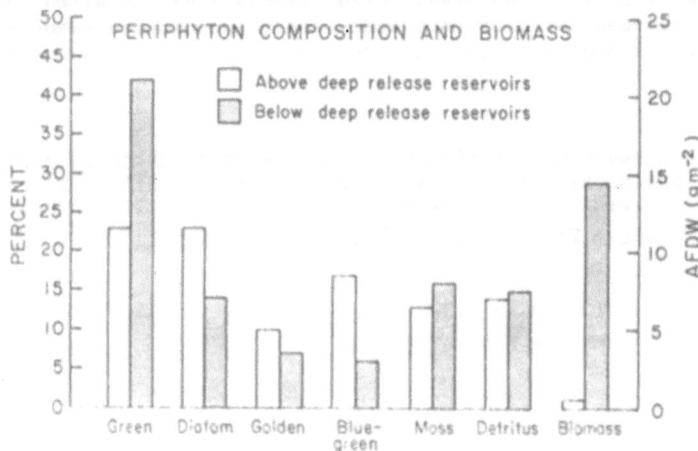

Figure 2. Mean percentage composition of major periphytic components and
mean biomass (ash-free dry weight) above and below deep-release
reservoirs.

Table 2. Selected examples of the responses of algal species to stream
         regulation.

   I. Found only at downstream sites

       A. Planktonic forms

              Asterionella formosa
              Cyclotella bodanica
              Melosira italica
              Stephanodiscus niagarae

       B. Non-planktonic forms

              Diatoma vulgare
              Gomphonema olivaceum
              Tolypothrix sp.

  II. Found only at upstream sites

              Nostoc parmeloides
              Achnanthes marginulata
              Navicula falaisensis var. *lanceola*

 III. Commonly found at downstream sites, rarely upstream

              Cladophora glomerata
              Microspora sp.

  IV. Commonly found at upstream sites, rarely downstream

              Achnanthes affinis
              Caloneis ventricosa var. *alpina*

---

a regulated site in a British river, and believed that the reservoir acted
as a barrier to downstream movement of inocula. Absence of certain diatom
species such as *Achnanthes marginulata* and *A. affinis* at many downstream
sites (Table 2) can most likely be attributed to the environmental
requirements of these species, rather than lack of opportunity to become
established. If reservoirs were barriers to dispersal, downstream
communities would be attentuated versions of those upstream, with taxa that
are rare above reservoirs most likely absent at downstream sites. This was
not the case in the present study.

     Biomass values were substantially higher below the reservoirs
(Figure 2). The principal factors responsible for the enhanced algal
standing crop at the regulated sites were increased nutrient loading,
temperature modification and flow regulation (Lowe 1979). While unregulated
stream segments were ice-covered for several months of the year,
temperatures below deep-release reservoirs never dropped to 0°C. At the
reference sites, spring runoff or sudden spates due to mountain storms
caused dramatic decreases in the algal standing crop. However, flow
regulation below the reservoirs eliminated sudden spates and droughts
(Zimmermann & Ward 1984) producing relatively uniform current and stable
substrates which allowed the development of dense layers of diatoms or
large mats of *Cladophora* with associated epiphytes and trapped detritus.
As a result, the diatom *Didymosphenia geminata* was able to produce dense,

Table 3. Shannon-Weaver diversity of epilithic algal communities above and below 9 reservoirs in 5 major drainage basins in Colorado.

Drainage Basin	Reservoir	Diversity Upstream	Diversity Downstream
South Platte	Eleven Mile	3.29	1.36
	Antero	0.99	2.59
Rio Grande	Rio Grande	2.00	2.90
	Mountain Home	1.09	3.25
Colorado	Williams Fork	2.37	2.54
	Rifle Gap	2.02	2.01
Yampa	Lake Catamount	3.05	2.25
San Juan	Lemon	1.47	2.06
	Vallecito	2.50	1.58

dendritic colonies below Lemon Reservoir, but was rare above it. *Cladophora glomerata* and *Ulothrix zonata* comprised the majority of the green algal biomass. While *U. zonata* was more prevalent at the upstream sites during spring and summer, *C. glomerata* had its best development immediately below the dams where it formed dense mats. However, *U. zonata* also grew with *C. glomerata* throughout the year at the regulated stations. Similar association of these algae was reported by Ward (1976) below Cheeseman Reservoir, Colorado. In addition, Spence and Hynes (1971) and Skulberg (1984) noted the abundant development of *U. zonata* in the regulated portions of the rivers they studied. Since Blum (1956) had previously indicated that *U. zonata* is a cold-water form, Spence and Hynes (1971) believed that lower water temperatures below the reservoir were responsible for the extensive growth of this alga in the Grand River, Ontario. However, *C. glomerata* is commonly found in eutrophic waters (Welch 1980). Lowe (1979) believed that increased nutrient load below deep-release reservoirs favors species such as *C. glomerata* which are normally associated with nutrient-rich habitats. Skulberg (1984) also reported quantitative and qualitative increases in green algal species with high saprobic affinities occurring below a deep-release reservoir on the River Daleai, Norway.

Although species diversity usually decreases with increasing fertility of the water (Wetzel 1975a), Shannon-Weaver diversity often increased below the reservoirs (Table 3). Modifications in temperature and flow which affected biomass were probably also responsible for increased diversity. Temperatures at the downstream stations with greater diversity ranged from 3-15°C favoring the growth of cold-water stenotherms (Patrick 1977; Ward 1974) which were also present above the reservoirs. However, factors such as winter ice cover, or high turbidity associated with low temperatures during spring runoff, limited community development at the upstream stations. Furthermore, spring runoff and spates at these sites caused the algal communities to shift to a pioneer or plagioclimax stage with corresponding decrease in diversity. Since mats of *Cladophora glomerata* expand the number of micro-habitats, thereby increasing the possibility for higher diversity, the presence of these mats at several of the regulated stations may help explain the comparably high diversity values at these locations.

Data from the three surface-release reservoirs (Brainard, Manitou and Estes Lakes) provided some interesting variations. Rather than eutrophic algae dominating the regulated stations, the flora at the sites above and below these reservoirs was almost identical. More notable was the absence of large mats of filamentous green algae at the regulated stations and the lack of dense stands of aquatic vascular plants and bryophytes which were present below some of the deep-release reservoirs.

CONCLUSIONS

The effects of deep-release reservoirs on the phytobenthos in this study are similar to those reported for other localities (Holmes & Whitton 1981; Skulberg 1984; Ward 1974; 1982; Ward & Stanford 1979; Wetzel 1975b). Algal biomass was dramatically higher, filamentous green algae formed dense mats and eutrophic taxa became more prevalent in the flora below the impoundments. The flora of reference and regulated sites of the surface-release reservoirs were similar in species composition. However, the algal biomass was significantly higher at the regulated sites. The increase in algal biomass below the deep-release reservoirs was believed to be caused by increased nutrient loading as well as high water clarity, relatively stable flow and water temperatures which favor cold-stenotherms, yet prevent ice formation.

ACKNOWLEDGEMENTS

Appreciation is extended to Mrs. Nadine Kuehl for typing the manuscript. Support was provided by a research grant to J. V. Ward from the Colorado Experiment Station.

REFERENCES

Blum, J. L. 1956. The ecology of river algae. Bot. Rev., **22**: 291-341.

Cholnoky, B. J. 1968. Die Okologie der Diatomeen in Binnengewässern. 699 pp. J. Cramer, Lehre.

Cline, L. D. and Ward, J. V. 1984. Biological and physiochemical changes downstream from construction of a subalpine reservoir, Colorado, U.S.A. In: Regulated Rivers. (Ed. by A. Lillehammer and S. J. Saltveit). pp. 233-243. Oslo Univ. Press, Oslo.

Holmes, N. T. H. and Whitton, B. A. 1981. Phytobenthos of the River Tees and its tributaries. Freshwat. Biol., **11**: 139-163.

Lawson, L. L. and Rushforth, S. R. 1975. The diatom flora of the Provo River, Utah, U.S.A. Bibliotheca Phycologica, **17**: 1-149.

Lowe, R. L. 1974. Environmental requirements and pollution tolerance of freshwater diatoms. 334 pp. EPA-670/4-74-005, Cincinnati.

Lowe, R. L. 1979. Phytobenthic ecology and regulated streams. In: The Ecology of Regulated Streams. (Ed. by J. V. Ward and J. A. Stanford). pp. 25-33. Plenum Press, New York.

Marcus, M. D. 1980. Periphytic community response to chronic nutrient enrichment by a reservoir discharge. Ecology, **61**: 387-399.

Parker, B. C., Samsel, G. L. and Prescott, G. H. 1973. Comparison of macrohabitats of macroscopic subalpine stream algae. Amer. Midl. Nat., **90**: 143-153.

Patrick, R. 1977. Ecology of freshwater diatoms and diatom communities. In: The Biology of Diatoms. (Ed. by D. Werner). pp. 284-332. Univ. of Calif. Press, Berkeley.

Prescott, G. W. 1970. How to Know the Freshwater Algae. 348 pp. Wm. C. Brown Co., Dubuque, Iowa.

Ross, L. E. and Rushforth, S. R. 1980. The effects of a new reservoir on
the attached diatom communities in Huntington Creek, Utah,
U.S.A. Hydrobiologia, **68**: 157-165.

Shannon, C. E. and Weaver, W. 1949. The Mathematical Theory of
Communication. 125 pp. Univ. of Illinois Press, Urbana.

Skulberg, O. M. 1984. Effects of stream regulation on algal vegetation.
In: Regulated Rivers. (Ed. by A. Lillehammer and S. J. Saltveit).
pp. 107-124. Oslo Univ. Press, Oslo.

Spence, J. A. and Hynes, H. B. N. 1971. Differences in benthos upstream and
downstream of an impoundment. J. Fish. Res. Board Can., **28**: 35-43.

Stanford, J. A. and Ward, J. V. 1984. The effects of regulation on the
limnology of the Gunnison River: A North American case history. In:
Regulated Rivers. (Ed. by A. Lillehammer and S. J. Saltveit).
pp. 467-480. Oslo Univ. Press, Oslo.

Stewart, W. D. P. 1973. Nitrogen fixation. In: the Biology of the
Blue-Green Algae. (Ed. by N. G. Carr and B. A. Whitton).
pp. 260-278. Blackwell, Oxford.

Ward, J. V. 1974. A temperature-stressed stream ecosystem below a
hypolimnial release mountain reservoir. Arch. Hydrobiol., **74**: 247-275.

Ward, J. V. 1976. Comparative limnology of differentially regulated
sections of a Colorado mountain river. Arch. Hydrobiol., **78**: 319-342.

Ward, J. V. 1982. Ecological aspects of stream regulation: Responses in
downstream lotic reaches. Wat. Poll. and Mgmt. Reviews, **2**: 1-26.

Ward, J. V. 1984. Stream regulation of the upper Colorado River: Channel
configuration and thermal heterogeneity. Verh. Internat. Verein.
Limnol., **22**: 1862-1866.

Ward, J. V. and Dufford, R. G. 1979. Longitudinal and seasonal distribution
of macroinvertebrates and epilithic algae in a Colorado springbrook-
pond system. Arch. Hydrobiol., **86**: 284-321.

Ward, J. V. and Short, R. A. 1978. Macroinvertebrate community structure of
four special lotic habitats in Colorado, U.S.A. Verh. Internat.
Verein. Limnol., **20**: 1382-1387.

Ward, J. V. and Stanford, J. A. 1979. Ecological factors controlling stream
zoobenthos with emphasis on thermal modification of regulated
streams. In: The Ecology of Regulated Streams. (Ed. by J. V. Ward
and J. A. Stanford). pp. 35-55. Plenum Press, New York.

Welch, E. B. 1980. Ecological Effects of Waste Water. 337 pp. Cambridge
Univ. Press, Cambridge.

Wetzel, R. G. 1975a. Limnology. 743 pp. W. B. Saunders Co., Toronto.

Wetzel, R. G. 1975b. Primary production. In: River Ecology. (Ed. by
B. A. Whitton). pp. 230-247. Univ. of Calif. Press, Berkeley.

Zimmermann, H. J. and Ward, J. V. 1984. A survey of regulated streams in
the Rocky Mountains of Colorado, U.S.A. In: Regulated Rivers. (Ed.
by A. Lillehammer and S. J. Saltveit). pp. 251-262. Oslo Univ. Press,
Oslo.

THE ECOLOGY OF REGULATED STREAMS:

PAST ACCOMPLISHMENTS AND DIRECTIONS FOR FUTURE RESEARCH

James V. Ward and Jack A. Stanford

Department of Zoology, Colorado State University
Fort Collins, Colorado 80523 U.S.A.
Flathead Lake Biological Station
University of Montana, Bigfork, Montana 59911 U.S.A.

INTRODUCTION

The control of stream discharge by man made dams has been practiced for millennia (Smith 1971), yet recognition of the often dramatic ecological changes in downstream lotic reaches has been slow. Until quite recently most research focused on the lentic water body behind the dam, despite the fact that virtually all of the world's river systems are regulated by dams. Reviews dealing with ecological effects of impoundment, although otherwise comprehensive, often give only cursory and general consideration to the receiving stream environment (Ridley & Steel 1975; Baxter 1977).

An analysis of publications dealing specifically with regulated streams reveals that the majority of papers and nearly all major works (books, comprehensive reviews) have appeared during the past decade (Figure 1). Books that focus on running waters below dams include the proceedings of the first (Ward & Stanford 1979) and second (Lillehammer & Saltveit 1984) international symposia on regulated streams, and a recent book by Petts (1984) that provides an ecological perspective for the management of impounded rivers. General reviews include articles concentrating on reservoirs that also provide an overview of downstream effects (Neel 1963; Ridley & Steel 1975; Baxter 1977), as well as those that focus on regulated lotic reaches (Brooker 1981; Ward & Stanford 1981; Ward 1982). Geographical reviews of stream regulation are available for Africa (Davies 1979), Australia (Walker 1979), Canada (Baxter & Glaude 1980), France (Décamps 1984a), Great Britain (Armitage 1979), North America (Stanford & Ward 1979), Norway (Lillehammer & Saltveit 1979), Spain (Garcia de Jalón 1984) and Sweden (Henricson & Müller 1979). In addition, a few of the site specific publications plotted in Figure 1 are based on very detailed and comprehensive studies (e.g., Pěnáz et al. 1968).

The purpose of this paper is three-fold: 1) to briefly summarize the major downstream effects of stream regulation, 2) to relate stream ecosystem concepts to regulated river systems, and 3) to propose directions for future research.

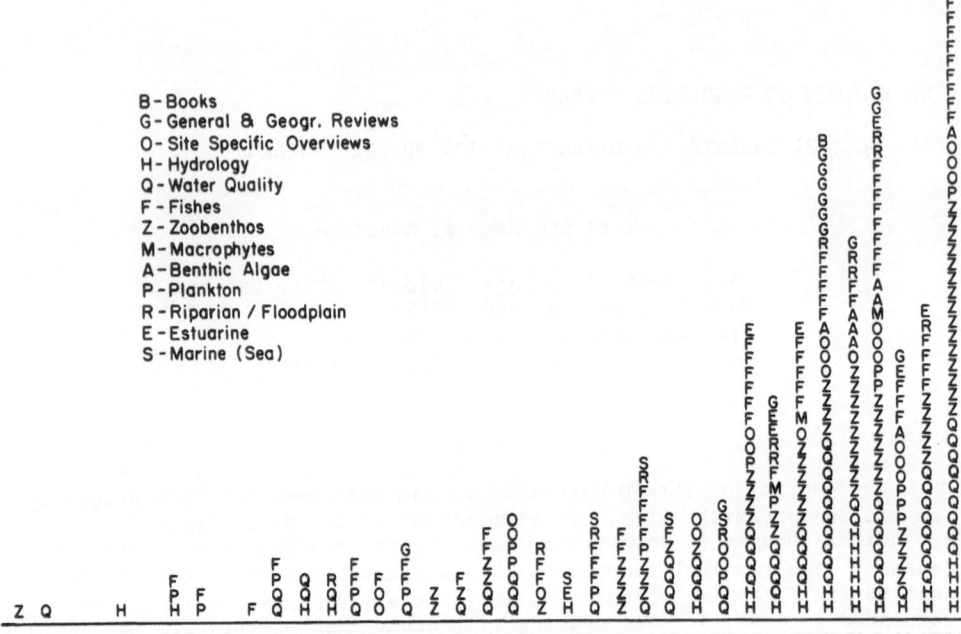

Figure 1. Numbers of publications per year (1948-1984) dealing with
downstream effects of stream regulation. Each publication
(except multiple-author books) is listed under only one
category. Compiled from literature citations (n=347) listed
in the International Newsletter on Regulated Stream Limnology
(excluding agency, governmental, or similar reports).

SYNTHESIS OF CURRENT KNOWLEDGE

This section provides a broad overview of the effects of regulation on
biotic and abiotic components in lotic reaches downstream from dams. Much
of the following discussion is based on Ward (1982) and Petts (1984).
Additional references are drawn largely from recent literature and review
papers. The approach is to focus on broad generalizations, while
recognizing the variable and individualistic responses elicited by stream
regulation.

The Flow Regime

Regulation may modify the flow regime in five major ways: 1) reduce
mean annual runoff, 2) reduce seasonal flow variability, 3) alter timing of
annual extremes, 4) reduce flood magnitudes, and 5) impose unnatural
pulses. These modifications are by no means mutually exclusive. For
example, a reduction of mean annual water yield (resulting from evaporative

loss from the reservoir) may be accompanied by extreme diel fluctuations in discharge below a hydroelectric dam. Such short-term flow fluctuations may be associated with reductions in annual discharge extremes. Most regulation schemes tend to reduce the magnitude and frequency of major flood events. Regulation may shift the timing of seasonal flow events. Release of irrigation water, for example, has changed segments of the Murray-Darling River system, Australia, from a summer-low to a summer-high flow regime (Cadwallader 1978).

Suspended Particles

Reservoirs alter the supply of organic and inorganic particles to downstream reaches; the regulated flow regime alters the downstream transport of available particles. Reservoirs generally act as settling basins, thereby truncating the downstream transport of sediment and detritus (Leopold et al. 1964; Stanford & Ward 1983; Kondratieff & Simmons 1984). This clarifying effect is in some cases countered by the export of reservoir plankton (Décamps et al. 1979) or by turbidity currents that maintain their integrity from the inlet stream to the reservoir outlet (Neel 1963).

The accumulated sediment is periodically flushed from some reservoirs, resulting in order of magnitude increases in downstream suspended solids concentrations (Gray & Ward 1982). The predominant effect of impoundment, however, is to increase markedly water clarity downstream.

Channel Morphology

Modification of the discharge and sediment regimes by stream regulation disrupts the hydrodynamic equilibrium. Adjustments in channel morphology occur as the altered system moves toward a new equilibrium state.

The erosive power of sediment-free water released from the dam may induce a variety of adjustments in channel morphology, including changes in cross-sectional area, downcutting, and lateral movements. Degradation (downcutting) is the predominant hydrodynamic tendency in river segments immediately below dams. However, if flood peaks are reduced this limits the competence of regulated streams to transport substrate particles so that the erosive potential may not be fully realized. Typically, finer particles are removed leaving an armor layer (see Simon 1979) of coarse particles on the stream bed that slows the degradation process. The degradation that does occur in the main channel reduces the base level of tributaries entering the regulated stream. This effectively rejuvenates tributaries, inducing headcutting, thereby increasing sediment input to the main channel. The downstream extent of channel adjustments is largely dependent on tributaries (non-regulated runoff) which add sediment to the main river and contribute to the recovery of natural flow conditions.

Chemical Conditions

Physical, chemical, and biological phenomena occurring within impoundments are responsible for differences between the chemistry of water entering the reservoir and that discharged from the dam (Hannan 1979). The most dramatic changes occur in deep, stratified reservoirs with long water retention times. Because stratified reservoirs exhibit vertical gradients in chemical conditions, the chemistry of water discharged from the dam varies as a function of release depth.

Reservoirs function as "sinks" for many substances, reducing their concentrations below dams compared to values in streams flowing into reservoirs (Table 1; Stanford & Ward 1983). For example, phosphorus retention within a large reservoir (Table 1), resulted primarily from sedimentation of particulate forms of phosphorus (Soltero et al. 1973). The net output of nitrogen from dams may also be less than nitrogen input to the reservoir, but nitrate-nitrogen concentrations in the effluent are often greater than in the influent. Mobilization of nitrate is attributed to mineralization of organic matter (nitrification) and nitrogen fixation within reservoirs (Rada & Wright 1979; Stanford & Ward 1983). Some compounds that are normally absent in the inflow (e.g., H_2S) may be produced within the impoundment and released downstream. Impoundment tends to alter seasonal patterns and to reduce natural temporal variability in the chemistry of water released from reservoirs.

Concentrations of dissolved gases in water released from dams may differ markedly from values in influent streams. Reported conditions range from anoxic to supersaturated, depending upon factors such as release depth, retention time, basin morphology, and the trophic status of the reservoir, in addition to operation and design variables at the dam. Equilibration with atmospheric conditions will eventually occur, but that requires some distance for regulated stream channels that are deep and non-turbulent. However, technologies are available to prevent the occurrence of dissolved gas concentrations that are inimical to lotic organisms (Weitkamp & Katz 1980). If the potential for severe oxygen deficits exists (e.g. deep-release dams impounding stratified eutrophic reservoirs), air drafts in the release tunnels will generally alleviate the problem (Neel 1963). Spillway deflectors and other devices are available to reduce gas supersaturation, a condition that is most pronounced in the tailwaters of high dams where falling water mixes with air that is subsequently dissolved under hydrostatic pressure in deep plunge basins (Weitkamp & Katz 1980).

Thermal Conditions

Three primary factors, hydrology, climate and insolation, interact to determine the thermal characteristics of running waters (Ward 1985). The thermal regime is a composite of 1) annual range, 2) diel range, 3) winter minimum, 4) summer maximum, and 5) periodicity patterns. All of these components are biologically significant and all may be altered by stream regulation.

The major thermal effects of surface-release reservoirs in the temperate zone are to elevate summer temperatures and delay vernal warming and autumnal cooling in the receiving stream. These alterations while not generally severe, can markedly influence lotic organisms (Fraley 1979).

The most dramatic deviation from natural thermal conditions are manifest in stream reaches below large, stratified deep-release impoundments of the temperate zone. Deep-release storage reservoirs that release water from the hypolimnion markedly decrease annual and diel ranges, produce winter warm and summer cool conditions, and disrupt thermal periodicity patterns in the receiving stream.

Large volumes of flowing water maintain their thermal integrity for greater distances downstream than small volumes of water. Therefore, reductions in discharge tend to increase summer temperatures and decrease winter temperatures in the receiving stream. Cool water discharged from deep-release hydroelectric dams during summer more rapidly equilibrates with ambient air temperatures during low flow (non-generation) periods. Diel fluctuations in discharge from hydroelectric dams may result in enhanced diel thermal fluctuations in downstream reaches during summer

Table 1. Percent change based on discharge – weighted chemical composition of the influent and effluent waters of Bighorn Lake, a large deep-release reservoir, Montana/Wyoming, U.S.A. (Based on 1968 data from Soltero et al. 1973).

	Percent Change
Ca^{2+} (m-eq 1^{-1})	– 1.3
Mg^{2+} (m-eq 1^{-1})	–11.9
Na^+ (m-eq 1^{-1})	–10.4
K^+ (m-eq 1^{-1})	0.0
HCO_3^- (m-eq 1^{-1})	– 1.3˙
Cl^- (m-eq 1^{-1})	– 9.1
SO_4^{2-} (m-eq 1^{-1})	– 0.2
NH_3-N (mg 1^{-1}N)	–20
NO_2^--N (mg 1^{-1}N)	–43
NO_3^--N (mg 1^{-1}N)	+11
Ortho-PO_4^{3-} (mg 1^{-1})	–68

(Ward & Stanford 1982a). A falling water level in a stratified reservoir with fixed outlet structures, or changing the depth from which water is released may produce abrupt temperature changes in the receiving stream. In addition, because of the relationship between temperature and discharge, any moderation of the natural year-to-year differences in annual discharge by stream regulation may be reflected by reductions in year-to-year variations in the temperature regime.

Stream Biota

The modified environmental conditions below dams alter structural and functional attributes of the biotic community. The following discussion will focus on direct sublethal effects of the altered environmental conditions on stream organisms. Predator-prey, host-parasite and competitive interactions are also altered by stream regulation, but few definitive data are available. For purposes of discussion it is assumed that complicating factors such as pollution and channelization do not occur.

Few studies have focused on the downstream effects of stream regulation on benthic algae; even fewer data are available for macrophytes. Nonetheless, some typical responses of aquatic vegetation are apparent, including increases in standing crop and total streambed coverage, invasion by angiosperms of reaches previously devoid of higher aquatic plants, and an enhancement of filamentous chlorophytes and bryophytes. One study documents a significant increase in periphyton production, which is largely attributed to nitrogen fixation in the upstream reservoir (Marcus 1980). Skulberg (1984) perceives stream regulation as "enriching" (sensu lato) environmental conditions for lotic algae in oligotrophic waters. Factors commonly occurring in regulated streams that generally favor plant growth include relatively uniform current, water level, and temperature; reductions in the frequency and severity of flood events; high water clarity; substrate stability; and greater nutrient availability. The detrimental effects of short-term flow fluctuations below hydroelectric dams are often more than countered by the greater substrate stability induced by long-term reductions in flood peaks, and by the cumulative effects of other favorable conditions.

More data are available regarding the responses of benthic invertebrates to stream regulation (reviewed by Armitage 1984), although precise interpretation is limited by a lack of basic knowledge (e.g. life histories, food habits, niche breadth) for many lotic species. In otherwise undisturbed regulated streams, reductions in species diversity, alterations of community composition and feeding guilds, and increases (constant flow) or decreases (fluctuating flow) in abundance, are characteristic zoobenthos responses.

The modified temperature regime below dams often does not supply sufficient thermal information for some species (Ward & Stanford 1982b). For example, species requiring winter chill to break egg diapause may be eliminated by the winter-warm conditions below deep-release dams. Summer cool conditions will eliminate species requiring a high absolute temperature, or the accumulation of a large number or degree days, for maturation or emergence. Altered thermoperiodicity patterns may also disrupt life cycle events. Species from spring brooks, in contrast, are adapted to constant thermal conditions and may invade stream reaches below deep-release dams, especially if flow amplitudes are reduced. Shallow reservoirs act as heat traps that may increase downstream summer temperatures to the detriment of the indigenous cold water fauna (Fraley 1979). Suboptimal thermal conditions also reduce fecundity and place species at a competitive disadvantage (Vannote & Sweeney 1980).

Constant flow and high substrate stability favor many sessile species. Sessile forms are often filter feeders (e.g., black flies, net spinning caddisflies) that are further enhanced by inputs of reservoir seston. Grazers are expected to increase as stable mineral particles develop dense growth of algae and where macrophytes provide surfaces for epiphytic algae. Coarse particle detritivores, however, are reduced because the upstream reservoir truncates detrital transport. Eliminating the flushing action of flood peaks may have deleterious effects on some stream systems, however, by allowing nuisance plant growth or failing to maintain silt-free substrate interstices.

Short-term flow fluctuations strand zoobenthos along shorelines and induce drift (Brusven 1984). Recolonization may not be sufficient to restore losses because the upstream reservoir blocks drift from the upper reaches, thereby eliminating a major recolonization pathway for lotic invertebrates. Because of differential tolerance to stranding and propensity to drift, certain groups are reduced or eliminated in regulated streams with extreme fluctuations in discharge. Short-term flow fluctuations also entrain detrital aggregates, eliminating the preferred microhabitat of some zoobenthos. Abrupt changes in discharge may, in some cases, result in extreme temperature fluctuations below hypolimnial-release reservoirs.

Although many species of benthic invertebrates are eliminated or reduced in abundance in regulated streams, a few taxa flourish under the altered environmental conditions. Some of those favored were present in lower numbers prior to impoundment, but others not normally occurring in unregulated streams invade the modified running water ecosystem.

Fishes have received more detailed study than any of the other biotic components of regulated streams. Anadromous migrants of commercial value (e.g. salmon) have been most intensively investigated in this regard.

Stream regulation poses many problems for adult fishes attempting to return to freshwater to spawn. Dams block upstream migration to spawning grounds that may have been inundated by a reservoir or rendered unsuitable for spawning because of changes in the substrate. Even if upstream passage is possible, stream regulation increases the time (and energy) required to reach the spawning grounds and may fail to provide appropriate temperature or flow stimuli. Weakened and disoriented fishes concentrated below dams may incur large predation losses (Raymond 1979). Gas supersaturation poses an additional danger.

Impoundments also increase the passage time of downstream migrating juveniles, which may also be subjected to turbine mortality, excessive predation, and gas supersaturation. Rapid drops in water level are especially detrimental to survival of juvenile salmonids (Saltveit & Styrvold 1984). Despite intensive management efforts it is not always possible to preserve the anadromous fishery of a regulated stream (Geen 1975; Mundie 1979).

The sublethal responses of resident fishes to stream regulation are in some respects analogous to those described for macroinvertebrates: species diversity is characteristically reduced, community structure is altered and standing crop can either increase or decrease. Exotic fishes such as rainbow trout (an introduced species throughout most of its present range) find suitable conditions below deep-release dams where indigenous species have been reduced or eliminated by the altered thermal, sediment and flow conditions (Holden 1979).

In some cases stream regulation improves conditions for the downstream fishery. Increases in population density and biomass following regulation can occur even if temperature limits individual growth rates (Crisp et al. 1983). However, winter-warm conditions below deep-release dams considerably extends the growth period in some regions.

Marked flow fluctuations decrease fish productivity (partly by reducing the invertebrate food base), disrupt spawning, and increase egg and larval mortality. The productive "put-and-take" recreational fisheries of some regulated streams are largely attributable to plankton and forage fish released from an upstream reservoir.

Some fish species exhibit considerable adaptation to the altered temperatures below dams, including genetic plasticity in response to thermal heterogeneity (Zimmerman 1984). Hybridization may also increase as altered environmental conditions cause simultaneous reproduction of species whose reproductive periods are temporally isolated in unregulated streams (Zimmerman & Wooten 1981).

INTEGRATION WITH STREAM ECOSYSTEM CONCEPTS

The preceding overview was intentionally presented with a narrow focus, namely the effects of a single dam on conditions within the stream channel immediately below that dam. Indeed, much of the research on regulated streams has used such an approach. In this section we draw upon

recent concepts of stream ecology in an attempt to spatially expand our
perception of regulated streams, to view them as ecosystems in a total
watershed context.

The Stream and its Valley

The theme of the Baldi Memorial Lecture presented at the nineteenth
International Congress of Limnology was "The stream and its valley" (Hynes
1975). In this seminal paper Prof. H. B. N. Hynes called for a more
holistic approach to the study of streams, an approach that views running
waters "as parts of the valleys that they drain." A similar theme was the
subject of the most recent Baldi Lecture, "Beyond the shoreline: A
watershed-ecosystem approach" (Likens 1984).

The watershed approach emphasizes energy, material, and nutrient
interactions between terrestrial and aquatic systems, meaning that stream
ecologists should be cognizant of the hydrological, geochemical,
meteorological, and vegetational setting. These watershed variables exert
major controls on water and sediment yields; the quality and quantity of
dissolved and particulate organic matter, nutrients and other ions entering
streams; and matter and energy transformations within the stream. As so
eloquently stated by Hynes (1975):

> "We may conclude then that in every respect the valley rules the
> stream. Its rocks determine the availability of ions, its soil,
> its clay, even its slope. The soil and climate determine the
> vegetation, and the vegetation rules the supply of organic
> matter. The organic matter reacts with the soil to control the
> release of ions, and the ions, particularly nitrate and phosphate,
> control the decay of the litter, and hence lie right at the root of
> the food cycle."

The River Continuum Concept

Whereas the primary emphasis of the watershed approach is to expand
the intellectual vision of stream ecologists laterally to include the
adjacent terrestrial system, the river continuum concept (Vannote et
al. 1980; Minshall et al. 1983) takes this one step further by focusing on
the longitudinal gradient of conditions that occur from the headwaters
to the mouth of river systems.

According to the river continuum concept (RCC), initially formulated
for deciduous forest streams, the headwaters are heavily canopied,
light-limited heterotrophic systems with coarse substrata. Fed largely by
groundwater, the headwaters exhibit low amplitude temperature and flow
regimes. The major energy source is leaf litter that enters the stream from
the terrestrial system.

As tributaries enter, the stream grows larger and the canopy opens. In
middle reaches the water is still relatively clear and shallow, but now the
stream bottom receives direct solar radiation. Aquatic angiosperms and
attached algae attain maximum development in middle reaches resulting in
autotrophic conditions. Temperature and discharge exhibit large temporal
fluctuations.

The light-limited heterotrophy in the lower reaches of river systems
results from quite different conditions than those prevailing in headwater
reaches. The greater depth, lower water clarity, and unstable substratum of
large rivers reduce their suitability for benthic plants, although
phytoplankton may now contribute to primary productivity. The large volume
of water resists short-term temperature changes and flow fluctuations are

moderated by the cumulative variations of many tributaries.

According to the RCC, biotic diversity attains the highest levels in middle reaches concomitant with naturally high levels of environmental heterogeneity (e.g., diel temperature fluctuations). Empirical evidence, based on species richness of benthic invertebrate assemblages, supports this tenet of the RCC (Minshall et al. 1985), although Statzner and Higler (1985) question whether it applies to all biotic components.

The fish fauna of the headwaters, according to the RCC, consists of cold stenothermal invertivores. In the middle reaches piscivores and invertivores predominate, whereas the fish fauna of the lower reaches consists largely of bottom feeders and planktivores.

Shredders, invertebrates that feed on coarse particulate organic matter (CPOM), predominate in the headwaters where allochthonous leaf litter comprises the major energy source, but decrease in relative abundance downstream as CPOM becomes less important. Collectors, species that feed on fine particulate organic matter (FPOM), are important in all reaches, but are virtually the only non-predaceous invertebrates in the lower reaches. Grazers, species that feed on algae attached to rocks and to the surfaces of higher plants, are best developed in the middle reaches where autotrophy predominates.

Although some aspects of these postulated longitudinal changes have been criticized or modified (Statzner & Higler 1985), and the applicability of the RCC to certain regions has been questioned (e.g. Winterbourn et al. 1981), it is clear that for a given hydrological, geochemical, meteorological, and vegetational setting, more-or-less predictable resource gradients do occur in river systems as they course from the headwaters to the sea. The biota respond to these longitudinal changes in environmental conditions.

Nutrient Spiralling

Because of the unidirectional flow of water, nutrient cycling in streams involves a longitudinal vector and so is more appropriately termed "nutrient spiralling" (Webster & Patten 1979). Spiralling length is the average downstream distance traveled by a nutrient "atom" as it completes one loop of an imaginary spiral (Newbold et al. 1981). Determined by radioassay techniques, spiralling length is the sum of 1) transport in the water column (dissolved), 2) transport in FPOM and CPOM (particulates), and 3) transport by consumers (drift).

Spiralling length is an index of the efficiency of utilization of nutrients supplied from the watershed, since it reflects the number of times a nutrient atom is recycled within a stream reach. In a small woodland stream in the southeastern U.S., the spiralling length of phosphorus was 193 m (Newbold et al. 1981).

Spiralling length is expected to vary as a function of position along the river continuum (Newbold this volume). It should be sensitive to watershed perturbations such as stream regulation.

The Serial Discontinuity Concept

If changes in structural and functional attributes occur along the river continuum, then the influence of stream regulation on these attributes should vary as a function of dam position (Figure 2). The serial discontinuity concept (SDC) was recently developed to provide a broad conceptual framework to deal with the disruptions (discontinuities) in stream ecosystem structure and function engendered by stream regulation (Ward & Stanford 1983). In its initial formulation the basic objectives of the SDC were four-fold: 1) to recognize the importance of dam position, 2) to propose discontinuity distance as a measure of the upstream or downstream shifts (e.g. in maximum autotrophy) that result from impoundment, 3) to promote a watershed approach for regulated stream research, and 4) to proffer a predictive conceptual model to be tested.

Although there have been some initial attempts to examine certain structural aspects of the SDC (Stanford & Ward 1984), the concept remains to be tested in a holistic way, and the processes and interactions involved remain to be elucidated. This will require intensive long-term investigations of a variety of regulated river systems. Nonetheless, the SDC has already been recognized as having contributed to stream ecosystem theory (Cummins et al. 1984).

PROSPECTS FOR FUTURE RESEARCH

As stated initially in this paper, our understanding of the ecology of regulated streams is still in its infancy. Many aspects remain largely unexplored. However, rather than listing specific topics that have received little attention (e.g. microbial ecology), the approach of this section is to deal briefly with a few broad areas that we believe provide opportunities to substantially contribute to our understanding of regulated lotic ecosystems.

Landscape Ecology

Landscape ecology resides at the intersection of several disciplines and "considers the development and dynamics of spatial heterogeneity, spatial and temporal interactions and exchanges across heterogeneous landscapes, influences of spatial heterogeneity on biotic and abiotic processes, and management of spatial heterogeneity" (Risser et al. 1984). It is concerned with the structure and dynamics of patches on the scale of kilometers (Forman & Godron 1981). For example, an agricultural landscape unit might be comprised of stands of upland and bottomland forest, fields of corn and alfalfa, an abandoned oil field, and a farm pond.

For stream ecologists landscape ecology is a logical extension of the recognition that the stream cannot be separated from its valley (Hynes 1975). Décamps (1984b), in a paper entitled "Towards a landscape ecology of river valleys," entreats stream ecologists to integrate the concept of the stream corridor (Forman & Godron 1981) with studies of riparian and floodplain vegetation, plant succession, and anthropogenic alterations, to provide needed insight into the structure and function of intensively managed rivers. A renewed interest in the functional relationships between running waters and the riparian zone has led to new insights, especially regarding the importance of alluvial forests to the lower reaches of aboriginal rivers (see Cummins et al. 1984).

Because the hydrologic regime influences the composition and dynamics of alluvial forests, it is not surprising that stream regulation induces changes in the riparian and floodplain vegetation (Décamps 1984a; Nilsson 1984; Petts 1984). For example, reduced flood peaks following regulation of the River Romanche, France (Décamps 1984a), allowed hardwoods (*Fraxinus excelsior*) to invade lower alluvial deposits below the dam which, prior to regulation, contained willows (*Salix eleagnos*) and alder (*Alnus incana*). Since the alluvial forests exert various controls on the adjacent river, stream regulation, by altering vegetation patterns, results in further

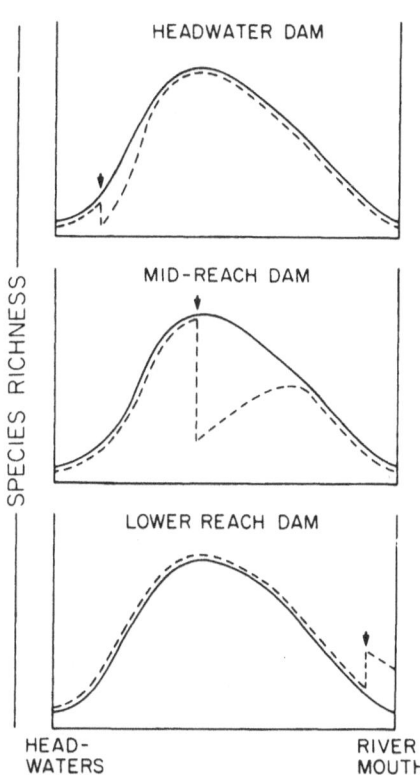

Figure 2. An example of the importance of dam position as prescribed by the serial discontinuity concept (Ward & Stanford 1983). The solid curve (repeated three times) shows the predicted pattern of species richness (benthic macroinvertebrates) along the course of a river system according to the river continuum concept (Vannote et al. 1980; Minshall et al. 1985). The dashed lines indicate the postulated responses of species richness to single deep-release dams located in the headwaters, middle, or lower reaches of a river system.

interactions at the aquatic-terrestrial interface (Grelsson 1984).

Limnologists have long recognized the potential importance of shoreline morphometry for lake productivity. Stream ecologists might profit by giving more attention to the configuration of the shoreline, including the development of backwaters, side channels, islands, and adjoining swamps, marshes and oxbow lakes. The water bodies of the floodplain are extensions of the river ecosystem. For example, they serve as nursery grounds for riverine fishes; such ecological interactions, however, depend on periodic inundation which may be eliminated by flow regulation.

Décamps (1984a; b) suggests that consideration of the natural and cultural landscape mosaics in river valleys will enable stream ecologists to gain a fuller understanding of lotic ecosystems, including the effects of stream regulation. A landscape ecology approach may prove to facilitate the development of a synthetic perspective of regulated streams as modified ecosystems. Indeed, we believe that the ecosystem concept should be extended to include the total landscape of a river basin, whether regulated or not.

Groundwater Ecology

Stream ecologists, especially in North America, have given but scant attention to groundwater systems. Most groundwater studies have been conducted by hydrologists and engineers. Just as stream ecologists should not ignore the adjacent terrestrial environment, a holistic approach to the understanding of running waters requires a more thorough consideration of groundwaters (Danielopol 1980; Hynes 1983).

Groundwater is an important source of organic matter for running waters, although its significance has probably been underestimated in studies of organic carbon budgets (Hynes 1983). The interface between the stream and groundwater systems is a biologically active zone that apparently removes a large proportion of the dissolved organic matter before it enters the stream. This is particularly significant because much of the water in stream channels has entered via groundwater pathways (i.e., contributions from overland flow are often not quantitatively important; Freeze & Cherry 1979). Hynes (1983) suggests that the interface serves as a sink for organic matter and may function to maintain the relative seasonal constancy of dissolved organic matter characterizing many lotic ecosystems.

This interface, the hyporheic zone, is best developed from a biological standpoint in gravel bed streams with relatively silt-free interstices (Stanford & Gaufin 1974). Fishery biologists have for some time recognized the importance of intergravel water movements and interstitial oxygen levels for the incubation of salmonid eggs and larvae (Vaux 1962).

The hyporheic zone serves as a refuge for lotic zoobenthos (Schwoerbel 1967), offering shelter from floods, drought, predators, and extreme temperatures, and providing suitable and predictable conditions for immobile stages such as eggs, pupae and diapausing nymphs. The hyporheos is a faunal reservoir capable of recolonizing the surface benthos, should the latter be depleted by adverse conditions.

Groundwater quality and quantity is an area of topical importance, yet has been given little consideration by stream ecologists with the exception of a few European investigations (e.g. Danielopol 1976; Mêstrov & Lattinger-Penko 1981).

Groundwater also influences the interactions between running waters and terrestrial vegetation referred to in the discussion of landscape ecology. Groundwater dynamics are critical determinants of the composition and density of the riparian and floodplain vegetation, and of the successional processes that structure the alluvial forest (e.g. Décamps 1984b).

The full extent to which stream regulation modifies groundwater ecology remains to be investigated. Regulation may raise or lower the water table of the floodplain, depending on whether the stream channel undergoes aggradation or degradation (Simons 1979; Petts 1984). Any changes in the level of the water table are known to markedly alter riparian plant communities (Pautou et al. 1979). Little or nothing is known, however, regarding the effects of stream regulation on groundwater quality, or the exchange of carbon at the hyporheic interface. We are unaware of any studies that have specifically examined the influence of stream regulation on the interstitial or hyporheic fauna.

Time Scales

Petts (1984; this volume) has pointed to the importance of employing a temporal perspective when interpreting the downstream effects of stream regulation. Because controlling factors operate at different frequencies, response times are dependent upon the component under consideration. For example, the downstream effects of increased diel flow fluctuations will be immediately apparent (e.g. increased invertebrate drift rates) when a hydroelectric dam commences operation, whereas increased substrate stability resulting from reductions in annual flood peaks will obviously manifest over a much longer period.

Ecosystems are characterized by a hierarchy of time scale responses. Chemical changes in the water discharged from the reservoir will generally stabilize within 5 to 10 years following dam closure (Petts this volume). The downstream temperature regime may readjust even more rapidly. However, it can take decades or even centuries for the channel morphology below the dam to attain a new equilibrium state. Because systems that have undergone complete physical readjustments have rarely been observed, ecologists may be examining transient biotic states. However, although few definitive data are available, it is likely that the most dramatic biotic responses to stream regulation occur within the first several years following dam closure, even though channel morphology continues to slowly re-equilibrate over a much longer period. Indeed, this would be especially true for organisms with short life cycles that are highly responsive to abiotic variables that undergo rapid readjustment (e.g. temperature). Nonetheless, even gradual physical readjustments will eventually lead to a marked deviation from the original equilibrium state; concomitant alterations in the biota, though subtle and occurring in small increments, may result in large cumulative changes in the community. Those studying regulated streams would be well advised to carefully view their findings within a temporal framework.

Estuarine and Marine Systems

It is perhaps not surprising that stream ecologists, being primarily concerned with freshwaters, have only rarely considered the influence of stream regulation on estuarine and marine systems. The available data show that the effects of regulation can be profound (e.g. Aleem 1972; Hassan 1975; Neu 1975; Sharaf El Din 1977; Corlett 1978; Connell et al. 1981; Berkes 1982).

Estuaries belong to what Odum (1971) calls "fluctuating water level ecosystems", that "are maintained in an early, relatively fertile stage by the tides, which provide the energy for rapid nutrient cycling." Most estuarine systems are, therefore, dependent upon a high level of environmental variability; their structural and functional attributes are seriously disrupted by stream regulation or any change that reduces or temporally displaces the natural environmental variability. Sediment and nutrient dynamics are altered by flow regulation. For example, eutrophication, held in check by light limitations, can result if stream regulation reduces suspended sediment levels in the estuary. Reduced river discharge allows upstream incursions of sea water, whereas unnaturally high discharge during the dry season constrains the normal upstream penetration of the salt wedge. Changes in the proportionate contribution of fresh and sea water may dramatically alter the temperature regime during certain seasons.

Many plants of the estuarine floodplain and delta depend on the cycle of flooding and exposure, and may not reproduce in the absence of large water level fluctuations. Many of the changes resulting from river regulation threaten the important role of estuaries as nursery grounds for marine fishes. Changes in the estuarine environment may disrupt the migration of anadromous and catadromous fishes. Many higher vertebrates are also dependent on estuaries and are adversely affected by any deviation

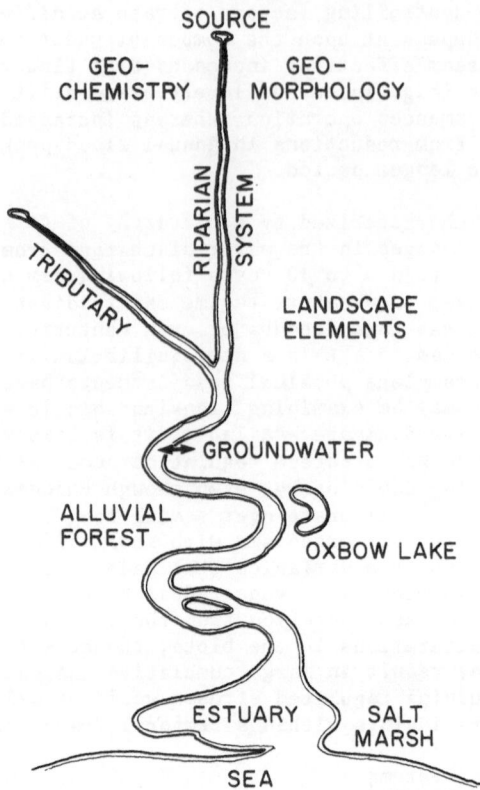

Figure 3.   Some elements to consider, along with climate, to provide the
broad spatiotemporal perspective necessary for a holistic
assessment of the potential effects of stream regulation on a
river system (see text).

from the natural cycle of water level changes (Davies 1979; Petts 1984).

Alterations attributable to stream regulation can also effect the
marine environment. Coastal erosion and deposition processes may be
disrupted by changes in freshwater outflow and sediment loads. Harnessing
the Nile floodwaters by Aswan Dam has decreased nutrient levels in the
coastal waters of the Mediterranean Sea, reduced plankton blooms, and
markedly reduced the commercial sardine harvest (Aleem 1972).

This brief overview of selected examples regarding interactions
between freshwater, estuarine, and marine environments, clearly indicates
that those of us striving toward a holistic perspective of regulated river
systems must consciously integrate estuaries and even marginal seas into
our conceptual framework.

Regulated Streams as Field Experiments

In a recent paper (Ward & Stanford 1984) we pointed out the value of
regulated streams as appropriate sites for testing ecological theory in the
context of running waters. Indeed, regulated streams have several
attributes making them especially attractive in this regard: they exhibit
distinctive and predictable spatial environmental gradients; their insular
nature (for obligatory rheophiles) provides suitable conditions for

examining colonization pathways; the abbreviated species complement is amenable to investigations of competitive exclusion and predator-prey interactions; and their manipulative capabilities provide control over downstream conditions.

Stream ecologists interested in the resiliency properties of lotic ecosystems, for example, should not overlook regulated streams as study sites. Conversely, researchers more directly interested in the effects of regulation, would profit by viewing their results in a broad ecological context. Limnologists have gained great insight into lentic systems by conducting "whole lake" experiments (Schindler 1980). It is not inconceivable that stream ecologists would considerably advance understanding of the structure and function of lotic ecosystems by managing regulated rivers for scientific purposes.

CONCLUSIONS

The principal theme of this paper is that stream ecologists should strive toward a more holistic perspective of regulated streams as modified ecosystems. What is presently known about biotic and abiotic conditions in regulated streams should be integrated with stream ecosystem concepts and placed within the context of modern ecological theory (see Figure 3). Regulated rivers are four-dimensional systems according to this viewpoint. The lateral dimension extends beyond the shoreline and considers the stream channel as but a part of the greater landscape. The longitudinal dimension extends from the headwaters to the estuary and marginal seas. Vertically the stream is viewed as an integral part of the groundwater system. The fourth dimension, time, provides the temporal framework.

ACKNOWLEDGEMENTS

We thank the members of the Local Organizing Committee for inviting us to contribute to the symposium, and gratefully acknowledge support provided by the Alberta Department of the Environment and the University of Alberta. Appreciation is extended to Mrs. Nadine Kuehl for typing the manuscript. This paper was written while JVW was supported by a research grant from the Colorado Experiment Station.

REFERENCES

Aleem, A. A. 1972. Effect of river outflow management on marine life. Mar. Biol., 15: 200-208.

Armitage, P. D. 1979. Stream regulation in Great Britain. In: The Ecology of Regulated Streams. (Ed. by J. V. Ward and J. A. Stanford). pp. 165-181. Plenum, New York.

Armitage, P. D. 1984. Environmental changes induced by stream regulation and their effect on lotic macroinvertebrate communities. In: Regulated Rivers. (Ed. by A. Lillehammer and S. J. Saltveit). pp. 139-165. Oslo Univ. Press, Oslo.

Baxter, R. M. 1977. Environmental effects of dams and impoundments. Ann. Rev. Ecol. Syst, 8: 255-283.

Baxter, R. M. and Glaude, P. 1980. Environmental effects of dams and impoundments in Canada: Experience and prospects. Can. Bull. Fish. Aquat. Sci., 205: 1-34.

Berkes, F. 1982. Preliminary impacts of James Bay Hydroelectric Project, Quebec, on estuarine fish and fisheries. Arctic, 35: 524-530.

Brooker, M. P. 1981. The impact of impoundments on the downstream fisheries
and general ecology of rivers. Adv. Appl. Biol., **6**: 91-152.

Brusven, M. A. 1984. The distribution and abundance of benthic insects
subjected to reservoir-release flows in the Clearwater River, Idaho,
USA. In: Regulated Rivers. (Ed. by A. Lillehammer and
S. J. Saltveit). pp. 167-180. Oslo Univ. Press, Oslo.

Cadwallader, P. L. 1978. Some causes of the decline in range and abundance
of native fish in the Murray-Darling River system. Proc. Roy. Soc.
Victoria, **90**: 211-224.

Connell, D. W., Bycroft, B. M., Miller, G. J. and Lather, P. 1981. Effects
of a barrage on flushing and water quality in the Fitzroy River
Estuary, Queensland. Aust. J. Mar. Freshwater Res., **32**: 57-63.

Corlett, J. 1978. Ecological implications of proposed water storage schemes
in British estuaries. Hydrobiol. Bull. (Amst.), **12**: 291-298.

Crisp, D. T., Mann, R. H. K. and Cubby, P. R. 1983. Effects of regulation
of the River Tees upon fish populations below Cow Green Reservoir.
J. Appl. Ecol., **20**: 371-386.

Cummins, K. W., Minshall, G. W., Sedell, J. R., Cushing, C. E. and
Petersen, R. C. 1984. Stream ecosystem theory. Verh. Int. Verein.
Limnol., **22**: 1818-1827.

Danielopol, D. L. 1976. The distribution of the fauna in the interstitial
habitats of riverine sediments of the Danube and the Piesting
(Austria). Int. J. Speleol., **8**: 23-51.

Danielopol, D. L. 1980. The role of the limnologist in ground water
studies. Int. Revue ges. Hydrobiol., **65**: 777-791.

Davies, B. R. 1979. Stream regulation in Africa: A review. In: The
Ecology of Regulated Streams. (Ed. by J. V. Ward and J. A. Stanford).
pp. 113-142. Plenum, New York.

Décamps, H. 1984a. Biology of regulated rivers in France. In: Regulated
Rivers. (Ed. by A. Lillehammer and S. J. Saltveit). pp. 495-514.
Oslo Univ. Press, Oslo.

Décamps, H. 1984b. Towards a landscape ecology of river valleys. In:
Trends in Ecological Research for the 1980s. (Ed. by J. H. Cooley and
F. B. Golley). pp. 163-178. Plenum, New York.

Décamps, H., Capblancq, J., Casanova, H. and Tourneq, J. N. 1979.
Hydrobiology of some regulated rivers in the southwest of France.
In: The Ecology of Regulated Streams. (Ed. by J. V. Ward and
J. A. Stanford). pp. 273-288. Plenum, New York.

Forman, R. T. T. and Godron, M. 1981. Patches and structural components for
a landscape ecology. BioScience, **31**: 733-740.

Fraley, J. J. 1979. Effects of elevated stream temperatures below a shallow
reservoir on a cold water macroinvertebrate fauna. In: The Ecology
of Regulated Streams. (Ed. by J. V. Ward and J. A. Stanford).
pp. 257-272. Plenum, New York.

Freeze, R. A. and Cherry, J. A. 1979. Groundwater. Prentice-Hall,
Englewood Cliffs. 604 pp.

García de Jalón, D. 1984. Stream regulation in Spain. In: Regulated
Rivers. (Ed. by A. Lillehammer and S. J. Saltveit). pp. 481-492.
Oslo Univ. Press, Oslo.

Geen, G. H. 1975. Ecological consequences of the proposed Moran Dam on the
Fraser River. J. Fish. Res. Board Can., **32**: 126-135.

Grelsson, G. 1984. Comparison of vegetation stability on two river banks,
subject to short-term water-level regulation, at the River Umealven in
Northern Sweden. In: Regulated Rivers. (Ed. by A. Lillehammer and
S. J. Saltveit). pp. 125-132. Oslo Univ. Press, Oslo.

Gray, L. J. and Ward, J. V. 1982. Effects of sediment releases from a
reservoir on stream macroinvertebrates. Hydrobiologia, **96**: 177-184.

Hannan, H. H. 1979. Chemical modifications in reservoir-regulated streams.
In: The Ecology of Regulated Streams. (Ed. by J. V. Ward and
J. A. Stanford). pp. 75-94. Plenum, New York.

Hassan, E. M. 1975. Some effects of river regulation on marginal seas. Ocean Manag., **2**: 333-344.

Henricson, J. and Müller, K. 1979. Stream regulation in Sweden with some examples from Central Europe. In: The Ecology of Regulated Streams. (Ed. by J. V. Ward and J. A. Stanford). pp. 183-199. Plenum, New York.

Holden, P. B. 1979. Ecology of riverine fishes in regulated stream systems with emphasis on the Colorado River. In: The Ecology of Regulated Streams. (Ed. by J. V. Ward and J. A. Stanford). pp. 57-74. Plenum, New York.

Hynes, H. B. N. 1975. The stream and its valley. Verh. Int. Verein. Limnol., **19**: 1-15.

Hynes, H. B. N. 1983. Groundwater and stream ecology. Hydrobiologia, **100**: 93-99.

Kondratieff, P. K. and Simmons, G. M. 1984. Nutritive quality and size fractions of natural seston in an impounded river. Arch. Hydrobiol., **101**: 401-412.

Leopold, L. B., Wolman, M. G. and Miller, J. P. 1964. Fluvial Processes in Geomorphology. Freeman, San Francisco. 522 pp.

Likens, G. E. 1984. Beyond the shoreline: A watershed-ecosystem approach. Verh. Int. Verein. Limnol., **22**: 1-22.

Lillehammer, A. and Saltveit, S. J. 1979. Stream regulation in Norway. In: The Ecology of Regulated Streams. (Ed. by J. V. Ward and J. A. Stanford). pp. 201-213. Plenum, New York.

Lillehammer, A. and Saltveit, S. J. (Eds.) 1984. Regulated Rivers. Oslo Univ. Press, Oslo. 540 pp.

Marcus, M. D. 1980. Periphytic community response to chronic nutrient enrichment by a reservoir discharge. Ecology, **61**: 387-399.

Mestrov, M. and Lattinger-Penko, R. 1981. Investigation of the mutual influence between a polluted river and its hyporheic. Int. J. Speleol., **11**: 159-171.

Minshall, G. W., Petersen, R. C., Cummins, K. W., Bott, T. L., Sedell, J. R., Cushing, C. E. and Vannote, R. L. 1983. Interbiome comparison of stream ecosystem dynamics. Ecol. Monogr., **53**: 1-25.

Minshall, G. W., Petersen, R. C. and Nimz, C. F. 1985. Species richness in streams of different size from the same drainage basin. Am. Nat., **125**: 16-38.

Mundie, J. H. 1979. The regulated stream and salmon management. In: The Ecology of Regulated Streams. (Ed. by J. V. Ward and J. A. Stanford). pp. 307-320. Plenum, New York.

Neel, J. K. 1963. Impact of reservoirs. In: Limnology in North America. (Ed. by D. G. Frey). pp. 575-593. Univ. Wisc. Press, Madison.

Neu, H. J. A. 1975. Runoff regulation for hydro-power and its effect on the ocean environment. Can. J. Civil Eng., **2**: 583-591.

Newbold, J. D. 1986. Applicability of the nutrient spiralling concept to river-reservoir systems. This volume.

Newbold, J. D., Elwood, J. W., O'Neill, R. V. and Van Winkle, W. 1981. Measuring nutrient spiralling in streams. Can. J. Fish. Aquat. Sci., **38**: 860-863.

Nillson, C. 1984. Effect of stream regulation on riparian vegetation. In: Regulated Rivers. (Ed. by A. Lillehammer and S. J. Saltveit). pp. 93-106. Oslo Univ. Press, Oslo

Odum, E. P. 1971. Fundamentals of Ecology. Saunders, Philadelphia.

Pautou, G., Girel, J., Lachet, B. and Ain, G. 1979. Recherches écologiques dans la vallée du Haut-Rhone francais. Doc. Cartogr. Ecol., **22**: 1-63.

Peñáz, M., Kubicek, F., Marvan, P. and Zelinka, M. 1968. Influence of the Vir Valley Reservoir on the hydrobiological and ichthyological conditions in the River Svratka. Acta Sci. Nat. Brno, **2**: 1-60.

Petts, G. E. 1984. Impounded Rivers. Wiley, Chichester. 326 pp.

Petts, G. E. 1986. Time-scales for ecological change in regulated rivers. This volume.

Rada, R. G. and Wright, J. C. 1979. Factors affecting nitrogen and phosphorus levels in Canyon Ferry Reservoir, Montana, and its effluent waters. Northwest Sci., **53**: 213-220.

Raymond, H. L. 1979. Effects of dams and impoundments on migration of juvenile chinook salmon and steelhead trout from the Snake River, 1966-1975. Trans. Am. Fish. Soc., **108**: 509-529.

Ridley, J. E. and Steel, J. A. 1975. Ecological aspects of river impoundments. In: River Ecology. (Ed. by B. A. Whitton). pp. 565-587. Blackwell, Oxford.

Risser, P. G., Karr, J. R. and Forman, R. T. T. 1984. Landscape Ecology - Directions and Approaches. Illinois Nat. Hist. Surv. Spec. Publ. No. 2, Champaign. 38 pp.

Saltveit, S. J. and Styrvold, J. O. 1984. Density of juvenile Atlantic salmon (*Salmo salar* L.) and brown trout (*Salmo trutta* L.) in two Norwegian regulated rivers. In: Regulated Rivers. (Ed. by A. Lillehammer and S. J. Saltveit). pp. 565-587. Oslo. Univ. Press, Oslo.

Schindler, D. W. 1980. Evolution of the Experimental Lakes Project. Can. J. Fish. Aquat. Sci., **37**: 313-319.

Schwoerbel, J. 1967. Das hyporheische Interstitial als Grenzbiotop zwischen oberirdischem und subteranem Ökosystem und seine Bedeutung für die Primar-Evolution von Kleinsthöhlenbewohnern. Arch. Hydrobiol. Suppl., **33**: 1-62.

Sharaf El Din, S. H. 1977. Effect of the Aswan High Dam on the Nile flood and on the estuarine and coastal circulation pattern along the Mediterranean Egyptian coast. Limnol. Oceanogr., **22**: 194-207.

Simon, D. B. 1979. Effects of stream regulation on channel morphology. In: The Ecology of Regulated Streams. (Ed. by J. V. Ward and J. A. Stanford). pp. 95-111. Plenum, New York.

Skulberg, O. 1984. Effect of stream regulation on algal vegetation. In: Regulated Rivers. (Ed. by A. Lillehammer and S. J. Saltveit). pp. 107-124. Oslo Univ. Press, Oslo.

Smith, N. 1971. A History of Dams. Peter Davies, London. 279 pp.

Soltero, R. A., Wright, J. C. and Horpestad, A. A. 1973. Effects of impoundment on the water quality of the Bighorn River. Water Res., **7**: 343-354.

Stanford, J. A. and Gaufin, A. F. 1974. Hyporheic communities of two Montana Rivers. Science, **185**: 700-702.

Stanford, J. A. and Ward, J. V. 1979. Stream regulation in North America. In: The Ecology of Regulated Streams. (Ed. by J. V. Ward and J. A. Stanford). pp. 215-236. Plenum, New York.

Stanford, J. A. and Ward, J. V. 1983. The effects of mainstream dams on the physicochemistry of the Gunnison River, Colorado. In: Aquatic Resources Management of the Colorado River Ecosystem. (Ed. by V. D. Adams and V. A. Lamarra). pp. 43-56. Ann Arbor Sci. Publ., Ann Arbor.

Stanford, J. A. and Ward, J. V. 1984. The effects of regulation on the limnology of the Gunnison River: A North American case history. In: Regulated Rivers. (Ed. by A. Lillehammer and S. J. Saltveit). pp. 467-480. Oslo Univ. Press, Oslo.

Statzner, B. and Higler, B. 1985. Questions and comments on the river continuum concept. Can. J. Fish. Aq. Sci., **42**: 1038-1044.

Vannote, R. L., Minshall, G. W., Cummins, K. W., Sedell, J. R. and Cushing, C. E. 1980. The river continuum concept. Can. J. Fish. Aquat. Sci., **37**: 130-137.

Vannote, R. L. and Sweeney, B. W. 1980. Geographic analysis of thermal equilibria; a conceptual model for evaluating the effect of natural and modified thermal regimes on aquatic insect communities. Am. Nat., **115**: 667-695.

Vaux, W. G. 1962. Interchange of stream and intergravel water in a salmon spawning riffle. Spec. Sci. Publ. U. S. Fish Wildl. Serv., Fish., **405**: 1-11.

Walker, K. F. 1979. Regulated streams in Australia. In: The Ecology of Regulated Streams. (Ed. by J. V. Ward and J. A. Stanford). pp. 143-163. Plenum, New York.

Ward, J. V. 1982. Ecological aspects of stream regulation: Responses in downstream lotic reaches. Wat. Poll. Mgmt. Reviews, **2**: 1-26.

Ward, J. V. 1985. Thermal characteristics of running waters. Hydrobiologia, **125**: 31-46.

Ward, J. V. and Stanford, J. A. (Eds.). 1979. The Ecology of Regulated Streams. Plenum, New York. 398 pp.

Ward, J. V. and Stanford, J. A. 1981. Tailwater biota: Ecological response to environmental alterations. In: Surface-water Impoundments. (Ed. by H. G. Stefan). pp. 1516-1525. Am. Soc. Civil Eng., New York.

Ward, J. V. and Stanford, J. A. 1982a. Effects of reduced and perturbated flow below dams on fish food organisms in Rocky Mountain trout streams. In: Allocation of Fishery Resources. (Ed. by J. H. Grover). pp. 493-506. FAO, Rome.

Ward, J. V. and Stanford, J. A. 1982b. Thermal responses in the evolutionary ecology of aquatic insects. Ann. Rev. Entomol., **27**: 97-117.

Ward, J. V. and Stanford, J. A. 1983. The serial discontinuity concept of lotic ecosystems. In: Dynamics of Lotic Ecosystems. (Ed. by T. D. Fontaine and S. M. Bartell). pp. 29-42. Ann Arbor Sci. Publ., Ann Arbor.

Ward, J. V. and Stanford, J. A. 1984. The regulated stream as a testing ground for ecological theory. In: Regulated Rivers. (Ed. by A. Lillehammer and S. J. Saltveit). pp. 23-38. Oslo Univ. Press, Oslo.

Webster, J. R. and Patten, B. C. 1979. Effects of watershed perturbation on stream potassium and calcium dynamics. Ecol. Monogr., **49**: 51-72.

Weitkamp, D. E. and Katz, M. 1980. A review of dissolved gas supersaturation literature. Trans. Am. Fish. Soc., **109**: 659-702.

Winterbourn, M. J., Rounick, J. S. and Cowie, B. 1981. Are New Zealand stream ecosystems really different? N. Z. J. Mar. Freshwater Res., **15**: 321-328.

Zimmerman, E. G. 1984. Genetic and physiological correlates in fish adapted to regulated streams. In: Regulated Rivers. (Ed. by A. Lillehammer and S. J. Saltveit). pp. 273-292. Oslo Univ. Press, Oslo.

Zimmerman, E. G. and Wooten, M. C. 1981. Allozymic variation and natural hybridization in sculpins, *Cottus confusus* and *Cottus cognatus*. Biochem. Syst. Ecol., **9**: 341-346.

SUMMARY AND CONCLUSIONS

J. F. Craig [1] and J. B. Kemper [2]

[1]Department of Fisheries and Oceans
Freshwater Institute, 501 University Crescent
Winnipeg, Manitoba R3T 2N6
[2]Planning Division, Alberta Environment
Oxbridge Place, 9820-106 Street
Edmonton, Alberta T5K 2J6

The papers presented within the section on streamflows and fisheries requirements reflect the increasing use of physical habitat simulation models to assess the effects of stream and river regulation on fish. Scott and Shirvell provide a critical review of the IFIM method and pose serious questions about eight of the basic assumptions underlying the method. In support of their cautionary tone they provide a summary of other studies, conducted worldwide, which show highly variable results when weighted usable area calculations are compared to the standing crop of fish. The submission by Geer describes a stepwise method of analysing the output from IFIM modelling and comparing recommended, optimal and historical flows and habitat conditions. Irvine experimented with a replicated series of five artificial streams in New Zealand. Compared to control streams, cyclic discharges were found to have very little effect on rainbow trout (*Salmo gairdneri*) fry biomass. In addition he found that a five-fold change in discharge did not produce any changes in the trout populations.

Beecher reports on a modification which simulates fish feeding stations near sheer zones to the habitat component of the PHABSIM model. Wesche et al. provide results of monitoring a steep gradient stream which was subjected to heavy inundation of sediment due to a construction accident. The authors, who had previously determined optimum flushing flows for the reach of stream, were able to follow the effects of four subsequent high flow events. These events exceeded the recommended flushing flow and thereby provided a unique test of the effects of the flushing flow predictions. A more general review of flushing flows from a biological and engineering perspective is provided in the paper by Reiser et al.

Several presentations at the conference which were supported by comments from the audience, stressed the importance of integrating the results of water quality and aquatic benthic (macrophyte and invertebrate) studies with the results obtained from physical fish habitat simulation modelling. Although IFIM has been well developed for fish little attention has been paid to macroinvertebrates. Invertebrates are less mobile than fish and take longer to reinvade a disturbed habitat and also they have a narrower range of tolerance. Gore has developed models which rely on the accurate measurements of basic stream parameters, current velocity, depth

and substrate characteristics. These models appear to successfully predict
density and biomass changes of benthic macroinvertebrates in response to
alterations in these parameters. However, Gore indicates that there is a
need for more instream flow studies to test these models. In a case study
on the Rhône River near Lyon, France, Gaschignard and Berly found that the
changes in hydraulics, sediments and organic matter from the normal river
load reduce the diversity of macroinvertebrates.

Saltveit et al. show that stoneflies (Plecoptera) alone may be good
indicators of water quality and environmental change, especially in cool
temperate regions. Current velocity is a major factor controlling the
distribution and abundance of stoneflies. Some species are better adapted
than others in periods of low flow. Temperature is also an essential
factor for consideration in that it controls physiological processes which
determine such features as egg, nymph and adult size and fecundity.
Armitage brings attention to the importance of temperature. He relates the
structure of macroinvertebrate faunas of streams below 29 reservoirs in the
UK to flow and other physical and chemical characteristics of the sites.
Armitage finds a positive relationship between the age of the reservoir and
the number of families of invertebrates in the receiving stream. However
the situation is complicated by the fact that the older reservoirs are at
lower latitudes. Is the relationship found by Armitage mainly one of
temperature? Armitage believes more studies are necessary to determine the
effects of temperature and sediments.

The relevance of temperature as a controlling factor is again
emphasised in a number of papers grouped under the heading of "Physical
Processes". Crisp indicates the importance of knowing the temperature
requirements at different life stages of fish. These may often have to be
determined under controlled laboratory conditions as is demonstrated for
whitefish (*Coregonus lavaretus*) egg development by Saltveit and Brabrand.
Crisp also stresses that requirements may change intraspecifically as well
as interspecifically. Jensen suggests that stock differences are the
reason that some sea trout (*Salmo trutta*) populations in Norway grow faster
than predicted from a model developed by Elliott (1975). Wilson et
al. have made some predictions on the effects of a planned large
hydroelectric project in Alaska on Pacific salmon species. These
predictions are based on an instream temperature model. They point out
that the major problems with using this model are the lack of good
long-term temperature data and the fish thermal tolerance criteria used are
based on salmonid species from locations south of Alaska (the criteria may
not apply to stocks in Alaska). In adjusting the flow from a reservoir to
cater for the requirements of the fish (or invertebrates) more information
is required about temperature equilibrium in relation to distance
downstream and rate of discharge (Crisp this volume).

Gilvear describes the potential of reservoir releases to remove the
fine superficial covering of material that accumulates during low flows.
Carling and McCahon point out that the unsteady and non-uniform flow
characteristics of most natural streams make it impossible to collect field
data which are good enough to apply to hydraulic predictive models.
However, field data are still needed to give scale for laboratory studies.
Bed topography is important in controlling water velocity and thus sediment
supply rate and potential deposition. Poorly graded stream bottoms with
the size range cut off at 2mm prevent excessive quantities of silt entering
the void spaces. Large peaks of water discharge by power plants cause bank
erosion and increase in the quantity of suspended solids and turbidity.
Alasaarela and Virtanen use a flow model and a sediment interaction model
to demonstrate that releasing water at several periods rather than at one
time over a 24 h period diminishes this erosion effect.

The fourth dimension, time, of the river continuum theory, necessary to an understanding of the regulated river system, is discussed by Petts. He finds spatial patterns of change within regulated rivers and divides them into three compartments, adjusted, changing and stable. The stable system has higher invertebrate density than the other compartments. This is similar to the findings of Armitage, already referred to, who finds that the diversity of invertebrate families is related to the age of the reservoir. The time scale involved in this stabilisation process is in decades.

The results of attempts to change riverine water quality are reported in three papers while a fourth submission describes the application of phosphorus nutrient spiralling concepts to various lotic systems (Newbold). Hauser and Ruane discuss a series of modelling studies which examine the feasibility and cost effectiveness of controlling BOD in the Holsten river, Kingsport, Tennessee. The model is supported by detailed measurements and calibration of DO and BOD in various components of the riverine system. The results suggest that despite many extensive management strategies tested DO levels would still fall below desired levels. The modelling exercise was judged to be a useful tool in promoting cooperation among agencies, industries and water managers.

The manipulation of river and reservoir oxygen levels through construction and manipulation of regulatory structures is the theme of the papers by Cassidy and Dunn, and Lakso. These authors report the results of expensive engineering which is of interest to the water managers. Cassidy and Dunn outline improvements in reservoir and downstream water quality through the use of selective water withdrawal within the reservoir.

The status of several massive interbasin transfer schemes in China, and the possible biological implications of the projects are presented by Wu. The inland and coastal fisheries impacts and the transfer of invertebrates, parasites and disease organisms between basins are considered.

More general effects of reservoir regulation on rivers are reviewed by Ward and Stanford with specific reference to the different categories of the lotic community. The authors call for a more holistic approach to the study of regulated streams.

The restorative capabilities of a reservoir on the Ter River, Spain, are described by Puig et al., while Dufford et al. report that marked increases in algal biomass are common downstream from day release reservoirs in the western parts of the USA.

The papers presented in this volume summarise the advances that have been made in the study of regulated streams since the first symposium held at Erie, Pennsylvania, USA, in 1979 (Ward & Stanford 1979) and the second meeting held at Oslo, Norway, in 1982 (Lillehammer & Saltveit 1984). Contributors were encouraged by the organisers to present papers which emphasised recent advances. An evident development in recent years has been the holistic approach to the study of the effects of regulated streams on the biota. A number of predictive models have been constructed but these models are only as good as the empirical data on which they are based. Since these data are often lacking or are not sound quantitatively, many of the models are somewhat tentative. However, the development of a model is probably better than no model and it will often point to areas of research which are needed for refinement of the model. It was apparent from the Edmonton meeting that this development is still in the embryonic stage although progress has been made since the late 1970s.

413

Recent work has tended to emphasise impact assessment studies which frequently involve the development or use of complex physical habitat simulation models for fish. Rarely are companion models for lotic invertebrate communities used in conjunction with the fisheries models. Consequently stream flow recommendations or impact predictions have tended to be based largely upon single resource criteria. The need for more integrated, holistic approaches was recognised and widely accepted at the Edmonton Symposium. Much of the previously collected data on lotic systems will not be useful in predictive models because of the way it was collected. Further assessment projects will be shirking their responsibilities if they fail to add new quantitative information on the effects of water quality, flows and assemblage changes in the riverine community. The editors believe that the development of more accurate models can only be effected through cooperative effort among water resource managers, scientists and engineers. The regulated stream conferences, held every three years have been, and will continue to be an important forum for the exchange and improvement of our knowledge of regulated streams.

REFERENCES

Elliott, J. M. 1975. The growth of brown trout, *Salmo trutta* L., fed on maximum rations. J. Anim. Ecol. 44: 805-821.

Lillehammer, A. and Saltveit, S. J. (Eds.). 1984. Regulated Rivers. Oslo Univ. Press, Oslo.

Ward, J. V. and Stanford, J. A. (Eds.). 1979. The Ecology of Regulated Streams. Plenum Press, New York.

CONTRIBUTORS

Erkki Alasaarela
Technical Research Centre of Finland/
Building Laboratory, P.O. Box 181
SF-90101 Oulu

J. Armengol
Departament d´Ecologia
Universitat de Barcelona, Spain

Patrick Armitage
Freshwater Biological Association
River laboratory
East Stoke, Wareham, Dorset BH20 6BB, U.K.

Hal A. Beecher
Washington Department of Game
600 N. Capitol Way
Olympia, Washington 985-4 USA

B. Bengeyfield
Global Fisheries Consultants Ltd.
13069 Marine Drive
Surrey, B.C. V4A 1E5

Alain Berly
Département de Biologie Animale et Ecologie
U.A. C.N.R.S. 367 "Ecologie des eaux douces"
Université Claude-Bernard, LYON I
69622 VILLEURBANNE-Cédex FRANCE

S. P. Blachut
Habitat Management Division
Department of Fisheries and Oceans
1090 West Pender St.
Vancouver, B.C. V6E 2P1

Age Brabrand
Laboratory of Freshwater Ecology and Inland Fisheries (LFI)
Zoological Museum, University of Oslo
Sarsgt. 1, 0562 Oslo 5, Norway

John E. Brittain
Zoological Museum
University of Oslo
Sars Gate 1, 0562 Oslo 5, Norway

P. A. Carling
Freshwater Biological Association
The Ferry House
Ambleside, Cumbria LA22 OLP U.K.

Richard A. Cassidy
U.S. Army Corps of Engineers
Portland District
Portland, Oregon U.S.A.

L. D. Cline
Department of Zoology
Colorado State University
Fort Collins, Colorado, U.S.A.

G. A. Cole
Institute of Hydrology
Wallingford, Oxfordshire, OX10 8BB, U.K.

J. F. Craig
Department of Fisheries and Oceans
Freshwater Institute
501 University Crescent
Winnipeg, Manitoba
R3T 2N6

D. T. Crisp
Freshwater Biological association
The Ferry House
Far Sawrey, Ambleside, Cumbria LA22 OLP U.K.

R. G. Dufford
Department of Botany
Colorado State University
Fort Collins, Colorado, U.S.A.

Patrick E. Dunn
U.S. Army Corps of Engineers
Portland District
Portland, Oregon U.S.A.

Odile Gaschignard
Département de Biologie Animale et Ecologie
U.A. C.N.R.S. 367 "Ecologie des eaux douces"
Université Claude-Bernard, LYON I
69622 VILLEURBANNE-Cédex FRANCE

William H. Geer
Utah Division of Wildlife Resources
Salt Lake City, Utah USA 84116

D. J. Gilvear
Department of Geography
University of Technology
Loughborough, Leicestershire, U.K.

G. Gonzalez
Departament d'Ecologia
Universitat de Barcelona, Spain

James A. Gore
Faculty of Biological Sciences
University of Tulsa
Tulsa, OK   USA   74104

A. Gustard
Institute of Hydrology
Wallingford, Oxfordshire, OX10 8BB, U.K.

V. R. Hasfurther
Wyoming Water Research Center
University of Wyoming
Laramie, Wyoming U.S.A.   82071

Gary E. Hauser
Water systems Development Branch
Tennessee Valley Authority
Norris, Tennessee USA

W. A. Hubert
Department of Zoology and Physiology
University of Wyoming
Laramie, Wyoming U.S.A.   82071

J. R. Irvine
Department of Zoology
University of Otago
P.O. Box 56
Dunedin, New Zealand
Present Address:
Department of Fisheries and Oceans
Fisheries Research Branch
West Vancouver Laboratory
4160 Marine Drive
West Vancouver, British Columbia, Canada
V7V 1N6

Arne Johan Jensen
Directorate for Nature Management
Tungasletta 2, N-7000 Trondheim Norway

Michael D. Kelly
Actic Environmental Information and Data Center
University of Alaska
707 A Street
Anchorage, Alaska   99501 USA

J.B. Kemper
Planning Division, Alberta Environment
Oxbridge Place, 9820-106 Street
Edmonton, Alberta
T5K 2J6

Esko Lakso
Tec.Lich, National Board of Waters
Finaldn, Kokkola Water District
Torikatu 40 B, KOKKOLA

Thomas R. Lambert
Pacific Gas and Electric Company
San Ramon, California

Albert Lillehammer
Zoological Museum
University of Oslo
Sars gate 1, 0562 Oslo 5, Norway

C. P. McCahon
Department of Applied Biology
U.W.I.S.T., King Edward VII Avenue
Cathays Park, Cardiff

Paul R. Meyer
Arctic Environmental Information and Data Center
University of Alaska
707 A Street
Anchorage, Alaska 99501 USA

J. Denis Newbold
Stroud Water Research Center
Academy of Natural Sciences of Philadelphia
Box 512, R.D. #1
Avondale, PA 19311

J. Penuelas
Departament d´Ecologia
Universitat de Barcelona, Spain

Geoffrey E. Petts
Department of Geography
University of Technology
Loughborough, Leicestershire, UK

M. A. Puig
Deparamento de Ecologia
Fas. Biologia. Univ. de Murcia. Spain

Michael P. Ramey
Bechtel Inc.
San Francisco, California

Dudley W. Reiser
Bechtel Inc.
San Francisco, California

Richard J. Ruane
Water Quality Branch
Tennessee Valley Authority
Chattanooga, Tennessee USA

F. Sabater
Departament d´Ecologia
Universitat de Barcelona, Spain

S. Sabater
Departament d´Ecologia
Universitat de Barcelona, Spain

Svein Jakob Saltveit
Zoological Museum
University of Oslo
Sars Gate 1, 0562 Oslo 5, Norway

D. Scott
Department of Zoology
University of Otago
Dunedin, New Zealand

C. S. Shirvell
Department of Zoology
University of Otago
Dunedin, New Zealand
Present Address:
Department of Fisheries and Oceans,
Room 207, 417-2nd Avenue West,
Prince Rupert, B.C., Canada V8J 1G8

Q. D. Skinner
Range Management Division
University of Wyoming
Laramie, Wyoming U.S.A. 82071

H. A. Smith
Environmental Services
B.C. Hydro
1265 Howe St.
Vancouver, B.C. V6Z 2C8

Jack A. Stanford
Flathead Lake Biological Station
University of Montana
Bigfork, Montana 59911 U.S.A.

Markku Virtanen
Technical Research Centre of Finland/
Building laboratory, P.O. Box 181
SF-90101 Oulu

J. V. Ward
Department or Zoology
Colorado State University
Fort Collins, Colorado, 80523 U.S.A.

T. A. Wesche
Wyoming Water Research Center
University of Wyoming
Laramie, Wyoming U.S.A. 82071

William J. Wilson
Arctic Environmental Information and Data Center
University of Alaska
707 A Street
Anchorage, Alaska 99501 USA

Huisheng Wu
Institute of Hydrobiology
Academia Sinica
Wuhan, Hebei Province, People's Republic of China

H. J. Zimmermann
Department of Zoology
Colorado State University
Fort Collins, Colorado U.S.A.

INDEX

Bear River, US, 2
Bed topography, 412
Bedload, 51, 54, 235, 241-243,
     284, 294, 300
Beecher, H.A., 71-82, 411
Bengeyfield, B., 289, 301
Benthos, 5, 8, 10, 20, 119, 124,
     245, 314, 315, 317, 321
  macrobenthos, 412
  zoobenthos, 383, 396, 402
Bere Stream, UK, 165
Berly, A., 145-157, 412
Bicarbonates, 376
Bighead, 367, 369, 370
Bighorn Lake, US, 395
Big Thompson River, US, 384
Biomass, 29, 36, 85, 104, 105,
     111, 112, 122, 386-389,
     397, 411-413
Biota, 4, 395
  aquatic, 4
Blachut, S.P., 289-301
Black Brows Beck, UK, 210
Blackfly, 110, 396
Blue Nile, 320
Blue River, US, 384
BMWP (biotic) score, 133, 135-137
BOD (see oxygen)
Boothwood, UK, 136, 140
Bosmina longirostris, 369
Boundary layer, 106, 109
Brabrand, A., 219-228, 412
Brachyptera risi, 120, 122
Brainard Reservoir, US, 383, 384
Brazos River, US, 118, 122
Bream, 370
British Columbia, Canada, 289-301
Brittain, J.E., 117-129
Bryophytes, 380, 389
Bubble disease, 367, 368
Burnhope, UK, 134
Buttonrock Reservoir, US, 384
Caddisflies, net spinning, 396
Caenis luctuosa, 379
  C. moesta, 265
Calcium, 303
California, US, 27
Calitriche sp., 379
Caloneis ventricosa var. *alpina*,
     387
Campostoma anomalum (Stone
     roller), 31
Canada, 27
Cannonsville Reservoir, US, 171
Canterbury, New Zealand, 32, 34
Canterbury Plains, New Zealand, 34
Capnia, 118
  C. atra, 119, 123
  C. limata, 120
Carangidae, 366

Carassius auratus (crucian carp),
     370
Carbon, 303, 402
Carbonaceous biochemical oxygen
     demand (CBOD), 331, 332,
     334
Carling, P.A., 229-244, 412
Carp,
  black, 367, 368, 370
  common, 369, 370
  grass, 367-371
  silver, 367-370
Cascade Mountains, US, 76
Cassidy, R.A., 339-357, 413
Castleshaw, UK, 136
Catadromous fishes, 403
Catostomids, 51
Centrarchids, 51
Ceraclea dissimilis, 151, 153, 155
  Ceraclea spp., 147, 148
Ceratium hirundinella, 369
Changjiang River (Yangtze), China,
     363-372
Channelization, 2
Channel bed, 261
Channel slope, 48, 49
Charr, 220
  arctic, 216
  sea, 212
Cheesman Reservoir, US, 122, 171,
     388
Cheumatopsyche lepida, 379
Chironomus sp., 368, 380
Chironomidae, 136, 141, 142, 151,
     262, 264, 265
Chironomids, 39
Chironomimi, 148, 155
Chloride, 137, 376
Chloroperla, 118, 124
  C. torrentium, 265
Chloroperlidae, 118, 151
Chlorophyll, 380
Chlorophyta, 385
Chlorophytes, 395
Chrysophyta, 385
Chrysophyte, 386
Cinclidotus fontinaloides, 379,
     380
Claassenia sabulosa, 119, 120,
     123, 125
Cladophora glomerata, 386-388
Clarity, 396, 398
Cline, L.D., 383-390
Clutha River, New Zealand, 37
Clyde, UK, 262
Cohort analysis, 108
Coilia brachygnathus, 370
  C. extensis, 370
  C. mystus (anchovy), 366
  Coilia spp. (anchovy), 366